Bio-manufactured Nanomaterials

Kaushik Pal
Editor

# Bio-manufactured Nanomaterials

## Perspectives and Promotion

Springer

*Editor*
Kaushik Pal
Laboratório de Biopolímeros e Sensores
Instituto de Macromoléculas
Universidade Federal do Rio de Janeiro
(LABIOS/IMA/UFRJ)–Cidade Universitária
Rio de Janeiro, Brazil

ISBN 978-3-030-67222-5    ISBN 978-3-030-67223-2 (eBook)
https://doi.org/10.1007/978-3-030-67223-2



This Springer imprint is published by the registered company Springer Nature Switzerland AG
The registered company address is: Gewerbestrasse 11, 6330 Cham, Switzerland

# Preface

*Bio-manufactured Nanomaterials: Perspectives and Promotion* has paved the way for the development of new and more capable novel bio-synthesis of nanomaterials, which has emerged as a potentially attractive "eco-friendly" alternative. Drawing upon case studies from bio-engineering and structural monitoring and involving chemical and bio-electronic sensors, this book suggests an approach that frames the relevant technical issues in such a way as to expedite the consideration of unique bio-inspired materials. Nano-biotechnology-based systems are vital to our awareness of our surroundings and provide safety, security, and surveillance, as well as enable monitoring of our healthcare treatment, environmental and plant biology, social, ethical, and regulatory implications of various aspects of green nanotechnologies, as well as eloquent foreseeable spectroscopic utilizations of some leading functional nanomaterials. A transformative advance in the field of nano/biotechnology has been the development of smart diversities of applications. Controllable novel designing features found in these biomaterials like hierarchical structure and composite nature and resorting to bio-inspired manufacturing processes like biomineralization and self-assembly could yield lab-made materials that are multifunctional, lightweight, benign, and recyclable. To be very specific, a brief outline of the discipline is followed by a discussion of general aspects related to the structure and synthesis of natural biomaterials. Next, the recent progress made in the development of bio-inspired materials with improved mechanical resistance, optical, self-cleaning, adhesiveness, anti-adhesion, and anti-microbial properties is reviewed with reference made to the most noteworthy examples. It enables a multidisciplinary approach for identifying opportunities and making realistic assessments of technical risk and could be used to guide relevant research and development in smart bioengineering technologies.

Recent years have witnessed the sheer growth of biomaterial fabrication concepts and nanotechnology-based innovations in bioengineering applications. This book is a collection of contributions from biochemists to biotechnologists to material scientists worldwide. The fabrication and characterization of nano-systems still exists as an evergreen domain in bioscience. The quality measurements of newly designed polymeric stuffs demand systematic and unexplored characterization protocols. As an outcome of this, the researches in this regard have attracted considerable attention of the scientific community worldwide. The fundamental idea of biosensor technology can not only provide interpretive power and customized outputs but also significantly improve bioelectronics system performance and capabilities. Nano-biotechnology is now emerging as a specific stream taught in universities and research institutes and is applied in industries for the attainment of new products with specific features. We hope that our new book entitled *Bio-manufactured Nanomaterials: Perspectives and Promotion* will surely be an asset to young scholars, scientists, engineers, and academicians working in the area of biogenic applications.

Next generation, industrial potential scale nano/biomaterials will possess embedded intelligence to provide the end user with critical data in a more rapid, reliable, robust, economical, and efficient manner with a seamless interface to applications. Recent breakthroughs mentioned in this book illustrate the novelty research finding, fabrication strategy, spectroscopic characterization, and detailed exploration and the utilization of prepared functional semiconductor nanomaterial samples for spectroscopic and microscopic investigations. The unique scientific challenges and significant outcome within this entire book will attribute for developing versatile applications of various morphologies novel synthetic strategy employed to smart technologies cost-effective pathways which are very promising to the reader. This book provides an overview of broadly provides detailed information on the emergence of different types of nanomaterials as transduction platforms utilized in the development of nano-biotechnology by distinguish editor **Prof. (Dr.) Kaushik Pal, Ph.D.; D.Sc.(Malaysia); Marie-Curie PDF (Europe); CAS Felw(China); R & D Scientist (Cheaptube, USA).**

This book contains **20 significant chapters**. Those chapters are categorized diversity of exploration undertaken following sub-chapters are mentioned below.

**Chapter#1**<Introduction to Nanobiotechnology: Novel and Smart Applications>; **Chapter #2** <Quantum Dot-Based Photo-electrochemical Biosensors: Principles, Fabrication, and Applications>; **Chapter #3**<Synthesis and Antimicrobial Abilities of Metal Oxide Nanoparticles>; **Chapter #4**<Emerging Nanomaterial-Based Medications: Key Challenges and Opportunities>; **Chapter #5**<Anti-biofilm Activities of Nanocomposites: Current Scopes and Limitations>; **Chapter #6**<Living Nano-factories: An Eco-friendly Approach Towards Medicine and Environment>; **Chapter #7**<Nanomaterial-Based Bioscaffolds for Enhanced Biomedical Applications>; **Chapter #8**<Highly Toxic Nanomaterials for Cancer Treatment>; **Chapter #9**<Applications of Nanomaterials in Tissue Engineering and Regenerative Medicine>; **Chapter #10**<Nanomaterials for the Management of Multidisciplinary Dental Sciences and Applications>; **Chapter #11**<Biosafety and Toxicity of Nanomaterials for the Management of Drug and Gene Delivery>; **Chapter #12**<Bio-inspired Materials in Nanobiotechnology Applications and Industrial Potential Scale>; **Chapter #13**<Phyto-fabricated Metal Oxide Nanoparticles as Promising Antibacterial Agents>; **Chapter #14**<A Unique Perspective in Precision of Nano-biotechnology for Sustainable Agricultural Fields>; **Chapter #15**<Key Changes and Scopes of Biomaterials Commercialization:

Therapeutic Delivery>; **Chapter #16**<Synthesis and Application of Nanomaterials for Biomedical Anticancer Therapy>; **Chapter #17**<Progress in Nanomaterial Self-Assembly for Bio-scaffolds: Exclusive Biomedical Applications>; **Chapter #18**<An Impact of Antibacterial Efficacy of Metal Oxide Nanoparticles: A Promise for Future>; **Chapter #19**<Structural Analysis and Thermal Properties of Graphene and Biocomposite Potential Application in Various Sensors>;**Chapter #20**<Conclusion, Outlook, Future Aspects, and Utilization of Functional Bio-engineered Nanomaterials>.

Nano/bioscience and novel sensor modulation technology has emerged as a hot topic from postgraduate to PhD levels and diversified industrial courses in polytechnic colleges, research institutes, and universities. Keeping the young researcher in mind, this book is edited to highlight the latest innovations and principles behind these findings, specific to nanostructured materials, hybrid nanocomposites, and unique sensor applications. This book is devoted to novel architectures at the nano-level with an emphasis on new synthesis and characterization methods. Special emphasis is given to new applications of nano-structures and nano-composites in various fields. Chapter-wise bibliographies have been introduced to further research on these topics. Recent breakthroughs in research finding, fabrication strategy, spectroscopic characterization, and detailed exploration and the utilization of prepared functional nanomaterial samples for spectroscopic and microscopic investigations are explored. The book is an outstanding resource reference for anyone involved in the field of semiconductor nanomaterials design for advanced technologies. The unique scientific challenges and significant outcome within this entire book will attribute for developing versatile applications of various synthetic strategy employed to smart technologies cost-effective pathways which are very promising to readers.

The unique scientific challenges and significant outcome within this entire book will attribute editorial efforts and enthusiasm of all the contributors for writing their chapters which are very promising to readers in current science and technology. Furthermore, without the tireless efforts we express our sincere gratitude to the creative environments of *Laboratório de Biopolímeros e Sensores, Instituto de Macromoléculas, Universidade Federal do Rio de Janeiro (LABIOS/IMA/UFRJ), Rio de Janeiro, Brazil* to our colleagues for their encouragement and support to success this project.

Rio de Janeiro, Brazil Kaushik Pal

# Contents

# About the Editor

**Kaushik Pal,** Ph.D.; D.Sc.(Malaysia); Marie Curie PDF (Europe); CAS Fellow (China), is Distinguished Professor and Group Leader, *Laboratório de Biopolímeros e Sensores, Instituto de Macromoléculas, Universidade Federal do Rio de Janeiro (LABIOS/IMA/UFRJ), Rio de Janeiro, Brazil.*

He received his Ph.D. from the *University of Kalyani (India)*, as well as an Honorís Causà DOCTOR OF SCIENCE (D.Sc.) from *IKTBN Sepang, Ministry of Education, and Govt. of Malaysia, as well as* the prestigious Marie Curie Postdoctoral Fellowship (Greece) offered by European Union; and the *Chinese Academy of Science (CAS)* Fellowship from Wuhan University, China. As a distinguished academician, Prof. Pal has held research and teaching positions at various top institutions, including *University Federal Rio de Jàneiro (Brazil), University of Maribör (Slovenia), and Yarmouk University (Jordan)*. He has published 85 articles in international peer-reviewed (SCI/Scopus) journals and edited or authored 25 books.

# Contributors

**Rao Zahid Abbas** Department of Parasitology, University of Agriculture, Faisalabad, Pakistan

**Gözde Koşarsoy Ağçeli** Department of Biology, Hacettepe University, Ankara, Turkey

**Kehkashan Alam** Department of Chemistry, Aligarh Muslim University, Aligarh, Uttar Pradesh, India

**Alaa A. A. Aljabali** Faculty of Pharmacy, Department of Pharmaceutics and Pharmaceutical Technology, Yarmouk University, Irbid, Jordan

**J. Anbu** Faculty of Pharmacy, Department of Pharmacology, Ramaiah University of Applied Sciences, Bangalore, Karnataka, India

**Amjad Islam Aqib** Department of Medicine, Cholistan University of Veterinary and Agriculture Sciences, Bahawalpur, Pakistan

**Dominic Augustine** Faculty of Dental Sciences, Department of Oral Pathology and Microbiology, M.S. Ramaiah University of Applied Sciences, Bangalore, India

**Rabia Aziz** Digital Health and Bioinformatic Engineering, University of Salerno, Fisciano, SA, Italy

**Arundhati Barik** Department of Chemistry, Ravenshaw University, Cuttack, India

**Mallikharjuna Reddy Bogala** Department of Materials Science, Central University of Tamil Nadu, Thiruvarur, TN, India

Department of Metallurgical and Materials Engineering, The University of Alabama, Tuscaloosa, AL, USA

**Subhendu Chakroborty** School of Sciences, Auro University, Surat, Gujarat, India

**P. K. Chauhan** Faculty of Applied Sciences and Biotechnology, Shoolini University, Solan, Himachal Pradesh, India

**Parveen Chauhan** Faculty of Applied Sciences and Biotechnology, Shoolini University, Solan, Himachal Pradesh, India

**R. Deveswaran** Pharmaceutics, Faculty of Pharmacy, Ramaiah University of Applied Sciences, Bangalore, Karnataka, India

**Kanika Dulta** Faculty of Applied Sciences and Biotechnology, Shoolini University, Solan, Himachal Pradesh, India

**Alka Gupta** Department of Materials Science and Metallurgical Engineering (MSME), University Institute of Engineering and Technology; Chatrapati Shahu Ji Maharaj Kanpur University Kalyanpur, Kanpur, UP, India

**Meenu Gupta** Department of Botany, J.D.W.C., Patna, India

**Vanishri C. Haragannavar** Faculty of Dental Sciences, Department of Oral Pathology and Microbiology, M.S. Ramaiah University of Applied Sciences, Bangalore, India

**Muhammad Imran** Department of Parasitology, University of Agriculture, Faisalabad, Pakistan

**Neha Jain** KLE Society's Institute of Dental Sciences, Bangalore, India

**Saranya Jayaram** Department of Life Sciences, CHRIST (Deemed to be University), Bangalore, India

**Arshi Khan** Department of Chemistry, Aligarh Muslim University, Aligarh, Uttar Pradesh, India

**Ishaat M. Khan** Department of Chemistry, Aligarh Muslim University, Aligarh, Uttar Pradesh, India

**Chhaya Kumar** Faculty of Dental Sciences, Department of Pedodontics and Preventive Dentistry, M.S. Ramaiah University of Applied Sciences, Bangalore, India

**K. Vineeth Kumar** Faculty of Dental Sciences, Department of Oral and Maxillofacial Surgery, RUAS, Bangalore, Karnataka, India

**P. R. Anil Kumar** Department of Applied Biology, Biomedical Technology Wing, Sree Chitra Tirunal Institute for Medical Sciences and Technology, Thiruvananthapuram, Kerala, India

**Bernadette K. Madathil** Advanced Center for Tissue Engineering, Department of Biochemistry, University of Kerala, Thiruvananthapuram, Kerala, India

**A. Manjceevan** Department of Chemistry, Faculty of Sciences, University of Jaffna, Jaffna, Sri Lanka

**Shruthi Nagaraja** Faculty of Dental Sciences, Department of Conservative Dentistry and Endodontics, M.S. Ramaiah University of Applied Sciences, Bangalore, India

**Praseetha P. Nair** Department of Chemical Engineering, Government Engineering College, Thrissur, Kerala, India

**Teresa Oliveira** Departamento de Medicina Veterinária, Escola de Ciências e Tecnologia, Universidade de Évora, Évora, Portugal

Mediterranean Institute for Agriculture, Environment and Development, Évora, Portugal

**Kaushik Pal** Laboratório de Biopolímeros e Sensores, Instituto de Macromoléculas, Universidade Federal do Rio de Janeiro (LABIOS/IMA/UFRJ)–Cidade Universitária, Rio de Janeiro, Brazil

Wuhan University, Wuchang Dist., Hubei, China

**Pravati Panda** Department of Chemistry, Ravenshaw University, Cuttack, India

**Rita Payan-Carreira** Departamento de Medicina Veterinária, Escola de Ciências e Tecnologia, Universidade de Évora, Évora, Portugal

Mediterranean Institute for Agriculture, Environment and Development, Évora, Portugal

**Maria del Pilar Rodriguez-Torres** Departamento de Ingeniería Molecular de Materiales, Centro de Física Aplicada y Tecnología Avanzada, Universidad Nacional Autónoma de México, Querétaro, Mexico

**Pushpalatha** Faculty of Dental Sciences, Department of Pediatric Dentistry, Ramaiah University of Applied Sciences, Bangalore, Karnataka, India

**C. Pushpalatha** Faculty of Dental Sciences, Department of Pedodontics and Preventive Dentistry, M.S. Ramaiah University of Applied Sciences, Bangalore, India

**Karthikeyan Rajendran** Department of Biotechnology, Mepco Schlenk Engineering College, Sivakasi, Tamil Nadu, India

**Joana Reis** Escola Superior Agrária de Ponte de Lima, Instituto Politécnico de Viana do Castelo, Viana do Castelo, Portugal

**Sandhya Kalathilparambil Santhosh** Department of Life Sciences, CHRIST (Deemed to be University), Bangalore, India

**R. Sarankumar** Department of Electronics and Communication Engineering, QIS Institute of Technology, Ongole, India

**Suma Sarojini** Department of Life Sciences, CHRIST (Deemed to be University), Bangalore, India

**Kumari Seema** Department of Chemistry, J.D.W.C., Patna, India

**Shrikanth** Mazumdar Shaw Medical Centre, Narayana Hrudayalaya Multispeciality Hospital, Bangalore, India

**N. Shruthi** Faculty of Dental Sciences, Department of Conservative and Endodontics, Ramaiah University of Applied Sciences, Bangalore, Karnataka, India

**V. Shwetha** Faculty of Dental Sciences, Department of Oral Medicine and Radiology, Ramaiah University of Applied Sciences, Bangalore, Karnataka, India

**A. Sivakami** Department of Physics, School of Sciences, Malla Reddy University, Hyderabad, India

**S. V. Sowmya** Faculty of Dental Sciences, Department of Oral Pathology and Microbiology, M.S. Ramaiah University of Applied Sciences, Bangalore, India

**Alisha Tahir** Department of Physical Chemistry, Quaid e Azam University, Islamabad, Pakistan

**Vinoy Thomas** Department of Materials Science and Engineering, Center for Nano-scale Materials & Biointegration (CNMB), University of Alabama at Birmingham (UAB), Birmingham, AL, USA

**Mridul Umesh** Department of Life Sciences, CHRIST (Deemed to be University), Bangalore, India

**M. V. Basavanag Unnamatla** Universidad autonoma del estado de Mexico, Toluca de Lerdo, Mexico

**Sugumari Vallinayagam** Department of Biotechnology, Mepco Schlenk Engineering College, Sivakasi, Tamil Nadu, India

**Elizabeth Abbu Varghese** Faculty of Dental Sciences, Department of Pedodontics and Preventive Dentistry, M.S. Ramaiah University of Applied Sciences, Bangalore, India

**Shiny Velayudhan** Division of Dental Products, Department of Biomaterial Science and Technology, Biomedical Technology Wing, Sree Chitra Tirunal Institute for Medical Sciences and Technology, Thiruvananthapuram, Kerala, India

**S. Vinodha** Department of Chemical Engineering, Sethu Institute of Technology, Virudhunagar, Tamil Nadu, India

**Faiz Warsi** Department of Chemistry, Aligarh Muslim University, Aligarh, Uttar Pradesh, India

**Iqra Zaheer** Department of Pathology, University of Agriculture, Faisalabad, Pakistan

**Tean Zaheer** Department of Parasitology, University of Agriculture, Faisalabad, Pakistan

**Mazhar Salim Al Zoubi** Faculty of Medicine, Department of Basic Medical Sciences, Yarmouk University, Irbid, Jordan

# Abbreviations

| | |
|---|---|
| 3T3 | 3-day transfer, inoculum $3 \times 10^5$ cells |
| A549 | An adenocarcinoma human alveolar basal epithelial cell |
| Ag NPs | Silver nanoparticles |
| ATP | Adenosine triphosphate |
| Au NPs | Gold nanoparticles |
| BALB/c | An albino mouse strain |
| BBB | Blood-brain barrier |
| BEAS-2B | An immortalized but non-tumorigenic epithelial cell line |
| $Ca^{++}$ | Calcium ions |
| $CeO_2$ NP | Cerium oxide nanoparticles |
| CNT | Carbon nanotube |
| CuO NPs | Copper oxide nanoparticles |
| DNA | Deoxyribonucleic acid |
| EMA | European Medicines Agency |
| FDA | United States Food and Drug Administration |
| $Fe_3O_4$ NPs | Iron oxide nanoparticles |
| FESEM | Field emission scanning electron microscopy |
| GDH | Glucose dehydrogenase |
| GMPs | Good manufacturing practices |
| GOx | Glucose oxidase |
| GQD | Graphene quantum dot |
| HEK293 | Human embryonic kidney 293 cells |
| HeLa | An immortal cell line |
| $HepG_2$ | A human liver cancer cell line |
| HNSCC | Head and neck squamous cell carcinoma |
| HRP | Horseradish peroxidase |
| IF-γ | Interferon-gamma |
| IMR-90 | A human diploid cell strain |
| IRv | Infrared |
| LC50 | Median lethal dose |
| LDH | Lactate dehydrogenase |

| | |
|---|---|
| LoVo | A colon cancer cell line |
| MCF7 | A breast cancer cell line |
| MNP | Metal nanoparticle |
| MONP | Metal oxide nanoparticle |
| MRI | Magnetic resonance imaging |
| MRSA | Methicillin-resistant *Staphylococcus aureus* |
| NAD | Nicotinamide adenine dinucleotide |
| NCBI | National Center for Biotechnology Information |
| NIH | National Institute of Health Swiss mouse embryo |
| NIH/3T3 | A continuous cell line of highly contact-inhibited cells |
| NP | Nanoparticle |
| OA | Osteoarthritis |
| p53 | Tumor suppressor protein |
| PBMC | Peripheral blood mononuclear cells |
| PEC | Photoelectrochemical |
| PEG | Polyethylene glycol |
| p-$H_2$AX | $H_2$A histone family member X |
| Pt NPs | Platinum nanoparticles |
| PubMed | Public/Publisher MEDLINE |
| Q-dot | Quantum dot |
| Rad51 | RAD51 recombinase |
| RBC | Red blood cells |
| RNS | Reactive nitrogen species |
| ROS | Reactive oxygen species |
| $SiO_2$ NPs | Silicon oxide nanoparticles |
| SOPs | Standard operating procedures |
| $TiO_2$ NPs | Titanium oxide nanoparticles |
| TLR-4 | Toll-like receptor type 4 |
| TNF-α | Tumor necrosis factor-alpha |
| $U_{937}$ | A model cell line |
| UV | Ultraviolet |
| UV-vis. | Ultraviolet visible |
| VSMCs | Vascular smooth muscle cells |
| ZnO NPs | Zinc oxide nanoparticles |

# Chapter 1
# Introduction to Nanobiotechnology: Novel and Smart Applications

**A. Sivakami, R. Sarankumar, and S. Vinodha**

## 1.1 Introduction

Nanobiotechnology is a new scientific technique involving materials and devices capable of altering the physical and chemical properties of a substance at the molecular level (Whitesides 2013). Nanobiotechnology (the Greek term nano means "dwarf") develops and uses materials, devices, and systems by manipulating matter on the scale of nanometers, i.e., at the level of atoms, molecules, and supramolecular structures (Jain 2005; Morrow et al. 2007).

Through this technique, atomic or molecular grade machines can be created by imitating or integrating biological systems, or by constructing small tools to analyze or modulate, on a molecular basis, the various properties of a biological system (Niemeyer and Mirkin 2004). Therefore, by incorporating cutting-edge applications of information technology (IT) and nanotechnology into current biological problems, nanobiotechnology will provide comfort in several paths of life sciences (Crean et al. 2011).

To some degree, this technology has the power to dissolve evident distinctions among physics, chemistry, and biology and to form the present thoughts and understanding. With this motivation, and through the extensive use of nanobiotechnology

A. Sivakami (✉)
Department of Physics, School of Sciences, Malla Reddy University, Hyderabad, India
e-mail: sivakami.phy@mallareddyuniversity.ac.in

R. Sarankumar
Department of Electronics and Communication Engineering, QIS Institute of Technology, Ongole, India

S. Vinodha
Department of Chemical Engineering, Sethu Institute of Technology, Virudhunagar, Tamil Nadu, India



K. Pal (ed.), *Bio-manufactured Nanomaterials*,
https://doi.org/10.1007/978-3-030-67223-2_1

over time, numerous different problems and guidelines will simultaneously arise in learning, science, and diagnostics.

Nanotechnology typically deals with the creation of materials, devices, or other structures ranging in size from 1 to 100 nm with at least one dimension. Meanwhile, biotechnology deals with biological topics, including microorganisms, metabolic, and other physiological processes. In evolving and applying numerous advantageous tools in the study of life, the interaction of these two technologies in nanobiotechnology may play a dynamic role (Morais et al. 2014; Chandrasekaran et al. 2011). The applications of nanobiotechnology in various sectors are shown in Fig. 1.1.

Nanobiotechnology is quite complex, commencing from extensions of traditional system physics to entirely different methods focused on molecular self-assembly, from the creation of fresh nanoscale dimensional materials to the investigation of whether atomic scale/level issues can be managed directly. This definition includes the use of scientific fields as varied as semiconductor physics, organic chemistry, surface science, molecular biology, and micro fabrication (Ratner et al. 2004; Peer et al. 2007).

The amalgamation of different kinds of Nanoparticles (NPs) in diverse dimensions, forms, and chemical compositions by means of exact differences is an important nanotechnology research. The development of new metal NPs, e.g., gold, silver, palladium, platinum, and zinc, from natural resources, has attained importance and attention in recent years owing to environmentally sustainable social material technologies that are essential to progress (Xu et al. 2009; Tiwari et al. 2012).

If NPs are biologically synthesized by naturally occurring organic compounds, namely carbohydrates, vitamins, botanical extracts, proteins, lipids, biologically degradable polymers, and microorganisms, the possible benefits of nanobiotechnology increase (Lehner et al. 2013; Sundar et al. 2010). The processing of a little

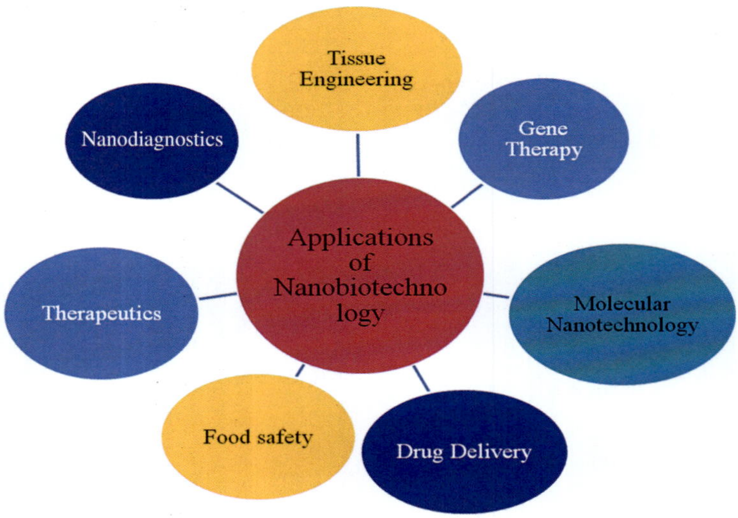


**Fig. 1.1** Applications of nanobiotechnology in different sectors

inorganic NPs, mainly metal NPs, some metal oxides, and salts, is induced by these innovations (Chen et al. 2003; Wang et al. 2012).

Many of the above raw materials shall be used to synthesize biologically on large scale NPs, including plant-based materials, which will be the best ecofriendly choice. Currently, numerous plant components, e.g., leaf, seed, core, latex, and stem, are used to synthesize biologically metal NPs.

## 1.2 Nanomaterials for Nanobiotechnology

In order to provide the groundwork for new developments in the field of manufacturing and medicine, new materials are being produced at an increasing rate. Nanoformulation depends on oral and current drug supply in biotechnology will induce molecular imaging across the blood brain barrier for brain tissue engineering applications. It is also involved in the development of CNS-focused neuro protection, minimum side effects, drug acceptance assistance and effective bio discovery (Kiessling et al. 2014; Prasad et al. 2016). The various biological syntheses of nanoparticles in the different nanobiotechnology sectors is shown in Fig. 1.2.

Biomaterials have been used throughout history to improve health and cure diseases. Leading biomaterials included gold for dental treatment, wooden teeth, and glass eyes. With the growth of science and technology, nanosized particles have also

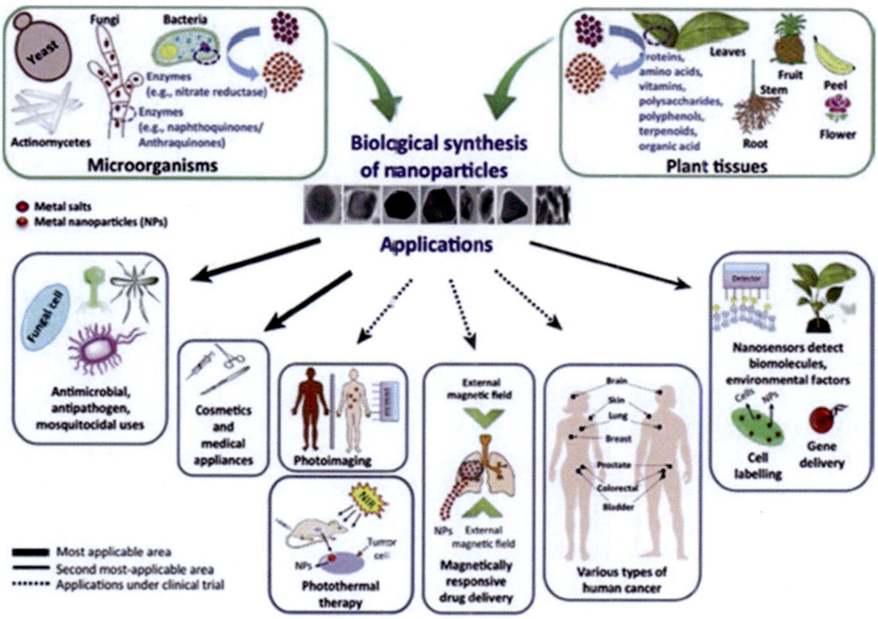


**Fig. 1.2** Different biological syntheses of nanoparticles used in biomedical research field. (He et al. 2019, J. of Food and Drug Analysis)

been developed. Nanoparticles are 1–100 nm in size and are capable of holding imaging agents, molecular medicines, nucleic acid, and many additional elements (He et al. 2019).

Nanotechnology was utilized by ancient Egyptians, but in the late 1900s, technological investigation and knowledge of nanomaterials advanced greatly. Materials can display different characteristics at nanoscale proportions than their counterparts at larger scales, enabling them to be used in biomedical engineering contexts, including drug delivery methods, deoxyribonucleic acid (DNA) labeling, and cancer treatment.

Intelligent nanomaterials were developed to provide the material with extra functionality and additional benefits under unique conditions (McNamara and Tofail 2013). Outside stimuli, including action in an aqueous environment, can cause quick and revocable modifications in the nanostructure of the particle. The nanoparticle has the ability to come back to its original state once the stimulus is removed (Ge et al. 2014). Thermosensitive and pH-responsive nanoparticles are commonly utilized. Nanoparticle conversion into nanocomposites also makes their use more effective in nanomedicine (Wei et al. 2015; Govindaraju et al. 2015). Ultimately, the addition of intelligent ingredients will change the nanoparticles into penetration enhancers, direct site-explicit gathering, and offer dynamic control. Stimuli-responsive polymers have attracted researchers in the areas of controlled drug-release, cell-adhesion mediators, and control of enzyme function and gene expression in medicine. For controlled drug delivery, solid lipid nanoparticles (SLNs) are utilized, originated by a conversion from alpha- to β-forms by way of a built-in trigger system (Zheng et al. 2015; Zhao and Nalwa 2007).

For vivo applications, lipid-based nanostructures have demonstrated the least toxicity. A variety of smart gels are established and are able to provide response to variations in the body such as glucose and those changes in pH, body temperature, and even explicit molecules (Petros and DeSimone 2010). In the field of endocrinology and in the handling of illnesses such as diabetes, such systems can be useful. Some, for example, provide straight reaction, allowing insulin to be supplied in connection to the extra glucose identified.

Healthcare is required to comprehensively study new materials being created. Intelligent nanomaterials are being investigated for applications in recognizing, investigating, imaging, and regenerative drugs (Lehner et al. 2013; Juan et al. 2009). Biologically decomposable intelligent nanomaterials are being investigated in lieu of biomedical applications in vivo together with momentary implantations, drug delivery transporters, and tissue-engineering scaffolds.

The permeability and retention advantages of polymer-based nanoparticles have been identified as making them a good choice for drug delivery systems. Polyde, L-actide-*co*-glycolide (PLGA), Poly Capro Lactone (PCL), Poly Lactic Acid (PLA), chitosan, and gelatin are polymeric systems. Form-memory metallic alloys, because of their use as sensors and actuators, are observed as hopeful intelligent materials (Stoeckel 2000).

They can be incorporated into textiles in many healthcare or traditional applications to provide multifunctional clothing. Nitinol wires, being super-elastic, are one such example and are utilized in the recovery of any form. Nitinol has also been investigated for use in medical application, biomedical engineering. and Micro Electromechanical Systems (MEMS) applications (Zhao et al. 2013; Wadood 2016). In addition to many other uses, these smart metals are known to be applied in orthodontic wires and for stents in cardiovascular contexts (Londe et al. 2008).

Moreover, buckminsterfullerene ($C_{60}$) and carbon nanotubes (CNTs) are the best choice to experience quick, revocable modifications in conformation and properties in connection to the small modifications in their close atmospheres. Carbon nanostructures offer biocompatibility benefits and little in the way of harmful side effects (Maiti et al. 2018). Hence, smart carbon nanomaterials are utilized in several areas such as biomedicine, bio-imaging, optical devices intended for energy transmission, and drug delivery (Kostarelos et al. 2015; Yamashita et al. 2012). Composite-nanostructure systems are an important field of growth at the present time. A typical, unique sample is the amalgamation of shape memory materials into other well-designed materials to form hybrid composites in which each individual material benefits. The shape memory alloys show substandard dynamic response and poor efficacy on their own. However, while their displacements are small and many of them are very brittle, ceramics including piezoelectric show a greater dynamic response. When associated with monolithic shape memory alloys, a scheme of shape memory alloys and piezoelectric ceramics are able to formulate smart hybrids that can produce a greater movement than traditional piezoelectric ceramics and process an enhanced dynamic response.

In order to achieve controlled functionality and biocompatibility, the design of these systems mandates enhanced specificity. Hybrid nanostructures have been recognized as intelligent stimulus-responsive systems as drug supply transporters, theranostics, and imaging agents for their use in nanomedicine. Researchers are concentrating and implementing the carbon based, lipid based, metallic, hybrid and composite nanostructures materials in medicine for focusing different applications.

## 1.3 Different Nanomaterials Used in the Nanobiotechnology Field

With the aid of nanotechnology, an extensive range of NPs with anticipated physical and chemical structures are synthesized, but they are divided into a few distinct groups. It is possible to classify nanoparticles into different groups. Nanoclusters are semi-crystalline and have a narrow size range and dimensions from 1 to 10 nm. Nanopowders are formed with the help of accumulation of non-crystalline nanomaterials ranging from 10 to 100 nm dimensions. The nanomaterials having the sizes of 100–1000 nm are nanocrystals, and are single crystalline (Su and Kang 2020). There are several other types of nanomaterials, depending on the morphology, such as nanowires, nanocups, nanostars, nanospheres, nanocapsules, and quantum dots (Lehner et al. 2013) as shown in Fig. 1.3.

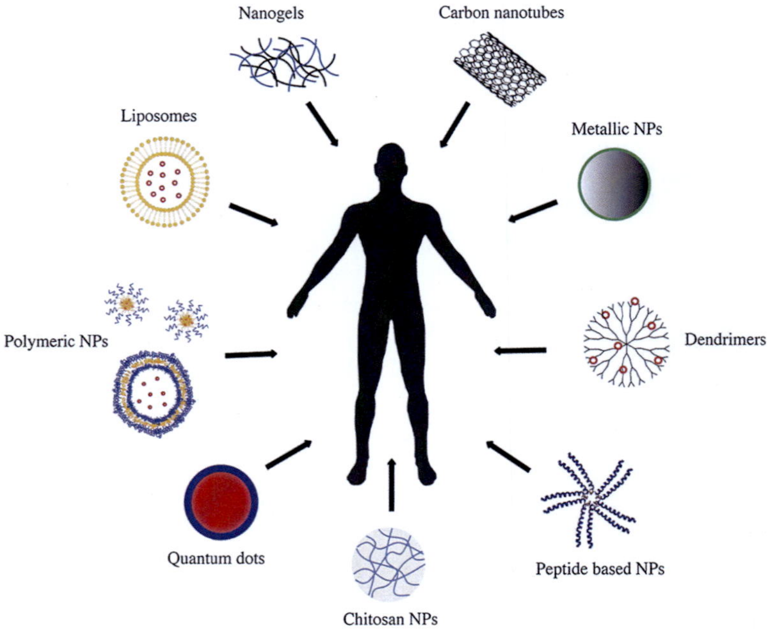


**Fig. 1.3** Various nanomaterials used in biotechnology. (Lehner et al. 2013, Biology and Medicine)

## *1.3.1 Liposomes*

Liposomes are vesicles with at least one lipid bilayer. They are prepared from biological membranes that are broken. Liposomes were investigated and regarded as nano-containers of biopolymers in the 1960s, then were developed for drug-carrying in the 1970s. In the chemistry domain, these materials are utilized as nano- and micro-reactors, in phase handover responses and in catalysis; they have been recorded as very beneficial. They improved the appreciation of the biological usefulness of proteins in the field of biochemistry, especially in excretion, signaling, and gene delivery (Daraee et al. 2016).

Liposomes are used as drug carriers in the pharmaceutical industry for various kinds of therapeutics in cancer treatment, vaccination, operation, and anti-fungal treatment. It is also utilized in the development of cosmetics, shampoos, and skin care products. Superparamagnetic Nanoparticles Ferromagnetic or ferromagnetic materials such as iron oxide are made of super paramagnetic nanoparticles. In addition to iron, these include nickel, cobalt, and manganese. In order to obtain desired properties, these NPs may be covered by inorganic materials such as gold, silica, and few certain organic materials, namely peptides, fatty acids, surfactants, polysaccharides, and polymers.

The induced magnetism that is an important aspect of these NPs is superparamagnetism, which distinguishes them from other NPs. These materials are pinched into a magnetic field, but after elimination of the applied field they maintain no residual magnetism. In magnetic resonance imaging (MRI magnetically guided directed drug supply, magnetic separation, magnetic hyperthermia, and magnetically supported gene transfection keen on cells) various bio-applications are found such as contrast agents (Gregoriadis 1995).

### 1.3.2 *Fullerenes*

Bucky balls and CNT Fullerenes are high-carbon hollow cylinders, tubes, or ellipsoids. Although CNTs are single-walled or multi-walled carbon atom tubes that can be open or closed at their ends, bucky balls are spheres (Bakry et al. 2007). The same kind of arrangement, consisting of layers of carbon atoms commonly connected by pentagonal, hexagonal, or heptagonal rings, is possessed by fullerenes and graphite.

CNT is thoroughly studied and identified to ensure special and remarkable mechanical and electronic behavior that make them greater to metals and it should be recognized as one of the lightest materials. In the field of nanotechnology, CNTs have made major advances as they hold strong possibility in the production of nanodevices including sensors, energy generation and storing, catalysis, computers, and electronics (Anilkumar et al. 2011). CNTs have also been used in other bio-applications, namely drug supply, bio-sensors, bio-electronics, and so on.

### 1.3.3 *Dendrimers*

Dendrimers are polymeric structures with normal-size branches, radial structure, and circular form in the solution. Dendrimers are designed by repeated reactions layer-by-layer from the center to the boundary, resulting in the development of covalent bonds. The thickness of dendrimers is increased because of geometric growth on branching points (Kalpana et al. 2018). Synthesis of dendrimer is probable by choosing a reagent with active surface of different groups.

The dendrimer molecule was first developed in the 1980s, but became popular in the twenty-first century owing to the huge quantity of applications discovered. In the area of nanomedicine, numerous labels for the proof of identity of enzymes, dyes, and several molecules have appeared as a valuable tool in the development of a high amount of molecular "hooks" available on their surface. Their fabrications can be very difficult and expensive, creating a hitch on a wide scale for their applications.

### *1.3.4 Quantum Dots*

Semiconductor nanoparticles are composed to formulate quantum dots (QD), which are nanocrystals that can emit light or fluoresce when electronically irradiated or excited (Smith and Nie 2009; Chen et al. 2017). Depending on the size of the nanocrystals, they are able to release light in very nearly any spectral color (Molaei 2019a, b). If it switches from the blue to the red end of the continuum, it is necessary to increase the size of the QDs, and vice versa (Molaei 2019a, b). The infrared (IR) or ultraviolet (UV) spectrum can also be modified outside the visible spectrum.

## 1.4 Advantages of Nanobiotechnology

Due to high affinity and easy compatible binding of biological samples with different nanomaterials, the biomedical research can be improved a lot using nanomaterials (Fig. 1.4).

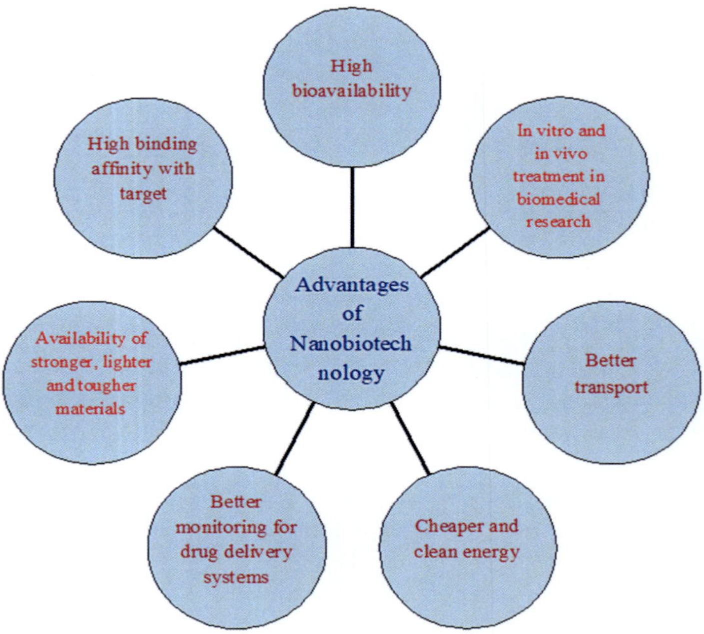


**Fig. 1.4** Factors involved in the advancement nanobiotechnology in various research areas

## 1.5 Nanobiotechnology in Medicine

Nanomedicine, the application of nanobiotechnology to medicine, is the use and development of resources and materials, novel devices, and systems through the control of matter on the nanometric level. Healthcare has already been impacted by nanobiotechnology. Many advancements have been made in all areas of the medical field. Health and medicine are thus the most exciting and promising spheres of influence in nanotechnology (Dash et al. 2020).

Potential developments have been offered in the pharmaceutical field, medical imaging and diagnosis, curing of cancer, implantable materials, and renewal of tissues and multifunctional platforms with the help of nanotechnology in combination with some of these modes of action. The scientific community has contributed to the achievement of tremendous accomplishments with significant opportunity for future prospects. These are biological fields influence by nanotechnology which then creates nano version of that field (Jain 2008) include biological treatment techniques, pharmaceuticals, ophthalmology, oncology, cardiology, neurology, orthopedics, pulmonology, dermatology, regenerative medicine, and tissue engineering.

In the late 1960s, ETH Zurich identified the initial outcome associated with the development of nanomedicine. Dramatic scientific, technical, and industrial advancements in nanomedicine are seen in more recent times, that is, the last couple of decades. In the beginning of 2003, the European Science Foundation initiated its "Forward look on nanomedicine," which acted as a boon for further expansion of the field. In addition to the application of nanotechnology in clinical arenas, nanomedicine is rooted in fundamental and essential theories and principles of nanotechnology; for instance, all the materials that hold nanoscale characteristics possess distinctive features, or else they are absent at a macroscopic level. Nanomedicine finds its advancements in a multidisciplinary nature, incorporating the principles and techniques of biology, chemistry and physics, similar to how nanotechnology sustains and benefits from mathematics and engineering.

The persistent accomplishments of nanotechnology in the healthcare division is determined by the ability to put efforts and work at different cellular mechanisms, organic molecules, and a wide range of biological processes. Owing to these grounds, the field of medicine has had a considerable look at nanotechnology and incorporating it as one of the ideal solutions for the identification and treatment of various infections and diseases. One among the countless applications of nanotechnology in medicine is in the field of drug delivery. Proper definition of the protocols and systems for synthesis, functionalization, and appropriate usage of synthesized nano-sized particles and nano-carriers has showered the clinical and scientific community with innovative and novel therapeutic approaches from molecular target to radiofrequency ablation and from personalized therapies to noninvasive practices (Hossen et al. 2019) as shown in Fig 1.5.

The key significant focus of nanomedicine comprise of supply of pharmaceuticals in vitro, on vivo and in vivo diagnostics, together with imaging, implanted devices, and regenerative medicine.

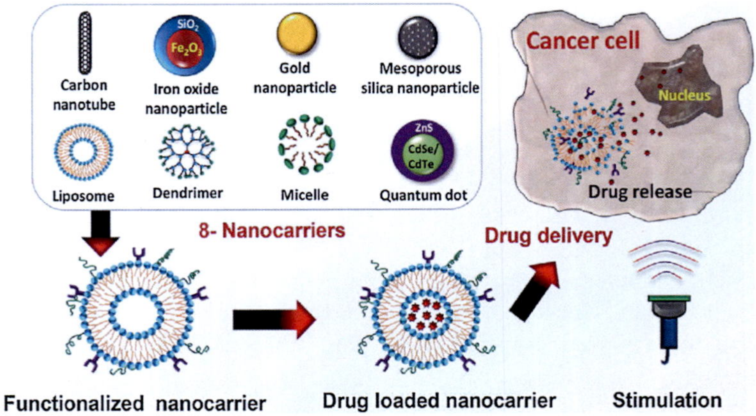


**Fig. 1.5** Nanomaterials involved in the field of drug delivery. (Hossen et al. 2019, J. of Adv. Res.)

### *1.5.1 Nano-pharmaceuticals*

Nanotechnology in nano-pharmaceuticals consists of inventing, improving, and releasing drugs. Screening and fixing drugs that can work the best occurs during the inventing step. It is also proved that the process of inventing drugs needs to be improved via nanotechnology in order to establish substantial groups of small spaces, which paves way for direct readings of the signals by means of micro-fluidics. The main aim in reducing the work area can be compared to computers that once took up an entire room, but can now be held in the palm of a hand. Hence to shape in accordance with today's technology and advancements in an efficient way, researchers are intending to switch laboratory experimentation to fit into a range within nanoscale for the screening of drugs, which would also make it possible to conduct thousands of trials at a time. Most popularly used nanoparticles used for the screening of drugs include nanoparticles of gold, quantum dots, nanolasers, and so on (Hossen et al. 2019).

### *1.5.2 Drug Discovery and Drug Development*

As already mentioned, nano-biotechnology is applied with a great effect in diagnostics and drug delivery. In recent years, research has added further improvements to the advantage of nanotechnology in drug invention. Nanomaterials such as nanocrystals and added nanoparticles such as colloids of gold, nanobodies, magnetic nanoparticles, nanoshells dendrimers, fullerenes, and nano-barcodes cover exceptional benefits in the invention of novel drugs; for example, magnetic nanoparticles or quantum dots are applied in the process of bar-coding of definite analytes that are the

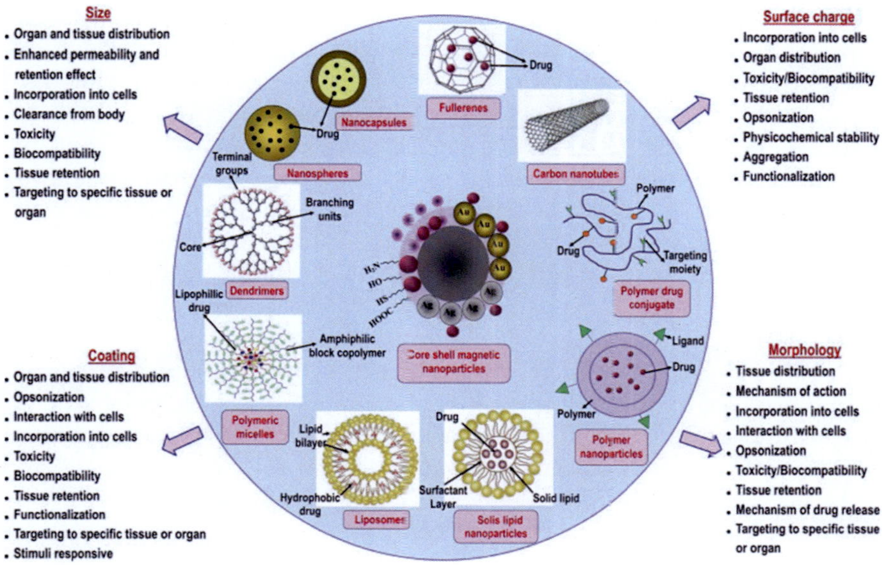


**Fig. 1.6** Illustration of different nanomaterials used in biomedical research. (Rana and Radhika 2019, Nanoscience and Nanotechnology in Drug Delivery Micro and Nano Technologies)

vital parts of the bio-barcode assay, which can then facilitate as a forthcoming substitute to the protein chain reaction (Rana and Radhika 2019) as shown in Fig. 1.6.

The name "Nanomolecular diagnostics" can be defined as the application of nanobiotechnology at molecular range for care and detaction of chronic diseases via widening the restrictions of molecular diagnostics on a nanoscale. The field of nano-molecular diagnostics has the potential to fulfill the present needs of the clinical laboratory, with due consideration of the cost factor as well.

Nanotechnology led to the development of variety nanoparticles in competent diagnostics and therapeutics possessing attributes at the nano-scale range. Specific ligands have the feasibility of immobilizing on the biological site surface, which can be turned out as ultimate aspirant for molecular susceptible finding, molecular imaging scheme, and innovative transporters for the medicines targeted, delivery of gene and drug, and photo-thermal therapy. The chosen nanoparticles which are in nano-vesicle form are supposed to be enclosed by a membrane or a layer and also encompass a suitable surface (sphere-shaped, cylinder-shaped, plate-like for molecular assembly of polymeric or inorganic drugs). The particle size and size distribution of the nanoparticle is an important parameter under critical consideration, thus performing a noteworthy task for pore structure diffusion of a cellular membrane. Competent fluorescent probes are required by the biomarkers for occupying distinguished color, which is solely dependent upon lean distribution of average-sized particles emitting a broad range of wavelengths. Several luminescent layers contained in the multifunctional nanoparticle core confirm to be similar to magnetic nanoparticles, thereby detecting and manipulating the particles. As a consequence, the nanoparticles provide a joint platform of nanotechnology, biotechnology, and information technology, making it possible to carry out biochemical, molecular, and biological processes like genetics, pharmacogenomics, and so on.

The ability to redesign the surface morphological characters of nanoparticles facilitates in the attainment of the objectives such as blood brain barrier and dermal tight junctions in a further effective and powerful manner through prevailing over the concern of drugs on the physiological barriers. A vascular structured nanoparticle has the tendency to enter into the lesion, which is of greater use in the case of malignant tumors. Such particles may perhaps restrain coating of an inorganic core of super-paramagnetic materials with natural and synthetic polymer.

Nanoparticles like fluorophores or semiconductor nanoparticles burn out or blink, whereas gold nanoparticles do not. Easy means of preparation, lower toxicity range, and the feasibility of getting bonded to the biological molecules are other added advantages of these particles. In addition, the frequency of the laser utilized for visualization of these particles selected that only negligible harm is caused to the biological tissues. Uni-molecular tracking of the biological samples and drug samples make use of this technology (Hossen et al. 2019).

Furthermore, quantum dots are being utilized for unimolecular tracking where brightness, stableness, and smaller size makes its more advantageous and superior for tracking compared to fluorescent dyes. Another feature is that they also sustain longer time than dyes—up to 40 min in contrast to 5 s. Quantum dots are also utilized for the detection mechanisms that can last even longer. A typical example is the receptors that build up a drug to fight against epilepsy and depression. Prolongation of Alzheimer's disease (AD) or Parkinson's disease (PD), and ill health caused by harmful emitted rays or chemical nerve agents, can be reduced by the use of drugs found through nano-lasers. The mitochondrion considered as a vital organelle of the cell is affected in AD and PD. Drug identification to safeguard the mitochondrion is very slow and complicated. Nano-lasers help to increase the speed of the process. Research provides evidence for the use of lasers, like those in compact disc players but combined with living cells, to sense the mitochondria fatality. Another laser technique allows a mitochondria flow within a solid-state microscopic laser, and it is also identified that the mitochondria have the capability of "lasing" itself. The rate of recurrence exposed by the mitochondria is a function of their health. A single color was identified using healthy mitochondria and a different color was seen with those that were swollen or dying. The death signal received from the PD or AD can be given to mitochondria with the help of this technology. As a result, thousands of drugs that inhibit the death signal can be tested at the same time. Cyclosporine is the most accepted and popularly used drug, but the immune system is found to get weakened by it. Jain (2008) proposes that the laser technology paves way for the researchers to discover a remedy using this drug that overcomes the existing limitations. Imaging at the nanoscale by and large uses atomic force microscopy (AFM), which can be used for the discovery of drugs as well. The tip of an AFM is connected to a ligand and thus biological systems can be investigated using the functionalized tip. The lively nature of the cells can be modified.

The reaction taking place at the functionalized tip may also be seen. Upon proceeding, diverse biological systems can be explored along with their chemical entities and surface properties. AFM is being used along with functionalizing the tip for AD. Aβ, which is a peptide, is implicated in the mechanism of AD and the only difficulty encompassed by the peptide is that it gets altered from the property of solubility to insolubility. AFM is applied for identifying the suitability of antibodies to hinder the morphology of the peptide. Previous works proved results through testing two antibodies (M266.2 and $m_3D_6$). Complete inhibition of morphology was observed using M266.2, whereas it was found that $m_3D_6$ could only retard the morphology of the peptide. Simultaneous experiments can be handled by devising a new device. This new device is around 1 $cm^2$ and contains more than thousands and up to millions of vessels with an open top and closed bottom resembling nano test tubes. Millions of drugs can be screened at a time using this chip, which has also been suggested for chemotherapy. Cancer cell membranes have pumps that allow the chemotherapy drug out.

### *1.5.3 Nanobiotechnology in Drug Delivery Systems*

As technologies evolve, researchers seek to demonstrate substantial significance in the application of nanobiotechnology to the pharmaceutical industry. Key relevance of nanobiotechnology is found and proven in drug invention, drug improvement, and drug delivery, an area collectively called nano-pharmaceuticals.

Predominant application of nanoparticles using these technologies has evolved in contributions to drug discovery and development through nanobiotechnology (Patra et al. 2018). Nanobiotechnology not only facilitates the discovery of drugs, but in addition a number of drugs are being developed from nanomaterials, including dendrimers, fullerenes, nano-bodies, and so on.

Recently, nanobiotechnology has been utilized in most of the biotechnological and pharmaceutical industries. In drug improvement, nanobiotechnology is used in all stages, from formulation to application of suitable dosages and appropriate delivery systems for proper administration and monitoring. Nanobiotechnology is applicable to the diagnostic process for many deadly diseases. It is predicted that in the near future nanobiotechnology, in connection with computer systems, may provide appropriate and absolute information about biological systems or an individual cell. The aid of this virtual representation might lead to the development of novel drugs of higher precision and accuracy without the need to carry out experimentation in live organisms.

The drug supply scientists are looking the task such as low solubility, higher molecular dimensions and bioavailability. Administering geriatric and pediatric drugs, delivery of protein and peptide drugs, and so on, are other hindrances in this field. The foremost need in the drug delivery field at present is the development of safe, noninvasive drug delivery techniques (Zhang et al. 2018).

The above-mentioned troubles in drug delivery can be overcome by the significant role played by the nanobiotechnology sector in providing the following solutions with regard to drug delivery problems: (a) This technology helps reduce the size of the drug particle to a range of nanometer size, thereby enhancing the plane area and in due course the rate of dissolution; (b) It improves the solubility owing to drug size of nano-meter range; (c) This technology aids the development of noninvasive drug administration practices, eliminating injectable drugs; (d) It has developed nanoparticle formulations as a superior alternative to nonstable formulations with shorter shelf lives; (e) It improves the solubility property of a weakly soluble drug and enhances the ability of absorption, improves bioavailability and discharge rate of bulky molecules, reduces the optimal dosage and enhances the safety by reduction of effects; (f) Principles of nanobiotechnology extend sustained and controlled release formulations with improved patient compliance.

### *1.5.4 Nanobiotechnology in Biomedical Imaging and Diagnostics*

The history of past decades shows that there has been great advancement in the areas of biotechnology and chemistry that has led to the emergence of new chemical and biological molecules for a wide range of application in human activities. Problems from the nonrational use of newly developed materials of higher technology, such as environmental degradation, toxicity of the ecosystem, high demand for the fabrication of cleansing units for the chemical and biological wastes, have accompanied these advances. The demand therefore rose for improvement in detection processes for the newly developed materials. Also, protein or nucleic acid bioanalytical assays detection became popular tools (Han et al. 2019).

Today, technological and scientific progression and added awareness lend a hand to the researchers and scientists in the development of a molecular tool for detection of toxic and contaminated material in daily life and bimolecular identification in a precise manner (e.g., proteins) that can cause various syndromes. The diagnosis with clinical application and the monitoring of human illnesses using of molecular technology is called molecular diagnostics. This is associated with the extended understanding of "DNA diagnostics" that is highly useful in protein technology, genes, and nucleic acid. Enzyme-Linked Immunosorbent Assay (ELISA) and Monoclonal antibodies (MAbs) are also considered as vital tools in biotechnology diagnostics. The term *genomic diagnostics* is widely applied in the studies pertaining to organism genes, its base configuration and sequence, regulation mechanisms, various interactions, and the final medicines via genomic molecular diagnostics. Nanobiotechnology also uses frequent terms from molecular diagnostics, which are added in the comprehensive type of biochips/microarrays. It is observed that the terms nanochips and nanoarrays is found more appropriate for approach (Cholkar et al. 2017). Nanotechnologies in a chip form are relevant to diverse applications and execute assays in nano volume capacities. Many nano devices are devised (enclosing nanostructures in a well-organized fashion) from inorganic materials to investigate the DNA sequence which is having wide range of medicinal and biological applications. These devised nano devices are portrayed as markers.

Nano devices like sensors in nano dimensions that are found to be precise during assays, i.e., living organisms and other receptors, are noticeable. Labeling systems that have the tendency to estimate the presence or value of products with prominent sensitivity and superior fastness in assays also fall under the category of nanoparticles. It is to be distinguished that biomaterials are considered marching toward excellence under nanoscale technologies. New approaches to the material science in innovative technological platforms (for commercial products), mainly in daily goods and therapeutics, are being provided by their properties.

Quantum nanodots, which can be categorized under spherical nanocrystals, are far and wide utilized as probes for imaging and as therapeutic products that are produced by means of metal elements possessing semiconductor properties (e.g., CdSe, CdTe, CdS, ZnS, PbS) and other metals (e.g., Au). Diameter of quantum nanodots ranges from 2 to 10 nm, corresponding to 10–50 atoms. In a general view, quantum nanodots consist of a semiconductor core usually covered with a shell such as ZnS to enhance the optical properties and buffer solubility.

The use of quantum nanodots (nanocrystals), semiconductor quantum dots (semiconductor nanocrystals), or simply quantum dots (QD) which are composed from stabilizers of metal ions and colloidal for making use in biological labeling and imaging, is still mounting (Siddique and Chow 2020). Tumor imaging agents, fluorescent labels for labeling of cells, intercellular biosensors for deep tissue, and sensitizers used in photodynamic therapy are other future applications. Further usage may not be limited in biological applications due to their surface chemistry and low cytotoxicity. Literature proves that there has been great interest in the study of nanoparticles used as nanoparticle mediated delivery of drugs due to the broad range of applications (Kumari et al. 2020; Wang et al. 2008).

## 1.6 Biosensors

The need for bioanalytical tools has led to biosensor technology advancement that also has increased in recent years due its ability in alteration and the interface of biological receptor together with the biomolecule that needs to be defined in an analytical grade. Biochemical sensors are those chemical sensors with a system dependent on biochemical mechanisms. Biosensor fabrication and development is applied for many areas and are persistently budding on a rise during the years. At present, target identification, assay development and validation, lead optimization and absorption, metabolism, distribution, excretion, and toxicity are major applications of biosensors. Other fields in which biosensors are widely applied include clinical diagnostics, quality control and pharmaceutical analysis, biomedicines, microbiological virus and bacteria detection, food production and quality check, veterinary, municipal, and industrial solid waste management, detecting toxic gas in industries and its prevention, and surveillance of ecological contaminations. The capability of these sensors to grant assay results in low-detection range and higher specialism in a very short period offers immense returns and benefits (Jianrong et al. 2004; Alejandro and Merkoci 2016). Techniques that are applied today for the biological and environmental sample studies usually take ample time—days and even weeks—to achieve their end result. Human lives are placed under severe risk due to the delay in initiating a remedy for illness due to the usage of certain improper measures (Mobed et al. 2020), as depicted in Fig. 1.7.

The effective use of developed biosensors is dependent on several factors based on the material used, i.e., mechanical support of the materials, surrounding substances or membranes, and active features of the biological systems. Supports for biosensors include various types, such as the sensor optical examination in which optical waveguides play a key role, dynamic support like fluorescent nanoparticles, metals spheres, metal films, and organic/inorganic nanoparticles and microparticles (Bagyalakshmi et al. 2020).

## 1.7 Other Applications of Nanobiotechnology

### *1.7.1 Tissue Engineering*

Tissue engineering is a fast-growing scientific area (in the medical field) focused on creating, repairing, and/or replacing cells, tissues, and organs using cells or combinations of cells with biomaterials together with dynamic biological molecules

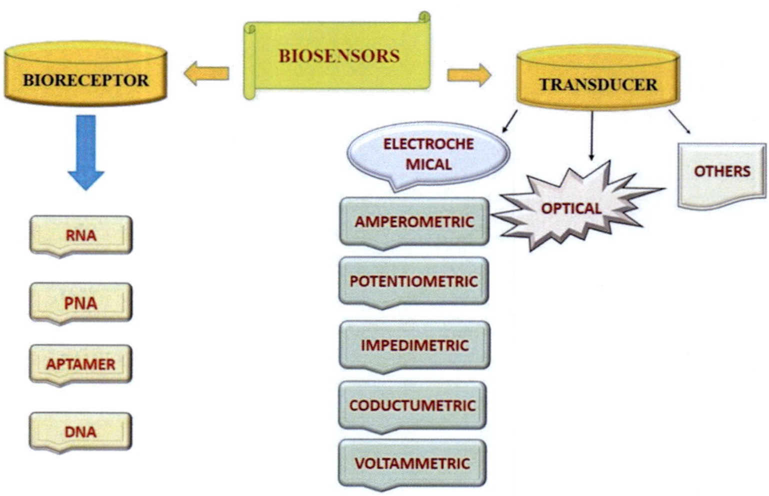


**Fig. 1.7** Different types of Biosensors used in various research fields. (Mobed et al. 2020, Analytical Methods)

helping in the formulation of materials that resemble the body's native tissue (Almouemen et al. 2019).

Nanofabrication practices in tissue engineering hold several benefits. Nanotechnology plays a vital role in the fabrication of nanofibers, nanopatterns, and controlled-release nanoparticles with relevance to tissue engineering (Dzobo et al. 2018). Since nanometer-sized biomaterials are to be developed like the extracellular fluids, cardiac tissues, bone marrow etc., native tissues may be mimicked.

### *1.7.2 Nanobiotechnology in Agriculture and Food Sectors*

The drastic revolution that took place in the agricultural and foodstuff sectors, predominantly the invention of smart and active packaging, has been achieved due to the attainment of fast growth in the field of nanotechnology with due facilitations in nano-sensors, nano-pesticides, and nano-fertilizers (Khan 2017). In addition, improvement in quality of food and safety, crop development and examination of the conditions of the environment, are due to the rise in novel nanomaterials. Nanotechnology thus offers surplus opportunity through sustainable and novel substitutes in the agricultural and food sectors (Shafiq et al. 2020; Chen et al. 2014) as show in Fig. 1.8.

The Nanomaterials grasp name is for well designing of taste with color additives, preservatives and carriers for foodstuff supplement (i.e., nano encapsulation and nano emulsion), together with the products for animal feed. Engineered nanomaterials possess certain unique characteristics that offer added benefits in the processing of food as chief components and supplement materials. Moreover, the U.S. FDA permits inorganic chemicals like MgO, $SiO_2$, and $TiO_2$ as anti-caking agents, carriers for food flavor, and foodstuff color additives. For example, $TiO_2$ is used widely as additive in candy, white sauces, puddings, and cake icing.

Nanomaterials that are specially designed for the packaging of food enjoy lot of advantages in comparison to conservative wrapping supplies. Mechanical, thermal, other barrier properties, and economic viability has led to nano-clay, which is used for food packaging, being one of the most widely used novel nanomaterials. For food preservation and storage, edible coating with nanomaterials has proved its potentiality. Lively investigations, research, and expansion paves the way for applications in foodstuff packaging including detection of pathogens, pesticides, and toxin identification owing to ultrasensitive characteristics of nanomaterials. Agriculture uses nanotechnology to increase foodstuff production with an equal or higher nutritional rate as well as improved safety and quality aspects. The most important way crop production can be improved is the efficient usage of pesticides, fertilizers, herbicides, and other factors or regulators that help in the regulation of plant growth. Controlled and appropriate release of herbicides, pesticides, and vital factors that assist plant growth can be achieved with nanocarriers. Protection of plants against pathogenic infections due to intrinsic toxicity is studied with the assistance of metal oxide nanomaterials like ZnO, $TiO_2$, and

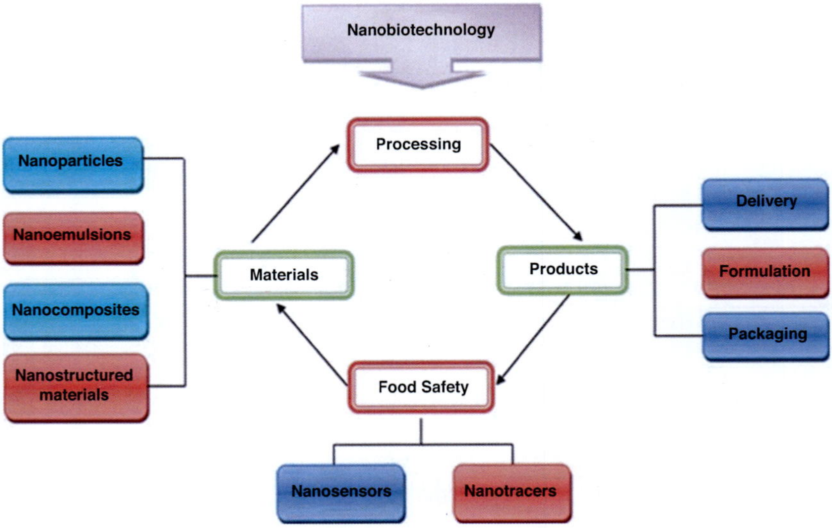


**Fig. 1.8** Nanobiotechnology in food sectors. (Shafiq et al. 2020, Foods)

CuO. Wireless nanosensors have also been developed to assess the nutrient efficiency in crops, retardation in crop growth due to infections, diseases and environmental conditions, and these demonstrate the beneficial aspects of nanomaterials. Newly devised nanosensors can detect pesticides and herbicides along with pathogens at meager quantities in foodstuff, which is notable. Thus, these real-time and in situ examination practices help to rectify the loss of potential crops, leading to the enhancement of crop production with appropriate use of nano-pesticide, nanofertilizer, and nano-herbicides (Zhao et al. 2020). A modern study reports that online monitoring of fungicide in aquatic environments (both fresh and salt water), with a lower detection range of about 1 mM, can be attained using copper-doped montmorillonite. Pathogens in wastewater can be detected with graphene nanomaterials, and purification for use as drinking water signifies prospective applications in aquaculture. Numerous added nanomaterials such as nanoparticles of copper, carbon nanotubes, gold, and silver are under consideration to be developed as nanosensors for real-time monitoring of crop health and growth and other environmental conditions.

Many diseases are not cured today but it may be cured by the rapid development in nanobiotechnology. The potential benefits and expectations of nanobiotechnology are endless in the nanomedicine still [Neeraja et al. (2017)]. The different agencies and governments are well-focused toward the development of nanomedicine to cure diseases that lack cures at present.

## 1.8 Conclusions, Outlook, and Future Prospects

Nanobiotechnology is showing great potential in the development of drug delivery and discovery, imaging (diagnostics), therapeutics, molecular imaging, regenerative medicine, tissue engineering, biomarkers, and biosensors fields. It promises a good future for the clinical diagnosis and drug delivery sectors. Due to continuous development of this field, there is the possibility of curing incurable diseases by using nanobiotechnology resources. It is showing great significance in manipulating the bio-functionalization at the atomic or molecular scale. The applications of nanobiotechnology depend on the physiochemical and size-dependent biological properties, surface, pH, ligand capacity, and toxicity. Several factors, such as synthesis methods, pH, temperature, pressure, particle, and pore size of nanoparticles, greatly influence the quality of bio-devices. The combination of biological molecules with different types of nanomaterials such as fibers, particles, and grains has led the challenge of developing biomedical devices in the healthcare and agriculture industries.

## References


Alejandro, C.-G., & Merkoci, A. (2016). Nanobiosensors in diagnostics. *Nanobiomedicine, 3*, 1–26.

Almouemen, N., Kelly, H. M., & O'Leary, C. (2019). Tissue engineering: Understanding the role of biomaterials and biophysical forces on cell functionality through computational and structural biotechnology analytical methods. *Computational and Structural Biotechnology Journal, 17*, 591–598.

Anilkumar, P., Lu, F., Cao, L., Luo, P. G., Liu, J.-H., Sahu, S., Tackett, K. N., Wang, Y., & Sun, Y.-P. (2011). Fullerenes for applications in biology and medicine. *Current Medicinal Chemistry, 18*, 2045–2059.

Bagyalakshmi, S., Sivakami, A., & Balamurugan, K. S. (2020). A Zno nanorods based enzymatic glucose biosensor by immobilization of glucose oxidase on a chitosan film. *Obesity Medicine, 18*, 100229.

Bakry, R., Vallant, R. M., Najam-ul-Haq, M., Rainer, M., Szabo, Z., Huck, C. W., & Bonn, G. K. (2007). Medicinal applications of fullerenes. *International Journal of Nanomedicine, 2*, 639–649.

Chandrasekaran, A. R., Venugopal, J., Sundarrajan, S., & Ramakrishna, S. (2011). Fabrication of a nanofibrous scaffold with improved bioactivity for culture of human dermal fibroblasts for skin regeneration. *Biomedical Materials, 6*, 015001.

Chen, Y., Mohanraj, V. J., & Parkin, J. E. (2003). Chitosan-dextran sulfate nanoparticles for delivery of an anti-angiogenesis peptide. *Letters in Peptide Science, 10*, 621–629.

Chen, H., Seiber, J. N., & Hotze, M. (2014). ACS select on nanotechnology in food and agriculture: A perspective on implications and applications. *Journal of Agricultural and Food Chemistry, 62*, 1209–1212.

Chen, F., Gao, W., Qiu, X., Zhang, H., Liu, L., PuLiao, F. W., & Luo, Y. (2017). Graphene quantum dots in biomedical applications: Recent advances and future challenges. *Frontiers in Laboratory Medicine, 1*, 192–199.

Cholkar K., Hirani N. D., Natarajan, C. (2017). Nanotechnology-Based Medical and Biomedical Imaging for Diagnostics. *Emerging Nanotechnologies for Diagnostics, Drug Delivery and Medical Devices, Micro and Nano Technologies*, 355–374.

Crean, C., Lahiff, E., Gilmartin, N., Diamond, D., & O'Kennedy, R. (2011). Polyaniline nanofibres as templates for the covalent immobilisation of biomolecules. *Synthetic Metals, 161*, 285–292.

Daraee, H., Etemadi, A., Kouhi, M., Alimirzalu, S., & Akbarzadeh, A. (2016). Application of liposomes in medicine and drug delivery. *Artificial Cells, Nanomedicine, and Biotechnology, 44*, 381–391.

Dash, D. K., Panik, R. K., Sahu, A. K., & Tripathi, V. (2020). *Role of nanobiotechnology in drug discovery, development and molecular diagnostic, applications of nanotechnology* (Chap. 1, pp. 113). London: Intech Open Access

Dzobo, K., Thomford, N. K., Senthebane, D. A., Shipanga, H., Rowe, A., Dandara, C., Pillay, M., Shirley Caroline, K., & Motaung, M. (2018). Advances in regenerative medicine and tissue engineering: Innovation and transformation of medicine. *Stem Cells International, 2018*, 1–24.

Ge, L., Li, Q., Wang, M., Ouyang, J., Li, X., & Xing, M. M. (2014). Nanosilver particles in medical applications: Synthesis, performance, and toxicity. *International Journal of Nanomedicine, 9*, 2399.

Govindaraju, K., Krishnamoorthy, K., Alsaga, S. A., Singaravelu, G., & Premanathan, M. (2015). Green synthesis of silver nanoparticles for selective toxicity towards cancer cells. *IET Nanobiotechnology, 9*, 325–330.

Gregoriadis, G. (1995). Engineering liposomes for drug delivery: Progress and problems. *Trends in Biotechnology, 13*, 527–537.

Han, X., Xu, K., Taratula, O., & Farsad, K. (2019). Applications of nanoparticles in biomedical imaging. *Nanoscale, 11*, 799–819.

He, X., Deng, H., & Hwang, H.-M. (2019). The current application of nanotechnology in food and agriculture. *Journal of Food and Drug Analysis, 27*, 1–21.

Hossen, S., Khalid Hossain, M., Basher, M. K., Mia, M. N., Rahman, M. T., & Jalal Udin, M. (2019). Smart nanocarrier-based drug delivery systems for cancer therapy and toxicity studies: A review. *Journal of Advanced Research, 15*, 1–18.

Jain, K. K. (2005). The role of nanobiotechnology in drug discovery. *Drug Discovery Today, 10*, 1435–1442.

Jain, K. K. (2008). Drug delivery systems - An overview: Drug delivery systems. In *Methods in molecular biology* (p. 437). Totowa: Humana Press.

Jianrong, C., Yuqing, M., Nongyue, H., Xiaohua, W., & Sijiao, L. (2004). Nanotechnology and biosensors. *Biotechnology Advances, 22*, 505–518.

Juan, J. S., Nó, M. L., & Schuh, C. A. (2009). Nanoscale shape-memory alloys for ultrahigh mechanical damping. *Nature Nanotechnology, 4*, 415–419.

Kalpana, N., Mohan, A., Thakur, S., & Kumar, P. (2018). Dendritic platforms for biomimicry and biotechnological applications. *Artificial Cells, Nanomedicine, and Biotechnology, 46*(sup1), 861–875.

Khan, A. S. (2017). Nanotechnology applications in food and agriculture. In *IEEE 17th International Conference on Nanotechnology (IEEE-NANO)*, Pittsburgh, PA, pp. 877–878

Kiessling, F., Mertens, M. E., Grimm, J., & Lammers, T. (2014). Nanoparticles for imaging: Top or flop. *Radiology, 273*, 10–28.

Kostarelos, K., Bianco, A., & Prato, M. (2015). The winding road for carbon nanotubes in nanomedicine. *Materails Today, 18*, 12–19.

Kumari, A., Pooja, S., & Acharya, A. (2020). Introduction to molecular imaging, diagnostics, and therapy. In *Nanomaterial - Based biomedical applications in molecular imaging, diagnostics and therapy*. New York: Springer.

Lehner, R., Wang, X., Marsch, S., & Hunziker, P. (2013). Intelligent nanomaterials for medicine: Carrier platforms and targeting strategies in the context of clinical application. *Nanomedicine: Nanotechnology, Biology, and Medicine, 9*, 742–757.

Londe, G., Han, A., & Cho, H. J. (2008). MEMS for nanotechnology: Top-down perspective. In S. Seal (Ed.), *Functional nanostructures. Nanostructure science and technology*. New York, NY: Springer.

Maiti, D., Tong, X., Mou, X., & Yang, K. (2018). Carbon-based nanomaterials for biomedical applications: A recent study. *Frontiers in Pharmacology, 19*, 1401–1416.

McNamara, K., & Tofail, S. A. M. (2013). Biomedical applications of nanoalloys. In C. Florent (Ed.), *Nanoalloys: From fundamentals to emergent applications* (Vol. 91, pp. 805–843). Elsevier Oxford Medical Clinics of North America.

Mobed, A., Hasanzadeh, M., Ahmadalipour, A., & Fakhari, A. (2020). Recent advances in the biosensing of neurotransmitters: material and method overviews towards the biomedical analysis of psychiatric disorders. *Analytical Methods, 12*, 557–575.

Molaei, M. J. (2019a). A review on nanostructured carbon quantum dots and their applications in biotechnology. *Sensors and Chemiluminescence Talanta, 196*, 456–478.

Molaei, M. J. (2019b). Carbon quantum dots and their biomedical and therapeutic applications: A review. *RSC Advances, 9*, 6460–6481.

Morais, M. G., Martins, V. G., Steffens, D., Pranke, P., & Vieira da Costa, J. A. (2014). Biological applications of nanobiotechnology. *Journal of Nanoscience and Nanotechnology, 14*, 1007–1017.

Morrow, K. J., Bawa, R., & Wei, C. (2007). Recent advances in basic and clinical nanomedicine. *Medical Clinics of North America, 91*, 805–843.

Neeraja, P., Benazir, F., & Muvvala, S. (2017). Role of nanobiotechnology in pharmacy and medicine: A review. *Journal of Pharmaceutical and Scientific Innovation, 6*, 31–36.

Niemeyer, C. M., & Mirkin, C. A. (2004). *Nanobiotechnology: Concepts, applications and perspectives*. New York: Wiley.

Patra, J. K., Das, G., Fraceto, L. F., Campos, E. V. R., Maria, R.-T., Laura, S., Luis, A., Grillo, R., Kumaraswamy, M., Sharma, S., Habtemariam, S., & Shin, H.-S. (2018). Nano based drug delivery systems: Recent developments and future prospects. *Journal of Nanobiotechnology, 16*, 71.

Peer, D., Karp, J. M., Hong, S., Farokhzad, O. C., Margalit, R., & Langer, R. (2007). Nanocarriers as an emerging platform for cancer therapy. *Nature Nanotechnology, 2*, 751–760.

Petros, R. A., & DeSimone, J. M. (2010). Strategies in the design of nanoparticles for therapeutic applications. *Natural Review of Drug Discovery, 9*, 615–627.

Prasad, R., Pandey, R., & Barman, I. (2016). Engineering tailored nanoparticles with microbes: Quo vadis. *Wiley Interdisciplinary Review of Nanomedicine and Nanobiotechnology, 8*, 316–330.

Rana, V., Radhika, S. (2019). Recent advances in development of nano drug delivery, applications of targeted nano drugs and delivery systems, *Nanoscience and nanotechnology in drug delivery micro and nano technologies*, 93–131.

Ratner, B. D., Hoffman, A. S., Schoen, F. J., & Lemons, J. E. (2004). *Biomaterials science: An introduction to materials in medicine*. Amsterdam: Academic Press.

Shafiq, M., Anjum, S., Hano, C., Anjum, I., & Abbasai, B. H. (2020). An overview of the applications of nanomaterials and nanodevices in the food industry. *Foods, 9*, 148–175.

Siddique, S., & Chow, J. C. L. (2020). Application of nanomaterials in biomedical imaging and cancer therapy. *Nanomaterials, 10*, 1700–1740.

Smith, A., & Nie, S. (2009). Next-generation quantum dots. *Nature Biotechnology, 27*, 732–733.

Stoeckel, D. (2000). Nitinol medical devices and implants. *Minimally Invasive Therapy & Allied Technologies, 9*, 81–88.

Su, S., & Kang, P. M. (2020). Systemic review of biodegradable nanomaterials in nanomedicine. *Nanomaterials, 10*, 656–677.

Sundar, S., Kundu, J., & Kundu, S. C. (2010). Biopolymeric nanoparticles. *Science and Technology of Advanced Materials, 11*, 014104.

Tiwari, A., Mishra, A. K., Kobayashi, H., & Turner, A. P. (2012). *Intelligent nanomaterials*. Wiley: Hoboken, NJ.

Wadood, A. (2016). Brief overview on nitinol as biomaterial. *Advances in Materials Science and Engineering, 2016*, 1–5.

Wang, X., Yang, L., Chen, Z., & Chin, D. M. (2008). Application of nanotechnology in cancer therapy and imaging. *CA: A Cancer Journal of Clinicians, 58*, 97–110.

Wang, A. Z., Langer, R., & Farokhzad, O. C. (2012). Nanoparticle delivery of cancer drugs. *Annual Review of Medicine, 63*, 185–198.

Wei, L., Lu, J., Xu, H., Patel, A., Chen, Z. S., & Chen, G. (2015). Silver nanoparticles: Synthesis, properties, and therapeutic applications. *Drug Discovery Today, 20*, 595–601.

Whitesides, G. M. (2013). The 'right' size in nanobiotechnology. *Nature Biotechnology, 21*, 1161–1165.

Xu, S., Luo, Y., Graeser, R., Warnecke, A., Kratz, F., Hauff, P., Licha, K., & Haag, R. (2009). Development of pH-responsive core–shell nanocarriers for delivery of therapeutic and diagnostic agents. *Bioorganic & Medicinal Chemistry Letters, 19*, 1030–1034.

Yamashita, T., Yamashita, K., Nabeshi, H., Yoshioka, T. Y., Tsunoda, S., & Sutsumi, Y.-T. (2012). Carbon nanomaterials: Efficacy and safety for nanomedicine. *Materials, 5*, 350–363.

Zhang, R., Li, J., Zhang, T., Amini, M. A., He, C., Lu, B., Ahmed, T., Lip, H., Rauth, A. M., & Yu, X. Y. (2018). Importance of integrating nanotechnology with pharmacology and physiology for innovative drug delivery and therapy – An illustration with firsthand examples. *Acta Pharmacology Sinica, 39*, 825–844.

Zhao, Y. L., & Nalwa, H. S. (2007). *Nanotoxicology: Interactions of nanomaterials with biological systems*. Los Angeles, CA: American Scientific Publishers.

Zhao, T., Li, Y., Xia, Y., Venkatraman, S. S., Xiang, Y., & Zhao, X. (2013). Formation of a nanopattering NiTi surface with Ni-depleted superficial layer to promote corrosion resistance and endothelial cell-material interaction. *Journal of Materials Science Materials in Medicine, 24*, 105–114.

Zhao, L., Lu, L., Wang, A., Zhang, H., Huang, M., Wu, H., Xing, B., Wang, Z., & Ji, R. (2020). Nano-biotechnology in agriculture: Use of nanomaterials to promote plant growth and stress tolerance. *Journal of Agricultural Food and Chemistry, 68*, 1935–1947.

Zheng, Y., Tang, Y., Bao, Z., Wang, H., Ren, F., Guo, M., Quan, H., & Jiang, C. (2015). FePt nanoparticles as a potential X-ray activated chemotherapy agent for HeLa cells. *International Journal of Nanomedicine, 10*, 6435–6444.

# Chapter 2
# Quantum-Dot-Based Photoelectrochemical Biosensors: Principles, Fabrication, and Applications

**A. Manjceevan**

## 2.1 Background of Materials and Discoveries

Discovery of photoelectric effect by Edmond Becquerel in 1839 was the starting point of the development of photoelectrochemistry. Research based on photoelectrochemistry became a hot topic due to substantial research scrutiny in multiple applications (Zhao et al. 2014). The basis of photoelectrochemistry is the conversion of photon energy into electrical energy. The absorbance of photon energy by absorber materials such as semiconductors produces electron-hole pairs, followed by charge separation and transfer process. Photoelectrochemistry is used in many applications such as sensitized solar cells (Manjceevan and Bandara 2018), photocatalysts (Zhang et al. 2011), artificial photosynthesis (Tsukamoto et al. 2018), photoelectrochemical water splitting (Berglund et al. 2016) and so on. Application of photoelectrochemistry in photoelectrochemical (PEC) sensors coupled with quantum dots (Q-dots) used in bioanalysis became an interesting topic in recent years (Ma et al. 2018).

### 2.1.1 *Properties of Quantum Dots (QDs)*

Synthesis and application of Q-dots semiconductor materials attracted considerable interest in various applications due to its characteristic features: efficient light harvesting properties, tunable band gap energy, multiple excitation generation, facile synthetic methods, and application in contexts such as solar cells, photocatalyst, photodetectors, lasers, and medical imaging and sensors (Tang et al. 2011; Wang

A. Manjceevan (✉)
Department of Chemistry, Faculty of Science, University of Jaffna, Jaffna, Sri Lanka
e-mail: manjceevan@univ.jfn.ac.lk


K. Pal (ed.), *Bio-manufactured Nanomaterials*,
https://doi.org/10.1007/978-3-030-67223-2_2

et al. 2016). The band gap energy of Q-dot materials can be tunable and can harvest a wide range of the solar spectrum (Pattantyus-Abraham et al. 2010; Do Sung et al. 2013; Wang et al. 2016). The novel properties of Q-dots could be achieved through variation of size, morphology, synthesis methods and conditions of Q-dots, and used in desired applications including q-dot-based photochemical sensors to detect various chemical and biological materials (Piliego et al. 2013). A sensor is a device that converts the physical input to data that can be handled easily by humans (McGrath and Scanaill 2013). In q-dot-based PEC sensors, a Q-dot molecule is attached with a conducting electrode and acts as a sensitizer to the light to improve the sensitivity of the sensor.

### *2.1.2 Quantum-Dot-Based Photoelectrochemical Sensor*

Photoelectrochemical technique possesses potentially higher sensitivity than the conventional electrochemical and optical methods (Wang et al. 2012; Zhang and Zhao 2013). Electrochemiluminescence (ECL) analysis is based on electroluminescence method detecting light intensity, while the PEC analysis detects the photocurrent in a direct or indirect way that uses light to stimulate the material to produce electron transfer (Zhang and Zhao 2013). Further, Q-dots are incorporated into photoelectrochemical technique to enhance the light absorbance (Zhu et al. 2018).

In Q-dot-based PEC sensors, quantum dots are attached to a conducting electrode by linker molecule. Due to the presence of an immobilized layer, under dark conditions it shows a small charge flow and lower signal. However, the presence of Q-dots in PEC sensors act as a photo-switchable layer (Khalid et al. 2011b). Under the illumination condition, the photoexcitation of electrons in Q-dots facilitates the redox reaction and electron transfer process. The current depends on the surrounding chemical environment adjacent to the Q-dot surface of the electrode. Q-dot in PEC sensors induce the charge transfer between conductive electrodes and redox electrolytes and gives measurable and high responses (Hun et al. 2017a, b; Zang et al. 2017). The basic mechanism of PEC sensors is oxidation and reduction reaction of electrolyte materials at the electrode-electrolyte interface. The smaller dark current of PEC sensors increases due to the photoinduced current while illuminated, and it facilitates the high sensitivity.

The q-dot-based PEC sensor can be used to detect chemical and biological substances. In q-dot-based PEC sensors the electronic coupling of Q-dots with biological material leads to detection of the chemical and biological material (Yue et al. 2013; Victorious et al. 2019). Further, by using this method there is a chance to detect materials which cannot be detected by other analytical techniques (Yue et al. 2013). Additionally, q-dot-based PEC sensors have many advantages: they are easy to operate, require no expensive instruments, are low cost (Zang et al. 2017), have high sensitivity and wide linearity range, achieve rapid detection, have a low background current, and so on (Yue et al. 2013). Q-dots have tunable bandgap and its bandgap can be tuned to absorb the light in th UV, visible, and near IR regions.

Therefore, the q-dot-based PEC sensor can be switched on even in white light (Yue et al. 2013). For these reasons, research about Q-dot-based photoelectrochemical sensors become a popular topic and more research was performed to understand the mechanism behind this technique to improve the selectivity and sensitivity of detection of biological and chemical materials.

This chapter discuss the principles of q-dot-based PEC sensors, the fabrication method used by the researchers to enhance the features of q-dot-based PEC biosensor, and the progress of recent applications.

## 2.2 Principle of Photoelectrochemical Sensor

In q-dot-based PEC sensors the Q-dot is attached to the conducting electrode by linker molecules like alkanedithiols (Hsiao et al. 2020), 1,4-dithiane (de Sousa et al. 2003), 1,6-hexanedithiol (Etorki et al. 2019), and so on, and cannot be moved from its position. Sensor application commonly uses a three-electrode system such as working electrode, reference electrode, and counter electrode. The q-dot-based photoanode, Ag/AgCl, saturated KCl, and platinum are used as working electrode, reference electrode, and counter electrode, respectively. The photoelectrochemical sensor performs under applied bias potential (positive or negative) under illumination condition.

Q-dots are small semiconductors, have tunable bandgap (shown in Fig. 2.1), and absorb the light in wide spectrum. Under illumination, the electron in the valence band absorb the photon energy and move to an excited state. It creates electron-hole pairs in the conduction and valence bands (shown in Fig. 2.2). The created hole regenerates by getting the electron from the electrolyte through the oxidation process of electrolyte and tunnels into the valence band. If the electrolyte is a strong reducing agent, the electrolyte can easily oxidize and facilitate the hole movement from Q-dot to electrolyte. The excited electron in the conduction band could be injected into the conducting substrate like FTO glass, metal such as gold or

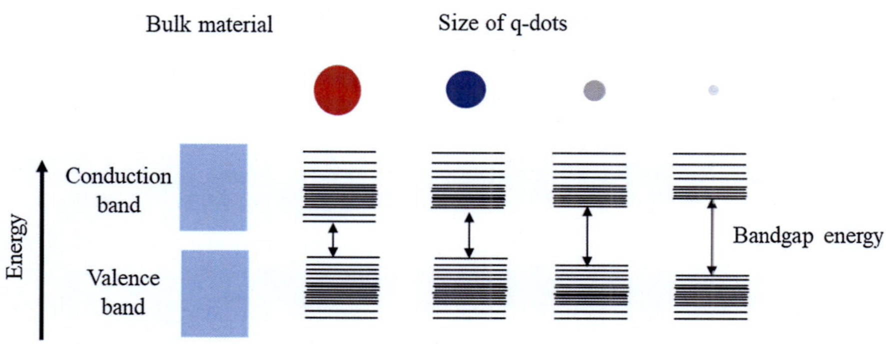


**Fig. 2.1** Size of band gap energy shows inverse correlation with particle size of Q-dots, which is known as quantum confinement effect

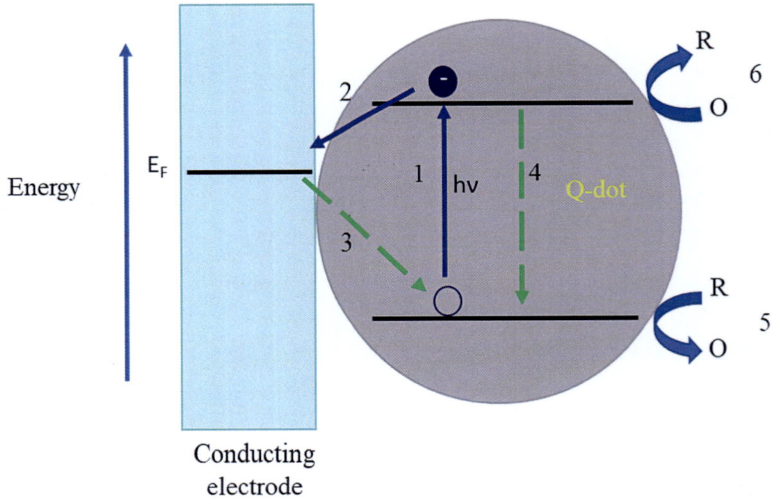


**Fig. 2.2** Charge transfer process after illumination. Excitation of electron from valence band to conduction band of Q-dot (1), injection of electron from Q-dot to conducting electrode (2), recombination of electron (3, 4), oxidation (5) and reduction (6) process at electrode-electrolyte interface (Yue et al. 2013)

aluminum, or it could be recombined with the electrolyte. The recombination of excited electron tunnels into the electrolyte and produces cathodic current, and the injection of electrons into the conducting substance produces anodic current. Reduction of electrolyte occurs due to excited electrons tunneling into the electrolyte. If the electrolyte is a strong oxidizing agent, then the excited electron easily tunnels into the electrolyte and produces cathodic current. The signal is produced as a result of net charge movement of anodic and cathodic current. If the electrolyte is a strong reducing agent, the excited electron directly injects from Q-dot to electrode. Therefore, more electron injects into the conducting substrate and produces more anodic current (Yue et al. 2013). Then if the electrolyte/analyte is strong reducing agent suppress the recombination of excited electron with electrolyte produces high anodic current and it could lead to better sensitivity of q-dot-based PEC sensor.

### *2.2.1 Traps State in Quantum Dots*

Defects available in the Q-dots known as mid-bandgap traps states may also affect the sensitivity of q-dot-based PEC sensor. The defects could be formed during the synthesis of Q-dot or post-synthesis period of Q-dots. Excited electrons under illumination condition occupy the traps states and facilitate recombination with a hole available in the valence band (Manjceevan and Bandara 2016). The recombination process decreases the sensitivity of q-dot-based PEC sensors. Therefore, quantum dots should be synthesized in an ideal condition to minimize the traps state and enhance the sensitivity of q-dot-based PEC sensor (Yue et al. 2013).

The anodic or cathodic current flow depends on several factors such as the properties of Q-dots, the valence band, conduction band position, Femi-level of conductive electrode, applied bias voltage, traps states, electron injection rates from Q-dot to conductive electrode or Q-dot to electrolyte, and connection between Q-dot and conductive electrode. If these properties are successfully controlled, then the concentration of analyte reflected by the charge flow rate could be measured from the amplitude of the current signal (Yue et al. 2013).

## 2.3 Fabrication of Q-Dot-Based Photoelectrochemical Sensor

Q-dot materials are small, typically around 1–10 nm, deposited on top of conducting substrate like gold, fluorine doped tin oxide (FTO) glass, indium doped tin oxide glass (ITO), and carbon substance, or on top of metal oxide semiconductor materials like $TiO_2$, ZnO, $SnO_2$, and so on to fabricate the q-dot-based PEC sensor (shown in Fig. 2.3) (Sheeney-Haj-Ichia et al. 2004; Chen et al. 2018). Common conducting substrates are selected as optically transparent materials and illuminate the electrode through transparent conducting substrates to avoid unwanted interactions with solutions (Yue et al. 2013).

### *2.3.1 Chemical Formulations and Structural Analysis of Linker Molecules*

The Q-dot materials immobilized on conducting substrate or connected through a wide bandgap semiconductor by using linker molecules (Miyake et al. 1999). Linker molecules have at least two functional groups and they improve the connection between quantum dots and the conducting substrates or oxide materials (Wang and Hu 2009). To connect the Q-dots and gold conducting substrates, several types of

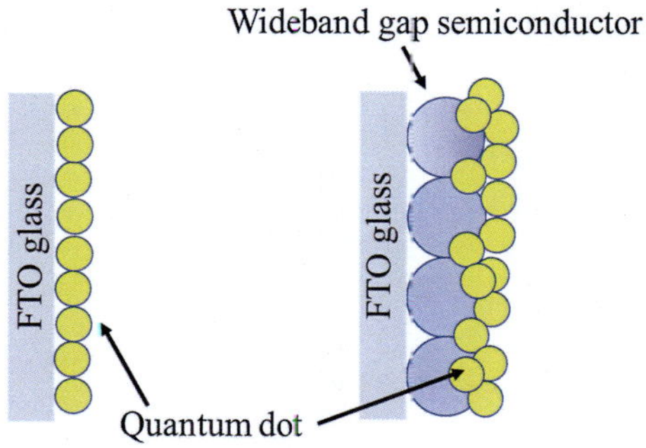


**Fig. 2.3** Q-dots attached with surface of FTO glass and on top of oxide semiconductor

linker molecules were used, including alkanedithiols (Hsiao et al. 2020), 1,4-dithiane (de Sousa et al. 2003), 1,4-benzenedithiol (Khalid et al. 2011a), 6-hexanedithiol (Nakanishi et al. 1998; Etorki et al. 2019), stilbenedithiol (Khalid et al. 2011a). These linker molecules have two thiol groups. Theoretically the one thiol group is attached to the Q-dots and another is bound with the electrode surface. In some instances, however, both thiol groups connected to the electrode may lead to inferior performance of electrodes, and this should be carefully avoided to get better output from devices (Yue et al. 2013). To connect the Q-dots with oxide semiconductors, use bifunctional linker molecules like 3-mercaptopropionic acid (MPA), thioglycolic acid (TGA) and cysteine (Cys) (shown in Fig. 2.4) (Sambur et al. 2010; Lai et al. 2014; Mashhadizadeh et al. 2017; Yang et al. 2018). These linker molecules have COOH and SH functional groups in which the COOH groups preferentially bind with $TiO_2$ materials and SH groups bind with Q-dots. Previous studies illustrate that MPA function well over Cys linker molecules due to higher loading of Q-dots observed in the presence of MPA than in others. However, in the presence of Cys linker molecules, the charge transfer rate is higher due to formation of zwitterionic structure. In zwitterion, the $NH_2$ group forms a positive charge and the COOH group forms a negative charge, and these functions facilitates the forward electron transfer from Q-dots to $TiO_2$ and suppresses the backward transfer of electrons (Margraf et al. 2013).

Linker molecules play a crucial role in the performance of q-dot-based PEC sensors. The linker molecules facilitate both the immobilization of the Q-dots (shown in Fig. 2.5) and the charge transfer process between Q-dots and the conducting electrode. In the increasing distance between the Q-dot and the conducting electrodes, strength of connection between them affects the output of the PEC sensor (Yue et al. 2013).

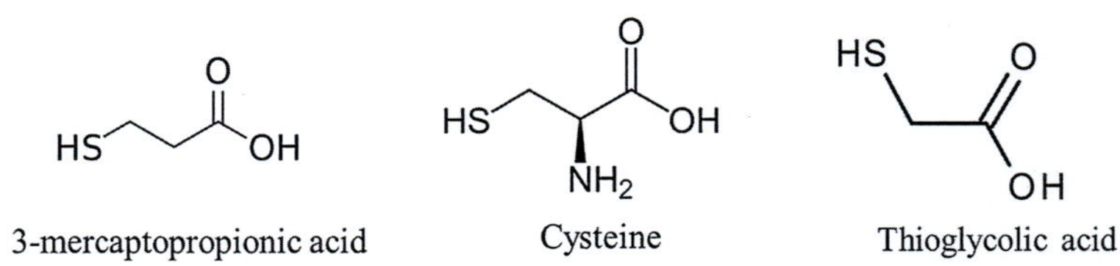


**Fig. 2.4** Chemical structures of linker molecules

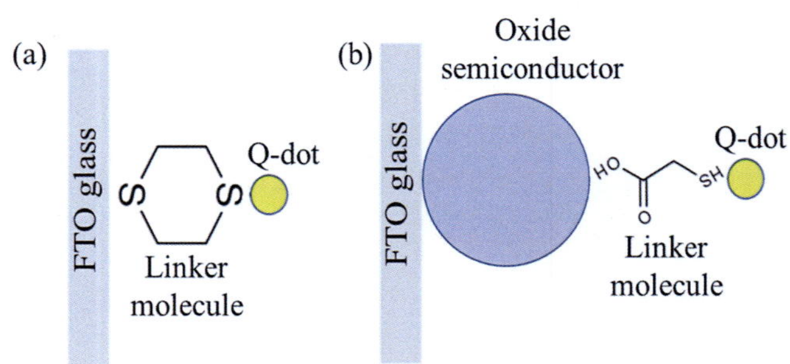


**Fig. 2.5** Linker molecules attached to the Q-dot on top of FTO glass (**a**); and on top of oxide semiconductor surface (**b**)

## 2.4 Importance of Linker Molecules in Photoelectrochemical Sensors

Disassembling Q-dot during the measurement in Q-dot-based PEC sensor may hinder the stability of photocurrent output. The strength of connection of Q-dots depends on the properties of the substrate, the type of linker molecules, and the Q-dots (Khalid et al. 2011a; Yue et al. 2013). Waqas Khalid et al. performed an analysis with different thiol linker molecules and different substrates. In their studies they used sintered substrates such as Au-coated glass slides, Au layers evaporated onto mica sheets, and Au films sputtered onto oxidized Si wafers, and they found no or little response of sensor without Q-dots. Further, they used thiol linker molecules in their studies, such as 1,4-benzenedithiol, 4-4′-biphenyldithiol, biphenyldithiol monoacetylated, *trans*-4,4′-stilbenedithiol, and *trans*-4,4′-stilbenedithiol monoacetylated, and connected the Q-dots with the substrates mentioned above. They analyzed the devices by using X-ray photoelectron spectroscopic (XPS) analysis and Signal-to-Noise and Drift Analysis of Photocurrent. They concluded that the high conductivity of linker molecules, better assembly of densely packed film, and strong attachment of the Q-dot layers are also important to the better performance of q-dot-based PEC sensor (Khalid et al. 2011a). If the Q-dots are connected with electrodes imperfectly, then the excited electron in the Q-dots will not effectively transfer from Q-dot to electrode. The excited electron accumulates in the Q-dot, and the charging and discharging process occurs and results in a lower performance of the Q-dot-based PEC sensor (Yue et al. 2013).

Q-dots could be connected with substrate by covalent and electrostatic interactions (Baron et al. 2005; Ma et al. 2015). Baron et al. prepared the electrode by connecting the CdS Q-dots with gold substrate by using barbiturate–triaminodiazine in which hydrogen bond interaction exists (Baron et al. 2005). Further, the hydrogen bond interaction exists in between guanine-cytosine, adenine-thymine in DNA connect immobilized the Q-dots with electrodes. Hydrogen bond interactions in DNA lead to better connection of Q-dots with the substrate and facilitates the electron transfer from Q-dots to conduction electrode. The usage of DNA as linker molecule has great benefits because the size of DNA could be altered and better control over the electron tunneling process between Q-dot and conducting electrode was possible (Yue et al. 2013). Electrostatic attraction between quantum dots with opposite charges attract each other and form heterojunction (Roy et al. 2020). Formation of heterojunction enhances the charge separation and amplitude of the signal.

## 2.5 Deposition Methods of Quantum Dots

Light-sensitized materials such as Q-dots could be deposited on conducting electrodes or oxide semiconductor materials as pre-synthesized Q-dots or by an in situ method at the electrode (Margraf et al. 2013). The pre-synthesized method of Q-dots could be performed by hot-injection methods, hydrothermal method, microwave assisted methods, so on. In the pre-synthesized method, the exact size and properties of Q-dots could be controlled. However, poor coverage of pre-synthesized Q-dots over conducting electrodes, or on top of $TiO_2$, leads to inferior performance of the device. In situ methods of deposition could be performed by chemical bath deposition (CBD) or successive ion layer adherence and reaction method (SILAR). The electrodes deposition performed by these methods shows well in different applications such as solar cells (Margraf et al. 2013) and PEC sensors due to higher loading of Q-dots. However, in the in situ method of deposition, the poor control of size and shape of the Q-dots leads to little control over the tuning of bandgap and shapes (Margraf et al. 2013).

Deposition of Q-dots on top of conducting electrodes could be performed by electrodeposition methods, spray pyrolysis method, evaporation methods, chemical bath deposition methods, and so on. The better coverage of Q-dots on the conducting surface shows higher absorbance of irradiated light. However, excessive Q-dot loading may lead to aggregation of Q-dots and increase the recombination of excited electrons (Manjceevan and Bandara 2016). As a result, it suppresses the amplitude of the signal. Therefore, the deposition of Q-dots should be optimized to achieve better amplitude of signal.

Absorbance of Q-dots and the amplitude of signal of q-dot-based PEC sensors could be increased by systematic stacking of multilayered Q-dots on top of a conducting substrate (Nakanishi et al. 1998; Manjceevan and Bandara 2018). Further, the valence band and conduction band positions can be arranged in a cascade manner. In multi-layered structure the excited electron-hole pair effectively separates and reduces the recombination of excited electrons (Sheeney-Haj-Ichia et al. 2004). The suppressed recombination of electron-hole pairs enhances the performance of current flow and enhances the sensitivity of the q-dot-based PEC sensor. However, the stacking of multi-layered Q-dots may affect the inner layer of Q-dots and hinder the performance (Manjceevan and Bandara 2018). The stacking process should be performed carefully to avoid the deleterious effects of deposition process (Fig. 2.6). Further, the deposition of Q-dots over metal oxide semiconductors like hybrid

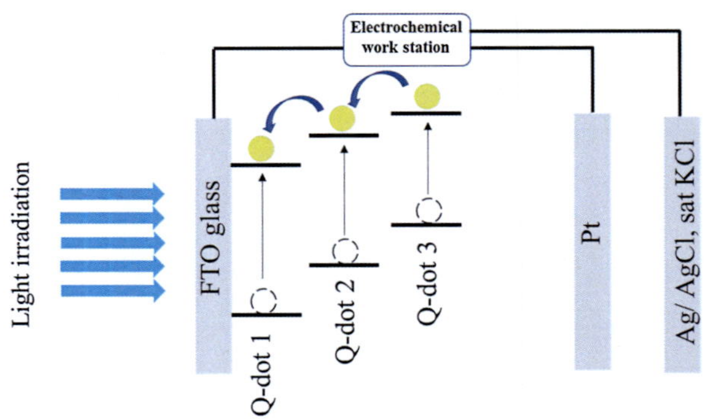


**Fig. 2.6** Schematic diagram of three electrode system of photoelectrochemical process of Q-dot sensor. Q-dots stacked on top of each other as cascade arrangement of band positions

structure enhances the charge carrier separation of excited electrons and leads to higher amplitude of signal in Q-dot-based PEC sensors. Similarly, the deposition of carbon nanotubes between Q-dot facilitates the charge transfer properties (Sheeney-Haj-Ichia et al. 2004).

## 2.6 Corrosion of Electrodes

Corrosion of electrodes, which affects stability, is one of the major problems in Q-dot-based electrodes (Yue et al. 2013; Manjceevan and Bandara 2016). In Q-dot-based PEC sensor applications, oxidation and reduction reaction occur at the electrodes-electrolyte interface. The photoexcitation of electrons creates electron hole pairs, where the electrons in conduction band and holes in valance band. Following this excitation, charge separation occurs effectively and redox reaction occurs at the electrodes-electrolyte interface. However, in the absence of redox reaction at electrode-electrolyte interface, charging and discharging of Q-dot in electrodes effect the photocurrent stability with time (Yue et al. 2013). To avoid such conditions, it is usual to add a different electron-hole donor such as ascorbic acid (Victorious et al. 2019), sodium sulfide (Yue et al. 2013), triethanolamine (Miyake et al. 1999), and so on.

## 2.7 Q-Dot Based Photoelectrochemical Biosensor

Biosensors consist of a biosensing element and transducer, in which the biosensing element senses the biological reaction or event and the transducer converts that into easily measurable output (Zhao et al. 2014). Biorecognition elements available on the surface of electrodes with quantum dots act as transducer in Q-dot-based photoelectrochemical sensors. Wideband gap semiconductor oxides such as $TiO_2$, ZnO, $SnO_2$ have many potential applications in biosensors as electron mediator, and they absorb the UV region of the spectrum. Deposition of Q-dots with wideband gap materials facilitates to increase the signal and obtain better sensitivity. There are several Q-dots used in PEC biosensor applications such as CdS (Golub et al. 2009; Khalid et al. 2011a; An et al. 2013), CdSe (Tyrakowski and Snee 2014), $CsPbBr_3$ (Zhu et al. 2018), graphene (An et al. 2013), and so on.

## 2.8 Applications of Q-Dot-Based Photoelectrochemical Sensor

The sensing property of Q-dot-based PEC sensors is based on the interaction of sensing materials with target substance as directly or indirectly categorized as direct or indirect sensor (Yue et al. 2013). These sensors are used in many applications such as detection of DNA, glucose, neuro transmitter, dihydronicotinamide adenine dinucleotide, and so on. In this chapter, we focus on some remarkable applications of Q-dots in the PEC biosensor.

### *2.8.1 Detection of DNA*

Q-dot-based PEC biosensing for detection of DNA perform with a three-electrode system in which the working electrode consists of Q-dots immobilized on a solid conductive substrate, with platinum and silver/silver chloride and saturated KCl as counter and reference electrodes. The working electrode consists of immobilized single-strand DNA on top of the Q-dot surface as probe and the biological process or biocatalytic processes based on DNA perform on the surface sense by the working electrode, and it records the photocurrent response before and after the process (shown in Fig. 2.7) (Zhao et al. 2014).

There are many interactions related to probe DNA, which is on Q-dots and could occur as hybridization, association with small molecules, interactions with proteins, aptamer–target Interactions, and DNA damage. The probe DNA are short oligonucleotides that can hybridize with target DNA in a specific region (Zhao et al. 2012, 2014). Pairing of target ssDNA with probe ssDNA is a bioprocess that occurs on the Q-dot surface and affects the current flow. In DNA adenosine pairing with thymine base by two bonds (A=T) and guanidine with cytosine C≡G by three bonds. Pairing of target ssDNA with probe ssDNA produce dsDNA. The change in current flow measurement due to conjugation of probe ssDNA with target ssDNA by pairing bases facilitates quantitative determination of the target DNA (Yue et al. 2013).

Zhao et. al prepared Q-dot-based PEC sensors for DNA detection by using CdS Q-dots and silver nanoparticles (Zhao et al. 2012). Silver nanoparticles connected with CdS Q-dots by DNA stimulate the exciton plasmon interactions (EPI) due to the natural overlap of absorbance band and suppress the charge transfer process (shown in Fig. 2.8) (Zhao et al. 2012). Therefore, the exciton plasmon interactions directly correlate with the concentration of target DNA. The photocurrent response decreases with logarithmic scale of target DNA concentration in a linear manner (Zhao et al. 2012).

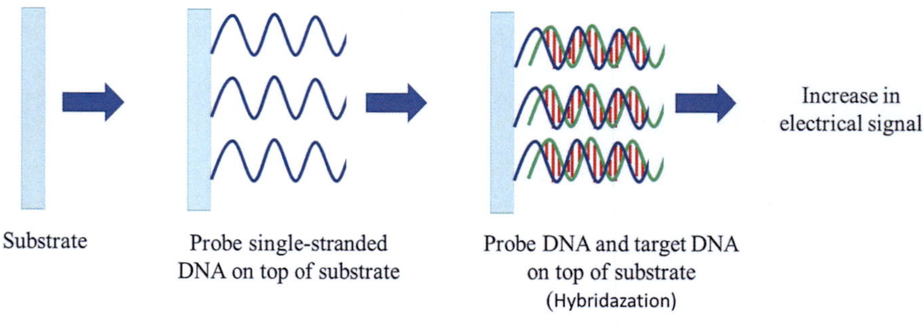


**Fig. 2.7** Schematic representation of PEC biosensor design (Zhao et al. 2014)

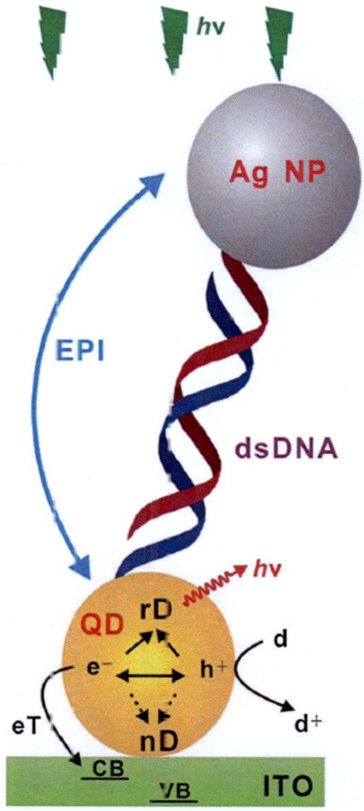


**Fig. 2.8** Energy transfer in Q-dot-based PEC biosensor. Illumination of quantum dot form the electron ($e^-$)-hole ($h^+$) pair. Excited electron injects into electrode and lose its energy by radiative decay (rD), non-radiative decay (nD) and exciton plasmon interaction (EPI) with silver nanoparticle connected by dsDNA. Reprinted with permission from Zhao et al. 2012. Copyright 2012, American Chemical Society

## 2.9 Detection of Gene Mutation

Q-dot-based PEC sensors for detecting DNA could be used to detect the target DNA quantitatively, which is available in the solution and used to predict gene mutations as well. The stability of dsDNA formed by the pairing of ssDNA on Q-dot as probe DNA and target DNA, which is available in solution, depends on the compatibility of those strands. However, even a single variation in target ssDNA could lead to poor compatibility and be energetically less favorable. It reflects in the output read-out of signal, and based on that output signal mutations in DNA could be identified.

Association of DNA non-covalently with small molecules could be performed in three main ways: intercalation, groove binding, and electrostatic interactions. Intercalation is the insertion of an intercalator, which is small organic molecule π-stack into the based pair of DNA. The process depends on the size and chemical nature of the intercalator and fits into the dsDNA (Liu and Sadler 2011; Zhao et al. 2014). Researchers introduce specific mutations by DNA intercalators and cross-ings in their study on DNA-related disease and diagnosis. Mutagens may be harmful to the gene function or kill it. DNA intercalators are used in treatments for rapid-growing cancer to inhibit the replication of DNA (Ashley and Poulton 2009). Intercalators are used in the biosensor for two main reasons: for signal reporter or mediator to shuttle electron (shown in Fig. 2.9). Under illumination the preferentially attached intercalators support generation of the photocurrent signal to monitor the DNA hybridization or damage (Zhao et al. 2014).

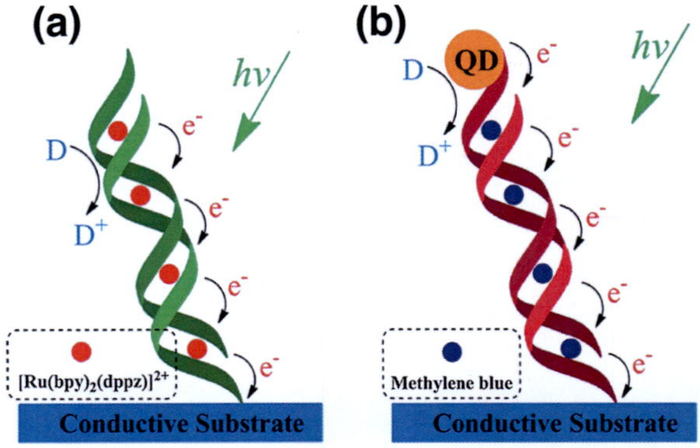


**Fig. 2.9** Diagrammatic representation of DNA intercalators act as (**a**) signal reporter with $[Ru(bpy)_2dppz]^{2+}$ and (**b**) electron mediator with methylene blue. Reprinted with permission from Zhao et al. 2014. Copyright 2014, American Chemical Society

## 2.10 Detection of Cocaine

PEC detection of cocaine could be performed by using a Q-dot signal reporter and supramolecular aptamer complexes. Aptamers are oligonucleotides that bind with the targets (shown in Fig. 2.10). An anticocaine aptamer subunit is attached to a conductive electrode surface and another subunit attached with Q-dots like CdS. CdS nanoparticles absorb the irradiated light and create an electron-hole pair and is followed by the hole transfer from Q-dot to electrolyte with a hole scavenger like triethanol amine creating the electron flow from Q-dots to electrode. Aptamers are small sized DNA or RNA and can join with specific target. Photocurrent produces only after joining the aptamer unit attached with CdS nanoparticles, and the aptamer subunit consists of cocaine attached with conductive electrode. Formation of the complex between cocaine and aptamer units is labeled by CdS Q-dot, changes the photocurrent response, and leads to quantitative detection of the cocaine. This method of detection has potent application, such as a lack of a background interfering signal (Golub et al. 2009; Yue et al. 2013).

## 2.11 Neurotransmitter Determination and Detection

Neurotransmitter substances are chemical products that play a key role in the nervous system, such as behavior and cognition functions in mammalians to conduct signals between neurons or neurons and non-neuron somatic cells (Tajik et al. 2020). There are many neurotransmitters available in which the biogenic amines are important, such as dopamine (Florescu and David 2017; Chen et al. 2018), epinephrine (Huang et al. 2020), norepinephrine (Tajik et al. 2020), and so on. Neurotransmitters in the central nervous system connect with various illnesses, such as Alzheimer's and Parkinson's diseases, heart failure, and so on. Monitoring the concentration of neurotransmitters in human fluid is important to diagnosis and treatment of disease (Tajik et al. 2020).

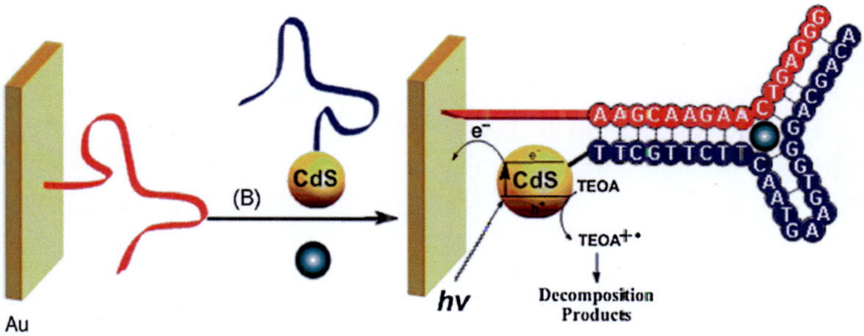


**Fig. 2.10** Self-Assembly of aptamer with CdS and aptamer unit with cocaine, functioning as a photoelectrochemical sensor. Reprinted with permission from Golub et al. 2009. Copyright 2009, American Chemical Society

There are many efforts made to detect the neurotransmitters by using photoelectrochemical, fluorimetry, chemiluminescence, chromatography, and other methods. Among them, the photoelectrochemical sensor is very cheap and provides a quick response and high sensitivity (Zhao et al. 2014; Tajik et al. 2020). To further improve the sensitivity of PEC sensors, many different types of Q-dots have been used in the Q-dot-based PEC sensor application to detect dopamine concentration. These include graphene quantum dot, SnSe (Hun et al. 2017a, b), Nano $MoS_2$ modified gold electrode (Hun et al. 2017a, b), perovskite quantum dot (Chen et al. 2018), and others. Graphene is two-dimensional material used in many applications due to its excellent properties (Hun et al. 2017a, b). SnSe is similar to graphene, like materials with group IV–VI elements. SnSe materials are also used to detect the dopamine concentration (Hun et al. 2017a, b).

Illumination of Q-dots on gold electrodes creates an electron-hole pair. The electrolyte consists of phosphate buffer and dopamine. The photo current response significantly higher with the dopamine in electrolyte, and the response increasing linearly with increasing dopamine concentration from the 0.01 to 10 μM concentration range. The applied potential, pH of electrolyte, and volume of Q-dots are important parameters to optimize and keep constant to determine the unknown concentration of dopamine. However, the coexistence of biological compounds such as ascorbic acid and uric acid in the samples interferes with the measurements. However, such interfering substances show a lower response at lower applied potentials (Hun et al. 2017a, b).

## 2.12 Glucose Biosensors

Diabetes mellitus is a chronic metabolic disorder that affects the lifetime of persons who suffer from it, but proper maintenance of the blood glucose level can reduce the harmfulness of the disease (Sabu et al. 2019). As glucose and insulin play a major role in maintaining the glucose level of blood, monitoring the level is essential.

Many tools, such as biosensors, are available to detect those levels. Those tools are based on enzymatic, non-enzymatic, electrochemical, and optical properties (Sabu et al. 2019). These biosensors are used to detect the serum glucose level accurately (Nikolelis and Nikoleli 2018). Glucose biosensors consist of a glucose recognition element, a transducer, and a data processor (Sabu et al. 2019). Q-dot materials used in biosensors are selected due to their characteristic features such as tunable band-gap energy and ability to absorb a wide range of the spectrum, which improves the sensitivity (Sabu et al. 2019).

Glucose could be detected indirectly by Q-dot-based PEC sensors combined with a suitable enzyme like glucose oxidase (GOx) or glucose dehydrogenase (GDH) (Wu et al. 2010; Razmi and Mohammad-Rezaei 2013; Xia et al. 2015). In the presence of glucose dehydrogenase enzyme, glucose produces NADH by electron transfer. The NADH could be detected by Q-dot-based PEC sensors under illumination conditions. Under illumination the electron in valence band goes to conduction band under illumination and creates electron-hole pair. NADH could be detected based on the current produced under illumination of Q-dots (Yue et al. 2013).

The oxidation of glucose by glucose oxidase produces $H_2O_2$. The production of $H_2O_2$ could be used to estimate blood glucose level. The production of $H_2O_2$ converts the phenol into quinone in the presence of horseradish peroxidase (HRP) (Wang et al. 2018). The production of quinone shows a quenching effect on the fluorescence of graphene Q-dots due to better electron transfer properties from quinone to graphene Q-dots (GQD).

Due to the drawbacks associated with enzyme-based PEC biosensors like pH and temperature dependency, poor reproducibility, high cost of enzymes, and a complicated process, more researchers tend to fabricate enzyme-free glucose biosensors by using metal nanoparticles and metal oxides (Xia et al. 2015; Hwang et al. 2018). An ultrasensitive electrochemical and photoelectrochemical Q-dot-based PEC biosensor to detect glucose was fabricated with enzyme-free CdS quantum dots modified CuO inverse opal photonic crystals (Xia et al. 2015). This method shows higher sensitivity, good stability, and reproducibility (Xia et al. 2015). The higher sensitivity of this method is due to the properties of CuO IOPCs and CdS Q-dots, such as large a surface area and uniform porous distribution of CuO IOPCs, surface passivation, and the protection of the potential barrier of CdS QDs (Xia et al. 2015). Likewise, there are many metal oxides, and enzyme-free sensors were reported by using direct oxidation of glucose such as $Co_3O_4$ (Zhang et al. 2016), $Co_3O_4/PbO_2$ core–shell nanorod (Chen et al. 2014), Ag/CuO (Zheng et al. 2014), $Cu@Cu_2O$/rGO modified glassy carbon electrode (Huo et al. 2014), NiO (Mishra et al. 2018) and so on.

## 2.13 Conclusion, Outlook, and Future Aspects

In this chapter, the general principles, fabrication methods of Q-dot-based photoelectrochemical biosensors, and their recent progress of applications are discussed. Q-dot-based PEC biosensors are widely used in various applications due to their high sensitivity, ease of fabrication, reproducibility, and low cost. These Q-dot-based PEC biosensors are used to detect the DNA, mutations, neurotransmitters, glucose, and so on. Further, there are many research projects scrutinizing other applications based on Q-dot PEC and to enhance the performances of such applications. Due to the rapid measurement, high sensitivity, and ease of fabrication of Q-dot PEC biosensors, if the active mechanisms behind the photoelectrochemical detection of specific DNA of pathogens are successfully identified, the research of Q-dot-based PEC systems could provide better solutions to the process of diagnosing disease, as it is capable of fast diagnosis minimizes the transmission of disease from one living being another.

**Acknowledgement** None declared.

**Conflicts of Interest** The authors declare no conflict of interest.

## References

An, X., Yu, X., Yu, J. C., & Zhang, G. (2013). CdS nanorods/reduced graphene oxide nanocomposites for photocatalysis and electrochemical sensing. *Journal of Materials Chemistry A, 1*(16), 5158–5164.

Ashley, N., & Poulton, J. (2009). Anticancer DNA intercalators cause p53-dependent mitochondrial DNA nucleoid re-modelling. *Oncogene, 28*(44), 3880–3891.

Baron, R., Huang, C. H., Bassani, D. M., Onopriyenko, A., Zayats, M., & Willner, I. (2005). Hydrogen-bonded CdS nanoparticle assemblies on electrodes for photoelectrochemical applications. *Angewandte Chemie International Edition, 44*(26), 4010–4015.

Berglund, S. P., Abdi, F. F., Bogdanoff, P., Chemseddine, A., Friedrich, D., & van de Krol, R. (2016). Comprehensive evaluation of CuBi2O4 as a photocathode material for photoelectrochemical water splitting. *Chemistry of Materials, 28*(12), 4231–4242.

Chen, T., Li, X., Qiu, C., Zhu, W., Ma, H., Chen, S., & Meng, O. (2014). Electrochemical sensing of glucose by carbon cloth-supported Co3O4/PbO2 core-shell nanorod arrays. *Biosensors and Bioelectronics, 53*, 200–206.

Chen, X., Li, D., Pan, G., Zhou, D., Xu, W., Zhu, J., Wang, H., Chen, C., & Song, H. (2018). All-inorganic perovskite quantum dot/TiO 2 inverse opal electrode platform: Stable and efficient photoelectrochemical sensing of dopamine under visible irradiation *Nanoscale, 10*(22), 10505–10513.

Do Sung, S., Lim, I., Kang, P., Lee, C., & Lee, W. I. (2013). Design and development of highly efficient PbS quantum dot-sensitized solar cells working in an aqueous polysulfide electrolyte. *Chemical Communications, 49*(54), 6054–6056.

Etorki, A., Awin, L., El Rais, M., & Shaban, I. (2019). Application of gold nanoparticles with 1,6-hexanedithiol modified screen-printed carbon electrode as a sensor for determination of arsenic in environmental samples. *Sensor Letters, 17*, 762–768.

Florescu, M., & David, M. (2017). Tyrosinase-based biosensors for selective dopamine detection. *Sensors, 17*(6), 1314.

Golub, E., Pelossof, G., Freeman, R., Zhang, H., & Willner, I. (2009). Electrochemical, photoelectrochemical, and surface plasmon resonance detection of cocaine using supramolecular aptamer complexes and metallic or semiconductor nanoparticles. *Anal Chem, 81*(22), 9291–9298.

Hsiao, S.-H., Wu, J.-X., & Chen, H.-I. (2020). High-selectivity NOx sensors based on an Au/InGaP Schottky diode functionalized with self-assembled monolayer of alkanedithiols. *Sensors and Actuators B: Chemical, 305*, 127269.

Huang, Q., Liu, Y., Zhang, C., Zhang, Z., Liu, F., & Peng, J. (2020). Au quantum dot/nickel tetraminophthalocyanaine-graphene oxide-based photoelectrochemical microsensor for ultrasensitive epinephrine detection. *ACS Omega, 5*(15), 8423–8431.

Hun, X., Wang, S., Mei, S., Qin, H., Zhang, H., & Luo, X. (2017a). Photoelectrochemical dopamine sensor based on a gold electrode modified with SnSe nanosheets. *Microchimica Acta, 184*(9), 3333–3338.

Hun, X., Wang, S., Wang, S., Zhao, J., & Luo, X. (2017b). A photoelectrochemical sensor for ultrasensitive dopamine detection based on single-layer NanoMoS2 modified gold electrode. *Sensors and Actuators B: Chemical, 249*, 83–89.

Huo, H., Guo, C., Li, G., Han, X., & Xu, C. (2014). Reticular-vein-like Cu@Cu2O/reduced graphene oxide nanocomposites for a non-enzymatic glucose sensor. *RSC Advances, 4*(39), 20459–20465.

Hwang, D.-W., Lee, S., Seo, M., & Chung, T. D. (2018). Recent advances in electrochemical non-enzymatic glucose sensors – A review. *Analytica Chimica Acta, 1033*, 1–34.

Khalid, W., El Helou, M., Murböck, T., Yue, Z., Montenegro, J.-M., Schubert, K., Göbel, G., Lisdat, F., Witte, G., & Parak, W. J. (2011a). Immobilization of quantum dots via conjugated self-assembled monolayers and their application as a light-controlled sensor for the detection of hydrogen peroxide. *ACS Nano, 5*(12), 9870–9876.

Khalid, W., Göbel, G., Hühn, D., Montenegro, J. M., Rivera Gil, P., Lisdat, F., & Parak, W. (2011b). Light triggered detection of aminophenyl phosphate with a quantum dot based enzyme electrode. *Journal of Nanobiotechnology, 9*, 46.

Lai, L.-H., Protesescu, L., Kovalenko, M. V., & Loi, M. A. (2014). Sensitized solar cells with colloidal PbS-CdS core-shell quantum dots. *Physical Chemistry Chemical Physics, 16*(2), 736–742.

Liu, H.-K., & Sadler, P. J. (2011). Metal complexes as DNA intercalators. *Accounts of Chemical Research, 44*(5), 349–359.

Ma, Z.-Y., Ruan, Y.-F., Zhang, N., Zhao, W.-W., Xu, J.-J., & Chen, H.-Y. (2015). A new visible-light-driven photoelectrochemical biosensor for probing DNA–protein interactions. *Chemical Communications, 51*(39), 8381–8384.

Ma, F., Li, C.-C., & Zhang, C.-Y. (2018). Development of quantum dot-based biosensors: Principles and applications. *Journal of Materials Chemistry B, 6*, 6173–6190.

Manjceevan, A., & Bandara, J. (2016). Robust surface passivation of trap sites in PbS q-dots by controlling the thickness of CdS layers in PbS/CdS quantum dot solar cells. *Solar Energy Materials and Solar Cells, 147*, 157–163.

Manjceevan, A., & Bandara, J. (2018). Systematic stacking of PbS/CdS/CdSe multi-layered quantum dots for the enhancement of solar cell efficiency by harvesting wide solar spectrum. *Electrochimica Acta, 271*, 567–575.

Margraf, J. T., Ruland, A. s., Sgobba, V., Guldi, D. M., & Clark, T. (2013). Quantum-dot-sensitized solar cells: Understanding linker molecules through theory and experiment. *Langmuir, 29*(7), 2434–2438.

Mashhadizadeh, M. H., Azhdeh, A., & Naseri, N. (2017). 3-Mercapto propionic acid self-assembled on gold nano-particles applied for modification of screen-printed electrode as a new digoxin electrochemical aptasensor using graphene oxide-based signal-on strategy. *Journal of Electroanalytical Chemistry, 787*, 132–138.

McGrath, M. J., & Scanaill, C. N. (2013). *Sensing and sensor fundamentals. Sensor technologies: Healthcare, wellness, and environmental applications* (pp. 15–50). Berkeley, CA: Apress.

Mishra, S., Yogi, P., Sagdeo, P. R., & Kumar, R. (2018). Mesoporous nickel oxide (NiO) nanopetals for ultrasensitive glucose sensing. *Nanoscale Research Letters, 13*(1), 16–16.

Miyake, M., Torimoto, T., Sakata, T., Mori, H., & Yoneyama, H. (1999). Photoelectrochemical characterization of nearly monodisperse CdS nanoparticles−Immobilized gold electrodes. *Langmuir, 15*(4), 1503–1507.

Nakanishi, T., Ohtani, B., & Uosaki, K. (1998). Fabrication and characterization of CdS-nanoparticle mono- and multilayers on a self-assembled monolayer of alkanedithiols on gold. *The Journal of Physical Chemistry B, 102*(9), 1571–1577.

Nikolelis, D. P., & Nikoleli, G. P. (2018). *Nanotechnology and biosensors*. Amsterdam: Elsevier.

Pattantyus-Abraham, A. G., Kramer, I. J., Barkhouse, A. R., Wang, X., Konstantatos, G., Debnath, R., Levina, L., Raabe, I., Nazeeruddin, M. K., & Gratzel, M. (2010). Depleted-heterojunction colloidal quantum dot solar cells. *ACS Nano, 4*(6), 3374–3380.

Piliego, C., Protesescu, L., Bisri, S. Z., Kovalenko, M. V., & Loi, M. A. (2013). 5.2% efficient PbS nanocrystal Schottky solar cells. *Energy & Environmental Science, 6*(10), 3054–3059.

Razmi, H., & Mohammad-Rezaei, R. (2013). Graphene quantum dots as a new substrate for immobilization and direct electrochemistry of glucose oxidase: Application to sensitive glucose determination. *Biosensors & Bioelectronics, 41*, 498–504.

Roy, P., Devatha, G., Roy, S., Rao, A., & Pillai, P. P. (2020). Electrostatically driven resonance energy transfer in an all-quantum dot based donor–Acceptor system. *The Journal of Physical Chemistry Letters, 11*(13), 5354–5360.

Sabu, C., Henna, T. K., Raphey, V. R., Nivitha, K. P., & Pramod, K. (2019). Advanced biosensors for glucose and insulin. *Biosensors & Bioelectronics, 141*, 111201.

Sambur, J. B., Riha, S. C., Choi, D., & Parkinson, B. A. (2010). Influence of surface chemistry on the binding and electronic coupling of CdSe quantum dots to single crystal TiO2 surfaces. *Langmuir, 26*(7), 4839–4847.

Sheeney-Haj-Ichia, L., Basnar, B., & Willner, I. (2004). Efficient generation of photocurrents by using CdS/carbon nanotube assemblies on electrodes. *Angewandte Chemie International Edition in English, 44*(1), 78–83.

de Sousa, J. R., Batista, A. A., Diógenes, I. C. N., Andrade, G. F. S., Temperini, M. L. A., Lopes, L. G. F., & de Sousa Moreira, Í. (2003). Characterization of a 1,4-dithiane gold self-assembled monolayer: An electrochemical sensor for the cyt-c redox process. *Journal of Electroanalytical Chemistry, 543*(1), 93–99.

Tajik, S., Dourandish, Z., Zhang, K., Beitollahi, H., Le, Q. V., Jang, H. W., & Shokouhimehr, M. (2020). Carbon and graphene quantum dots: A review on syntheses, characterization, biological and sensing applications for neurotransmitter determination. *RSC Advances, 10*(26), 15406–15429.

Tang, J., Kemp, K. W., Hoogland, S., Jeong, K. S., Liu, H., Levina, L., Furukawa, M., Wang, X., Debnath, R., & Cha, D. (2011). Colloidal-quantum-dot photovoltaics using atomic-ligand passivation. *Nature materials, 10*(10), 765–771.

Tsukamoto, T., Shimada, T., & Takagi, S. (2018). Artificial photosynthesis model: Photochemical reaction system with efficient light-harvesting function on inorganic nanosheets. *ACS Omega, 3*(12), 18563–18571.

Tyrakowski, C. M., & Snee, P. T. (2014). Ratiometric CdSe/ZnS quantum dot protein sensor. *Analytical Chemistry, 86*(5), 2380–2386.

Victorious, A., Saha, S., Pandey, R., Didar, T. F., & Soleymani, L. (2019). Affinity-based detection of biomolecules using photo-electrochemical readout. *Frontiers in Chemistry, 7*, 617–617.

Wang, F., & Hu, S. (2009). Electrochemical sensors based on metal and semiconductor nanoparticles. *Microchimica Acta, 165*(1), 1–22.

Wang, W., Bao, L., Lei, J., Tu, W., & Ju, H. (2012). Visible light induced photoelectrochemical biosensing based on oxygen-sensitive quantum dots. *Anal Chim Acta, 744*, 33–38.

Wang, H., Wang, Y., He, B., Li, W., Sulaman, M., Xu, J., Yang, S., Tang, Y., & Zou, B. (2016). Charge carrier conduction mechanism in PbS quantum dot solar cells: Electrochemical impedance spectroscopy study. *ACS Applied Materials & Interfaces, 8*, 28.

Wang, B., Shen, J., Huang, Y., Liu, Z., & Zhuang, H. (2018). Graphene quantum dots and enzyme-coupled biosensor for highly sensitive determination of hydrogen peroxide and glucose. *International Journal of Molecular Sciences, 19*(6), 1696.

Wu, P., He, Y., Wang, H.-F., & Yan, X.-P. (2010). Conjugation of glucose oxidase onto Mn-Doped ZnS quantum dots for phosphorescent sensing of glucose in biological fluids. *Analytical Chemistry, 82*(4), 1427–1433.

Xia, L., Xu, L., Song, J., Xu, R., Liu, D., Dong, B., & Song, H. (2015). CdS quantum dots modified CuO inverse opal electrodes for ultrasensitive electrochemical and photoelectrochemical biosensor. *Scientific Reports, 5*, 10838.

Yang, Y., Pan, D., Zhang, Z., Chen, T., Xie, H., Gao, J., & Guo, X. (2018). Ag2Se quantum dots for photovoltaic applications and ligand effects on device performance. *Journal of Alloys and Compounds, 766*, 925–932.

Yue, Z., Lisdat, F., Parak, W. J., Hickey, S. G., Tu, L., Sabir, N., Dorfs, D., & Bigall, N. C. (2013). Quantum-dot-based photoelectrochemical sensors for chemical and biological detection. *ACS Applied Materials & Interfaces, 5*(8), 2800–2814.

Zang, Y., Lei, J., & Ju, H. (2017). Principles and applications of photoelectrochemical sensing strategies based on biofunctionalized nanostructures. *Biosensors and Bioelectronics, 96*, 8–16.

Zhang, Z.-X., & Zhao, C.-Z. (2013). Progress of photoelectrochemical analysis and sensors. *Chinese Journal of Analytical Chemistry, 41*(3), 436–444.

Zhang, N., Zhang, Y., Pan, X., Fu, X., Liu, S., & Xu, Y.-J. (2011). Assembly of CdS nanoparticles on the two-dimensional graphene scaffold as visible-light-driven photocatalyst for selective organic transformation under ambient conditions. *The Journal of Physical Chemistry C, 115*(47), 23501–23511.

Zhang, E., Xie, Y., Ci, S., Jia, J., & Wen, Z. (2016). Porous Co3O4 hollow nanododecahedra for nonenzymatic glucose biosensor and biofuel cell. *Biosensors and Bioelectronics, 81*, 46–53.

Zhao, W.-W., Yu, P.-P., Shan, Y., Wang, J., Xu, J.-J., & Chen, H.-Y. (2012). Exciton-plasmon interactions between CdS quantum dots and Ag nanoparticles in photoelectrochemical system and its biosensing application. *Analytical Chemistry, 84*(14), 5892–5897.

Zhao, W.-W., Xu, J.-J., & Chen, H.-Y. (2014). Photoelectrochemical DNA biosensors. *Chemical Reviews, 114*(15), 7421–7441.

Zheng, B., Liu, G., Yao, A., Xiao, Y., Du, J., Guo, Y., Xiao, D., Hu, Q., & Choi, M. (2014). A sensitive AgNPs/CuO nanofibers non-enzymatic glucose sensor based on electrospinning technology. *Sensors and Actuators B: Chemical, 195*, 431–438.

Zhu, Y., Tong, X., Song, H., Wang, Y., Qiao, Z., Qiu, D., Huang, J., & Lu, Z. (2018). CsPbBr3 perovskite quantum dots/ZnO inverse opal electrodes: Photoelectrochemical sensing for dihydronicotinamide adenine dinucleotide under visible irradiation. *Dalton Transactions, 47*(30), 10057–10062.

# Chapter 3
# Synthesis and Antimicrobial Abilities of Metal Oxide Nanoparticles

**Pravati Panda, Arundhati Barik, M. V. Basavanag Unnamatla, and Subhendu Chakroborty**

## 3.1 Introduction

The rising population is under the radar of several types of infections, which ultimately led to a high fatality rate of many people worldwide (Silver 2011). Among the different types of infections, bacterial-borne diseases are the most common, which arises from the overuse and abuse of antibiotics. Gram-positive as well as gram-negative microbes are equally found to have pronounced impacts on the lives of humans and animals, thereby continuing to a worrisome situation (Davies and Davies 2010). From ancient times, antibiotics are being used for the treatment of various bacterial- and fungal-borne diseases. Almost 100,000 tons of antibiotics are produced every year globally to win the battle against these deadly diseases (Davies and Davies 2010). However, misuses of antibiotics as well as the lack of scientific tools and capabilities have led to the emergence of bacterial cell mutations and the generation of microbes that are resistant to such therapeutics, ultimately leading to multidrug resistance (MDR) (Wright and Sutherland 2007). Microorganisms are generally found to show MDR by developing new mechanisms inside their cell with time. These mechanisms mainly include the application of resistance plasmids and multidrug efflux pumps. Another factor of resistance is the creation of continual biofilm matrices against available antimicrobial therapy (González-Bello 2017) (Fig. 3.1). It is a self-defensive, rigid, and permeable slimy layer formed by the extracellular polymeric substances of bacteria that adhere to biotic and abiotic

P. Panda · A. Barik
Department of Chemistry, Ravenshaw University, Cuttack, India

M. V. B. Unnamatla
Universidad autonoma del estado de Mexico, Toluca de Lerdo, Mexico

S. Chakroborty (✉)
School of Sciences, Auro University, Surat, Gujarat, India
e-mail: subhendu.chakroborty@aurouniversity.edu.in


K. Pal (ed.), *Bio-manufactured Nanomaterials*,
https://doi.org/10.1007/978-3-030-67223-2_3

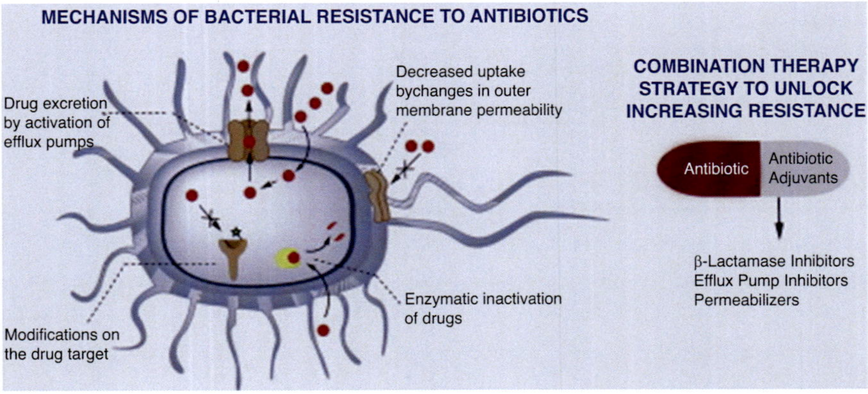


**Fig. 3.1** Mechanism of bacterial resistance to antibiotics (Bioorg Med Chem Lett 2017, 27(18): 4221–4228)

surfaces and makes them gain more resistance against a wide range of antibiotics cleaning and eradication than planktonic cells. Formations of the biofilm network contributed to chronic infections such as periodontitis, cystic fibrosis, and hence inactivate the traditional antibiotic therapy (De Vita et al. 2016). About 25% of infections are associated with biofilm formation in medical devices, which causes huge morbidity and mortality rates in patients and increases costs due to their long-term treatments (Mah and O'Toole 2001). In order to eradicate the bacterial-borne infections and to inhibit biofilm formation, it is of high demand to search for the new efficient treatment approach to disrupt the multicellular structures of the biofilm, thereby limiting the spread of pathogens.

Nowadays, nanotechnology creates an enormous interest among researchers by offering its vast application to tackle the diverse aspects of the eco-system as well as allowed to cross the boundaries of a specialized area of studies than known established one (Elfeky et al. 2020). It is the multidisciplinary science dealing with the creation and modification of nanosize materials by reducing the particle size from micrometer to nanometer (1–100 nm) (Liu et al. 2011). Reducing the particle size provides them a unique potential feature, which makes them wonderful alternatives in the area of health, industry, and medicine by offering a new vision in medical treatment (Mirza and Siddiqui 2014). From the last years, metal oxide nanoparticles (MONPs) are highly emphasizing their effective antimicrobial abilities against drug-resistant pathogenic microbes, which can combat bacteria and easily kill them. MONPs are effective due to their ionic charge, small particle size, and high surface to volume ratio, which make great candidates to unlock many possibilities in the area of nano-biofields (Salata 2004). The positively charged metal oxide nanoparticles have been demonstrated to get attached with cellular proteins present on the bacterial cell membrane firmly via electrostatic attraction that seems to be so strong that the particles might stick there for a very long hour (Kumar and Das 2017). This causes oxidative stress leading to the formation of ROS (reactive oxygen species), impair membrane function, protein dysfunction, genotoxicity, interference with nutrient uptake, and consequently inhibiting their growth up to the cell's death (Kumar and Das 2017). Literature survey includes various techniques for the synthesis of MONPs following simple, cost-effective, and greener protocol, among them double precipitation method, hydrothermal method, electrochemical method, and sol-gel techniques are important (Raghunath and Perumal 2017; Mahapatra et al. 2008). Scientific communities also analyzed the antimicrobial abilities of MONPs such as FeO, CuO, ZnO, $CeO_2$, HgO, and $MnO_2$ against different kinds of bacteria (Amna et al. 2015; Lemire et al. 2013; Baek and An 2011). This chapter summarizes the recent advances made so far in the area towards the synthesis of different metal oxide nanoparticles and its application as antimicrobial agents against MDR pathogens.

## 3.2 Synthesis and Fabrication of Nanoparticles

### *3.2.1 Synthesis of Iron Oxide Nanoparticles*

Iron oxide (FeO) nanoparticles were prepared from the peel extract of *Punica granatum* following environmentally benign and sustainable green technology (Irshad et al. 2017). In order to get small particle size and better morphology of nanoparticles, the authors monitored the reaction at different peel extract concentrations, i.e., 20, 40, and 60 ml, using 150 ml of 0.15 M $FeCl_3 \cdot 6H_2O$ solution. A 40-ml peel extract was found to be the optimal reaction condition for the synthesis of nanoparticles. When the color of the solution changed from brown to black, the solution was autoclaved at 200 °C for 5 h. The formed FeO nanoparticles were centrifuged at 10,000 rpm and were dried for 4 days. The size, crystallinity, and morphology were determined by different techniques, i.e., UV-Vis, X-ray, FT-IR, SEM, and EDX spectroscopy (Fig. 3.2).

Following ultrasound-assisted green strategy, $Fe_3O_4$ nanoparticles were prepared from the leaf extracts of *Artemisia haussknechtii Boiss* (Alavi and Karimi 2019).

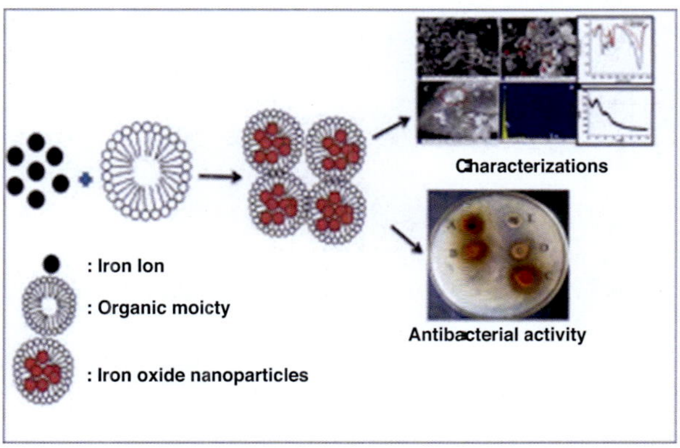


**Fig. 3.2** Synthesis of iron oxide nanoparticles (J Photochem Photobiol B 2017, 170: 241–246)

The NPs were synthesized by mixing up 0.2 M $FeCl_3{\cdot}6H_2O$, 0.1 M $FeCl_2{\cdot}4H_2O$, aqueous *A. haussknechtii*, and 0.1 M NaOH solution. The resulting solution was stirred under ultrasonic irradiation (power: 50 W) with a frequency of 40 kHz for 2 h. The NPs were purified by centrifuging for 5–10 min at 4000 rpm and further washed with 1:1 mixture of $CH_3OH:H_2O$ several times and incubated at 50 °C for 48 h to get the powder form of NPs. The synthesized NPs were successfully characterized by using different spectroscopic analysis, i.e., UV-Vis, EDX, RT-IR, SEM, XRD, and AFM (atomic force microscopic).

Magnetite ($Fe_2O_3$) nanoparticles were prepared following co-precipitation and hydrothermal techniques and further functionalized by oleic acid/propolis (El-Guendouz et al. 2016). The functionalization of iron oxide nanoparticles was carried out by the suspension of the synthesized NPs into the solution of oleic acid/propolis. The resulting mixture was stirred and heated for 60 min at 80 °C. Excess oleic acid/propolis was extracted by successive washing with water in a dropwise manner. The generated functionalized oleic acid/propolis-coated iron oxide nanoparticles were recrystallized by washing with ethanol and dried. The nanofabricated metal oxides were successfully analyzed by FT-IR, TEM, XRD, and Mossbauer spectral studies.

#### 3.2.1.1 Synthesis of Copper Oxide Nanoparticles

Copper oxide (CuO) nanoparticles were prepared from *hc*-pCUR-NS (highly cross-linked poly(curcumin) nanospheres) following base-free precipitation methodology (Masoule et al. 2019). A mixture of *hc*-pCUR-NS and $Cu(OAc)_2$ solution was sonicated and then refluxed for 24 h. The metal oxides were prepared by the chelation of Cu with the dicarbonyl group of curcumin. The CuO NPs were separated by centrifugation (8000 rpm) and washed, recrystallized by distilled water, and further dried in a desiccator to get the powder form. The SEM, XRD, X-ray, UV-Vis, FT-IR, and zeta potential of the synthesized NPs were determined. Negatively charged nanocomposites bearing small spherical CuO NPs were found to have the particle size 34 and 246 nm.

Copper oxide (CuO) nanoparticles were prepared following a solvothermal manner (AlYahya et al. 2018). A mixture of copper chloride (0.04 M), deionized water, EtOH, PVA, and 1 M NaOH was stirred for 30 min. The pH of the mixture solution was maintained to 10 throughout the reaction. The mixture was autoclaved in Teflon-lined stainless steel and heated at 160 °C for 16 h. The reaction mixture was cooled, and the nanoparticles generated solvothermally were filtered and recrystallized by distilled water and EtOH. The synthesized nanoparticles were further dried at 80 °C for 6 h inside a hot air oven. Particle size, morphology, and crystallinity nature of NPs were confirmed by SEM, XRD, PL studies, Raman studies, and FT-IR spectroscopy. The metal-oxygen bond vibration appeared at 510 $cm^{-1}$ in FT-IR data.

Iron-doped copper oxide nanoparticles were synthesized using sol-gel technique (Pugazhendhi et al. 2018). The mixture of $Cu(NO_3)_2{\cdot}3H_2O$ and $CH_3OH$ was allowed to dissolve completely by stirring for 1 h and then kept for 2 days to assist the gel formation. The developed gel was collected and dried in a hot air oven at 200 °C for 2 h and again calcined for 1 h at 300 °C. The synthesized CuO NPs were doped by $Fe(NO_3)_3 \cdot 9H_2O$ to get iron-doped copper oxide NPs. Synthesized NPs were successfully analyzed by using XRD, SEM, EDAX, and FT-IR data. The photocatalytic activities of the NPs were studied by FL (fluorescent light spectroscopic) and UV-Vis analysis. NPs were found to possess an average particle size of 21 nm.

Ramzan et al. (2020) brought forward the synthesis of copper oxide (CuO) nanoparticles from the aqueous extracts of *Cedrus deodara* as a safe and compatible reducing agent using a facile, non-toxic, and low-cost approach by taking copper sulfate pentahydrate as the starting material. Synthesized NPs were analyzed through XRD, TEM, SEM, UV-Vis, and FT-IR. Monoclinic nature of CuO NPs was depicted by XRD data. Along with this, spherical morphology and uniform symmetry were confirmed by SEM and TEM analysis.

Rafique et al. (2020) suggested the biosynthesis of CuO NPs from the leaf extracts of *Citrus aurantifolia* following a greener protocol. XRD analysis exposed the crystalline nature of CuO NPs with an average particle size of 22 nm and bandgap 3.48–3.51 eV. The authors also verified the photocatalyst abilities of the metal oxide NPs.

#### 3.2.1.2 Synthesis of Zinc Oxide Nanoparticles

Zinc oxide (ZnO) nanoparticles were prepared from biocompatible polyvinylalcohol (PVA) as a stabilizing agent under ultrasonication (Gedanken et al. 2016). The 1:15 ratio mixture of Zn(II)acetate dihydrate and PVA was subjected to ultrasonication, followed by the addition of NaOH. The pH of the solution was maintained to 8. The mixture was then removed and cooled in an ice bath to get colloidal NPs. The authors synthesized both ZnO without PVA and ZnO-PVA nanoparticles under ultrasonic irradiation. The synthesized nanoparticles were characterized by ICP (inductive coupled plasma) analysis, XRD, UV-Vis, ESR, and HR-TEM analysis.

Silver (Ag)- and iodide (I)-doped zinc oxide (ZnO) nanoparticles were synthesized following the solvothermal method (Karami Xie et al., 2020). A mixture of ethylene glycol containing $Zn(OAc)_2$ was stirred at 60 °C for 45 min. The reaction mixture was autoclaved inside a Teflon-lined vessel at 170 °C for 18 h. After completion of the reaction, ZnO NPs were collected by centrifugation and washed multiple times with EtOH and $H_2O$ to get the pure form. The sample was dried at 80 °C for 12 h to get the powder form. In order to synthesize Ag-doped ZnO NPs, the authors took the mixture of zinc nitrate hexahydrate, silver nitrate, CTAB, and EtOH and stirred the reaction mixture at room temperature for 1 h 15 min followed by the addition of NaOH. After this step, the authors follow the same preparation steps as followed for the synthesis of ZnO NPs. Also, the preparation procedure for I-doped ZnO is the same as that of the Ag-doped one with a little variation that here HI was added with previously synthesized ZnO NPs. The synthesized NPs were analyzed using XRD, XPS, UV-Vis, photoluminescence, TEM, and SEM.

Curcumin-doped ZnO NPs were prepared using the surfactant-free synthesis method (Pourhajibagher et al. 2019). The mixture of zinc (II) sulfate heptahydrate and NaOH was stirred for 15 min. The mixture was kept in a microwave for 1 min. The synthesized NPs were recrystallized by deionized water and dried at 60 °C for 2 h. The curcumin-doped ZnO NPs were prepared by the dropwise addition of ethanol-dissolved curcumin solution to the prepared ZnO NPs, and the resulting mixture was kept under sonication (power: 100 W) for 4 h with a frequency of 30 kHz. After solid development, it was centrifuged to separate solid from a liquid. The obtained NPs were dried at 25–27 °C.

Ag@ZnO-coated cotton fabrics (CFs) were prepared (Manna et al. 2015). Initially, the authors synthesized ZnO NPs by mixing poly(allylamine) (PAA) with $Zn(NO_3)_2 \cdot 6H_2O$. The reaction mixture was heated for 2 h at 60 °C. The generated NPs were separated from the aqueous solution by centrifugation and dried at room temperature. Second, the authors synthesized Ag@ZnO NPs by adding an $AgNO_3$ solution at a different concentration to the freshly prepared ZnO components. Third, the authors prepared ZnO@CF by impressing the CF to the aqueous solution of PAA and zinc (II) nitrate hexahydrate. Finally, the Ag@ZnO@CF was prepared by dipping the prepared ZnO@CF into $AgNO_3$ solution and stirring overnight. The synthesized metal oxide-coated CFs were analyzed by UV-Vis, XRD, HRTEM, and FESEM to confirm the coating of Ag@ZnO NPs on the cotton fabrics. The presence of Ag in ZnO-coated CFs not only enables them to exhibit improved photocatalytic property but also allowed for visible light-driven activities. UV-Vis spectra studies suggested a bandgap of 3.17 eV for the synthesized metal oxide NPs.

Ag@ZnO and ZnO NPs have been soaked over well porous activated carbon (AC) (Prashantha Kumar et al. 2015). Initially, the authors synthesized ZnO@AC by dissolving HMTA to the aqueous solution of zinc acetate. The pH of the mixture was retained to 8 by the addition of liq. $NH_3$. AC was added to the resulting mixture of the solution and stirred at 70 °C for 24 h. The generated NPs were filtered, washed, and dried for 24 h at 90 °C. Second, they prepared Ag@ZnO@AC NPs by adding $AgNO_3$ solution to the freshly prepared ZnO@AC NPs and allowed the reaction mixture to stir at 40 °C for 24 h. The resulting mixture was cooled for 15 min followed by the slow addition of $NaBH_4$ with continuous stirring for 24 h at room temperature. The generated NPs were filtered, washed, and dried for 24 h at 80 °C. The formation of nanohybrids on the activated carbon surface was confirmed by XRD and TEM.

Ag-doped ZnO nanoparticles were synthesized following arginine-assisted immobilization method (Agnihotri et al. 2015). The presence of arginine stabilized ZnO NPs by deterring the suspension of zinc under alkaline conditions.

Zinc oxide (ZnO) nanoparticles were prepared by Dhandapani et al. (2020) from the leaf extracts of *Melia azedarach* using commercially available zinc nitrate as the starting material following the sol-gel technique. The particle size and morphology of the synthesized nanoparticles were successfully determined by UV-Vis, FT-IR, XRD, TEM, SEM, and EDAX spectral analysis. In the UV-Vis spectra, an absorption peak appears at 372 nm which is the distinct feature of ZnO NPs. Nanocrystal morphology with the hexagonal and spherical shape of the metal oxides was established by XRD, TEM, and SEM analysis. Metal oxide NPs were found to have an average particle size ranges between 33 and 96 nm.

Pillai et al. (2020) prepared zinc oxide (ZnO) nanoparticles from different plant extracts such as *Cinnamomum tamala, Beta vulgaris, Brassica oleracea var. Italica*, and *Cinnamomum verum* following a rapid, cost-effective, and eco-friendly method. Synthesized NPs were analyzed by SEM, FT-IR, XRD, and TEM analysis.

In another study, Pauzi et al. (2020) synthesized ZnO NPs from gum arabic (a natural stabilizing agent) and zinc nitrate under microwave irradiation method. The aqueous mixture was exposed to microwave irradiation at 450 W for 2 min with subsequent addition of aq. NaOH to maintain the pH of the mixture to 10. The mixture was further irradiated for 5 min at 450 W. The prepared gum arabic ZnO NPs were successfully analyzed by UV-Vis, XRD, and RTIR analysis.

#### 3.2.1.3 Cerium Oxide Nanoparticle Synthesis

Cerium oxide ($CeO_2$) nanoparticles were prepared by mixing ethylene glycol to $Ce(NO_3)_3 \cdot 6H_2O$ following the precipitation method (Bellio et al. 2018). The reaction mixture was subjected to stir for 30 min, followed by the addition of $NH_4OH$. The synthesized $CeO_2$ NPs were calcined for 8 h at 500 °C in an air furnace. The electronic and structural properties of $CeO_2$ nanoparticles were successfully determined by XRD and XPS (X-ray photoelectron spectroscopy) technique.

#### 3.2.1.4 Mercury Oxide Nanoparticle Synthesis

Mercuric oxide (HgO) nanoparticles were prepared from the flower extracts of *Callistemon viminalis* following green technology by heating the mixture of flower extract and mercuric acetate at 60 °C with 150 rpm (Das et al. 2015). The plant protein-coated HgO NPs showed an absorption maxima at 243 nm, which was confirmed by UV-Vis spectroscopy. The average diameters of the synthesized particles were found in the range of 2–4 nm. The particle size and morphology were confirmed by XRD, TGA, TEM, and FT-IR spectroscopy. HgO NPs were found to have an optical band gap of 2.48 eV, which suggested its further use in metal oxide semiconductor-based photovoltaic cells.

#### 3.2.1.5 Manganese Dioxide Nanoparticle Synthesis

$MnO_2$ NPs were prepared from the extracts of *Cussonia zuluensis* (Mahlangeni et al. 2020). The metal oxide NPs were prepared by adding the methanolic solution of the extract with $KMNO_4$ solution in the dropwise manner. The synthesized NPs were analyzed by SEM, TEM, XRD, UV-Vis, and FT-IR analysis. When capped with biomolecules from the extract, $MnO_2$ NPs were found to be presented as ultrathin nanoflakes with green morphology with a diameter ranging from 11 to 29 nm, whereas the nanospheres surrounded with nanosheet morphology with particle size 6.99–16.57 nm was found when capped by aralia cerebroside.

#### 3.2.1.6 Synthesis of Copper Oxide and Zinc Oxide Nanoparticles

ZnO and CuO NPs were prepared following double-precipitation strategies (Varaprasad 2017). The procedure for the preparation involves the addition of a mixture of $Zn(NO_3)_2 \cdot 6H_2O$ and $NH_4OH \cdot HCl$ to distilled water. Then, CuO at different concentrations (0.1, 0.25, 0.5 g) was added to this solution. The prepared reaction mixture was stirred for 30 min, and pH of the solution was maintained at 12, and the precipitation was generated by the addition of NaOH solution in the dropwise manner. The resulting precipitate was filtered and dried for 2 h at 100 °C. Morphology and the structure of the NPs were analyzed by XRD, TEM, SEM, UV-Vis, FT-IR, PL, and TGA analysis. UV-Vis study suggested that increasing the CuO concentration leads to a decrease in the intensity of the emission band of ZnO NPs. Order of decreasing intensity of emission band was $ZnO > ZnO\text{-}CuO_{0.1} > ZnO\text{-}CuO_{0.5} > CuO$.

ZnO and CuO NPs were prepared from *Penicillium chrysogenum* (Mohamed et al. 2020). A mixture of zinc (II) acetate and copper (II) acetate was mixed with the biomass filtrate *P. chrysogenum* and allowed to stir (150 rpm) at 30 °C for 48 h. The generated metal oxides were separated and dried at 80 °C for 48 h. Synthesized metal oxides were successfully analyzed through XRD, UV-Vis, TEM, SEM, and FT-IR analysis.

CuO and ZnO NPs were prepared using their respective Schiff's base ligand by calcination for 5 h at 600 °C in good yield with unique morphology and high crystallinity (Alothman and Albaqami 2020). Synthesized metal oxides were analyzed by SEM, XRD, ADX, TEM, and FT-IR analysis. The average particle size of synthesized CuO and ZnO was 42.68 and 62.43 nm, with bandgap 2.38 and 3.34 eV, respectively.

#### 3.2.1.7 Synthesis of Cobalt Ferrite Nanoparticles

Cobalt ferrite nanoparticles were prepared using a simple and cost-effective sol-gel auto-combustion method by mixing cobalt nitrate hexahydrate and ferric nitrate with simultaneous addition of citric acid as a reducing agent (Sharma et al. 2019). The pH of the solution was retained at 7 using liq. $NH_3$. The prepared solution was stirred for 3 h at 80–90 °C until the development of gel. The obtained gel was burned to ash, which was annealed for 5 h at 500 °C and grinded in mortar and pestle to get cobalt ferrite NPs in powder form. The synthesized NPs were found to have a surface area of 22.15 $m^2/g$.

#### 3.2.1.8 Synthesis of Cerium Oxide and Yttrium Oxide Nanoparticles

Hetero-structured $CeO_2$-$Y_2O_3$ NPs were prepared following the hydrothermal method using cerium nitrate, yttrium nitrate, and NaOH as reducing agent (Magdalane et al. 2017). The synthesized metal oxide NPs were studied for their photocatalytic ability against the degradation of a synthetic dye, i.e., Rhodamine B (RhB). The particle size and morphology were confirmed by XRD, SEM, TEM, EDX, and FT-IR analysis. Temperature and pressure played a pivotal role in controlling the shape and size of the synthesized NPs. The oxygen vacancy and lattice constant parameters of NPs also found to depend upon the concentration $OH^-$ ion, which provided better morphology at low pressure and temperature. Using the metal oxide as catalyst, *p*-aminophenol was prepared conveniently by the hydrogenation/reduction of *p*-nitrophenol. To enhance the photocatalytic degradation (PCD) efficiency, the authors used $H_2O_2$, which improved the formation of ROS species.

#### 3.2.1.9 Copper Oxide, Zinc Oxide, and Magnesium Oxide Nanoparticle Synthesis

Different shapes of CuO, ZnO, and MgO nanoparticles were prepared using sol-gel method and hydrothermal method (EL-Mekkawi et al. 2018). Crystal properties, type, and purity of the synthesized metal oxides were determined by using the XRD analysis. Particle sizes and shapes were examined by TEM. Surface properties of the prepared metal oxides were analyzed through XRD.

#### 3.2.1.10 Copper Oxide, Nickel Oxide, and Magnesium Oxide Nanoparticle Synthesis

Copper oxide, nickel oxide, and magnesium oxide nanoparticles were prepared from *Cynomorium coccineum* extract (Jabli et al. 2020). Metal oxide NPs were prepared by mixing $CuSO_4$, $NiSO_4$, and $MgSO_4$ to the extract solution of *Cynomorium coccineum*. The reaction mixture was stirred for 2 h at 70 °C. The developed precipitate was filtered, washed, and dried. The structure and morphology were confirmed by XRD, TEM, SEM, and FT-IR analysis.

## 3.3 Antimicrobial Activity of Metal Oxide Nanoparticles

### *3.3.1 Iron Oxide Nanoparticles as Antimicrobial Agents*

After the successful synthesis of iron oxide nanoparticles following the concepts of green strategies, the authors further studied the antibacterial activities of biogenic FeO nanoparticles using the agar well diffusion method (Irshad et al. 2017). It was found that as compared to FeO nanoparticles prepared by using 20 ml [zone of inhibition: 18 (±0.4) mm] and 60 ml [zone of inhibition: 14 (±0.3) mm] extract solution, FeO nanoparticles prepared by using 40 ml [zone of inhibition: 22 (±0.5) mm] peel extract solution showed strongest antibacterial activity against *P. aeruginosa*. The

MIC (minimum inhibitory concentration) for the antibacterial activity of FeO was found to be 0.062 mg/mL. FeO NPs showed the strongest antibacterial activity by producing reactive oxygen species (ROS) that interact and damage the cellular components like a cell membrane, DNA, and other vital enzymes, thereby ultimately causing bacterial cell death. Synthesized NPs were found to have no hemolytic activity.

Using disc diffusion assay, antimicrobial abilities of the prepared magnetite ($Fe_3O_4$) nanoparticles were carried out against three pathogenic microbes, i.e., *S. marcescens* ATCC 13880, *S. aureus* ATCC 43300, and *E. coli* ATCC 25922 (Alavi and Karimi 2019). The MIC of the synthesized NPs towards the studied bacteria was found in the range 12.5–50 μg/mL with minimum bactericidal concentration (MBC) 50–100 μg/mL. Furthermore, as compared to *S. marcescens* ATCC 13880 and *E. coli* ATCC 25922, the NPs were found to be more efficient towards *S. aureus*. The average diameter of NPs for antibiofilm activities was found to be 83.4 nm. It also acts as antispreading agent by decreasing the colony expansion against *S. aureus*.

The antimicrobial abilities of functionalized nanoparticles towards the methicillin-resistant strains of *S. aureus* (MRSA) were evaluated by El-Guendouz et al. (2016). The NPs exhibited potent antibacterial activity by inhibiting the biofilm formation in catheters. The potency of NPs against *S. aureus* was found to be dependent upon the mode of preparation. Mainly, particles prepared by the co-precipitation method using $Na_2SO_3$ and $Fe^{3+}$ solution and derivatized by oleic acid/propolis were found to be more effective against all MRSA strains by impairing the biofilm formation on catheters ($p < 0.001$).

### 3.3.2 Copper Oxide Nanoparticle as Antimicrobial Agents

The authors performed the antimicrobial abilities of the prepared CuO NPs towards both the microbial strains i.e., *Enterococcus faecalis* (gram positive) and *Pseudomonas aeruginosa* (gram negative) using CFU and disc diffusion method (Masoule et al. 2019). Compared to gram-negative bacteria, the CuO-linked poly(curcumin) nanospheres were found to have higher bactericidal activity towards gram-positive bacteria. The increasing antibacterial potency of nanocomposites might be due to the release of $Cu^{2+}$ ions in the aqueous state. The released $Cu^{2+}$ ions selectively bound with the various functional moieties of the proteins and enzymes and brought about the inhibition of cellular processes.

After fruitful synthesis of copper oxide nanoparticles, the authors evaluated their antibacterial property using the traditional agar well disc diffusion method (AlYahya et al. 2018). The synthesized particles with particle size 2 μm were found to have enhanced antibacterial properties and rupture the bacterial cell wall both in *B. thuringiensis* (gram-positive) and *P. aeruginosa* (gram-negative) bacteria. The zone of inhibition increases with a decrease in the particle size of nanoparticles. The nanoparticle adsorbed in the cell membrane of the bacteria and causes the leakage of the proteins and minerals and hence ultimately leads to bacterial cell death.

In vitro antibacterial and anti-biofilm properties of iron-doped copper oxide nanoparticles were demonstrated against two pathogenic bacteria, i.e., *S. aureus* and *S. epidermidis* (Pugazhendhi et al. 2018). The synthesized particles were found to be ineffective towards gram-positive pathogens, i.e., *S. aureus,* which might be due to their thicker wall. The thin wall of gram-negative microbes (*S. epidermidis*) allows the easy penetration of Fe-doped CuO nanoparticles and causes bacterial cell death and hence ultimately acts as an antibiofilm agent.

Synthesized CuO NPs were studied for their in vitro antimicrobial abilities towards *E. coli* and *S. aureus* following the disc diffusion technique (Ramzan et al. 2020). The results showed that metal oxides NPs were exhibiting highly potential antibacterial abilities towards *E. coli* with the maximum zone of inhibition 29 nm.

The biosynthesized CuO NPs were evaluated for their antibacterial properties in order to apply it for wastewater purification generated from the living places and industries (Rafique et al. 2020). CuO NPs were found to have 91% Rhodamine B (RhB) dye removal ability. Also, the NPs act as a potential antibacterial agent against *S. aureus* and *E. coli.*

### 3.3.3 Zinc Oxide Nanoparticles as Antimicrobial Agents

Antimicrobial abilities of both ZnO without PVA, as well as ZnO-PVA nanoparticles, were screened against two pathogenic bacteria, i.e., *S. aureus* and *E. coli* (Gedanken et al. 2016). NPs with particle size 5 nm produced increased the amounts of ROS, which ruptures the bacterial cell wall and hence causes bacterial cell death. As compared to ZnO without PVA, ZnO, with PVA nanoparticles showed enhanced antibacterial properties against both the bacterial strains; reducing the particle size enhances the antibacterial activity.

Karami et al. deliberated the antibacterial abilities of Ag- and I-doped ZnO nanoparticles against *S. aureus*: USA 300 and *E. coli*: MG1655 pathogens (Karami et al. 2020). The study revealed that the synthesized Ag:I:ZnO NPs had pronounced activity against both the bacterial strains. The generation of ROS and $Ag^+$ ions mainly plays a major role in enhancing the bactericidal activity of the NPs both in the presence of visible light and dark conditions. The generated Ag + ion binds with the lipopolysaccharides present on the bacteria cell membrane through electrostatic interaction and creates holes on them, thereby penetrating inside the cell, binding with the DNA, and eventually leading to bacterial death. The antibacterial activities of the NPs were found a little higher towards gram-negative bacteria due to its thinner cell wall.

In another report, Pourhajibagher et al. explored the antimicrobial activity as well biofilm inhibitory ability of curcumin-doped ZnO nanoparticles by using the disc diffusion method (Pourhajibagher et al. 2019). The authors found that photo activated 7.5 wt% cCur/ZnO NPs showed antibacterial potency by preventing biofilm formation with $p < 0.05$.

Antibacterial activity of the synthesized Ag@ZnO-coated cotton fabrics were explained against *S. aureus* and *P. aeruginosa* bacterial strains by Manna et al. (2015). The coated CFs showed excellent antibacterial activities towards both the bacterial strains.

In an extensive study, antimicrobial abilities of the prepared ZnO and Ag@ZnO@AC NPs were studied using disc diffusion technique against both G+ and G− pathogenic bacteria (Prashantha Kumar et al. 2015). Results of the study indicated that Ag@ZnO@AC NPs exhibited maximum zone of inhibition and killing kinetics with MIC value 0.20 mg/ml against *E. coli* and 0.30 mg/ml against *S. aureus*. The MBD (minimum bactericidal concentration) of the NPs was found as 0.35 mg/ml towards *E. coli* and 0.60 mg/ml against *S. aureus*. The bactericidal potency of Ag@ZnO@AC NPs was found a little higher than ZnO@AC NPs.

Agnihotri et al. (2015) augmented the potential bactericidal ability of Ag@ZnO hybrid nanoparticles leading to 100% bacteria-killing ability was studied against *E. coli* and *B. subtilis*. The hybrid nature enhances the bactericidal performance as compared to their individual nanocomponents (Ag or Zn) by direct contact kill of bacteria. The potency of NPs was not significantly declined though, after used by several times. The enhanced bactericidal efficacy and good biocompatibility would further make it feasible to utilize this immobilization approach for synthesizing new-generation antibacterial coatings.

The antimicrobial abilities of the prepared ZnO NPs were determined towards various pathogenic strains such as *P. aeruginosa*, *E. coli*, *S. aureus*, *S. thalpophilum*, *B. subtilis,* and *K. Pneumoniae* (Dhandapani et al. 2020). Synthesized NPs exhibited excellent potency both against gram-positive and gram-negative bacterial strains.

Evaluations of the antibacterial abilities of the synthesized NPs were demonstrated both towards *S. aureus* (gram-positive) and *E. coli* (gram-negative) (Pillai et al. 2020). All the synthesized NPs showed excellent antibacterial potency towards both G+ and G− bacterial strains, whereas ZnO nanoparticles synthesized from *Beta vulgaris* were found to be inactive against *S. aureus*.

Antimicrobial and antibiofilm abilities of the synthesized gum arabic ZnO NPs were studied towards *E. coli* and *S. aureus* (Pauzi et al. 2020). The MIC and MBC of the metal oxide NPs for *S. aureus* and *E. coli* were found to be 31.25, and 62.5 μg/mL, respectively. The observed results suggested that the synthesized NPs will serve as a natural new-generation antibacterial agent.

### 3.3.4 Cerium Oxide Nanoparticles as Antimicrobial Agents

Antibacterial activities of the synthesized $CeO_2$ nanoparticles were evaluated against the gram-negative bacterial strain (Bellio et al. 2018). The cerium oxide nanoparticle acts as potent antibacterial agent by attaching themselves to the bacterial cell membrane through electrostatic interaction. Due to no-toxic nature of $CeO_2$, it remains attached to the bacterial cell membrane for a long hour and generates reactive oxygen species (ROS), thereby disrupting the outer cell membrane of bacteria allowing the leakage of cellular components.

### 3.3.5 Mercury Oxide Nanoparticles as Antimicrobial Agents

The antibacterial activity of plant-coated mercuric oxide nanoparticles were investigated against *Escherichia coli* (MTCC 1687) using the disc diffusion method (Das et al. 2015). HgO nanoparticles were stabilized by the high potential of *Callistemon viminalis,* thereby enhancing the antibacterial activity, which not only finds application in medicine but also is found to have profound use in food as well as cosmetic industries.

Antibacterial activities of the synthesized $MnO_2$ NPs were demonstrated against the pathogenic bacteria (Mahlangeni et al. 2020). Unfortunately, the synthesized NPs did not meet the expectation, i.e., the antibacterial activity of the synthesized NPs was found to be negligible.

### 3.3.6 Copper Oxide and Zinc Oxide Nanoparticle as Antimicrobial Agents

After the successful synthesis of CuO and ZnO nanoparticles, the author investigated their antimicrobial ability against *E. coli* (Varaprasad 2017). NPs (ZnO-$CuO_{0.5}$) were found to have the highest zone of inhibition towards *E. coli* at both low and high concentration levels as compared to their individual components (ZnO or CuO). Co-assembly of different nanoparticles enhances the production of reactive oxygen species (ROS), which helps in rupturing the bacteria cell wall causing bacterial cell death. The zone of inhibition was found in the range of 9–22 mm. The synthesized nanoparticles would be further applied as potential agents for industry and advanced biomedical purposes.

Very recently, Mohamed et al. (2020) exploited the antibacterial and antibiofilm potency of the synthesized metal oxides towards both gram-positive and gram-negative bacterial strains. From the biology point of view, it was observed that the bio-secreted proteins CapE and reduce the CuO and ZnO to hexagonal and spherical shape with particle size 10.5–59.7 and 9.0–35.0 nm, respectively. Both the metal oxide NPs exhibited higher potency against two pathogenic bacteria strains. Moreover, it was found that as compared to ZnO NPs, CuO NPs were found to be more effective towards *S. aureus* (MIC: 1.0 mg/mL) and, therefore, can be applied in various medicinal purposes. The metal oxides bind with the biomolecules present on the bacterial cell membrane, leading to the inactivation of the bacterial proteins and ultimately cause bacterial death.

In another study, Alothman and Albaqami (2020) studied the antibacterial abilities of the synthesized metal complexes against *E. coli* and *S. aureus*. The results revealed that the metal oxide NPs exhibited potential antibacterial activity as compared to the reference drug HPIMC.

### *3.3.7 Cobalt Ferrite Nanoparticles as Antimicrobial Agents*

The Authors envisaged the in vitro antimicrobial activities of synthesized cobalt ferrite NPs towards gram-positive (*S. aureus*, *B. subtilis*) and gram-negative (*P. aeruginosa*, *E. coli*) microbes (Sharma et al. 2019). The cobalt ferrite NPs were found to exhibit optimum potency against *E. coli* with zone of inhibition of 15 mm. Also, the particles were demonstrated to have the ability to destroy the bacterial membrane, leading to cytoplasmic leakage, which enhances the anti-bactericidal ability.

### *3.3.8 Cerium Oxide and Yttrium Oxide Nanoparticles as Antimicrobial Agents*

Antimicrobial abilities of the synthesized NPs were evaluated, and it was found that the nanoparticles displayed excellent antimicrobial properties by inhibiting the growth of bacteria cell wall both in G+ and G− pathogenic species (Magdalane et al. 2017). The generated $Ce^{4+}$ ions and $Y^{3+}$ ions from the metal oxide NPs interacted with the bacterial cell wall, thereby causing pits and holes on it. As a result, the metal ions easily penetrate inside the cell, target bacterial DNA, and hence leading to bacterial cell death (Fig. 3.3).

### *3.3.9 Copper Oxide, Zinc Oxide, and Magnesium Oxide Nanoparticles as Antimicrobial Agents*

Antimicrobial abilities of the prepared metal oxides (CuO, ZnO, and MgO) were studied against G+ (*S. aureus* and *B. subtilis*) and G− (*E. coli*) pathogens using agar diffusion method (EL-Mekkawi et al. 2018). Both ZnO and CuO NPs showed high antimicrobial abilities. However, MgO NPs were found to have significantly low antimicrobial activities. The physicochemical properties of ZnO and CuO nanoparticles significantly influenced their antimicrobial performance. For instance, thin CuO nanoparticles possessed higher antimicrobial abilities than thick CuO nanoparticles which might be due to the easy penetration of thin CuO nanoparticles into bacterial membranes. Due to the lowest particle size and the highest total pore volume of hexagonal ZnO sample, it exhibited the highest antimicrobial abilities towards all bacteria under examination.

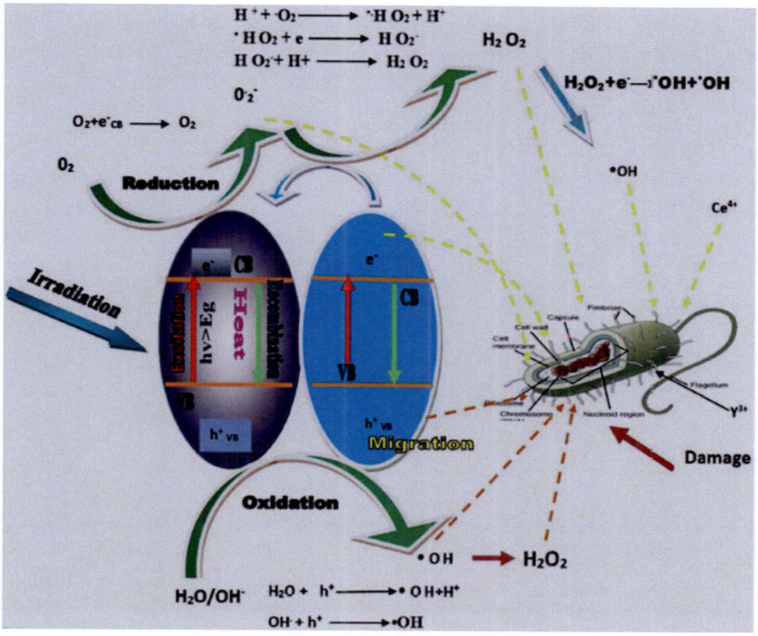


**Fig. 3.3** Plausible mechanism for antimicrobial abilities of $CeO_2/Y_2O_3$ nanoparticles (J Photochem Photobiol B 173:23–34)

### *3.3.10 Copper Oxide, Nickel Oxide, and Magnesium Oxide Nanoparticles as Antimicrobial Agents*

Antibacterial abilities of prepared nickel oxide, copper oxide, and magnesium oxide nanoparticles were studied against *S. aureus*, *S. typhi*, and *C. albicans* (Jabli et al. 2020). Cotton printed with nickel oxide nanoparticles proved high antimicrobial activity towards *S. typhi*.

## 3.4 Conclusion, Outlook, and Future Prospects

In an era of increasing bacterial resistance to classical antibiotics, nanotechnology brings a new ray of hope through nanoantibiotics to sustain the battle against the pathogenic bacteria causing infections. Particularly metal oxide nanoparticles such as iron oxide NPs, copper oxide NPs, zinc oxide NPs, and cerium oxide NPs, along with their various metal oxide hybrids, were found to have excellent ability to fight the war against the pathogenic strains. Due to their small particle size, it remains sticked to the bacteria cell wall for a long period of time. It punctures the bacteria cell wall by generating reactive oxygenated species (ROS) leading to the leakage of the cytoplasmic components, targets the bacterial DNA, thereby causing bacterial death. However, excess release of metal ions leads to toxicity, which can be minimized by functionalization, ion doping, and polymer conjugating the nanoparticles.

From the environment and biological point of view, synthesis of various metal oxide nanoparticles following green and sustainable strategies is highly demanding due to the major advantages associated with this, i.e., atom economic synthesis, easy work-up, cost-effective, less toxicity, etc. Therefore, this chapter will help the future researchers to synthesize various metal oxide nanoparticles and to study their anti-microbial abilities against various microbes which can be considered as suitable alternatives against currently used antibiotics and also will be beneficial for future invention of promising pharmaceuticals.

**Conflicts of Interest** All the authors have declared that there is no conflicts of interest.

## References

Agnihotri, S., Bajaj, G., Mukherji, S., & Mukherji, S. (2015). Arginine-assisted immobilization of silver nanoparticles on ZnO nanorods: An enhanced and reusable antibacterial substrate without human cell cytotoxicity. *Nanoscale, 7*, 7415–7429. https://doi.org/10.1039/C4NR06913G.

Alavi, M., & Karimi, N. (2019). Ultrasound assisted-phytofabricated $Fe_3O_4$ NPs with antioxidant properties and antibacterial effects on growth, biofilm formation, and spreading ability of multidrug resistant bacteria. *Artificial Cells, Nanomedicine and Biotechnology, 47*(1), 2405–2423. https://doi.org/10.1080/21691401.2019.1624560.

Alothman, A. A., & Albaqami, M. D. (2020). Nano-sized Cu(II) and Zn(II) complexes and their use as a precursor for synthesis of CuO and ZnO nanoparticles: A study on their sonochemical synthesis, characterization, and DNA-binding/cleavage, anticancer, and antimicrobial activities. *Applied Organometallic Chemistry, 34*(10), e5827. https://doi.org/10.1002/aoc.5827.

AlYahya, S., Rani, B. J., Ravi, G., Yuvakkumar, R., Arun, A., Ameen, F., & AlNadhary, S. (2018). Size dependent magnetic and antibacterial properties of solvothermally synthesized cuprous oxide ($Cu_2O$) nanocubes. *Journal of Materials Science: Materials in Electronics, 29*, 17622–17629. https://doi.org/10.1007/s10854-018-9865-7.

Amna, S., Shahrom, M., Azman, S., Noor, H., Mohamad, K., Ling, C., Siti, K., Mohd, B., Habsah, H., & Dasmawati, M. (2015). Review on zinc oxide nanoparticles: Antibacterial activity and toxicity mechanism. *Nano-Micro Letters, 7*(3), 219–242. https://doi.org/10.1007/s40820-015-0040-x.

Baek, Y. W., & An, Y. J. (2011). Microbial toxicity of metal oxide nanoparticles (CuO, NiO, ZnO, and $Sb_2O_3$) to *Escherichia coli, Bacillus subtilis*, and *Streptococcus aureus*. *Science of the Total Environment, 409*(8), 1603–1608. https://doi.org/10.1016/j.scitotenv.2011.01.014.

Bellio, P., Luzi, C., Mancini, A., Cracchiolo, S., Passacantando, M., Pietro, L. D., Perilli, M., Amicosante, G., Santucci, S., & Celenza, G. (2018). Cerium oxide nanoparticles as potential antibiotic adjuvant. Effects of $CeO_2$ nanoparticles on bacterial outer membrane permeability. *Biochimica et Biophysica Acta-Biomembranes, 1860*(11), 2428–2435. https://doi.org/10.1016/j.bbamem.2018.07.002.

Das, A. K., Marwal, A., Sain, D., & Pareek, V. (2015). One-step green synthesis and characterization of plant proteincoated mercuric oxide (HgO) nanoparticles: Antimicrobial studies. *International Nano Letters, 5*, 125–132. https://doi.org/10.1007/s40089-015-0144-9.

Davies, J., & Davies, D. (2010). Origins and evolution of antibiotic resistance. *Microbiology and Molecular Biology Reviews, 74*(3), 417–433. https://doi.org/10.1128/MMBR.00016-10.

De Vita, D., Simonetti, G., Pandolfi, F., Costi, R., Di Santo, R., & D'Auria, F. D. (2016). Exploring the anti-biofilm activity of cinnamic acid derivatives in *Candida albicans*. *Bioorganic & Medicinal Chemistry Letters, 26*(24), 5931–5935. https://doi.org/10.1016/j.bmcl.2016.10.091.

Dhandapani, K. V., Anbumani, D., Gandhi, A. D., Annamalai, P., Muthuvenkatachalam, B. S., Kavitha, P., & Ranganathan, B. (2020). Green route for the synthesis of zinc oxide nanoparticles from *Melia azedarach* leaf extract and evaluation of their antioxidant and antibacterial activities. *Biocatalysis and Agricultural Biotechnology, 24*, 101517. https://doi.org/10.1016/j.bcab.2020.101517.

Elfeky, A. S., Salem, S. S., Elzaref, A. S., Owda, M. E., Eladawy, H. A., Saeed, A. M., Awad, M. A., Abou-Zeid, R. E., & Fouda, A. (2020). Multifunctional cellulose nanocrystal/metal oxide hybrid, photo-degradation, antibacterial and larvicidal activities. *Carbohydrate Polymers, 230*, 115711. https://doi.org/10.1016/j.carbpol.2019.115711.

El-Guendouz, S., Aazza, S., Lyoussi, B., Bankova, V., Lourenço, J. P., Costa, A. M. R., Mariano, J. F., Miguel, M. G., & Faleiro, M. L. (2016). Impact of biohybrid magnetite nanoparticles and moroccan propolis on adherence of methicillin resistant strains of *Staphylococcus aureus*. *Molecules, 21*(9), 1208. https://doi.org/10.3390/molecules21091208.

EL-Mekkawi, D. M., Selim, M. M., Hamdi, N., Hassan, S. A., & Ezzat, A. (2018). Studies on the influence of the physicochemical characteristics of nanostructured copper, zinc and magnesium oxides on their antibacterial activities. *Journal of Environmental Chemical Engineering, 6*(4), 5608–5615. https://doi.org/10.1016/j.jece.2018.08.044.

Gedanken, A., Nagvenkar, A., Deokar, A., & Perelshtein, I. (2016). A one-step sonochemical synthesis of stable ZnO–PVA nanocolloid as a potential biocidal agent. *Journal of Materials Chemistry B, 4*, 2124–2132. https://doi.org/10.1039/C6TB00033A.

González-Bello, C. (2017). Antibiotic adjuvants–A strategy to unlock bacterial resistance to antibiotics. *Bioorganic & Medicinal Chemistry Letters, 27*(18), 4221–4228. https://doi.org/10.1016/j.bmcl.2017.08.027.

Irshad, R., Tahir, K., Li, B., Ahmad, A., Siddiqui, A. R., & Nazir, S. (2017). Antibacterial activity of biochemically capped iron oxide nanoparticles: A view towards green chemistry. *Journal of Photochemistry and Photobiology. B, 170*, 241–246. https://doi.org/10.1016/j.jphotobiol.2017.04.020.

Jabli, M., Al-Ghamdi, Y. O., Sebei, N., Almalki, S. G., Alturaiki, W., Khaled, J. M., Mubarak, A. S., & Algethami, F. K. (2020). Green synthesis of colloid metal oxide nanoparticles using *Cynomorium coccineum*: Application for printing cotton and evaluation of the antimicrobial activities. *Materials Chemistry and Physics, 249*, 123171. https://doi.org/10.1016/j.matchemphys.2020.123171.

Karami, A., Xie, Z., Zhang, J., Kabir, M. S., Munroe, P., Kidd, S., & Zhang, H. (2020). Insights into the antimicrobial mechanism of Ag and I incorporated ZnO nanoparticle derivatives under visible light. *Materials Science and Engineering: C, 107*, 110220. https://doi.org/10.1016/j.msec.2019.110220.

Kumar, M., & Das, A. (2017). Emerging nanotechnology based strategies for diagnosis and therapeutics of urinary tract infections: A review. *Advances in Colloid and Interface Science, 249*, 53–65. https://doi.org/10.1016/j.cis.2017.06.010.

Lemire, J. A., Harrison, J. J., & Turner, R. J. (2013). Antimicrobial activity of metals: Mechanisms, molecular targets and applications. *Nature Reviews. Microbiology, 11*, 371–384. https://doi.org/10.1038/nrmicro3028.

Liu, J., Qiao, S. Z., & Hu, Q. H. (2011). Magnetic nanocomposites with mesoporous structures: Synthesis and applications. *Small, 7*(4), 425–443. https://doi.org/10.1002/smll.201001402.

Magdalane, C. M., Kaviyarasu, K., Vijayad, J. J., Siddhardhae, B., & Jeyarajf, B. (2017). Facile synthesis of heterostructured cerium oxide/yttrium oxide nanocomposite in UV light induced photocatalytic degradation and catalytic reduction: Synergistic effect of antimicrobial studies. *Journal of Photochemistry and Photobiology. B, 173*, 23–34. https://doi.org/10.1016/j.jphotobiol.2017.05.024.

Mah, T. F. C., & O'Toole, G. A. (2001). Mechanisms of biofilm resistance to antimicrobial agents. *Trends in Microbiology, 9*(1), 34–39. https://doi.org/10.1016/S0966-842X(00)01913-2.

Mahapatra, O., Bhagat, M., Gopalakrishnan, C., & Arunachalam, K. D. (2008). Ultrafine dispersed CuO nanoparticles and their antibacterial activity. *Journal of Experimental Nanoscience, 3*(3), 185–193. https://doi.org/10.1080/17458080802395460.

Mahlangeni, N. T., Magura, J., Moodley, R., Baijnath, H., & Chenia, H. (2020). Biogenic synthesis, antioxidant and antimicrobial activity of silver and manganese dioxide nanoparticles using *Cussonia zuluensis* Strey. *Chemical Papers, 74*, 4253–4265. https://doi.org/10.1007/s11696-020-01244-9.

Manna, J., Goswami, S., Shilpa, N., Sahu, N., & Rana, R. K. (2015). Biomimetic method to assemble nanostructured Ag@ZnO on cotton fabrics: Application as self-cleaning flexible materials with visible-light photocatalysis and antibacterial activities. *ACS Applied Materials & Interfaces, 7*(15), 8076–8082. https://doi.org/10.1021/acsami.5b00633.

Masoule, S. F., Pourhajibagher, M., Safari, J., & Khoobi, M. (2019). Base-free green synthesis of copper(II) oxide nanoparticles using highly cross-linked poly(curcumin) nanospheres: Synergistically improved antimicrobial activity. *Research on Chemical Intermediates, 45*, 4449–4462. https://doi.org/10.1007/s11164-019-03841-0.

Mirza, A. Z., & Siddiqui, F. A. (2014). Nanomedicine and drug delivery: A mini review. *International Nano Letters, 4*(1), 94. https://doi.org/10.1007/s40089-014-0094-7.

Mohamed, A. A., Abu-Elghait, M., Ahmed, N. E., & Salem, S. S. (2020). Eco-friendly mycogenic synthesis of ZnO and CuO nanoparticles for *in vitro* antibacterial, antibiofilm, and antifungal applications. *Biological Trace Element Research*. https://doi.org/10.1007/s12011-020-02369-4.

Pauzi, N., Zain, N. M., Kutty, R. V., & Ramli, H. (2020). Antibacterial and antibiofilm properties of ZnO nanoparticles synthesis using gum Arabic as a potential new generation antibacterial agent. *Materials Today*. https://doi.org/10.1016/j.matpr.2020.06.359.

Pillai, A. M., Sivasankarapillai, V. S., Rahdar, A., Joseph, J., Sadeghfar, F., Anuf, A. R., Rajesh, K., & Kyzas, G. Z. (2020). Green synthesis and characterization of zinc oxide nanoparticles with antibacterial and antifungal activity. *Journal of Molecular Structure, 1211*, 128107. https://doi.org/10.1016/j.molstruc.2020.128107.

Pourhajibagher, M., Vaziri, A. S., Takzaree, N., & Ghorbanzadeh, R. (2019). Physico-mechanical and antimicrobial properties of an orthodontic adhesive containing cationic curcumin doped zinc oxide nanoparticles subjected to photodynamic therapy. *Photodiagnosis and Photodynamic Therapy, 25*, 239–246. https://doi.org/10.1016/j.pdpdt.2019.01.002.

Prashantha Kumar, T. K. M., Mandlimath, T. R., Sangeetha, P., Sakthivel, P., Revathi, S. K., Ashok Kumar, S. K., & Sahoo, S. K. (2015). Highly efficient performance of activated carbon impregnated with Ag, ZnO and Ag/ZnO nanoparticles as antimicrobial materials. *RSC Advances, 5*, 108034–108043. https://doi.org/10.1039/C5RA19945J.

Pugazhendhi, A., Kumar, S. S., Manikandan, M., & Saravanan, M. (2018). Photocatalytic properties and antimicrobial efficacy of Fe doped CuO nanoparticles against the pathogenic bacteria and fungi. *Microbial Pathogenesis, 122*, 84–89. https://doi.org/10.1016/j.micpath.2018.06.016.

Rafique, M., Tahir, M. B., Irshad, M., Nabi, G., Gillani, S. S. A., Iqbal, T., & Mubeen, M. (2020). Novel Citrus aurantifolia leaves based biosynthesis of copper oxide nanoparticles for environmental and wastewater purification as an efficient photocatalyst and antibacterial agent. *Optik, 219*, 165138. https://doi.org/10.1016/j.ijleo.2020.165138.

Raghunath, A., & Perumal, E. (2017). Metal oxide nanoparticles as antimicrobial agents: A promise for the future. *International Journal of Antimicrobial Agents, 49*(2), 137–152. https://doi.org/10.1016/j.ijantimicag.2016.11.011.

Ramzan, M., Obodo, R. M., Mukhtar, S., Ilyas, S. Z., Aziz, F., & Thovhogi, N. (2020). Green synthesis of copper oxide nanoparticles using Cedrus deodara aqueous extract for antibacterial activity. *Materials Today*. https://doi.org/10.1016/j.matpr.2020.05.472.

Salata, O. V. (2004). Applications of nanoparticles in biology and medicine. *Journal of Nanbiotechnology, 2*(1), 3. https://doi.org/10.1186/1477-3155-2-3.

Sharma, R. P., Raut, S. D., Mulani, R. M., Kadam, A. S., & Mane, R. S. (2019). Sol–gel autocombustion mediated cobalt ferrite nanoparticles:a potential material for antimicrobial applications. *International Nano Letters, 9*, 141–147. https://doi.org/10.1007/s40089-019-0268-4.

Silver, L. L. (2011). Challenges of antibacterial discovery. *Clinical Microbiology Reviews, 24*, 71–109. https://doi.org/10.1128/CMR.00030-10.

Varaprasad, K. (2017). Co-assembled ZnO (shell) – CuO (core) nano-oxide materials for microbial protection. *Phosphorus, Sulfur, and Silicon and the Related Elements, 193*(2), 74–80. https://doi.org/10.1080/10426507.2017.1417301.

Wright, G. D., & Sutherland, A. D. (2007). New strategies for combating multidrug-resistant bacteria. *Trends in Molecular Medicine, 13*, 260–267. https://doi.org/10.1016/j.molmed.2007.04.004.

# Chapter 4
# Emerging Nanomaterial-Based Medications: Key Challenges and Opportunities

**Joana Reis, Teresa Oliveira, and Rita Payan-Carreira**

## 4.1 An Overview of Nanomaterial-Aggregate Medications

Nanoparticles (NPs) are gaining rising projection as therapeutic agents with increased efficacy and diminished systemic drug side effects. Nanomaterials may also be engineered to be multifunctional, combining therapeutic, targeting and diagnostic features (Seleci et al. 2016; Park et al. 2009). The term nanoparticle refers mostly to materials within a size range from 1 to 100 nm. NPs close to 100 nm have been extensively used to favour drug accumulation, uptake and therapeutic response. Nanomaterial-based systems may be classified into three categories: inorganic, organic and hybrid (organic-inorganic) (Fig. 4.1) (Peer et al. 2007). Inorganic nanomaterials for theranostics (a term that derives from the combination of therapy and diagnostics) include metallic NPs such as gold, silver, iron, and cobalt, but also carbon-based nanomaterials (as carbon nanotubes (CNTs), fullerene, graphene oxide and quantum dots), and ceramic NPs including mesoporous silica NPs, aluminium oxide or titanium dioxide (Wang et al. 2020a; Baig et al. 2019; Maiti et al. 2018; Slowing et al. 2007; Zang et al. 2019). Organic NPs include synthetic and natural polymer nanoparticles, liposomes, niosomes, solid lipid NPs, dendrimers and bio-macromolecule-based NPs (Mitragotri and Stayton 2014; Biswas et al. 2019).

J. Reis (✉)
Escola Superior Agrária de Ponte de Lima, Instituto Politécnico de Viana do Castelo, Viana do Castelo, Portugal
e-mail: jmargaridareis@esa.ipvc.pt

T. Oliveira · R. Payan-Carreira
Departamento de Medicina Veterinária, Escola de Ciências e Tecnologia, Universidade de Évora, Évora, Portugal

Mediterranean Institute for Agriculture, Environment and Development, Évora, Portugal


K. Pal (ed.), *Bio-manufactured Nanomaterials*,
https://doi.org/10.1007/978-3-030-67223-2_4

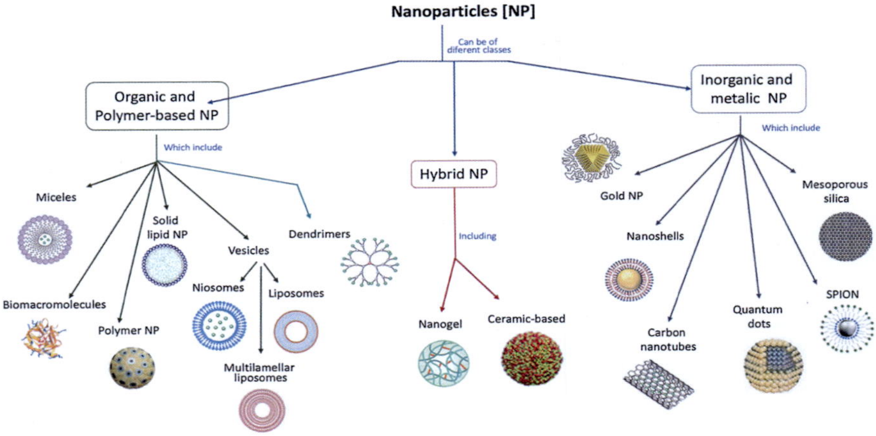


**Fig. 4.1** Classification of NPs according to its type (*NP* nanoparticles)

Hybrid nanoparticles include metal–organic frameworks, nanoscale coordination polymers and protein–ceramic hybrid NPs, for example (He et al. 2015; Vega-Chacón et al. 2020; Nguyen and Zhao 2015). The term hybrid has also been applied to NPs that combine at least two different materials in their structure, even if both are organic or inorganic. Hybrid NPs combine the properties of the different constituents to improve biocompatibility, stability, targeting efficiency and different roles in a diagnostic/therapeutic sequential approach. NPs have also been classified as soft, hard and other (Schütz et al. 2013).

NPs present very distinctive properties and behaviour when interacting with biological systems, compared to those of bulk material. Due to their surface-to-volume ratio and minute size, NPs hold singular physical and chemical features (Schwirn et al. 2014). Likewise, NPs based on the same material present distinguishing behaviour depending on size, shape, synthesis method, surface charge, surface modifications and functionalization, amongst other variables (Reis et al. 2011). One could say, for example, that not all gold nanoparticles are created equal (or induce equal biological response); the same is applicable to NPs in general. This uniqueness, associated with the submicrometric scale, hinders the prediction and assessment of their biological effects. Comparing the results from studies resorting to different methodologies, however minor may the variances seem, is also difficult.

Irrespective of its classification and of being highly promising, nanomaterial-based medications pose unique challenges and offer exceptional opportunities, which are discussed in this chapter from the perspective of clinical applications, with special emphasis in green nanomedicine approaches. Green synthesis of nanoparticles for medical application reduces the use of hazardous materials in the synthesis process and the deleterious health impacts of pharmaceutics and environmental deterioration. The green synthesis processes resort to simple, cost-effective, sustainable methods and raw materials, avoid toxic chemicals and by-products and reduce the number of steps in the procedures, promoting simultaneously a safer laboratory environment (Lam et al. 2017; Kanwar et al. 2019).

## 4.2 Key Challenges

### 4.2.1 *How Do NPs Interact with Biological Systems?*

In the organism, NPs contact with extracellular proteins and other biomolecules. When encountering a solid surface, proteins suffer significant conformational changes (Nakanishi et al. 2001; Gray 2004), leading to the formation of a dynamic "protein corona" on the NP surface. This protein corona is composed of a "hard corona," integrating the longer-lived and more stable NP protein coating, and a "soft corona," more changeable. However, it has been recognized that the composition of the hard corona in terms of the nature and amount of the proteins present depends not only on the nanoparticle but also from the protein concentration in the medium, with implications for prognostication of in vivo response based on the in vitro studies results (Monopoli et al. 2011). The protein corona is a major parameter for determining the identity of NPs and their interaction with cells and the immune system (Durán et al. 2015; Escamilla-Rivera et al. 2019; Zhang et al. 2019). Protein corona depends on the medium composition and pH, incubation time, shear stresses and temperature, amongst others, and on key NP features such as size, curvature, surface charge and functionalization (Durán et al. 2015; Lee 2020; Marichal et al. 2020). However, adsorption to the NP surface not only alters the nanomaterial surface presented to the biological system but also changes the morphology and function of the adsorbed molecules, such as enzymatic activity and transport ability (Devineau et al. 2017; Johnson et al. 2014; Breger et al. 2019; Ding et al. 2015; Salvati et al. 2013). The observed changes in conformation and function are dependent on the interaction between the biomolecule and the nanoparticle and do not depend solely on the protein nature (Marichal et al. 2020; Kokkinopoulou et al. 2017). Particle curvature largely affects the protein's secondary structure. Larger diameter particles drive more extensive particle–protein interaction, leading to more pronounced changes in its secondary structure; contrastingly, the NPs' curvature does not affect the protein's tertiary structure (Lundqvist et al. 2004). The NPs' shape also determines the amount of adsorbed protein: several studies suggest that for NPs with similar chemistry, porosity, surface potential and size in the *y*-dimension, a significant larger amount of proteins attached to rod-like structures when compared to spheres (Gagner et al. 2011; Madathiparambil Visalakshan et al. 2020). Observed shape effects of NPs may not be due to a physical effect but rather to the different chemistry associated with different crystal facet surface areas, as it happens in silver NPs (Pal et al. 2007). The shape can control the nanoparticle in vivo behaviour, setting its biodistribution, intravascular and transvascular transport (Toy et al. 2014) and immune response, as suggested by in vivo studies (Pujari-Palmer et al. 2016; Da Silva-Candal et al. 2019). The protein adsorption behaviour on the surface of NPs is affected by media pH, since electrostatic interaction occurs depending on the charges of the adsorbent and the adsorbate, according to the protein isoelectric point, even if other variables such as size might interfere (Reis et al. 2011; Zhang et al. 2019; Saikia et al. 2019). Protein to protein interactions and competition also influence the composition and stability of the protein layer adsorbed to the NPs (Lee 2020).

NP size strongly influences the biological response, as it is determinant for the protein corona formation, bioavailability, cell internalization, immunogenicity and cytotoxicity (Breger et al. 2019; Ivask et al. 2014). Particles < 10 nm undergo renal clearance whilst larger particles (>10 nm) are often rapidly sequestered from the blood, cleared through the liver and mononuclear phagocyte system and may accumulate in blood, spleen and liver, amongst other organs (Tang et al. 2016; Mahmoud et al. 2020; Jindal 2017). Once the NPs are systemically administered, they begin a complex cascade of interactions with the host extracellular medium and cells. Nanoparticle size is a critical factor for its circulation in the blood and lymph, tumour accumulation, tumour retention and drug release, as it is well known when considering liposomes and other drug nanocarriers (Oussoren and Storm 2001; Dong et al. 2019; Danaei et al. 2018; Nagayasu et al. 1999). Figure 4.2 presents a summary of some in vitro methods that may be used for NP characterization (Jain and Thareja 2019; Savage et al. 2019; Ke et al. 2015).

Each application poses its specific set of challenges and distinct aims. Intracavitary chemotherapy (e.g. intraperitoneal and intrapleural) aims at high local drug concentrations for an extended period and low absorption and systemic distribution. Systemic therapy targeting metastatic disease must face poor tumour perfusion, permeability and penetration; simultaneously, it must be able to be delivered to and enter target cells, bypass the endo-lysosomal compartment and allow appropriate intracellular processing and transport. This must be achieved with minimal systemic toxicity. For many applications, it is undesirable to have a high uptake by macrophage-type cells such as the liver Kupffer cells because it may result in an

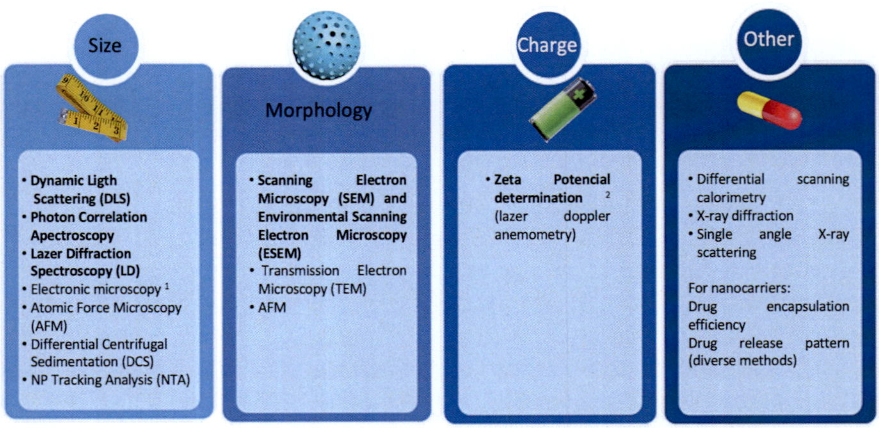


**Fig. 4.2** Some frequently used methods for NP characterization.[1]Complementary to light scattering-based techniques, allows for individual particle measurement; there are limitations when NPs are very small and in the presence of agglomeration.[2]Gives information about the net charge of NPs, strongly related to NP dispersion and stability when in suspension. The dispersion liquid must simulate the biological systems for appropriate prediction

early removal from circulation before therapeutic concentrations are achieved in target tissues and organs. However, other uses may aim at eliciting strong cellular immunity and phagocytosis (De Brito et al. 2019).

The huge potential for varied designs when considering nanomaterial-based medications can only be fully taken advantage of if the multiple variables concerning physicochemical properties and interactions with the host are fully understood and predictive models developed (Nguyen and Lee 2017; Andraos et al. 2020).

### 4.2.2 *Toxicity and Toxicity Assessment*

The assessment of NP toxicity poses difficult challenges. Interferences with in vitro assays may arise due to the nano-related optical, fluorescent, surface, oxidative and catalytic properties of NPs. For example, NP interfere with cell viability assays due to adsorption of lactate dehydrogenase (LDH) to the surface, with enzyme inactivation, and consequent result distortion of the most common tetrazolium salt-based tests but also may interfere with the reduced form of nicotinamide adenine dinucleotide (NADH) by oxidation, interfering with non-salt fluorometric methods (Savage et al. 2019; Andraos et al. 2020; Ong et al. 2014; Pem et al. 2018).

The size of NPs is also responsible for unique optical, magnetic, and electrical properties that interfere in the tests used for the assessment of their effects. When the material dimensions are much smaller than those of the wavelength of the electromagnetic radiation, the interaction with matter is altered. Thus, NPs do not absorb, scatter and reflect electromagnetic radiation in the same way as bulk materials and present fluorescent, magnetic and electrical capacities diverse from their conventional chemistry (Flory et al. 2011). For example, bulk metals absorb electromagnetic radiation whilst thin metal films may partially transmit and become somewhat transparent. Metal NPs present electromagnetic resonances originated by plasmons, quanta of plasma oscillations. Light-excited plasmons strongly scatter light and absorption, as well as enhance local electromagnetic field. In the case of metals (gold and silver), plasmon modes fall in the visible spectral region (Orendorff et al. 2006). In semiconductor quantum dots, another phenomenon known as quantum confinement effect causes the optical absorption and emission shift to higher energies with size reduction (Reithmaier et al. 1997; Orendorff et al. 2006). These aspects must be pondered when conducting in vitro assays since these aim at providing information on NP cytotoxicity, mechanisms of biological activity and cellular uptake. Figure 4.3 presents a summary of some of the in vitro eassays that may be used for NP toxicity assessment (Jain and Thareja 2019; Savage et al. 2019; Norris et al. 1993; Reithmaier et al. 1997).

However, it must be noted that each type of NP poses its unique set of possible interferences and that there is not a standardized one-size-fits-all solution when choosing the experimental protocols (Andraos et al. 2020; Pem et al. 2018); therefore, safety assessment should consider the type of NP, be stepwise, evaluate both material properties and effects with growing biological complexity and fostering non-animal methods [61].

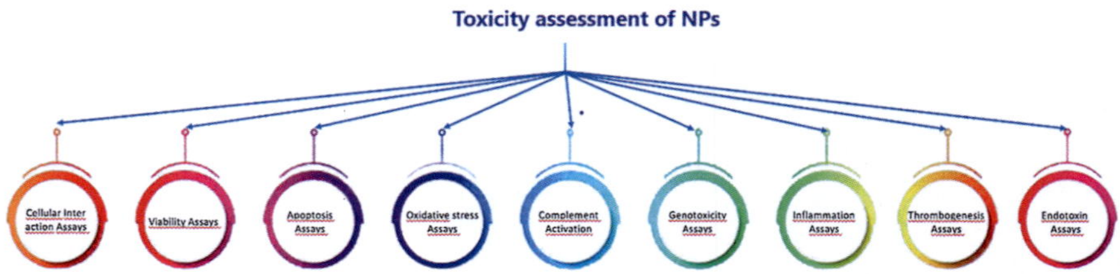


**Fig. 4.3** Possible NP toxicity assessment assays, dependent on the application and administration route

In vitro studies have lower related costs, are more easily replicated and pose none of the ethical issues associated with in vivo studies (Landsiedel et al. 2017; Swierczewska et al. 2018; Hartung 2018). However, they have limitations of their own. In vitro results do not always translate into similar in vivo observations, as it is the case of reported genotoxicity for some NPs, due to the diversity of cells and particles tested, methods and exposure scenarios (Price and Gesquiere 2020; Marquis et al. 2009; Maser et al. 2015). Often concentrations of NPs and routes of exposure do not sufficiently replicate "real life" situations. Comparison between studies is often difficult due to incomplete or distinct methodologies in characterization and assessment.

### *4.2.3 Nanoparticles and Oxidative Stress: The Face of Janus?*

The use of nanoparticles (NPs) in pharmaceutical products originates several intense and involuntary reactions (Wang et al. 2020b). One main mechanism for nanoparticle toxicity is the production of reactive species (ROS and RNS). Damage usually derives from an imbalance between its production and the scavenging activities set in place to avoid its accumulation in tissues (Magdolenova et al. 2014). Free radicals and other reactive species (ROS and RNS) are continuously generated as byproducts of normal cellular metabolism (Fig. 4.4).

In normal physiological conditions, the levels of oxidants are regulated by antioxidant defence mechanisms. These mechanisms control the amount of reactive species through specific scavenger reactions and detoxification pathways, which control the flux of ROS and act to inhibit and/or reduce the damage caused by deleterious reactive species (Prajitha et al. 2019). Disturbance of such balance originates an array of physiopathologic outcomes (e.g., DNA damage, inflammation, cell death fibrosis, metaplasia or carcinogenesis), which may be exponentially increased, if the situation sustains, by the involvement of immune cells and immunomodulators (Ferreira et al. 2018). NP-mediated ROS/RNS responses tend to spiralize, the direct consequences of ROS/RNS production triggering in themselves new sources for accrue generation of free radicals and oxidative species (Santos and Payan-Carreira 2018). The mechanisms underlying ROS generation by different NP are

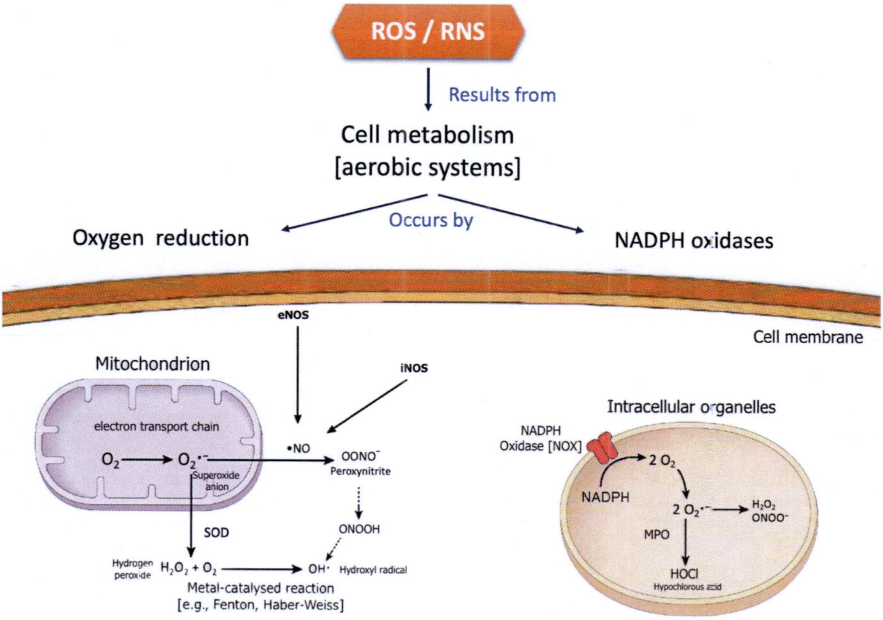


**Fig. 4.4** Source of reactive oxygen species (ROS) in normal cells. *SOD* superoxide dismutase, *iNOS* inducible nitric oxide synthase, *eNOS* endothelial nitric oxide synthase, *MPO* myeloperoxidase, *NADPH* reduced form of nicotinamide adenine dinucleotide phosphate, *ONOOH* peroxynitrous acid

still incompletely elucidated, particularly on what respects to the generation of RNS. NPs are usually first adsorbed on the cell surface. NP uptake into the cell occurs by diverse mechanisms (e.g., pinocytosis; caveolae, lipid raft and clathrin-dependent endocytosis; phagocytosis (Yu et al. 2020)—Fig. 4.5).

Once inside the cell, NPs may trigger oxidative stress through different mechanisms. Changes in redox may lead to oxidative modification of biomacromolecules such as proteins and lipids, with damage to the integrity of cell and mitochondrial membranes, impaired mitochondrial respiration and reduced membrane potentials, cellular enzyme deactivation, membrane structure disruption, depleted redox potential levels, along with DNA damage and the mechanism balancing cell death/proliferation. All these reactions aggravate the intracellular ROS accumulation; the expression of oxidative stress-related genes is compromised, interfering with the scavenging mechanisms that could re-balance the tissue and cell oxidative stress, driving an escalation of the situation. Figure 4.6 summarizes the proposed mechanism of NP-induced oxidative stress (Magdolenova et al. 2014; Ferreira et al. 2018; Santos and Payan-Carreira 2018; Manke et al. 2013; Sun et al. 2017; Fu et al. 2014).

The oxidative stress response to NPs has been graded according to three magnitude levels, which were related to the signalling cascades triggered by oxidative radicals generated in response to NPs toxicity. These levels relate to the sequence of events that follow the increase in cellular oxidative stress (Fig. 4.7). At low levels,

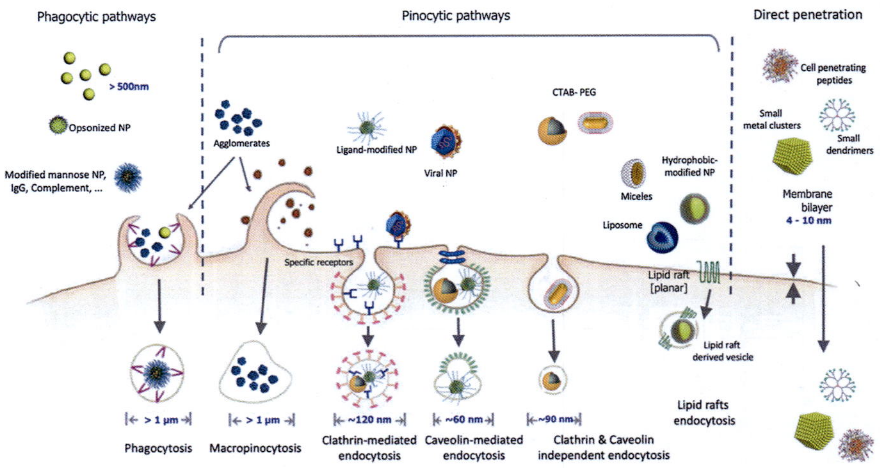


**Fig. 4.5** Cellular mechanisms of NP uptake, covering phagocytosis, endocytosis and transmembrane migration (adapted from Yu et al. 2020)

an attempt is made to correct the excessive levels of free radicals that accumulate in cells, but when the scavenger mechanisms are depleted, inflammation occurs that tend to aggravate the magnitude of installed oxidative stress, which then leads to cell death.

The kinetics and regulation of NP uptake depend on the cell characteristics and the NP size, the surface-to-volume ratio, surface features, zeta potential and NP resistance against aggregation (Liu and Shi 2019), amongst others (Fig. 4.8). These and other factors will also explain why NPs present very different ROS/RNS potential.

Interestingly, some NPs (both of inorganic and biological origin) also act as exogenous antioxidants (Tomankova et al. 2015). These characteristics may turn NPs into a useful tool in the medical area, particularly in the therapeutic approach to cancer, ageing-related and degenerative diseases, amongst others. Albeit NPs offer several advantages over traditional antioxidant delivery methods, they also have major hurdles, namely regarding the achievement of high specificity against ROS and a detailed knowledge on their way of integration into disease mechanisms. Additional studies are therefore foreseen to clarify the antioxidant properties, side effects and safety of NPs before any future biomedical application.

### *4.2.4 The Gap Between the Lab Bench and the Bedside: From Challenge to Opportunity*

The development of nanomaterial-based approaches has been swift and globally expanding since the introduction of the first FDA-approved doxorubicin liposomal formulation in 1995 (Khalil et al. 2019). However, regardless of the investments in nanomedicine research, only a few nano-formulations have entered the market as

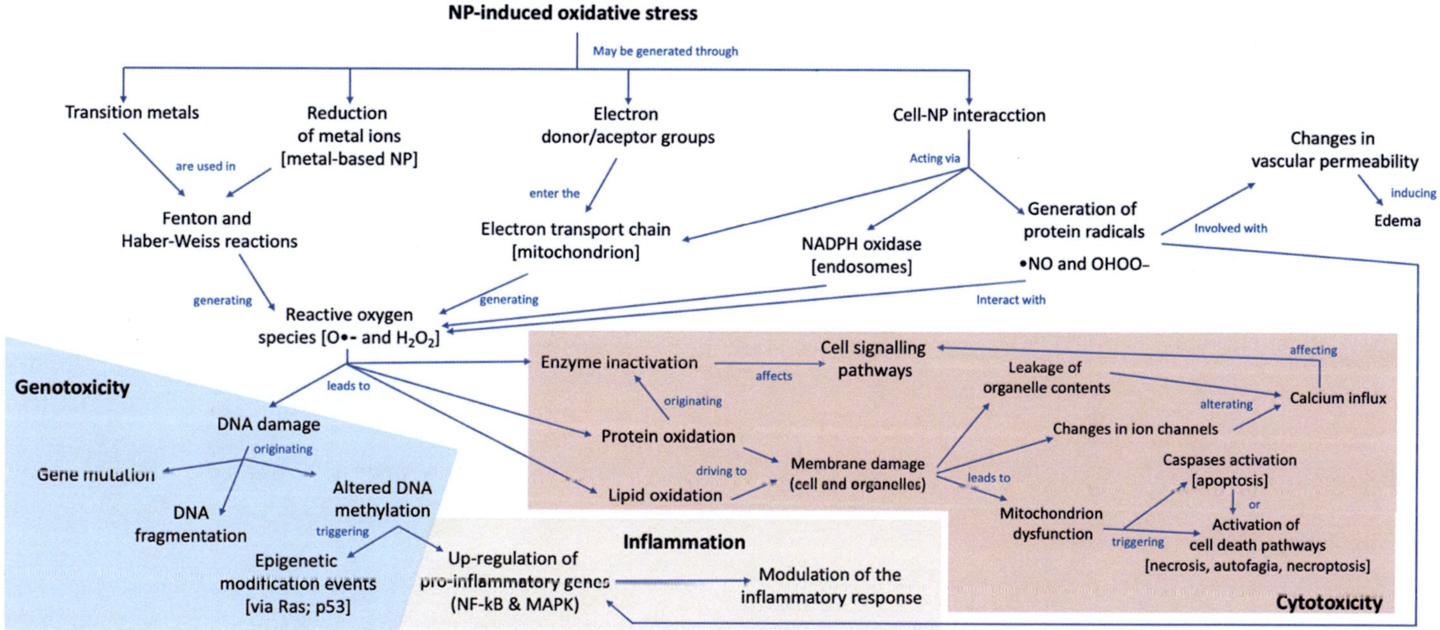


**Fig. 4.6** Proposed mechanisms of NP-induced oxidative stress. *NF-kB* nuclear factor-kB, *MAPK* mitogen-activated protein kinase, *˙NO* nitric oxide, $OONO^-$ peroxynitrite, $H_2O_2$ hydrogen peroxide; $O_2^{\cdot -}$ superoxide radical, *OH˙* hydroxyl radical

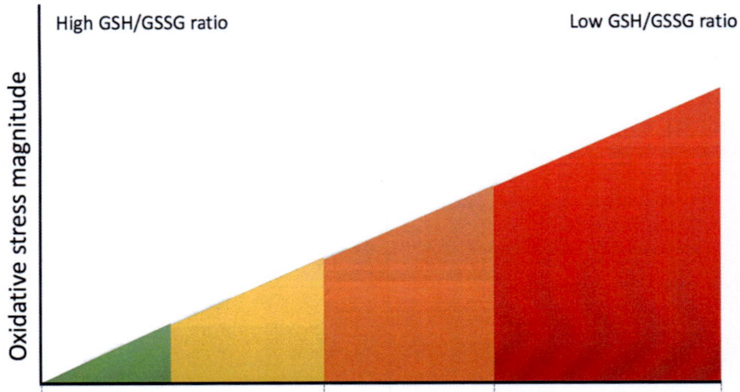


| | | | | |
|---|---|---|---|---|
| Cell response pathway | Physiological | Adaptation | Inflammation | Cytotoxicity |
| Modulators | Anti-oxidant molecules and endogenous scavengers | | MAPK & NF-kB signalling pathways | Caspases activation Deplection of ATP |
| Outcome | Cytoprotection | | Cytokines Chemokines Adesion molecules | Cell death [necrosis, autophagy, apoptosis] |

**Fig. 4.7** Magnitude of oxidative stress and corresponding main events. *MAPK* mitogen-activated protein kinase, *NF-kB* nuclear factor-kB

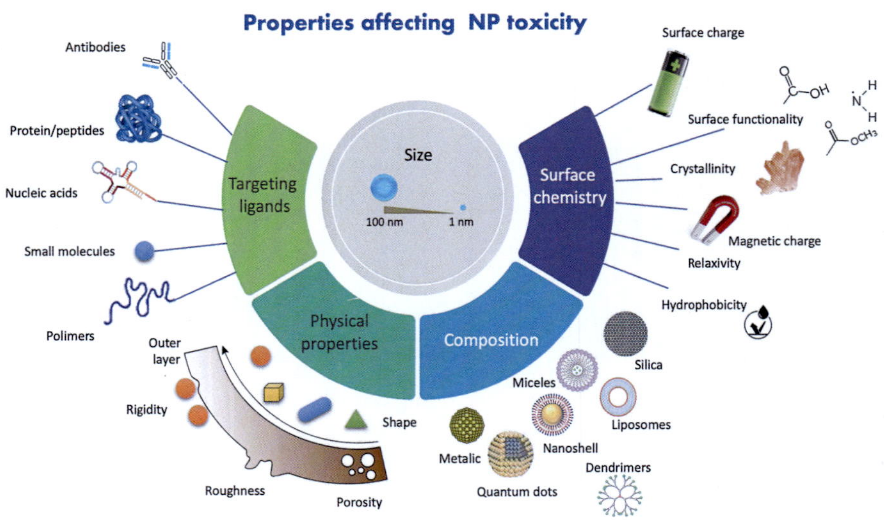


**Fig. 4.8** Determinants of NP toxicity

clinically approved products, and the gap between early development and clinical application remains a major challenge. One of the limitations is that laboratory animal models fail to accurately represent the diversity of the patients' population and even when using multiple animal models; with the consequent effects on increased animal use, the complexity of spontaneous disease and clinical settings is not well represented. Many drugs that are promising in rodent models are not successful when applied to human patients. However, natural disease in companion animals may help filling this gap, with advantage for both humans and animals, as long as a careful preclinical NP evaluation is performed and risk–benefit analysis done in an objective and careful way. Companion animals are exposed to the same environment of their human carers, and spontaneous disease better reflects the genetic, environmental and physiological variations observed in human disease, as explored in the One Health concept (Bobo et al. 2016; Henry 2014). Many types of cancer, musculoskeletal, neurological and genetic disorders, as well as cardiovascular, allergic disease, autoimmune and inflammatory diseases present common features in companion animals and humans (Kol et al. 2015; Reis et al. 2020; Hakanen et al. 2018; Jin et al. 2018; Fernández-Trapero et al. 2017). Progress in the knowledge of pathophysiology and applied therapeutics has the potential to benefit humans and animals, eliminating the moral dilemma of causing suffering to one being to potentially benefit another (or not, as often happens) and help overcoming the gap between early development and clinical application.

## 4.3 Biomedical Scopes and Opportunities

In this section, novel developments and opportunities presented by nanomaterial-based medications for the treatment and prevention of several medical conditions, with distinct aetiologies and physiopathology, are summarized.

### *4.3.1 Oncologic Disease*

Various new therapeutic and theranostics nanotechnology-based approaches are being explored to circumvent the challenges posed by conventional approaches to oncological disease. Table 4.1 summarizes some of the reported developments since 2018 but is by no means a thorough review considering the exponential growth in this area of research; potential applications are multiple and NP design possibilities unending. However, most have only been tested in vitro, often in mono-type cell cultures. Further advancement depends on more extensive testing, including organ and mini-organism on-a-chip and in vivo studies.

Bio-manufactured NPs have been tested as drug carriers, for their intrinsic anticancer effects, as sensitizers for chemotherapy, radiotherapy and phototherapy and as immunomodulators of host response. Metal NPs (e.g. silver, gold and palladium

**Table 4.1** A few recent nanomaterial-based medications for cancer treatment

| Nanomedicine | Details | Obs. | Refs. |
|---|---|---|---|
| Curcumin-based | Micellar NPs, acid-labile methotrexate and curcumin loaded for carcinoma treatment | In vitro<br>In vivo | Xie et al. (2018) |
| | TPGS/curcumin NPs for colon cancer treatment | In vitro<br>In vivo | Li et al. (2019) |
| | Human serum albumin/curcumin NPs for breast adenocarcinoma treatment | In vitro | Matloubi and Hassan (2020) |
| | Dextran/curcumin NPs, methotrexate delivery for breast cancer treatment | In vitro | Curcio et al. (2019) |
| | Curcumin-loaded silk fibroin NPs for hepatocarcinoma and neuroblastoma treatment | In vitro | Montalbán et al. (2018) |
| | Curcumin-loaded solid lipid NP for radiosensitization in breast cancer treatment | In vitro | Minafra et al. (2019) |
| | Phenyl boronic acid-conjugated and pH-responsive ZnO nanoparticles for targeted delivery of curcumin for breast cancer treatment | In vitro<br>In vivo | Kundu et al. (2019) |
| | Folic acid conjugated curcumin loaded biopolymeric gum acacia microsphere for triple negative breast cancer therapy | In vitro<br>In vivo | Pal et al. (2019) |
| | Complementary therapy to gemcitabine with phytosome complex of curcumin for pancreatic cancer treatment | Phase II trial | Pastorelli et al. (2018) |
| Metal and nonmetal NPs | Green AuNPs synthetized using a *Corchorus olitorius* leaf extract with cytotoxic activity in vitro in hepatocarcinoma, breast and colon carcinomas | In vitro | Ismail et al. (2018) |
| | Selenium nanoparticles by chitosan and *Pleurotus ostreatus*-fermented fenugreek against Erlich ascites and colon carcinomas | In vitro | El-Batal et al. (2020) |
| | AuNPs synthetized from *Abies spectabilis* extract and its anticancer activity on bladder cancer T24 cells | In vitro | Wu et al. (2019) |
| | Green synthesized silver NPs using extract of *Nepeta deflersiana* against human cervical cancer cells | In vitro | Al-Sheddi et al. (2018) |
| Albumin | Piceatannol nanoencapsulated in albumin effects on colon cancer cells | In vitro<br>In vivo | Aljabali et al. (2020) |
| Polymeric | Phloretin-loaded chitosan nanoparticles on oral cancer cells | In vitro | Mariadoss et al. (2019) |
| | Quercetin loaded PLGA microspheres in breast cancer treatment | In vitro | Karthick et al. (2019) |
| | Catechol-functionalized alginate nanoparticles as mucoadhesive carriers for intravesical chemotherapy | In vitro<br>Ex vivo | Sahatsapan et al. (2020) |

NPs) may be obtained from natural extracts through green chemistry processes and have wide potential medical application (Si et al. 2020).

Cancer possesses complex characteristics and tissue structure, distinctive from that of normal organs and tissues. Nanomedicine-based medications may use the unique NP features for more effective delivery of drugs with poor bioavailability (such as curcumin or piceatannol; cf. Batra et al. 2019), safer treatments with drugs with high rates of adverse reactions (such as paclitaxel or doxorubicin; cf. Chou et al. 2020; Chhikara et al. 2019), smart drug delivery by taking advantage of ease of interaction with microenvironment and cells and local pH changes and ROS for NP activation (cf. Zhang et al. 2020a, b; Haddada et al. 2018), tumour microenvironment modulation (Liu et al. 2018) and enhanced cancer immunotherapy (cf. Irvine and Dane 2020). Oncologic disease is multifactorial, triggered by the combination of genetic and environmental factors. In addition to therapeutic and theranostics applications, nanotechnology may assist in cancer prevention, not only through vaccines that stimulate the immune response but also by improving the bioavailability of compounds known for their preventive properties such as polyphenols (Arora et al. 2019), flavonoids (Khan et al. 2019), carotenoids (Zare et al. 2020), selenium species (Tan et al. 2018).

### *4.3.2 Degenerative Diseases*

Likewise, the potential use of nanotechnology-based approaches in the diagnosis and treatment of several degenerative diseases has been thoroughly developed. For instance, ROS and RNS play an essential role in the development of osteoarthritis (OA), a devastating condition affecting more than 400 million people worldwide.

As reviewed in Table 4.2, some of the nano-based approaches thus focus on ROS scavenging properties, whilst others modulate the expression of targeted genes, either reducing expression of proinflammatory factors or promoting regeneration. Nano-formulation also allows better more traditionally used drug availability in situ, with less secondary systemic effects.

Like OA, several other debilitating conditions can benefit from the use of nanomedicine. Some of these include other orthopaedic (e.g. rheumatoid arthritis) and non-orthopaedic (as retinal degenerative diseases, like age-related macular degeneration and diabetic retinopathy) (cf. Wang et al. 2018; De Matteis and Rizzello 2020), neurodegenerative disorders (e.g. Alzheimer, Parkinson and amyotrophic lateral sclerosis), amongst others.

Again, a major limitation is that several of these nanoparticles have only been tested in vitro. Whenever available, most in vivo studies were carried out in rodents, requiring further translational validation in other, for example models for spontaneous diseases.

**Table 4.2** Review of nanomaterial-based therapies for degenerative diseases

| Nanomedicine | Details | Obs. | Refs. |
|---|---|---|---|
| Dexamethasone | Dexamethasone-loaded ROS-responsive NPs with anti-inflammatory effects for the treatment of rheumatoid arthritis | In vitro In vivo | Ni et al. (2020) |
| Dopamine-melanin | Artificially melanin nanospheres (prepared by polymerization of dopamine) as a novel type of radical scavenger for OA therapy | In vitro In vivo | Zhong et al. (2019) |
| FT-C60 | Targeted and systemic FT-C60NPs developed to treat degenerative disc diseases by significantly attenuating mRNA expressions of proinflammatory factors | In vitro Ex vivo In vivo | Xiao et al. (2019) |
| Gold NPs | Antibody-AuNPs for the selective delivery of fluorescently tagged siRNAs or FITC-dextran dye into retinal cells | In vivo | Wilson et al. (2018) |
| Iron oxide NPs | Iron oxide NPs combined with nerve growth factor (NGF) and quercetin for the treatment of neurodegenerative diseases | In vitro | Katebi et al. (2019) |
| Lipid nanoparticle (LNP) | Lipid nanoparticle (LNP)-siRNA delivery system for the treatment of cartilage diseases by knocking down specific genes, providing a significantly chondroprotective effect | In vitro Ex vivo In vivo | Rai et al. (2019) |
| Non-viral NPs | LNPs for the delivery of mRNA to the back of the eye for the treatment of monogenic retinal degenerative disorders of the RPE | In vivo | Patel et al. (2019) |
| Polydopamine | Anti-angiogenic protein-loaded polydopamine (PDA) nanoparticles for the enhanced treatment of age-related macular degeneration (AMD) | In vitro Ex vivo | Jiang et al. (2020) |

### *4.3.3 Pain Control*

Nanotechnology presents outstanding opportunities for pain control, namely in pain refractory to conventional treatment, such as neuropathic and chronic pain. Adding to their abilities as drug carriers, NPs may also be developed to present intrinsic analgesic activity, act by reducing tolerance to opioids and as ROS scavengers, crossing the blood–brain barrier and being targeted at specific sites and cells to reduce interference of NPs with healthy tissue. The pro-inflammatory microglial activation with abnormal ROS generation in the spinal cord is key in the development of neuropathic pain (Da Silva-Candal et al. 2019; Kuthati et al. 2020; An et al. 2020). The therapeutic application of nanotechnologies is described for drugs like morphine, bupivacaine, baclofen, amitriptyline, doxepin and imipramine. The analgesic effects of these drugs are greatly improved using nanocarriers formulated with liposomes, PLGA, and polyesters (Kuthati et al. 2020). As referred previously in this chapter, nanotechnology may also contribute for the diagnosis of the underlaying conditions that cause severe pain, such as degenerative and oncologic diseases.

### 4.3.4 *Infectious and Parasitic Diseases*

Nanomaterial activity against several human pathogens, including parasites, fungi, bacteria, and viruses, has been studied. Biosynthesized NPs may have enhanced activity. Silver NPs synthesized using plant extracts exhibit a potent antibacterial effect, enhanced by using medicinal plants. They may potentially be used in replacement of antibiotics or in synergy (cf. Kotcherlakota et al. 2019). Antiviral activities of NPs maybe exerted through different mechanisms, namely from the combination of: (1) the NP surface charge, assisting cellular internalization and transport, (2) small particle size, enabling drug transfer to targeted locations, (3) large surface area, enabling large drug payloads, and (4) intrinsic antiviral assets, for example dendrimers and silver NPs (cf. Nasrollahzadeh et al. 2020). NPs may also assist in the combat against parasitic disease such as leishmaniasis and malaria that remain grave and difficult infections with major health impact. NPs may be used for more effective, less toxic, targeted drug delivery, in vaccine development and in vector control (Borgheti et al. 2020; Singh et al. 2019).

### 4.3.5 *Vaccines, Regenerative Medicine and Genetic Therapy*

Nanostructures allow the delivery of viral antigens in a controlled manner. By resembling the viruses' structure and size, without triggering a real infection, NPs are capable of activating follicular dendritic cells or B cells, promoting antigen presentation, inducing a humoral/cellular immune response (Nasrollahzadeh et al. 2020).

NP-based systems may be used in tissue engineering by integrating scaffolds, allowing the delivery of multiple growth factors, providing contrast for imaging and controlling scaffold properties (Fathi-Achachelouei et al. 2019). NPs may also be used to direct cell fate (Wu et al. 2018). Those nanomaterials that are derived from bacterial cultures or bio-molecules isolated from living organisms may be helpful in the prevention and management of osteogenic disorders, when compared to conventionally synthetized NPs, especially regarding improved cell attachment and proliferation, and their capability to avoid bacterial adhesion (Medina-Cruz et al. 2020). Non-viral NPs have been investigated for gene delivery and show promise as vectors for gene therapy in oncologic and neurodegenerative disease (Binda et al. 2020; Luly et al. 2020), although concerns on cytotoxicity, neurotoxicity, effects on normal cells function, complement activation, apoptosis induction, etc. remain.

## 4.4 Conclusions, Biomedical Aspects and Future Remarks

Nanomaterial-based therapeutic agents represent an immense clinical application potential, especially those that take a biogenic and biomimetic approach.

It is unlikely that widespread benefits will be readily apparent unless a deeper knowledge on NP interaction with media, cells and assessment methods is gained and consistently applied in research. Despite the profusion of publications in this area, most of the published studies rely on somewhat limited in vitro assessment, not always considering NP interference with standard cell culture response evaluation techniques. On the other hand, in vivo rodent models pose ethical issues and are limited in their representation of human natural disease.

Companion animals may help fill the gap between the development and application but only after a thorough characterization is done and suitable predictive models are developed and applied.

Current challenges may turn into opportunities since mechanisms such as oxidative stress and decreased pH associated with inflammation may be used to trigger NP activation and allow effective, targeted action. Interactions with biological molecules and immune system cells may be used as an advantage rather than being an obstacle.

Successful and timely development of biomedical applications will depend on a multidisciplinary and committed approach, with enhanced communication and cooperation between experts from different areas of knowledge.

**Acknowledgement** This work has been partially supported by the European Commission through Portugal 2020/Alentejo 2020, grant Mulbioimplant POCI-01-0145-FEDER-032486.

**Conflicts of Interest** The authors declare there are no conflicts of interest concerning this publication.

Ethical Issues There are no ethical issues associated with the present work.

## References

Aljabali, A. A. A., Bakshi, H. A., Hakkim, F. L., et al. (2020). Albumin nano-encapsulation of Piceatannol enhances its anticancer potential in colon cancer via downregulation of nuclear p65 and HIF-1α. *Cancers, 12*, 113. https://doi.org/10.3390/cancers12010113.

Al-Sheddi, E. S., Farshori, N. N., Al-Oqail, M. M., et al. (2018). Anticancer potential of green synthesized silver nanoparticles using extract of Nepeta deflersiana against human cervical Cancer cells (HeLA). *Bioinorganic Chemistry and Applications, 2018*, 9390784. https://doi.org/10.1155/2018/9390784.

An, Z., Yan, J., Zhang, Y., & Pei, R. (2020). Applications of nanomaterials for scavenging reactive oxygen species in the treatment of central nervous system diseases. *Journal of Materials Chemistry. B, Materials for Biology and Medicine, 8*, 8748. https://doi.org/10.1039/d0tb01380c.

Andraos, C., Yu, I. J., & Gulumian, M. (2020). Interference: A much-neglected aspect in high-throughput screening of nanoparticles. *International Journal of Toxicology, 39*, 397–421. https://doi.org/10.1177/1091581820938335.

Arora, I., Sharma, M., & Tollefsbol, T. O. (2019). Combinatorial epigenetics impact of polyphenols and phytochemicals in cancer prevention and therapy. *International Journal of Molecular Sciences, 20*. https://doi.org/10.3390/ijms20184567.
Baig, B., Halim, S. A., Farrukh, A., et al. (2019). Current status of nanomaterial-based treatment for hepatocellular carcinoma. *Biomedicine & Pharmacotherapy, 116*, 108852. https://doi.org/10.1016/j.biopha.2019.108852.
Batra, H., Pawar, S., & Bahl, D. (2019). Curcumin in combination with anti-cancer drugs: A nanomedicine review. *Pharmacological Research, 139*, 91–105. https://doi.org/10.1016/j.phrs.2018.11.005.
Binda, A., Murano, C., & Rivolta, I. (2020). Innovative therapies and nanomedicine applications for the treatment of Alzheimer's disease: A state-of-the-art (2017-2020). *International Journal of Nanomedicine, 15*, 6113–6135. https://doi.org/10.2147/IJN.S231480.
Biswas, A., Shukla, A., & Maiti, P. (2019). Biomaterials for interfacing cell imaging and drug delivery: An overview. *Langmuir, 35*, 12285–12305. https://doi.org/10.1021/acs.langmuir.9b00419.
Bobo, D., Robinson, K. J., Islam, J., et al. (2016). Nanoparticle-based medicines: A review of FDA-approved materials and clinical trials to date. *Pharmaceutical Research, 33*, 2373–2387. https://doi.org/10.1007/s11095-016-1958-5.
Borgheti, L. N., San Anselmo, M., & Lantero, E. (2020). Promising nanomaterials in the fight against malaria. *Proceedings of the Institutuion of Mechanical Engineering, Part L Journal of Materials Design and Applications*. https://doi.org/10.1039/D0TB01398F.
Breger, J. C., Oh, E., Susumu, K., et al. (2019). Nanoparticle size influences localized enzymatic enhancement—A case study with Phosphotriesterase. *Bioconjugate Chemistry, 30*, 2060–2074. https://doi.org/10.1021/acs.bioconjchem.9b00362.
Chhikara, B. S., Rathi, B., & Parang, K. (2019). Critical evaluation of pharmaceutical rational design of nano-delivery systems for doxorubicin in cancer therapy: Rationale of designing of nanoparticular delivery systems and impact of chemistry used with doxorubicin for anti-cancer therapy. *Journal of Material Nanoscience, 6*, 47–66. http://thesciencein.org/journal/index.php/jmns/article/view/83.
Chou, P.-L., Huang, Y.-P., Cheng, M.-H., et al. (2020). Improvement of paclitaxel-associated adverse reactions (ADRs) via the use of Nano-based drug delivery systems: A systematic review and network meta-analysis. *International Journal of Nanomedicine, 15*, 1731–1743. https://doi.org/10.2147/IJN.S231407.
Curcio, M., Cirillo, G., Tucci, P., et al. (2019). Dextran-curcumin nanoparticles as a methotrexate delivery vehicle: A step forward in breast cancer combination therapy. *Pharmaceuticals, 13*, 2. https://doi.org/10.3390/ph13010002.
Da Silva-Candal, A., Brown, T., Krishnan, V., et al. (2019). Shape effect in active targeting of nanoparticles to inflamed cerebral endothelium under static and flow conditions. *Journal of Controlled Release, 309*, 94–105. https://doi.org/10.1016/j.jconrel.2019.07.026.
Danaei, M., Dehghankhold, M., Ataei, S., et al. (2018). Impact of particle size and polydispersity index on the clinical applications of lipidic nanocarrier systems *Pharmaceutics, 10*. https://doi.org/10.3390/pharmaceutics10020057.
De Brito, R. C. F., Cardoso JM de, O., Reis, L. E. S., et al. (2019). Synthetic peptides elicit strong cellular immunity in visceral Leishmaniasis natural reservoir and contribute to long-lasting Polyfunctional T-cells in BALB/c mice. *Vaccine, 7*, 162. https://doi.org/10.3390/vaccines7040162.
De Matteis, V., & Rizzello, L. (2020). Noble metals and soft bio-inspired nanoparticles in retinal diseases treatment: A perspective. *Cell, 9*, 679. https://doi.org/10.3390/cells9030679.
Devineau, S., Zargarian, L., Renault, J. P., & Pin, S. (2017). Structure and function of adsorbed hemoglobin on silica nanoparticles: Relationship between the adsorption process and the oxygen binding properties. *Langmuir, 33*, 3241–3252. https://doi.org/10.1021/acs.langmuir.6b04281.
Ding, S., Cargill, A. A., Medintz, I. L., & Claussen, J. C. (2015). Increasing the activity of immobilized enzymes with nanoparticle conjugation. *Current Opinion in Biotechnology, 34*, 242–250. https://doi.org/10.1016/j.copbio.2015.04.005.

Dong, Y.-D., Tchung, E., Nowell, C., et al. (2019). Microfluidic preparation of drug-loaded PEGylated liposomes, and the impact of liposome size on tumour retention and penetration. *Journal of Liposome Research, 29*, 1–9. https://doi.org/10.1080/08982104.2017.1391285.

Durán, N., Silveira, C. P., Durán, M., & Martinez, D. S. T. (2015). Silver nanoparticle protein corona and toxicity: A mini-review. *J Nanobiotechnology, 13*, 55. https://doi.org/10.1186/s12951-015-0114-4.

El-Batal, A. I., Mosallam, F. M., Ghorab, M. M., et al. (2020). Factorial design-optimized and gamma irradiation-assisted fabrication of selenium nanoparticles by chitosan and Pleurotus ostreatus fermented fenugreek for a vigorous in vitro effect against carcinoma cells. *International Journal of Biological Macromolecules, 156*, 1584–1599. https://doi.org/10.1016/j.ijbiomac.2019.11.210.

Escamilla-Rivera, V., Solorio-Rodríguez, A., Uribe-Ramírez, M., et al. (2019). Plasma protein adsorption on Fe3O4-PEG nanoparticles activates the complement system and induces an inflammatory response. *International Journal of Nanomedicine, 14*, 2055–2067. https://doi.org/10.2147/IJN.S192214.

Fathi-Achachelouei, M., Knopf-Marques, H., Ribeiro da Silva, C. E., et al. (2019). Use of nanoparticles in tissue engineering and regenerative medicine. *Frontiers in Bioengineering and Biotechnology, 7*, 113. https://doi.org/10.3389/fbioe.2019.00113.

Fernández-Trapero, M., Espejo-Porras, F., Rodríguez-Cueto, C., et al. (2017). Upregulation of CB2 receptors in reactive astrocytes in canine degenerative myelopathy, a disease model of amyotrophic lateral sclerosis. *Disease Models & Mechanisms, 10*, 551–558. https://doi.org/10.1242/dmm.028373.

Ferreira, C. A., Ni, D., Rosenkrans, Z. T., & Cai, W. (2018). Scavenging of reactive oxygen and nitrogen species with nanomaterials. *Nano Research, 11*, 4955–4984. https://doi.org/10.1007/s12274-018-2092-y.

Flory, F., Escoubas, L., & Berginc, G. (2011). Optical properties of nanostructured materials: A review. *The Journal for Nurse Practitioners, 5*, 052502. https://doi.org/10.1117/1.3609266.

Fu, P. P., Xia, Q., Hwang, H.-M., et al. (2014). Mechanisms of nanotoxicity: Generation of reactive oxygen species. *Journal of Food and Drug Analysis, 22*, 64–75. https://doi.org/10.1016/j.jfda.2014.01.005.

Gagner, J. E., Lopez, M. D., Dordick, J. S., & Siegel, R. W. (2011). Effect of gold nanoparticle morphology on adsorbed protein structure and function. *Biomaterials, 32*, 7241–7252. https://doi.org/10.1016/j.biomaterials.2011.05.091.

Gray, J. J. (2004). The interaction of proteins with solid surfaces. *Current Opinion in Structural Biology, 14*, 110–115. https://doi.org/10.1016/j.sbi.2003.12.001.

Haddada, M. B., Ben Haddada, M., Gerometta, E., et al. (2018). Endemic plants: From design to a new way of smart hybrid nanomaterials for green nanomedicine applications. *Journal of Nanomedicine and Nanotechnology, 09*, 2–9. https://doi.org/10.4172/2157-7439.1000518.

Hakanen, E., Lehtimäki, J., Salmela, E., et al. (2018). Urban environment predisposes dogs and their owners to allergic symptoms. *Scientific Reports, 8*, 1585. https://doi.org/10.1038/s41598-018-19953-3.

Hartung, T. (2018). Perspectives on in vitro to in vivo extrapolations. *Applied In Vitro Toxicology, 4*, 305–316. https://doi.org/10.1089/aivt.2016.0026.

He, C., Liu, D., & Lin, W. (2015). Nanomedicine applications of hybrid nanomaterials built from metal-ligand coordination bonds: Nanoscale metal-organic frameworks and nanoscale coordination polymers. *Chemical Reviews, 115*, 11079–11108. https://doi.org/10.1021/acs.chemrev.5b00125.

Henry, C. J. (2014). Unleashing the power of comparative oncology models in nanomedicine research. *Eur J Nanomed, 7*, 129–133. https://doi.org/10.1515/ejnm-2014-0046.

Irvine, D. J., & Dane, E. L. (2020). Enhancing cancer immunotherapy with nanomedicine. *Nature Reviews. Immunology, 20*, 321–334. https://doi.org/10.1038/s41577-019-0269-6.

Ismail, E. H., Saqer, A. M. A., Assirey, E., et al. (2018). Successful green synthesis of gold nanoparticles using a Corchorus olitorius extract and their antiproliferative effect in cancer cells. *International Journal of Molecular Sciences, 19*, 2612. https://doi.org/10.3390/ijms19092612.

Ivask, A., Kurvet, I., Kasemets, K., et al. (2014). Size-dependent toxicity of silver nanoparticles to bacteria, yeast, algae, crustaceans and mammalian cells in vitro. *PLoS One, 9*, e102108. https://doi.org/10.1371/journal.pone.0102108.

Jain, A. K., & Thareja, S. (2019). In vitro and in vivo characterization of pharmaceutical nanocarriers used for drug delivery. *Artificial Cells in Nanomedicine and Biotechnology, 47*, 524–539. https://doi.org/10.1080/21691401.2018.1561457.

Jiang, P., Choi, A., & Swindle-Reilly, K. E. (2020). Controlled release of anti-VEGF by redox-responsive polydopamine nanoparticles. *Nanoscale, 12*, 17298–17311. https://doi.org/10.1039/d0nr03710a.

Jin, L., Balian, G., & Li, X. J. (2018). Animal models for disc degeneration-an update. *Histology and Histopathology, 33*, 543. https://doi.org/10.14670/HH-11-910.

Jindal, A. B. (2017). The effect of particle shape on cellular interaction and drug delivery applications of micro- and nanoparticles. *International Journal of Pharmaceutics, 532*, 450–465. https://doi.org/10.1016/j.ijpharm.2017.09.028.

Johnson, B. J., Russ Algar, W., Malanoski, A. P., et al. (2014). Understanding enzymatic acceleration at nanoparticle interfaces: Approaches and challenges. *Nano Today, 9*, 102–131. https://doi.org/10.1016/j.nantod.2014.02.005.

Kanwar, R., Rathee, J., Salunke, D. B., & Mehta, S. K. (2019). Green nanotechnology-driven drug delivery assemblies. *ACS Omega, 4*, 8804–8815. https://doi.org/10.1021/acsomega.9b00304.

Karthick, V., Panda, S., Kumar, V. G., et al. (2019). Quercetin loaded PLGA microspheres induce apoptosis in breast cancer cells. *Applied Surface Science, 487*, 211–217. https://doi.org/10.1016/j.apsusc.2019.05.047.

Katebi, S., Esmaeili, A., Ghaedi, K., & Zarrabi, A. (2019). Superparamagnetic iron oxide nanoparticles combined with NGF and quercetin promote neuronal branching morphogenesis of PC12 cells. *International Journal of Nanomedicine, 14*, 2157–2169. https://doi.org/10.2147/IJN.S191878.

Ke, X., Bittencourt, C., & Van Tendeloo, G. (2015). Possibilities and limitations of advanced transmission electron microscopy for carbon-based nanomaterials. *Beilstein Journal of Nanotechnology, 6*, 1541–1557. https://doi.org/10.3762/bjnano.6.158.

Khalil, I., Yehye, W. A., Etxeberria, A. E., et al. (2019). Nanoantioxidants: Recent trends in antioxidant delivery applications. *Antioxidants, 9*, 24. https://doi.org/10.3390/antiox9010024.

Khan, H., Ullah, H., Martorell, M., et al. (2019). Flavonoids nanoparticles in cancer: Treatment, prevention and clinical prospects. *Seminars in Cancer Biology*. https://doi.org/10.1016/j.semcancer.2019.07.023.

Kokkinopoulou, M., Simon, J., Landfester, K., et al. (2017). Visualization of the protein corona: Towards a biomolecular understanding of nanoparticle-cell-interactions. *Nanoscale, 9*, 8858–8870. https://doi.org/10.1039/c7nr02977b.

Kol, A., Arzi, B., Athanasiou, K. A., et al. (2015). Companion animals: Translational scientist's new best friends. *Science Translational Medicine, 7*, 308ps21. https://doi.org/10.1126/scitranslmed.aaa9116.

Kotcherlakota, R., Das, S., & Patra, C. R. (2019). Chapter 16 - Therapeutic applications of green-synthesized silver nanoparticles. In A. K. Shukla & S. Iravani (Eds.), *Green synthesis, characterization and applications of nanoparticles* (pp. 389–428). Amsterdam: Elsevier. https://doi.org/10.1016/B978-0-08-102579-6.00017-4.

Kundu, M., Sadhukhan, P., Ghosh, N., et al. (2019). pH-responsive and targeted delivery of curcumin via phenylboronic acid-functionalized ZnO nanoparticles for breast cancer therapy. *Journal of Advertising Research, 18*, 161–172. https://doi.org/10.1016/j.jare.2019.02.036.

Kuthati, Y., Navakanth Rao, V., Busa, P., et al. (2020). Scope and applications of nanomedicines for the management of neuropathic pain. *Molecular Pharmaceutics, 17*, 1015–1027. https://doi.org/10.1021/acs.molpharmaceut.9b01027.

Lam, P.-L., Wong, W.-Y., Bian, Z., et al. (2017). Recent advances in green nanoparticulate systems for drug delivery: Efficient delivery and safety concern. *Nanomedicine, 12*, 357–385. https://doi.org/10.2217/nnm-2016-0305.

Landsiedel, R., Ma-Hock, L., Wiench, K., et al. (2017). Safety assessment of nanomaterials using an advanced decision-making framework, the DF4nanoGrouping. *Journal of Nanoparticle Research, 19*, 171. https://doi.org/10.1007/s11051-017-3850-6.

Lee, H. (2020). Effects of nanoparticle electrostatics and protein–protein interactions on Corona formation: Conformation and hydrodynamics. *Small, 16*, 1906598. https://doi.org/10.1002/smll.201906598.

Li, H., Yan, L., Tang, E. K. Y., et al. (2019). Synthesis of TPGS/curcumin nanoparticles by thin-film hydration and evaluation of their anti-Colon Cancer efficacy in vitro and in vivo. *Frontiers in Pharmacology, 10*, 769. https://doi.org/10.3389/fphar.2019.00769.

Liu, Y., & Shi, J. (2019). Antioxidative nanomaterials and biomedical applications. *Nano Today, 27*, 146–177. https://doi.org/10.1016/j.nantod.2019.05.008.

Liu, J., Chen, Q., Feng, L., & Liu, Z. (2018). Nanomedicine for tumor microenvironment modulation and cancer treatment enhancement. *Nano Today, 21*, 55–73. https://doi.org/10.1016/j.nantod.2018.06.008.

Luly, K. M., Choi, J., Rui, Y., et al. (2020). Safety considerations for nanoparticle gene delivery in pediatric brain tumors. *Nanomedicine, 15*, 1805–1815. https://doi.org/10.2217/nnm-2020-0110.

Lundqvist, M., Sethson, I., & Jonsson, B.-H. (2004). Protein adsorption onto silica nanoparticles: Conformational changes depend on the particles' curvature and the protein stability. *Langmuir, 20*, 10639–10647. https://doi.org/10.1021/la0484725.

Madathiparambil Visalakshan, R., González García, L. E., Benzigar, M. R., et al. (2020). The influence of nanoparticle shape on protein Corona formation. *Small, 16*, e2000285. https://doi.org/10.1002/smll.202000285.

Magdolenova, Z., Collins, A., Kumar, A., et al. (2014). Mechanisms of genotoxicity. A review of in vitro and in vivo studies with engineered nanoparticles. *Nanotoxicology, 8*, 233–278. https://doi.org/10.3109/17435390.2013.773464.

Mahmoud, N. N., Albasha, A., Hikmat, S., et al. (2020). Nanoparticle size and chemical modification play a crucial role in the interaction of nano gold with the brain: Extent of accumulation and toxicity. *Biomaterials Science, 8*, 1669–1682. https://doi.org/10.1039/c9bm02072a.

Maiti, D., Tong, X., Mou, X., & Yang, K. (2018). Carbon-based nanomaterials for biomedical applications: A recent study. *Frontiers in Pharmacology, 9*, 1401. https://doi.org/10.3389/fphar.2018.01401.

Manke, A., Wang, L., & Rojanasakul, Y. (2013). Mechanisms of nanoparticle-induced oxidative stress and toxicity. *BioMed Research International, 2013*, 942916. https://doi.org/10.1155/2013/942916.

Mariadoss, A. V. A., Vinayagam, R., Senthilkumar, V., et al. (2019). Phloretin loaded chitosan nanoparticles augments the pH-dependent mitochondrial-mediated intrinsic apoptosis in human oral cancer cells. *International Journal of Biological Macromolecules, 130*, 997–1008. https://doi.org/10.1016/j.ijbiomac.2019.03.031.

Marichal, L., Degrouard, J., Gatin, A., et al. (2020). From protein Corona to colloidal self-assembly: The importance of protein size in protein–nanoparticle interactions. *Langmuir, 36*, 8218–8230. https://doi.org/10.1021/acs.langmuir.0c01334.

Marquis, B. J., Love, S. A., Braun, K. L., & Haynes, C. L. (2009). Analytical methods to assess nanoparticle toxicity. *Analyst, 134*, 425–439. https://doi.org/10.1039/b818082b.

Maser, E., Schulz, M., Sauer, U. G., et al. (2015). In vitro and in vivo genotoxicity investigations of differently sized amorphous SiO2 nanomaterials. *Mutation Research, Genetic Toxicology and Environmental Mutagenesis, 794*, 57–74. https://doi.org/10.1016/j.mrgentox.2015.10.005.

Matloubi, Z., & Hassan, Z. (2020). HSA-curcumin nanoparticles: A promising substitution for curcumin as a cancer chemoprevention and therapy. *Daru, 28*, 209–219. https://doi.org/10.1007/s40199-020-00331-2.

Medina-Cruz, D., Mostafavi, E., Vernet-Crua, A., et al. (2020). Green nanotechnology-based drug delivery systems for osteogenic disorders. *Expert Opinion on Drug Delivery, 17*, 341–356. https://doi.org/10.1080/17425247.2020.1727441.

Minafra, L., Porcino, N., Bravatà, V., et al. (2019). Radiosensitizing effect of curcumin-loaded lipid nanoparticles in breast cancer cells. *Scientific Reports, 9*, 11134. https://doi.org/10.1038/s41598-019-47553-2.

Mitragotri, S., & Stayton, P. (2014). Organic nanoparticles for drug delivery and imaging. *MRS Bulletin, 39*, 219–223. https://doi.org/10.1557/mrs.2014.11.

Monopoli, M. P., Walczyk, D., Campbell, A., et al. (2011). Physical-chemical aspects of protein corona: Relevance to in vitro and in vivo biological impacts of nanoparticles. *Journal of the American Chemical Society, 133*, 2525–2534. https://doi.org/10.1021/ja107583h.

Montalbán, M. G., Coburn, J. M., Lozano-Pérez, A. A., et al. (2018). Production of curcumin-loaded silk fibroin nanoparticles for cancer therapy. *Nanomaterials, 8*, 126. https://doi.org/10.3390/nano8020126.

Nagayasu, A., Uchiyama, K., & Kiwada, H. (1999). The size of liposomes: A factor which affects their targeting efficiency to tumors and therapeutic activity of liposomal antitumor drugs. *Advanced Drug Delivery Reviews, 40*, 75–87. https://doi.org/10.1016/s0169-409x(99)00041-1.

Nakanishi, K., Sakiyama, T., & Imamura, K. (2001). On the adsorption of proteins on solid surfaces, a common but very complicated phenomenon. *Journal of Bioscience and Bioengineering, 91*, 233–244. https://doi.org/10.1016/S1389-1723(01)80127-4.

Nasrollahzadeh, M., Sajjadi, M., Soufi, G. J., et al. (2020). Nanomaterials and nanotechnology-associated innovations against viral infections with a focus on coronaviruses. *Nanomaterials, 10*, 1072. https://doi.org/10.3390/nano10061072.

Nguyen, V. H., & Lee, B.-J. (2017). Protein corona: A new approach for nanomedicine design. *International Journal of Nanomedicine, 12*, 3137–3151. https://doi.org/10.2147/IJN.S129300.

Nguyen, K. T., & Zhao, Y. (2015). Engineered hybrid nanoparticles for on-demand diagnostics and therapeutics. *Accounts of Chemical Research, 48*, 3016–3025. https://doi.org/10.1021/acs.accounts.5b00316.

Ni, R., Song, G., Fu, X., et al. (2020). Reactive oxygen species-responsive dexamethasone-loaded nanoparticles for targeted treatment of rheumatoid arthritis via suppressing the iRhom2/TNF-α/BAFF signaling pathway. *Biomaterials, 232*, 119730. https://doi.org/10.1016/j.biomaterials.2019.119730.

Norris, D. J., Nirmal, M., Murray, C. B., et al. (1993). Size dependent optical spectroscopy of II--VI semiconductor nanocrystallites (quantum dots). *Zeitschrift fur Physik D: Atoms, Molecules and Clusters, 26*, 355–357.

Ong, K. J., MacCormack, T. J., Clark, R. J., et al. (2014). Widespread nanoparticle-assay interference: Implications for nanotoxicity testing. *PLoS One, 9*, e90650. https://doi.org/10.1371/journal.pone.0090650.

Orendorff, C. J., Sau, T. K., & Murphy, C. J. (2006). Shape-dependent plasmon-resonant gold nanoparticles. *Small, 2*, 636–639. https://doi.org/10.1002/smll.200500299.

Oussoren, C., & Storm, G. (2001). Liposomes to target the lymphatics by subcutaneous administration. *Advanced Drug Delivery Reviews, 50*, 143–156. https://doi.org/10.1016/s0169-409x(01)00154-5.

Pal, S., Tak, Y. K., & Song, J. M. (2007). Does the antibacterial activity of silver nanoparticles depend on the shape of the nanoparticle? A study of the gram-negative bacterium Escherichia coli. *Applied and Environmental Microbiology, 73*, 1712–1720. https://doi.org/10.1128/AEM.02218-06.

Pal, K., Roy, S., Parida, P. K., et al. (2019). Folic acid conjugated curcumin loaded biopolymeric gum acacia microsphere for triple negative breast cancer therapy in invitro and invivo model. *Materials Science & Engineering. C, Materials for Biological Applications, 95*, 204–216. https://doi.org/10.1016/j.msec.2018.10.071.

Park, K., Lee, S., Kang, E., et al. (2009). New generation of multifunctional nanoparticles for cancer imaging and therapy. *Advanced Functional Materials, 19*, 1553–1566. https://doi.org/10.1002/adfm.200801655.

Pastorelli, D., Fabricio, A. S. C., Giovanis, P., et al. (2018). Phytosome complex of curcumin as complementary therapy of advanced pancreatic cancer improves safety and efficacy of gemcitabine: Results of a prospective phase II trial. *Pharmacological Research, 132*, 72–79. https://doi.org/10.1016/j.phrs.2018.03.013.

Patel, S., Ryals, R. C., Weller, K. K., et al. (2019). Lipid nanoparticles for delivery of messenger RNA to the back of the eye. *Journal of Controlled Release, 303*, 91–100. https://doi.org/10.1016/j.jconrel.2019.04.015.

Peer, D., Karp, J. M., Hong, S., et al. (2007). Nanocarriers as an emerging platform for cancer therapy. *Nature Nanotechnology, 2*, 751–760. https://doi.org/10.1038/nnano.2007.387.

Pem, B., González-Mancebo, D., Moros, M., et al. (2018). Biocompatibility assessment of up-and down-converting nanoparticles: Implications of interferences with in vitro assays. *Methods and Applications in Fluorescence, 7*, 014001. https://doi.org/10.1088/2050-6120/aae9c8.

Prajitha, N., Athira, S. S., & Mohanan, P. V. (2019). Bio-interactions and risks of engineered nanoparticles. *Environmental Research, 172*, 98–108. https://doi.org/10.1016/j.envres.2019.02.003.

Price, E., & Gesquiere, A. J. (2020). Animal simulations facilitate smart drug design through prediction of nanomaterial transport to individual tissue cells. *Science Advances, 6*, eaax2642. https://doi.org/10.1126/sciadv.aax2642.

Pujari-Palmer, S., Chen, S., Rubino, S., et al. (2016). In vivo and in vitro evaluation of hydroxyapatite nanoparticle morphology on the acute inflammatory response. *Biomaterials, 90*, 1–11. https://doi.org/10.1016/j.biomaterials.2016.02.039.

Rai, M. F., Pan, H., Yan, H., et al. (2019). Applications of RNA interference in the treatment of arthritis. *Translational Research, 214*, 1–16. https://doi.org/10.1016/j.trsl.2019.07.002.

Reis, J., Capela-Silva, F., Potes, J., et al. (2011). Carbon nanotubes – Interactions with biological systems. In M. Naraghi (Ed.), *Carbon nanotubes - growth and applications* (pp. 473–492). London: InTech. https://doi.org/10.5772/16651.

Reis, J., Oliveira, T., Pereira, A., et al. (2020). Microtomographic characterization of calcifications in canine mammary tumours. *Veterinary and Comparative Oncology, 18*, 281–291. https://doi.org/10.1111/vco.12545.

Reithmaier, J. P., Röhner, M., Zull, H., et al. (1997). Size dependence of confined optical modes in photonic quantum dots. *Physical Review Letters, 78*, 378–381. https://doi.org/10.1103/PhysRevLett.78.378.

Sahatsapan, N., Ngawhirunpat, T., Rojanarata, T., et al. (2020). Catechol-functionalized alginate nanoparticles as Mucoadhesive carriers for Intravesical chemotherapy. *AAPS PharmSciTech, 21*, 212. https://doi.org/10.1208/s12249-020-01752-7.

Saikia, D., Deka, J. R., Wu, C.-E., et al. (2019). pH responsive selective protein adsorption by carboxylic acid functionalized large pore mesoporous silica nanoparticles SBA-1. *Materials Science & Engineering. C, Materials for Biological Applications, 94*, 344–356. https://doi.org/10.1016/j.msec.2018.09.043.

Salvati, A., Pitek, A. S., Monopoli, M. P., et al. (2013). Transferrin-functionalized nanoparticles lose their targeting capabilities when a biomolecule corona adsorbs on the surface. *Nature Nanotechnology, 8*, 137–143. https://doi.org/10.1038/nnano.2012.237.

Santos, D., & Payan-Carreira, R. (2018). The role of Superoxid dismutase in endometrial functional integrity. In S. Magliozzi (Ed.), *Superoxide dismutase: Structure, synthesis and applications* (pp. 1–44). New York: Nova Science. ISBN: 978-1-53613-081-2.

Savage, D. T., Hilt, J. Z., & Dziubla, T. D. (2019). In vitro methods for assessing nanoparticle toxicity. *Methods in Molecular Biology, 1894*, 1–29. https://doi.org/10.1007/978-1-4939-8916-4_1.

Schütz, C. A., Juillerat-Jeanneret, L., Mueller, H., et al. (2013). Therapeutic nanoparticles in clinics and under clinical evaluation. *Nanomedicine, 8*, 449–467. https://doi.org/10.2217/nnm.13.8.

Schwirn, K., Tietjen, L., & Beer, I. (2014). Why are nanomaterials different and how can they be appropriately regulated under REACH? *Environmental Sciences Europe, 26*, 4. https://doi.org/10.1186/2190-4715-26-4.

Seleci, M., Seleci, D. A., Joncyzk, R., et al. (2016). Smart multifunctional nanoparticles in nanomedicine. *BioNanoMaterials, 17*, 33–41. https://doi.org/10.1515/bnm-2015-0030.

Si, A., Pal, K., Kralj, S., et al. (2020). Sustainable preparation of gold nanoparticles via green chemistry approach for biogenic applications. *Materials Today Chemistry, 17*, 100327. https://doi.org/10.1016/j.mtchem.2020.100327.

Singh, O. P., Gedda, M. R., Mudavath, S. L., et al. (2019). Envisioning the innovations in nanomedicine to combat visceral leishmaniasis: For future theranostic application. *Nanomedicine, 14*, 1911–1927. https://doi.org/10.2217/nnm-2018-0448.

Slowing, I. I., Trewyn, B. G., Giri, S., & Lin, V. S.-Y. (2007). Mesoporous silica nanoparticles for drug delivery and biosensing applications. *Advanced Functional Materials, 17*, 1225–1236. https://doi.org/10.1002/adfm.200601191.

Sun, H., Yan, G., & Zhou, H. (2017). Nanoparticles-induced oxidative stress. In B. Yan, H. Zhou, & J. L. Gardea-Torresdey (Eds.), *Bioactivity of engineered nanoparticles* (pp. 53–79). Singapore: Springer. https://doi.org/10.1007/978-981-10-5864-6_4.

Swierczewska, M., Crist, R. M., & McNeil, S. E. (2018). Evaluating nanomedicines: Obstacles and advancements. *Methods in Molecular Biology, 1682*, 3–16. https://doi.org/10.1007/978-1-4939-7352-1_1.

Tan, H. W., Mo, H.-Y., Lau, A. T. Y., & Xu, Y.-M. (2018). Selenium species: Current status and potentials in cancer prevention and therapy. *International Journal of Molecular Sciences, 20*. https://doi.org/10.3390/ijms20010075.

Tang, S., Peng, C., Xu, J., et al. (2016). Tailoring renal clearance and tumor targeting of Ultrasmall metal nanoparticles with particle density. *Angewandte Chemie (International Ed. in English), 55*, 16039–16043. https://doi.org/10.1002/anie.201609043.

Tomankova, K., Horakova, J., Harvanova, M., et al. (2015). Reprint of: Cytotoxicity, cell uptake and microscopic analysis of titanium dioxide and silver nanoparticles in vitro. *Food and Chemical Toxicology, 85*, 20–30. https://doi.org/10.1016/j.fct.2015.10.012.

Toy, R., Peiris, P. M., Ghaghada, K. B., & Karathanasis, E. (2014). Shaping cancer nanomedicine: The effect of particle shape on the in vivo journey of nanoparticles. *Nanomedicine, 9*, 121–134. https://doi.org/10.2217/nnm.13.191.

Vega-Chacón, J., Tarhini, M., Lebaz, N., et al. (2020). Protein-silica hybrid submicron particles: Preparation and characterization. *Chemistry Africa., 3*, 793. https://doi.org/10.1007/s42250-020-00138-3.

Wang, Y., Rajala, A., & Rajala, R. V. S. (2018). Nanoparticles as delivery vehicles for the treatment of retinal degenerative diseases. *Advances in Experimental Medicine and Biology, 1074*, 117–123. https://doi.org/10.1007/978-3-319-75402-4_15.

Wang, J., Wu, X., Shen, P., et al. (2020a). Applications of inorganic nanomaterials in photothermal therapy based on combinational cancer treatment. *International Journal of Nanomedicine, 15*, 1903–1914. https://doi.org/10.2147/IJN.S239751.

Wang, Y., Zhang, H., Shi, L., et al. (2020b). A focus on the genotoxicity of gold nanoparticles. *Nanomedicine, 15*, 319–323. https://doi.org/10.2217/nnm-2019-0364.

Wilson, A. M., Mazzaferri, J., Bergeron, É., et al. (2018). In vivo laser-mediated retinal ganglion cell optoporation using KV1.1 conjugated gold nanoparticles. *Nano Letters, 18*, 6981–6988. https://doi.org/10.1021/acs.nanolett.8b02896.

Wu, C., Shen, Y., Chen, M., et al. (2018). Recent advances in magnetic-nanomaterial-based mechanotransduction for cell fate regulation. *Advanced Materials, 30*, e1705673. https://doi.org/10.1002/adma.201705673.

Wu, T., Duan, X., Hu, C., et al. (2019). Synthesis and characterization of gold nanoparticles from Abies spectabilis extract and its anticancer activity on bladder cancer T24 cells. *Artificial Cells in Nanomedicine and Biotechnology, 47*, 512–523. https://doi.org/10.1080/21691401.2018.1560305.

Xiao, L., Huang, R., Zhang, Y., et al. (2019). A new formyl peptide Receptor-1 antagonist conjugated fullerene nanoparticle for targeted treatment of degenerative disc diseases. *ACS Applied Materials & Interfaces, 11*, 38405–38416. https://doi.org/10.1021/acsami.9b11783.

Xie, J., Fan, Z., Li, Y., et al. (2018). Design of pH-sensitive methotrexate prodrug-targeted curcumin nanoparticles for efficient dual-drug delivery and combination cancer therapy. *International Journal of Nanomedicine, 13*, 1381–1398. https://doi.org/10.2147/IJN.S152312.

Yu, Z., Li, Q., Wang, J., et al. (2020). Reactive oxygen species-related nanoparticle toxicity in the biomedical field. *Nanoscale Research Letters, 15*, 115. https://doi.org/10.1186/s11671-020-03344-7.

Zang, S., Chang, S., Shahzad, M. B., et al. (2019). Ceramics-based drug delivery system: A review and outlook. *Reviews on Advanced Materials Science, 58*, 82–97. https://doi.org/10.1515/rams-2019-0010.

Zare, M., Norouzi Roshan, Z., Assadpour, E., & Jafari, S. M. (2020). Improving the cancer prevention/treatment role of carotenoids through various nano-delivery systems. *Critical Reviews in Food Science and Nutrition, 61*, 1–13. https://doi.org/10.1080/10408398.2020.1738999.

Zhang, T., Tang, M., Yao, Y., et al. (2019). MWCNT interactions with protein: Surface-induced changes in protein adsorption and the impact of protein corona on cellular uptake and cytotoxicity. *International Journal of Nanomedicine, 14*, 993–1009. https://doi.org/10.2147/IJN.S191689.

Zhang, Y., Yang, L., Yang, C., & Liu, J. (2020a). Recent advances of smart acid-responsive gold nanoparticles in tumor therapy. *Wiley Interdisciplinary Reviews Nanomedicine and Nanobiotechnology, 12*, e1619. https://doi.org/10.1002/wnan.1619.

Zhang, M., Chen, X., Li, C., & Shen, X. (2020b). Charge-reversal nanocarriers: An emerging paradigm for smart cancer nanomedicine. *Journal of Controlled Release, 319*, 46–62. https://doi.org/10.1016/j.jconrel.2019.12.024.

Zhong, G., Yang, X., Jiang, X., et al. (2019). Dopamine-melanin nanoparticles scavenge reactive oxygen and nitrogen species and activate autophagy for osteoarthritis therapy. *Nanoscale, 11*, 11605–11616. https://doi.org/10.1039/c9nr03060c.

# Chapter 5
# Anti-Biofilm Activities of Nanocomposites: Current Scopes and Limitations

**Sandhya Kalathilparambil Santhosh, Suma Sarojini, and Mridul Umesh**

## 5.1 Introduction to Biofilms

Bacterial infections are becoming a huge threat to mankind due to the emergence of multi-drug-resistant bacteria. These organisms deploy several strategies to defend the effects of drugs by possessing R plasmids, multi-drug efflux pumps, integrons, transposons or type IV secretion systems (Alavi and Karimi 2018). Another predominant reason that helps a few bacterial species to develop resistance against several drugs is the ability to form biofilms. A biofilm is a group of cells held together by a mesh-like framework given by proteins, exopolysaccharides, DNA and lipopeptides in the matrix (Nirwati et al. 2019). Some of the most common organisms that have the ability to form biofilms and are extensively responsible in causing nosocomial infections in patients are Gram-negative organisms such as *Pseudomonas aeruginosa, Klebsiella pneumoniae* and *Escherichia coli* and Gram-positive organisms such as *Staphylococcus aureus* and *Streptococcus pyogenes* (Khan et al. 2017). When bacteria are protected within a biofilm, they tolerate the antibiotics, host defence system and any form of stress quite easily (Sharma et al. 2019). Depletion or unavailability of nutrients, slow growth rate, low penetration power of antibiotics and dormancy of cells are few characteristic features seen in a biofilm that impart resistance against various drugs (Stewart 2002). The most common infections caused by biofilm formers are sinusitis, cystic fibrosis, chronic obstructive pulmonary disease, otitis media and chronic wounds (van Tilburg Bernardes et al. 2015). Cells in a biofilm have been observed to be 10–1000 times more resistant to antibiotics when compared to the planktonic cells (Sharma et al. 2019). When a biofilm is disrupted, the cells are dispersed to their original planktonic state and become susceptible to the same antibiotics that they were resistant

S. K. Santhosh (✉) · S. Sarojini · M. Umesh
Department of Life Sciences, CHRIST (Deemed to be University), Bangalore, India
e-mail: sandhya.ks@res.christuniversity.in


K. Pal (ed.), *Bio-manufactured Nanomaterials*,
https://doi.org/10.1007/978-3-030-67223-2_5

to, when they were enclosed in the biofilm (Berlanga et al. 2017). The scenario gets challenging when biofilms are established on medical devices such as catheters and implants which prove that bacteria can form biofilms on both biotic and abiotic surfaces. They are able to stay together as a community predominantly because of the presence of appendages like pili and flagella or in some other cases, electrostatic attraction or van der Waal's forces (Paluch et al. 2020). The first step in biofilm formation is attachment of cells to either the host tissue or any abiotic surface. Secondly, they are capable of producing certain signalling molecules that further alter gene expression in the favour of the bacterial species (Ivanova et al. 2020).

However, biofilms have positive aspects too. For example, they act as biological controls against pathogens infecting plants, as biofertilizers for improved production of crops, to mitigate hazardous pollutants from the environment, for treating waste water and to prevent corrosion. Although biofilms are helpful in agricultural and industrial domains, research is predominantly being done to combat the problems caused by biofilm-forming bacteria in the medical field. Biofilms are considered to be a master plan strategized by bacteria to survive in environments that give them too little to sustain (Muhammad et al. 2020). They achieve this by mere cell-to-cell communication with the neighbouring bacteria. This communication is scientifically termed as quorum sensing.

### *5.1.1 Quorum Sensing*

Bacterial cells make use of a phenomenon called quorum sensing to detect the number of organisms in the surrounding. They produce chemical molecules called auto inducers (AIs) that facilitate in measuring the cell density in a given site. In Gram-negative organisms, AIs are acyl homoserine lactones (AHLs), and Gram-positive organisms on the contrary produce short oligopeptides as their AIs. These signalling molecules diffuse from the cell to the exterior in a natural process. When many cells are present in a specific location, the concentration of AIs in the surrounding is more than the concentration of AIs within the cells. Therefore, to strike a balance, some AIs diffuse back into the cells and thereby increase the concentration of AIs within the cells to collectively affect the transcription of few genes that are responsible for virulence, sporulation, biofilm formation, bioluminescence, antibiotic production or competence in bacteria (Rutherford and Bassler 2012).

Biofilms are indirectly responsible for promoting multiple drug resistance in several bacterial species as they bring cells close to each other, making it extremely easy for conjugation to occur. R plasmids present in a specific strain/species can be passed on to the surrounding cells via horizontal gene transfer (HGT), thereby making them resistant to various drugs even in the planktonic stage (Madsen et al. 2012). These are the main reasons which makes it imperative to find a novel solution to tackle the menace created by biofilm formers.

Instead of disrupting the bacterial biofilm, scientists have been working towards preventing the communication between bacteria. This has been a practical move and has been found to be efficient due to molecules known as quorum quenchers. The target molecules here are the AIs without a doubt. They work in one or more ways against AIs. Enzymes like AHL lactonase and AHL acylases specifically hydrolyse homoserine lactone rings and amide bonds, respectively. Enzymes like oxidoreductases work a little differently by modifying the AHL molecules, so that they lose their function. The third strategy is to use inductor antagonists that can compete with AHLs for the binding site on the receptor or can non-competitively bind to the receptor to block the signal cascade reactions lined up (Paluch et al. 2020). For example, halogenated furanones isolated from marine red algae have structural similarity to the signalling molecules and have been classified under quorum sensing inhibitors (QSIs) (Hayat et al. 2019). In the last two cases, the auto inducers are not destroyed, but the ultimate goal of preventing biofilm formation is achieved.

## 5.2 Nanoparticles

The field of nanotechnology started establishing itself when inventions such as the scanning tunnel microscope and atomic force microscopes were made, as they play a major role in imaging surfaces at the atomic level (Ferdous and Nemmar 2020). Gold, platinum and silver nanoparticles (AgNPs) have gained utmost importance in the field of medical research. Specifically pertaining to silver, a lot of experiments are being done as it has been effective against a broad spectrum of pathogens since ancient times. Silver has been a part of several ointments and other topical agents to prevent any forms of bacterial infections on open wounds and burns (Mohanta et al. 2020). Considering their efficacy, nanotechnology is undoubtedly one of the best ways to target biofilms with the help of AgNPs or nanocomposites. The advantage nanoparticles have when compared to any other method used against biofilms is their tremendously small size that provides a large surface area, high reactivity and the ability to easily penetrate through the biofilm matrix. Their large surface area makes them a very good drug carrier (Qayyum and Khan 2016). Numerous physical and chemical methods to synthesize these particles had gained popularity until scientists analysed the harmful impact on the environment. In the recent past, eco-friendly methods have been identified to produce AgNPs. Once synthesized, they are also combined with polymers to form nanocomposites (Awad et al. 2015).

### *5.2.1 Synthesis of Nanoparticles*

Mostly, silver nitrate ($AgNO_3$) solution is used as the precursor molecule of silver. Rarely, silver wires and silver sulphate are in use as well. A reducing agent and a stabilizing agent are mandatory to synthesize AgNPs. The reducing agent is added to $AgNO_3$ solution to convert silver ions into elemental silver. Once, elemental silver starts forming, there is a visible colour change in the solution as well as agglomeration of these particles. Agglomeration reduces the surface exposure of nanoparticles drastically, and therefore, to prevent it from happening, it is a must to make use of a stabilizing agent. But the biggest disadvantage of using these reducing agents and stabilizers is their deleterious effects on the environment. Hence, scientists have recently started using greener protocols to synthesize nanoparticles (Iravani et al. 2014).

#### 5.2.1.1 Green Synthesis of Nanoparticles

Greener alternatives for the synthesis of AgNPs focus on the use of actinomycete (Hamed et al. 2020), bacteria, fungi or plants as the source of enzymes required for the reduction of silver. Fungi are more beneficial when compared to bacteria as they produce extracellular enzymes in bulk amounts which helps in reducing the steps involved in downstream processing. For example, the white rot fungi *Pycnoporus* sp. can be grown in malt extract for about 5 days at 32 ° C in an orbital shaker incubator. After 5 days of incubation, the mycelium should be filtered out using a Whatman filter paper, and equal amounts of $AgNO_3$ solution should be added at 1 mM concentration to the filtrate. The solution must be further kept in dark conditions inside a shaker incubator and later characterized using UV-Vis spectroscopy. The exact mechanism behind this method is not very clear. However, it is assumed that the fungal cell produces an extracellular enzyme called NADH dependent nitrate reductase which reduces the silver ions when they come in contact with the fungal cell walls. To be sure of the synthesis of AgNPs, a positive control ($AgNO_3$ solution and deionized water) and a negative control ($AgNO_3$ solution) can be kept in the same conditions (Gudikandula and Charya Maringanti 2016).

Plants are another source of multiple secondary metabolites that can be used for the production of AgNPs. Different parts of the plants can be procured, dried and powdered using a blender. The coarse powder must then be added into deionized water and sonicated. The solution should be filtered and about 10 ml of the filtrate should be added to 90 ml of $AgNO_3$ solution. This solution is incubated overnight at 60 °C. Gradually a visible colour change is observed which indicates the formation of AgNPs (Mohanta et al. 2020).

### *5.2.2 Characterization of Nanoparticles*

There are umpteen number of methods used to characterize silver nanoparticles. The parameters taken into consideration for their characterization are size, surface charge, shape and distribution (Carvalho et al. 2018). A surface plasmon band is a characteristic feature of silver nanoparticles that can be observed using UV-Vis spectrophotometer. In an experiment conducted by Krishna Gudikandula et al., a comparison was drawn between the surface plasmon band obtained with chemically synthesized silver nanoparticles and biologically synthesized silver nanoparticles to give an absorption peak at 430 and 420 nm, respectively. These readings were typical of silver nanoparticles (Gudikandula and Charya Maringanti 2016). Dynamic light scattering (DLS) spectroscopy is another method used to confirm the synthesis of nanoparticles. When light is incident on a particle, the light reflects at a specific angle and the angle between the incident light and reflected light can be calculated over time. Bigger particles reflect the incident light slowly, whereas smaller particles such as the nanoparticles reflect the incident light so quickly that it gets difficult to calculate the photon correlation with respect to time. This is repeated several times to get a precise value (Carvalho et al. 2018).

## 5.3 Nanocomposites

The solutions to various problems existing in the medical field today are nanocomposites. Nanocomposites are substances made of materials that possess multiple phases, and each of those phases has up to three dimensions in nanometre size. Nanocomposites can be broadly classified into three types: metal matrix nanocomposite (MMNC), ceramic matrix nanocomposite (CMNC) and polymer matrix nanocomposite (PMNC) (Omanović-Mikličanin et al. 2020).

Metal matrix nanocomposites (MMNCs): Metal matrix nanocomposites comprise ductile metal in which nanosized reinforcement is fixed. Due to its high ductility, strength, and toughness, they are extensively used in the aerospace industries.

Ceramic matrix nanocomposites (CMNCs): In ceramic matrix composites, one or more ceramic phases are added to augment the chemical stability and resist wear and tear. However, ceramic matrix composites have a disadvantage. They are extremely brittle and therefore not in much demand in the industrial domain. To overcome this, CMNCs were developed which were tougher than ceramic matrix composites. Polymer matrix nanocomposites (PMNCs): PMNCs are constructed with the help of fillers which are also called nanofillers. These nanofillers are broadly divided into 1D (linear), 2D (layered) and 3D (powdered) forms. The best example of 1D, 2D and 3D forms are carbon nanotubes, montmorillonite and silver nanoparticles (Omanović-Mikličanin et al. 2020).

### *5.3.1 Preparation of Nanocomposites*

During the synthesis of nanocomposites, at least one dimension of the various layers of inorganic or organic material are fillers. These fillers must be less than 100 nm in size (Fawaz and Mittal 2014). There are several methods by which nanocomposites can be synthesized. In situ polymerization method is mainly used to synthesize nanocomposites made of graphite. Graphite does not have charged groups naturally present on its surface, and therefore ionic interaction between graphite and the polymer would be difficult. However, the expanded graphite has pores big enough (2–10μm) such that the polymer material remains embedded within the pores even after the solvent has been extracted (Fu et al. 2019).

In the solution mixing method, the nanofillers are allowed to swell up in a solvent in which the polymer is soluble. The solvents used can be water, chloroform, toluene, etc. The nanofillers in use are most of the time layered silicates. When these silicates are mixed with polymers vigorously, the polymers intercalate between the layered silicates by displacing the solvent. Perfect mixing is obtained by magnetic stirring or ultrasonication followed by slow evaporation of the solvent and casting of the nanocomposite (Rane et al. 2018).

Sol-gel technique is used extensively in ceramic engineering (Shahjahan 2017). A widely used substance in this technique is tetra ethyl ortho silicate (TEOS) as it is highly efficient in forming networks (Owens et al. 2016). Sol is a substance that has solid materials distributed in a solution. They are in a colloidal state. They undergo slow hydrolysis reactions and thereafter polymerize to form a gel. The crystals grow which allows the polymers to seep between the various layers, thus forming nanocomposites (Khan et al. 2016).

### 5.3.2 Characterization of Nanocomposites

Various methods like FT-IR, TEM, TGA and XRD are used to characterize nanocomposites. FT-IR spectra reveal a lot about the functional groups present in a given sample based on the bond stretching observed as peaks. The peaks of AgNPs should not be showing any peaks at the carbonyl frequency region assuring the absence of the stabilizing agents like acetate added during the synthesis of nanocomposites. TEM has the ability to image the synthesized materials on a nanometre scale. The dispersion quality of nanoparticles can also be observed. It is important to confirm that a homogenous mixture is obtained such that there are no agglomerates of nanoparticles. These factors can be clearly visualized using a TEM image (Puggal et al. 2016). XRD makes use of wide-angle X-ray diffraction to check the crystalline nature, exfoliation and intercalation of polymers between nanoclay layers and dispersion of the nanoparticles within the polymer matrix. These parameters can be calculated using Bragg's law and Scherrer's law (El-Sheikhy and Al-Shamrani 2015). TGA can also be used to characterize NC. When polymers contain substances such as nanotubes and montmorillonite, the temperature at which thermal degradation happens is increased, i.e. thermal stability is enhanced. This is widely seen in polymethyl methacrylate (PMMA), polydimethylsiloxane (PDMS), polyamide and polypropylene. The reason for greater thermal stability is because of char formation as suggested by many researchers. The permeability is reduced, and the char thus formed, blocks the outward movement of the decomposed products on degradation. This quality is a characteristic feature of nanocomposites when compared to polymers without nanoparticles (Corcione and Frigione 2012).

### 5.3.3 Antibiofilm Activity of Nanocomposites

Two materials that are gaining a lot of importance in the field of nanotechnology are polyvinyl alcohol (PVA) and chitosan (CS) for their very good level of biodegradability and biocompatibility. In a study done by Omnia M Abdallah et al., PVA and

CS were mixed with biologically synthesized AgNPs to test their antimicrobial and antibiofilm activities. Both the solutions had 0.1% of AgNPs. Once the solutions were thoroughly mixed, they were casted into a petri dish each and further placed in a desiccator. To remove any further residual water or solvent molecules, the plates were kept at 60 °C. These nanofilms thus formed were used to check their antimicrobial, antibiofilm and cytotoxic effects. The films without the addition of AgNPs were set as controls (Abdallah et al. 2020).

#### 5.3.3.1 Application in Biomedical Devices

Titanium is a material that has been extensively used in implants. Likewise, infections have been unavoidable even after sterilizing and disinfecting the implant before surgeries (Corrêa et al. 2015). In a study conducted by Secenti et al., 20 New Zealand rabbits were deliberately injected with bacteria at the surgical site on iliac crests. One group had screws coated with AgNPs using the sol gel technique, and the other group had titanium screws without AgNPs. After a duration of 28 days, the rabbits were sacrificed, following which the screws and adjacent bones were tested for biofilm formation with the help of TEM and SEM. Observations concluded that AgNP-coated screws did not entertain biofilm formation, whereas the uncoated screws favoured biofilm formation drastically (Sivolella et al. 2012).

Another important medical scenario observed in the field of dentistry is stomatitis. It is a condition majorly caused by the organism *Candida albicans* that colonizes the rough edges of the inner surfaces of complete dentures. Predominantly, geriatric prosthetic wearers succumb to stomatitis because of reduced motor dexterity, memory loss, and cognitive impairment. However, to overcome this problem various experiments were carried out to modify the material used for dentures to enhance antimicrobial activity as most antifungal agents are not very effective against cells in a biofilm. Poly methyl methacrylate acrylic (PMMA) resin that comprises 1 μg/ml AgNP has been found to reduce the adherence of *Candida* spp. on the denture and thereby inhibit biofilm formation. In addition, the modified denture did not show any forms of cytotoxic or genotoxic effects (Corrêa et al. 2015).

There have been various such materials used as nanocomposites that have been successful against biofilm-forming pathogens. A few of them have been summarized in Fig. 5.1.

### 5.3.4 *Scope of Nanocomposites as Biofilm Disrupting Agents*

The multifunctional properties of nanocomposites make them ideal candidates for sustainable therapeutic agents against bacterial biofilms either directly or by conjugating with antimicrobial agents. Conjugation of polymeric nanomaterials in drug delivery has been prevalent in the medical field over the past few decades (Kumar et al. 2018; Umesh et al. 2018). Nanocomposites can be successfully employed for the delivery of phytochemical compounds specifically to biofilms, thereby solving the issue associated with hydrophobicity that limits their accessibility to the

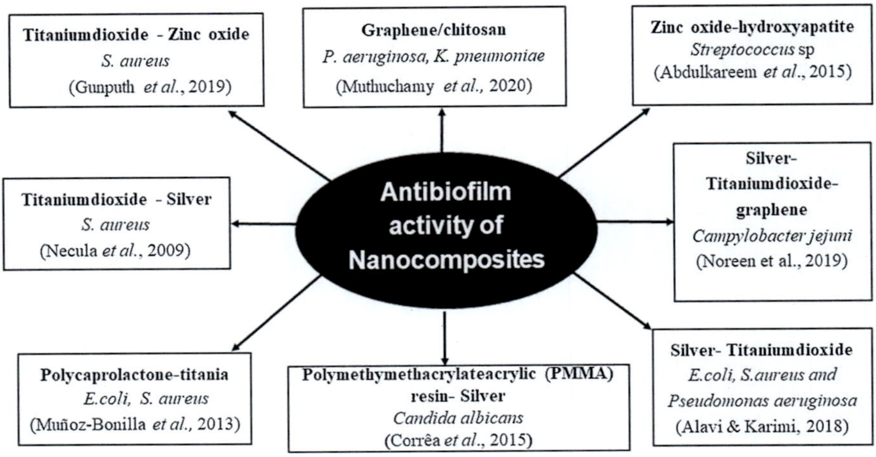


**Fig. 5.1** Biofilm disrupting activities of various nanocomposites

biofilms (Barros and Casey 2020). Enhancement of antibiofilm activity of nanocomposites can be carried out through surface functionalization with active metabolites through either covalent or noncovalent conjugation (Pathak 2019). Nanocomposites represent multiphase systems and are visualized as alternatives to overcome the limitations of microcomposites and monolithics and have the potential to become the materials of the future (Omanović-Mikličanin et al. 2020). Although the in-depth understanding regarding the effect of nanocomposites on human health and environment is not fully explored, recent research works suggest that functionalized nanocomposites can be less toxic to immune cells and thus offer a hope to be used as antibiofilm agents in in vivo conditions. In spite of having potential applications as antibiofilm agents in laboratory studies, the application of nanocomposites as therapeutic agents against biofilm-forming bacteria still faces a lot of hurdles. Translation potential of these nanocomposites from a lab scale to real-life therapeutic application is extremely challenging. Most of the reported applications of antibiofilm properties of nanocomposites were related to oral biofilm or periodontitis, the possibility of extending nanocomposites to hinder the biofilm-forming bacteria in respiratory and urogenital infection needs more focus. Another major constraint to solve is the production cost associated with nanocomposite synthesis (Ramasamy and Lee 2016). Green synthesis of nanocomposites with plant extracts was reported to have high antimicrobial activity along with the reduction in synthesis cost. This method also is eco-friendly as it reduces the usage of solvents, and the synthesized composites are likely to be biocompatible (Mondal et al. 2020). As the field of nanocomposite is relatively new and a multidisciplinary field encompassing science, technology and engineering, significant research and development in these sectors can truly revolutionize the application of nanocomposites against biofilm-forming pathogens.

### 5.3.5 *Limitations of Nanocomposites as Biofilm Disrupting Agents*

Most of the NC works by generating reactive oxygen species (ROS) which can hamper bacterial growth and multiplication. But it has been observed that ROS generation does not always directly cause cell death. Gene expression analysis has shown that ZnO NP can even inhibit the expression of oxidative stress genes (Kadiyala et al. 2018). The advantages of NC include their large surface area and high reactivity. These positive aspects can also lead to some side effects. Also chemically synthesized NPs can have toxicity issues due to the use of dangerous compounds during their synthesis some of which may remain in trace amounts in the NP and cause undesirable effects (Römling et al. 2005). In order to circumvent this problem, an environmentally friendly and less toxic approach called the "green synthesis" can be resorted to (Salem and Fouda 2020). Another aspect to be looked at is the mutagenicity of these nanostructures. Inadequate levels of NPs may not destroy the biofilms, but may induce mutations, which can lead to the emergence of "super mutants". Tungsten oxide NP has been shown to directly interact with DNA and cause single-strand breaks. So even though the majority of the bacterial cells were killed, few remaining were found to be mutants (Thongkumkoon et al. 2014). The degree of horizontal gene transfer was also more pronounced with the use of aluminium oxide NPs with the evidence of a bacterium being conjugated to many other bacteria (Qiu et al. 2012). Despite the several promises they offer; nanocomposites have some limitations as well. Our primary intention in using nanocomposites is to eradicate the pathogenic microbes. But the non-specificity of various NC may lead to elimination of symbiotic organisms as well (Qayyum and Khan 2016). This can lead to disruptions in the normal microflora composition.

## 5.4 Conclusions and Outlook

Biofilms have been a serious problem in the health sector for a while. Bacteria evolving themselves into superbugs that are resistant to over 20 different drugs have come into existence, and it is definitely the need of the hour to find novel solutions. The use of nanoparticles and nanocomposites in the field of medicine is a promising tool. However, the application of these methods is currently restricted to only implants or other medical devices and topical agents. A solution must be found to deal with biofilm-forming pathogens that cause severe lung infections or urinary tract infections. A lot is yet to be unravelled in this emerging field of nanotechnology to overcome this challenge of multiple drug resistance completely.

**Conflicts of Interest** All the authors have declared that there is no conflict of interest for publishing this work.

## References


Abdallah, O. M., EL-Baghdady, K. Z., Khalil, M. M. H., El Borhamy, M. I., & Meligi, G. A. (2020). Antibacterial, antibiofilm and cytotoxic activities of biogenic polyvinyl alcohol-silver and chitosan-silver nanocomposites. *Journal of Polymer Research, 27*(3), 74. https://doi.org/10.1007/s10965-020-02050-3.

Abdulkareem, E. H., Memarzadeh, K., Allaker, R. P., Huang, J., Pratten, J., & Spratt, D. (2015). Anti-biofilm activity of zinc oxide and hydroxyapatite nanoparticles as dental implant coating materials. *Journal of Dentistry, 43*(12), 1462–1469. https://doi.org/10.1016/j.jdent.2015.10.010.

Alavi, M., & Karimi, N. (2018). Antiplanktonic, antibiofilm, antiswarming motility and antiquorum sensing activities of green synthesized Ag–$TiO_2$, $TiO_2$–Ag, Ag–Cu and Cu–Ag nanocomposites against multi-drug-resistant bacteria. *Artificial Cells, Nanomedicine, and Biotechnology, 46*(Suppl 3), S399–S413. https://doi.org/10.1080/21691401.2018.1496923.

Awad, M. A., Mekhamer, W. K., Merghani, N. M., Hendi, A. A., Ortashi, K. M. O., Al-Abbas, F., & Eisa, N. E. (2015). Green synthesis, characterization, and antibacterial activity of silver/polystyrene nanocomposite. *Journal of Nanomaterials, 2015*, 1–6. https://doi.org/10.1155/2015/943821.

Barros, C. H. N., & Casey, E. (2020). A review of nanomaterials and technologies for enhancing the antibiofilm activity of natural products and phytochemicals. *ACS Applied Nano Materials, 3*(9), 8537–8556. https://doi.org/10.1021/acsanm.0c01586.

Berlanga, M., Gomez-Perez, L., & Guerrero, R. (2017). Biofilm formation and antibiotic susceptibility in dispersed cells versus planktonic cells from clinical, industry and environmental origins. *Antonie Van Leeuwenhoek, 110*(12), 1691–1704. https://doi.org/10.1007/s10482-017-0919-2.

Carvalho, P. M., Felício, M. R., Santos, N. C., Gonçalves, S., & Domingues, M. M. (2018). Application of light scattering techniques to nanoparticle characterization and development. *Frontiers in Chemistry, 6*, 237. https://doi.org/10.3389/fchem.2018.00237.

Corcione, C., & Frigione, M. (2012). Characterization of nanocomposites by thermal analysis. *Materials, 5*(12), 2960–2980. https://doi.org/10.3390/ma5122960.

Corrêa, J. M., Mori, M., Sanches, H. L., da Cruz, A. D., Poiate, E., & Poiate, I. A. V. P. (2015). Silver nanoparticles in dental biomaterials. *International Journal of Biomaterials, 2015*, 1–9. https://doi.org/10.1155/2015/485275.

El-Sheikhy, R., & Al-Shamrani, M. (2015). On the processing and properties of clay/polymer nanocomposites CPNC. *Latin American Journal of Solids and Structures, 12*(2), 385–419. https://doi.org/10.1590/1679-78251399.

Fawaz, J., & Mittal, V. (2014). Synthesis of polymer nanocomposites: Review of various techniques. In V. Mittal (Ed.), *Synthesis techniques for polymer nanocomposites* (pp. 1–30). New York: Wiley-VCH Verlag. https://doi.org/10.1002/9783527670307.ch1.

Ferdous, Z., & Nemmar, A. (2020). Health impact of silver nanoparticles: A review of the biodistribution and toxicity following various routes of exposure. *International Journal of Molecular Sciences, 21*(7), 2375. https://doi.org/10.3390/ijms21072375.

Fu, S., Sun, Z., Huang, P., Li, Y., & Hu, N. (2019). Some basic aspects of polymer nanocomposites: A critical review. *Nano Materials Science, 1*(1), 2–30. https://doi.org/10.1016/j.nanoms.2019.02.006.

Gudikandula, K., & Charya Maringanti, S. (2016). Synthesis of silver nanoparticles by chemical and biological methods and their antimicrobial properties. *Journal of Experimental Nanoscience, 11*(9), 714–721. https://doi.org/10.1080/17458080.2016.1139196.

Gunputh, U., Le, H., Besinis, A., Tredwin, C., & Handy, R. (2019). Multilayered composite coatings of titanium dioxide nanotubes decorated with zinc oxide and hydroxyapatite nanoparticles: Controlled release of Zn and antimicrobial properties against Staphylococcus aureus. *International Journal of Nanomedicine, 14*, 3583–3600. https://doi.org/10.2147/IJN.S199219.

Hamed, A. A., Kabary, H., Khedr, M., & Emam, A. N. (2020). Antibiofilm, antimicrobial and cytotoxic activity of extracellular green-synthesized silver nanoparticles by two marine-derived actinomycete. *RSC Advances, 10*(17), 10361–10367. https://doi.org/10.1039/C9RA11021F.

Hayat, S., Muzammil, S., Shabana, A. B., Siddique, M. H., Saqalein, M., & Nisar, M. A. (2019). Quorum quenching: Role of nanoparticles as signal jammers in Gram-negative bacteria. *Future Microbiology, 14*(1), 61–72. https://doi.org/10.2217/fmb-2018-0257.

Iravani, S., Korbekandi, H., Mirmohammadi, S. V., & Zolfaghari, B. (2014). Synthesis of silver nanoparticles: Chemical, physical and biological methods. *Research in Pharmaceutical Sciences, 9*(6), 385–406.

Ivanova, A., Ivanova, K., Tied, A., Heinze, T., & Tzanov, T. (2020). Layer-by-layer coating of aminocellulose and quorum quenching acylase on silver nanoparticles synergistically eradicate bacteria and their biofilms. *Advanced Functional Materials, 30*(24), 2001284. https://doi.org/10.1002/adfm.202001284.

Kadiyala, U., Turali-Emre, E. S., Bahng, J. H., Kotov, N. A., & VanEpps, J. S. (2018). Unexpected insights into antibacterial activity of zinc oxide nanoparticles against methicillin resistant *Staphylococcus aureus* (MRSA). *Nanoscale, 10*(10), 4927–4939. https://doi.org/10.1039/C7NR08499D.

Khan, W. S., Hamadneh, N. N., & Khan, W. A. (2016). Polymer nanocomposites – Synthesis techniques, classification and properties. In *Science and applications of tailored nanostructures*. Portland: One Central Press (OCP).

Khan, H. A., Baig, F. K., & Mehboob, R. (2017). Nosocomial infections: Epidemiology, prevention, control and surveillance. *Asian Pacific Journal of Tropical Biomedicine, 7*(5), 478–482. https://doi.org/10.1016/j.apjtb.2017.01.019.

Kumar, S., Sarita Nehra, M., Dilbaghi, N., Tankeshwar, K., & Kim, K.-H. (2018). Recent advances and remaining challenges for polymeric nanocomposites in healthcare applications. Progress in Polymer Science, 80, 1–38.

Madsen, J. S., Burmølle, M., Hansen, L. H., & Sørensen, S. J. (2012). The interconnection between biofilm formation and horizontal gene transfer. *FEMS Immunology & Medical Microbiology, 65*(2), 183–195. https://doi.org/10.1111/j.1574-695X.2012.00960.x.

Mohanta, Y. K., Biswas, K., Jena, S. K., Hashem, A., Abd-Allah, E. F., & Mohanta, T. K. (2020). Anti-biofilm and antibacterial activities of silver nanoparticles synthesized by the reducing activity of phytoconstituents present in the Indian medicinal plants. *Frontiers in Microbiology, 11*, 1143. https://doi.org/10.3389/fmicb.2020.01143.

Mondal, P., Anweshan, A., & Purkait, M. K. (2020). Green synthesis and environmental application of iron-based nanomaterials and nanocomposite: A review. *Chemosphere, 259*, 127509. https://doi.org/10.1016/j.chemosphere.2020.127509.

Muhammad, M. H., Idris, A. L., Fan, X., Guo, Y., Yu, Y., Jin, X., Qiu, J., Guan, X., & Huang, T. (2020). Beyond risk: Bacterial biofilms and their regulating approaches. *Frontiers in Microbiology, 11*, 928. https://doi.org/10.3389/fmicb.2020.00928.

Muñoz-Bonilla, A., Cerrada, M., Fernández-García, M., Kubacka, A., Ferrer, M., & Fernández-García, M. (2013). Biodegradable polycaprolactone-titania nanocomposites: Preparation, characterization and antimicrobial properties. *International Journal of Molecular Sciences, 14*(5), 9249–9266. https://doi.org/10.3390/ijms14059249.

Muthuchamy, M., Govindan, R., Shine, K., Thangasamy, V., Alharbi, N. S., Thillaichidambaram, M., Khaled, J. M., Wen, J.-L., & Alanzi, K. F. (2020). Anti-biofilm investigation of graphene/chitosan nanocomposites against biofilm producing P. aeruginosa and K. pneumoniae. *Carbohydrate Polymers, 230*, 115646. https://doi.org/10.1016/j.carbpol.2019.115646.

Necula, B. S., Fratila-Apachitei, L. E., Zaat, S. A. J., Apachitei, I., & Duszczyk, J. (2009). In vitro antibacterial activity of porous TiO2–Ag composite layers against methicillin-resistant Staphylococcus aureus. *Acta Biomaterialia, 5*(9), 3573–3580. https://doi.org/10.1016/j.actbio.2009.05.010.

Nirwati, H., Sinanjung, K., Fahrunissa, F., Wijaya, F., Napitupulu, S., Hati, V. P., Hakim, M. S., Meliala, A., Aman, A. T., & Nuryastuti, T. (2019). Biofilm formation and antibiotic resistance of Klebsiella pneumoniae isolated from clinical samples in a tertiary care hospital, Klaten, Indonesia. *BMC Proceedings, 13*(S11), 20. https://doi.org/10.1186/s12919-019-0176-7.

Noreen, Z., Khalid, N. R., Abbasi, R., Javed, S., Ahmad, I., & Bokhari, H. (2019). Visible light sensitive Ag/TiO2/graphene composite as a potential coating material for control of campylobacter jejuni. *Materials Science and Engineering: C, 98*, 125–133. https://doi.org/10.1016/j.msec.2018.12.087.

Omanović-Mikličanin, E., Badnjević, A., Kazlagić, A., & Hajlovac, M. (2020). Nanocomposites: A brief review. *Health and Technology, 10*(1), 51–59. https://doi.org/10.1007/s12553-019-00380-x.

Owens, G. J., Singh, R. K., Foroutan, F., Alqaysi, M., Han, C.-M., Mahapatra, C., Kim, H.-W., & Knowles, J. C. (2016). Sol–gel based materials for biomedical applications. *Progress in Materials Science, 77*, 1–79. https://doi.org/10.1016/j.pmatsci.2015.12.001.

Paluch, E., Rewak-Soroczyńska, J., Jędrusik, I., Mazurkiewicz, E., & Jermakow, K. (2020). Prevention of biofilm formation by quorum quenching. *Applied Microbiology and Biotechnology, 104*(5), 1871–1881. https://doi.org/10.1007/s00253-020-10349-w.

Pathak, Y. V. (Ed.). (2019). *Surface modification of nanoparticles for targeted drug delivery* (1st ed.). New York: Springer. https://doi.org/10.1007/978-3-030-06115-9.

Puggal, S., Dhall, N., Singh, N., & Litt, M. S. (2016). A review on polymer nanocomposites: Synthesis, characterization and mechanical prop. *Indian Journal of Science and Technology, 9*(4). https://doi.org/10.17485/ijst/2016/v9i4/81100.

Qayyum, S., & Khan, A. U. (2016). Nanoparticles *vs.* biofilms: A battle against another paradigm of antibiotic resistance. *MedChemComm, 7*(8), 1479–1498. https://doi.org/10.1039/C6MD00124F.

Qiu, Z., Yu, Y., Chen, Z., Jin, M., Yang, D., Zhao, Z., Wang, J., Shen, Z., Wang, X., Qian, D., Huang, A., Zhang, B., & Li, J.-W. (2012). Nanoalumina promotes the horizontal transfer of multiresistance genes mediated by plasmids across genera. *Proceedings of the National Academy of Sciences of the United States of America, 109*(13), 4944–4949. https://doi.org/10.1073/pnas.1107254109.

Ramasamy, M., & Lee, J. (2016). Recent nanotechnology approaches for prevention and treatment of biofilm-associated infections on medical devices. *BioMed Research International, 2016*, 1–17. https://doi.org/10.1155/2016/1851242.

Rane, A. V., Kanny, K., Abitha, V. K., & Thomas, S. (2018). Methods for synthesis of nanoparticles and fabrication of nanocomposites. In *Synthesis of inorganic nanomaterials* (pp. 121–139). Amsterdam: Elsevier. https://doi.org/10.1016/B978-0-08-101975-7.00005-1.

Römling, U., Gomelsky, M., & Galperin, M. Y. (2005). C-di-GMP: The dawning of a novel bacterial signalling system: C-di-GMP signalling in bacteria. *Molecular Microbiology, 57*(3), 629–639. https://doi.org/10.1111/j.1365-2958.2005.04697.x.

Rutherford, S. T., & Bassler, B. L. (2012). Bacterial quorum sensing: Its role in virulence and possibilities for its control. *Cold Spring Harbor Perspectives in Medicine, 2*(11), a012427–a012427. https://doi.org/10.1101/cshperspect.a012427.

Salem, S. S., & Fouda, A. (2020). Green synthesis of metallic nanoparticles and their prospective biotechnological applications: An overview. *Biological Trace Element Research, 199*, 344. https://doi.org/10.1007/s12011-020-02138-3.

Shahjahan, M. (2017). Synthesis and characterization of silver nanoparticles by sol-gel technique. *Nanoscience and Nanometrology, 3*(1), 34. https://doi.org/10.11648/j.nsnm.20170301.16.

Sharma, D., Misba, L., & Khan, A. U. (2019). Antibiotics versus biofilm: An emerging battleground in microbial communities. *Antimicrobial Resistance & Infection Control, 8*(1), 76. https://doi.org/10.1186/s13756-019-0533-3.

Sivolella, S., Stellini, E., Brunello, G., Gardin, C., Ferroni, L., Bressan, E., & Zavan, B. (2012). Silver nanoparticles in alveolar bone surgery devices. *Journal of Nanomaterials, 2012*, 1–12. https://doi.org/10.1155/2012/975842.

Stewart, P. S. (2002). Mechanisms of antibiotic resistance in bacterial biofilms. *International Journal of Medical Microbiology, 292*(2), 107–113. https://doi.org/10.1078/1438-4221-00196.

Thongkumkoon, P., Sangwijit, K., Chaiwong, C., Thongtem, S., Singjai, P., & Yu, L. D. (2014). Direct nanomaterial-DNA contact effects on DNA and mutation induction. *Toxicology Letters, 226*(1), 90–97. https://doi.org/10.1016/j.toxlet.2014.01.036.

van Tilburg Bernardes, E., Lewenza, S., & Reckseidler-Zenteno, S. (2015). Current research approaches to target biofilm infections. *Postdoc Journal: A Journal of Postdoctoral Research and Postdoctoral Affairs, 3*(6), 36–49. https://doi.org/10.14304/surya.jpr.v3n6.5.

Umesh, M., Priyanka, K., Thazeem, B., & Preethi, K. (2018). Biogenic PHA nanoparticle synthesis and characterization from *Bacillus subtilis* NCDC0671 using orange peel medium. *International Journal of Polymeric Materials and Polymeric Biomaterials, 67*(17), 996–1004. https://doi.org/10.1080/00914037.2017.1417284.

# Chapter 6
# Living Nano-factories: An Eco-friendly Approach Towards Medicine and Environment

**Meenu Gupta and Kumari Seema**

## 6.1 Overview

Nanotechnology is a nano-scale-based science which brings about synthesis of functional materials of size ranges from 0.1 to 100 nm scales. It is commercially most successful technology in the twenty-first century. Nanoparticles (1–100 nm dimensions) have unique optical, physiochemical, and biological properties. These highly reactive NPs can be suitably manipulated for desired applications.

Initial idea of nanotechnology was given by Richard Feynman (Nobel laureate) in the 1960s, but it gained considerable interest after the discovery of fullerene C60 in 1985 by Kroto et al., afterwards many nano-structured materials are produced globally (Singh and Nalwa 2011).

Extensive researches have been performed on large surface-to-volume ratios of NPs in comparison to other bulk metals and atoms. Optical properties of MNPs are responsible for their unique therapeutic potential, illustrated by surface plasmon resonance (Lin et al. 2014). Thus, MNPs play a pivotal role in fields like fluorescence-based biosensor, photonics (Doria et al. 2012), and bio-labeling (Zhang et al. 2012).

M. Gupta (✉)
Department of Botany, J.D.W.C., Patna, India
e-mail: botany@jdwcpatna.com

K. Seema
Department of Chemistry, J.D.W.C., Patna, India


K. Pal (ed.), *Bio-manufactured Nanomaterials*,
https://doi.org/10.1007/978-3-030-67223-2_6

## 6.2 Nanotechnology and Its Applications

Fabrication of nanomaterials consists of two approaches, i.e., top-down and bottom-up. In bottom-up approaches, synthesis of nanomaterials takes place through the assembly of molecular components, whereas top-down approaches include size reduction of bulk materials through material removal or division (Kralj and Makovec 2015). With the help of these approaches, nanomaterials of different morphologies like particles, rods, tubes, wire, sphere, and cubes have been made (Fig. 6.1). Unique properties of NPs made them attractive material for various fields. Medical applications of NPs include their use as nano-carrier for gene, drug and vaccine delivery. Another sparkling aspect is use of nano-scale biochips to monitor metabolic changes in humans and transmits information. DNA chips, carbohydrate chips, protein chips, and MEMs/NEMS are being utilized for supersensitive detection and diagnostics (Singh and Nalwa 2011).

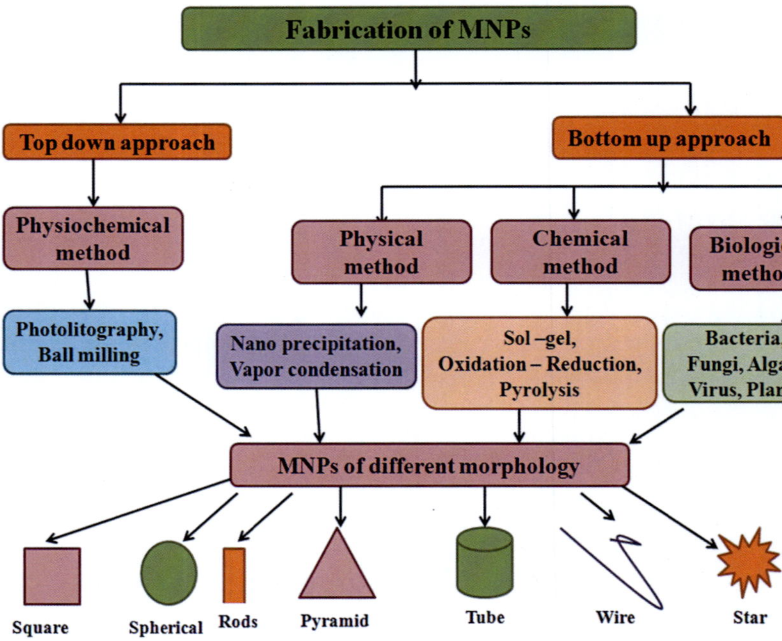


**Fig. 6.1** Diagrammatic representation of approaches used for the fabrication of MNPs

Drug delivery at nano-scale includes liposome, lipid NPs, polymeric nanomicelles, etc. and the routes for nano drug delivery can be oral, ocular, dermal, and pulmonary. NPs of the size $10^{-9}$ are able to cross the blood–brain barrier and thus effectively exploited to cure various types of cancer including brain cancer, nervous system diseases, etc. (Malam et al. 2011). MNPs are also being explored to find possibilities as alternative sources of renewable energy, required to improve the performance of bioprocesses. Various MNPs have been examined in bio-fuel operations like biogas, bio-ethanol, bio-diesel and bio-hydrogen generation for increasing yield of the process. Recently, MNPs are effectively used as asphaltenes adsorbents. Thirteen metals Cu, Cd, Fe, V, Mo, Pb, Ca, P, Mn, Ti, Si, Ni, and Co were found effective (Mazloom et al. 2020). Outline of various applications of MNPs is given in Table 6.1.

**Table 6.1** Diverse applications of MNPs

| Metal nanoparticles (MNPs) | Application | References |
|---|---|---|
| Gold nanoparticles (AuNPs) and bioconjugates of AuNPs | Cancer, HIV, TB treatment, peculiar drug delivery of doxorubicin, methotrexate, and paclitaxel | Rai et al. (2015) |
| | Cancer detection, photo acoustic imaging | Singh and Nalwa (2011) |
| | HIV treatment and TB detection | Rai et al. (2015) |
| | Effective against *Aspergillus flavus*, *Aspergillus niger*, *Puccinia graminis*, *Staphylococcus aureus*, *Staphylococcus epidermidis*, *Salmonella typhi*, *Staphylococcus aureus*, *Vibrio cholerae*, and HeLa cells | Pestovsky and Martínez-Antonio (2017), Ravishankar and Jamuna (2011) |
| | Radiotherapy/cancer treatment | Rai et al. (2015) |
| | Increase yield of bio-hydrogen production using *Clostridium butyricum* and artificial wastewater | Patrick et al. (2019) |
| | Enhance shoot to root ratio and length in *Lactuca sativa*, rapid germination of seeds in *Brassica juncea* | Pestovsky and Martínez-Antonio (2017) |
| | Effective against HIV and influenza virus | Ravishankar and Jamuna (2011) |
| AuNP-loaded PEGylated dendrimer | X-ray computed tomography contrast agent | Singh and Nalwa (2011) |

(continued)

**Table 6.1** (continued)

| Metal nanoparticles (MNPs) | Application | References |
|---|---|---|
| Silver nanoparticles (AgNPs) and bioconjugates of AgNPs | Activity against leukemia | Rai et al. (2015) |
| | As vehicle for specific drug delivery in cancer therapy, imaging of human oral cancer | |
| | Activity against human breast cancer (MCF-7) | |
| | Activity against murine lymphoma tumor L5178Y | |
| | Potent anti-HIV agent | |
| | Effective against HSV-2 virus | Rai et al. (2015) |
| | Activity against MDA-MB-231 breast cancer cells | |
| | Activity against cervical carcinoma and malignant melanoma | |
| | Effective against H1N1 influenza A virus | |
| | Activity against adenocarcinoma cell line of lung epithelium (A549) | |
| | Activity against cell line of cervical cancer SiHa | |
| | Effective against HSV-1, HSV-2, and HPIV-3 viruses | |
| | Activity against tumor B16F10 melanoma | |
| | Effective against adenovirus type 3 virus | |
| | Effective against *E. coli, Enterobacter aerogenes, P. aeruginosa, P. aeruginosa, Streptococcus pyogenes*, MRSA | Rai et al. (2015) |
| | *Neiserria gonorrhoeae* inhibition | Rai et al. (2015) |
| | *Staphylococcus aureus* and *E. coli* inhibition | Ravishankar and Jamuna (2011) |
| | Effective against *S. aureus* (methicillin resistant) | |
| | Activity against *M. tuberculosis* (MDR)and *E. coli* | |
| | Activity against *K. pneumoniae, S. aureus*, and *Bacillus* sp. inhibition | |
| | Dose-dependent activity against *Hippobosca maculate, Haemaphysalis bispinosa, Curvularia lunata, Aspergillus niger, Rhizoctonia solani, Botrytis cinerea, Sclerotinia sclerotiorum, Alternaria* sp., *Penicillium expansum, Rhizopus, Aspergillus terreus, Macrophomina phaseolina*, and *Pseudomonas aeruginosa* | Pestovsky and Martínez-Antonio (2017) |
| | Antifungal activity against *Trichophyton* and pathogenic *Candida* sp. | Kim et al. (2009) |
| | Antibacterial activity against *Bacillus thuringiensis, Xanthomonas oryzae, Pseudomonas syringae, Nitrosomonas europaea, Bacillus megaterium,* and *Burkholderia glumae* | Pestovsky and Martínez-Antonio (2017) |
| | Rapid germination of seeds in *Triticum aestivum,* increase leaf number, plant height, leaf length and inflorescence in *Borago officinalis,* increase root length in *Raphanus sativus* and *Arabidopsis* | Pestovsky and Martínez-Antonio (2017) |
| | Antifungal activity against *Penicillium* sp., *Macrophomina phaseolina, Aspergillus niger, Aspergillus tamarii, Aspergillus versicolor*, and *Aspergillus flavus* | Jogee et al. (2017) |
| | Inhibitory activity against *E. coli, Staphylococcus aureus, Staphylococcus epidermidis, Enterococcus faecium, Klebsiella pneumonia, Streptococcus pyogenes*, and *Salmonella typhi* | Ravishankar and Jamuna (2011) |
| | Increase yield of bio-hydrogen production using *Clostridium butyricum* and inorganic salts | Patrick et al. (2019) |
| | Check CD4-dependent binding of virion and infection. Act as mighty virucidal agent against free and cell-linked virus | Ravishankar and Jamuna (2011) |
| | Effective against respiratory syncytial virus, herpes simplex virus, HIV-1, monkey pox virus, and influenza virus | |

**Table 6.1** (continued)

| Metal nanoparticles (MNPs) | Application | References |
|---|---|---|
| Platinum nanoparticles (PtNPs) and bioconjugates of PtNPs | Diminish oxidative stress of cell | Rai et al. (2015) |
| | Provide bond strength between tooth forms | |
| | Poisonous effect on bacteria | |
| | Induce cytotoxicity in cancer cells | |
| | Treatment of Parkinson's disease | Rai et al. (2015) |
| | Bone allograft | |
| | Effective against *Cladosporium fulvum* and *Colletotrichum acutatum* | Pestovsky and Martínez-Antonio (2017) |
| | Detection of cancer cells | Rai et al. (2015) |
| | Prevention from bone deprivation | |
| Magnesium oxide nanoparticles (MgONPs) | Antibacterial activity against gram-negative, gram-positive, and spore-producing strains, e.g., *Bacillus subtilis, E. coli,* and *Bacillus megaterium* | Ravishankar and Jamuna (2011) |
| Zinc oxide nanoparticles (ZnONPs) | Biocidal activity against *Pseudomonas fluorescens, Staphylococcus aureus, Salmonella typhimurium, Listeria monocytogenes,* and *Salmonella enteritidis* | |
| | Effective against *Aspergillus flavus, Fusarium culmorum, A. niger, F. oxysporum, A. fumigatus, Rhizopus stolonifer, Trichoderma harzianum, Pseudomonas aeruginosa, Pythium, Botrytis cinerea, Penicillium expansum,* and *Phytophthora capsici* | Pestovsky and Martínez-Antonio (2017) |
| Sulfur nanoparticles | Inhibit growth of *Aspergillus niger, Fusarium oxysporum, Rhizopus oryzae, Penicillium italicum, Malus domestica,* and *Actinidia deliciosa* | Pestovsky and Martínez-Antonio (2017) |
| Aluminum nanoparticles (AlNPs) | Antimicrobial activity against *Escherichia coli* | Ravishankar and Jamuna (2011) |
| Copper nanoparticles/ copper oxide (CuO) NPs | Antimicrobial activity against *Pseudomonas aeruginosa, Proteus* sp., and *B. subtilis* | Ravishankar and Jamuna (2011) |
| | Increase shoot to root length in *Lactuca sativa,* more seed production in *Helianthus annuus* | Pestovsky and Martínez-Antonio (2017) |
| | Increase yield of bio-hydrogen production using *Clostridium acetobutylicum/Enterobacter cloacae* and glucose | Patrick et al. (2019) |
| | Enhanced seed production in *Helianthus annuus,* enhanced gibberellin, cytokinin, and indole acetic acid levels in *Vicia sativa* and *Triticum aestivum* | Pestovsky and Martínez-Antonio (2017) |
| | Inhibit *Pythium, Alternaria alternata, Aspergillus flavus, Fusarium solani, Curvularia lunata, Phoma destructiva,* and *Penicillium chrysogenum* | Pestovsky and Martínez-Antonio (2017) |

(continued)

**Table 6.1** (continued)

| Metal nanoparticles (MNPs) | Application | References |
|---|---|---|
| Titanium dioxide nanoparticles ($TiO_2NPs$) | Biocidal activity against *Escherichia coli, Listeria monocytogenes* and *S. aureus*<br>Photo-catalytic activity against bacteria and fungi | Ravishankar and Jamuna (2011) |
| | Photo-catalytic hydrogen production | Patrick et al. (2019) |
| | Potential application in MRI | Singh and Nalwa (2011) |
| | Increase the germination index, photosynthetic rate in *Spinacia oleracea* | Pestovsky and Martínez-Antonio (2017) |
| | Increase nitrate and protein content in *Spinacia oleracea* seedlings, protect *Spinacia oleracea* chloroplasts from aging | |
| | Cause mortality of *Rhipicephalus Boophilus microplus* larvae as well as adults of *Haemaphysalis bispinosa,* inhibit *Fusarium solani* and *Venturia inaequalis* | Pestovsky and Martínez-Antonio (2017) |
| | Antibacterial activity against *Bacillus subtilis, Pseudomonas fluorescens,* and *Escherichia coli* | Ravishankar and Jamuna (2011) |
| | Increase yield of bio hydrogen production using *Rhodobacter sphaeroides* and Sistrom medium | Patrick et al. (2019) |
| Carbon nanostructures gadofullerenes and gadonanotubes | For efficient performance of MRI | Singh and Nalwa (2011) |
| | Blood-pool, molecular-imaging, and guided therapy probes | |
| Diamond nanoparticles | Labeling agent for cellular imaging | Singh and Nalwa (2011) |
| Nickel nanoparticles | Increase yield of bio-hydrogen production using granular sludge and inorganic salts | Patrick et al. (2019) |
| | Active against *Rhipicephalus_Boophilus microplus, Hyalomma anatolicum* mosquito *Anopheles subpictus, Culex quinquefasciatus* and *Culex gelidus* | Pestovsky and Martínez-Antonio (2017) |
| Polyethylene glycol nanoparticles | Induce mortality in *Spodoptera litura* and tea spider mites (*Oligonychus coffeae*) | Pradhan et al. (2013) |
| Magnetic nanoparticles | Used in imaging, sensing, and drug delivery systems, human-induced pluripotent stem (IPS) cells with fluorescent magnetic nanoparticles (FMNPs) | Singh and Nalwa (2011) |
| | Biodiesel production | Patrick et al. (2019) |
| | Production of bio-ethanol employing magnetic NPs and immobilized *S. cerevisiae* cells | |
| | Increase yield of bio-hydrogen production using anaerobic sludge and potato starch/*Enterobacter cloacae* and inorganic salts | |
| | Increase biodiesel production using soybean oil and *Pseudomonas cepacia*/canola oil and *Thermomyces lanuginosus/Candida antarctica/Rhizomucor miehei* | |
| | Antifungal activity against *Alternaria tenuissima, Fusarium,* and *Bipolaris* sp. | Pestovsky and Martínez-Antonio (2017) |
| | Increase activity of methanogens, reduce sulfur reduction, decrease inhibition during biogas production using waste-activated sludge and anaerobic granular sludge | Patrick et al. (2019) |

**Table 6.1** (continued)

| Metal nanoparticles (MNPs) | Application | References |
|---|---|---|
| Chitosan-coated superparamagnetic iron oxide nanoparticles | Stem cell labeling, in vivo MRI tracking | Singh and Nalwa (2011) |
| Manganese nanoparticles ($MnO_2NPs$) | Bio-ethanol production using Baker's yeast and sugarcane leaves | Patrick et al. (2019) |
| | Enhanced nitrogen metabolism by enhancing the activity of several enzymes during seed division<br>Increase levels of chlorophyll *a* and *b* as well as of carotenoid and CP43 protein contents in *Vigna radiata* | Pestovsky and Martínez-Antonio (2017) |
| Silica nanoparticles ($SiO_2$) | Fast seedling growth of *Lycopersicon esculentum,* enhanced shoot to root ratio in *Lactuca sativa,* increase the percentage of filled spikelet in *Oryza sativa,* used against *Sitophilus oryzae,* protect insecticide abamectin from photolysis | Pestovsky and Martínez-Antonio (2017) |
| Selenium nanoparticles | Increase crop yield in *Zea mays* | Pestovsky and Martínez-Antonio (2017) |
| ZnO, $SiO_2$ nanoparticle conjugates | Inhibit growth of *Aspergillus niger* and *Fusarium oxysporum* | Mitra et al. (2015) |

### *6.2.1 Cytotoxicity of MNPs*

Many investigations have shown that MNPs exert cytotoxicity inside the living system. These MNPs either are absorbed (due to the high permeability of plasma membrane) in cells/tissues/organs/bone marrow or circulate in the blood stream. This absorption of MNPs leads to conformational changes in protein, cellular apoptosis, oxidative stress, and platelet activation. Clarence of MNPs from living system is also a huge problem because these MNPs are non-biodegradable and can aggregate in spleen and liver. Gold colloids (20 nm) induce autophagy in lung fibroblasts of human (Azharuddin et al. 2019).

#### 6.2.1.1 Phytotoxicity of MNPs

Agricultural applications of MNPs also include phytotoxicity. Besides this potential, eco-toxicity should also be considered. Intrinsic antimicrobial, antifungal properties of MNPs can harm nitrogen fixing free bacteria of lithosphere and can disturb natural micro flora and mutual symbiotic associations. Some of the MNPs transferred to the food web due to their longer residence. MNPs also interact with organic matter and affect their availability, mobility, and aggregation. In *Arabidopsis thaliana* 150 M concentration of tetrachloroauric ions in germination medium is proved to be sub-lethal. When *Triticum aestivum* seeds were treated with silver NPs at 10 mg/l, it results in diminishing length of root and shoot. Silver

NPs at concentration between 0.5 and 3 mg/l can cause inhibition of root elongation in *Arabidopsis thaliana* (Pestovsky and Martínez-Antonio 2017). Proteomic studies showed the expression of cellular detoxifying enzymes and modification proteins as a result of oxidative stress. ZnO NPs inhibit seed germination in *Brassica oleracea* at concentrations between 0.01 and 1000 mg/l (Pestovsky and Martínez-Antonio 2017). Fragmentation of DNA was observed in *Vicia narbonensis* after treatment with $TiO_2$ NPs (Ruffini Castiglione et al. 2014). Irrespective of the chemical nature of MNPs, they have phytotoxicity at high concentrations and toxicity for non-target organisms. MNPs in many cases cannot be completely removed from the agricultural products and cause toxicity in humans. Thus it is necessary to develop an eco-friendly, cost-effective, green process for the synthesis of less or non-toxic MNPs.

## 6.3 Biological Approach for Eco-friendly Synthesis of MNPs

Living organisms are fascinating nature's secrets; they have inherent capacity to cope up with changing environment and stress conditions. Microorganisms and plants have immense potential for cost-effective, eco-friendly, and non-toxic physiochemical synthesis of MNPs which require less energy input and avoid harsh chemicals, not even in lab but also in natural conditions of their habitat. Metabolism in living system proceed at optimum conditions of pH and temperature, thus these processes are energy efficient and do not require any input of high temperature (Mann 1993). For example, bio-mineralization process in prokaryotes and eukaryotes results in the production of various inorganic materials with different morphologies at both nano and macro scales (Gadd 2007).

Biological synthesis of MNPS have been considered as bottom-up approach in which the responsible oxidation/reduction reactions involve various active natural bio-molecules such as flavonoids, polyphenols, proteins, sugars, alkaloids, co-enzymes, terpenoids, and vitamins in eukaryotic as well as prokaryotic systems (Pal et al. 2019). These active substances work as stabilizing, reducing, and capping agent in the living system for the fabrication of MNPs.

### *6.3.1 Microbial (Prokaryote and Eukaryote) Fabrication of MNPs*

Synthesis of MNPs employing microbes is considered as green, sustainable method which links nanotechnology with application of microbiology. Microorganisms have many enzymes, proteins, metal-resistant genes, reducing co-factors which act as reducing agents in the synthesis of MNPs.

#### 6.3.1.1 Bacteria as Source for Biogenic Green Synthesis of MNPs (Prokaryotic Synthesis)

Microorganisms have to cope with harsh, stressful, and noxious environmental conditions. To adapt in such conditions, bacteria develop certain defense mechanisms which include accumulation of heavy metals from surroundings, efflux-pumping, and bio-sorption (Iravani 2014). For example, *Magnetospirillum magneticum* has been reported for the synthesis of magnetite ($Fe_3O_4$) and greigite ($Fe_3S_4$) (Dudek et al. 2017). Thus these mechanisms can be taken into consideration for green synthesis of MNPs.

Many bacteria have been found to have potential for the production of MNPs (Table 6.2). Synthesis of MNPs using prokaryotic bacterial system (bacteria/archaea) does not require toxic external chemicals as reducing agent, capping agent, and stabilizer which leads to the production of more biocompatible MNPs with very little or no toxicity (Khan et al. 2018).

**Table 6.2** List of bacteria and actinomycetes used for biological synthesis of MNPs

| Microorganism | Type of MNPs | Site | Applications | References |
|---|---|---|---|---|
| *Weissella oryzae* | Ag spherical | Intracellular | Activity against microbes and bio-film | Singh et al. (2016) |
| *Brevibacterium frigoritolerans* | Ag spherical | Extracellular | Antimicrobial | |
| *Pseudomonas deceptionensis* | Ag spherical | Extracellular | Activity against microbes and bio-film | |
| *Bacillus methylotrophicus* | Ag spherical | Extracellular | Activity against microbes | |
| *Bhargavaea indica* | Au anisotropic, flower | Extracellular | Antimicrobial | |
| *Bacillus amyloliquefaciens* | CdS cubic/ hexagonal | Extracellular | | |
| *Bacillus licheniformis* | Ag triangular | Extracellular | Activity against virus and bacteria | |
| *Bacillus persicus* | Ag triangular | Extracellular | Activity against virus and bacteria | |
| *Listeria monocytogenes* | Ag anisotropic and Au | Extracellular | Activity against microbes and mosquito | Soni and Prakash (2015) |
| *Pseudomonas aeruginosa* | Au | Extracellular | | Husseiney et al. (2007) |

(continued)

**Table 6.2** (continued)

| Microorganism | Type of MNPs | Site | Applications | References |
|---|---|---|---|---|
| *Bacillus subtilis* | Ag hexagonal and spherical | Extracellular | Activity against microbes and mosquito | Soni and Prakash (2015) |
| *Streptomyces anulatus* | Ag anisotropic and Au | Extracellular | Activity against microbes and mosquito | Soni and Prakash (2015) |
| *Escherichia coli* | Palladium spherical | Extracellular | | Deplanche et al. (2010) |
| *Aeromonas hydrophila* | Zn oval/spherical | Extracellular | Antimicrobial | Khan and Lee (2020) |
| *Ochrobactrum* sp. | Selenium and tellurium spherical/rod | Extracellular | | |
| *Rhodococcus* sp. | Au spherical | Intracellular | | |
| *Pseudomonas aeruginosa* | Ag spherical | Extracellular | Antimicrobial | |
| *Shewanella loihica PV-4* | Platinum, palladium, and gold spherical | Extracellular | Effective against bio-film and catalytic activities | Ahmed et al. (2018) |
| *Stenotrophomonas maltophilia* | Au spherical | Extracellular | Antimicrobial | Gahlawat and Choudhury (2019) |
| *Kocuria flava* | Cu spherical | Extracellular | | |
| *Acidithiobacillus ferrooxidans* | $Fe_2O_3$ nano-rods | Extracellular | Lithium ion battery anodes | Xiao et al. (2019) |
| *Bacillus cereus* | Ag spherical | Extracellular | Antibacterial activity | Gahlawat and Choudhury (2019) |
| *Alteromonas macleodii* | Ag spherical | Extracellular | | Gahlawat and Choudhury (2019) |
| *Alcaligenes faecalis* | Ag spherical | Extracellular | Antimicrobial and anti-bio-film activity | Divya et al. (2019) |
| *Deinococcus radiodurans* | Ag spherical | | Activity against bacteria, bio-fouling, and cancer | Gahlawat and Choudhury (2019) |
| *Pseudomonas aeruginosa* | CdS spherical | Extracellular | Eliminate cadmium from polluted water | |
| Actinomycetes | | | | |
| *Rhodococcus* NCIM 2891 | Ag spherical | Extracellular | Effective against microbes | Gahlawat and Choudhury (2019) |
| *Streptomyces* sp. LK3 | Ag spherical | Extracellular | Effective against microbes | |
| *Streptacidiphilus durhamensis* | Ag spherical | | Effective against cancer and bacteria | |
| *Streptomyces* sp. LK3 | Ag spherical | | Acaricidal | Singh et al. (2016) |

MNPs synthesized using actinomycetes have good stability and remarkable biocidal activity towards broad spectrum of pathogens. Reductase enzyme plays a significant role in the reduction of metal salts to MNPs (Karthik et al. 2014) (Table 6.2).

Bacteria possesses both extracellular and intracellular mechanisms for the synthesis of MNPs.

### Extracellular Fabrication of MNPs

Many prokaryotes including bacteria and archaea have been studied for physiochemical synthesis of MNPs through extracellular as well as intracellular pathways. Among all these methodologies, extracellular mechanism has attracted researchers because it does not require downstream processing for MNP recovery like cell wall degradation, sonication, several centrifugation and washing steps which are required in intracellular methodologies. Extracellular formation of MNPs takes place through bio-mineralization which includes precipitation, complex formation, efflux-pumping, and bio-sorption (Sengani et al. 2017). Detailed mechanism of extracellular MNP synthesis using bacteria is given in Fig. 6.2 (Singh et al. 2016).

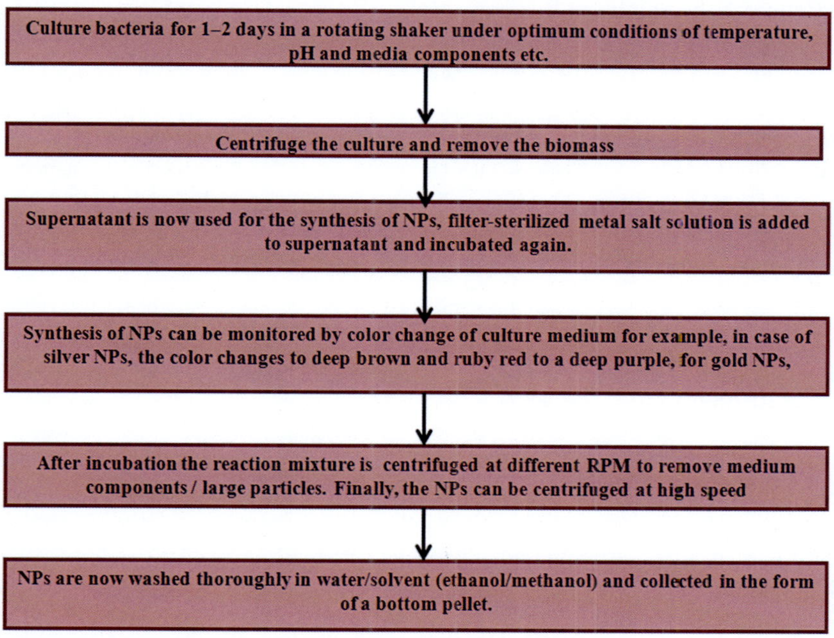


**Fig. 6.2** Mechanism of extracellular synthesis of MNPs in bacteria

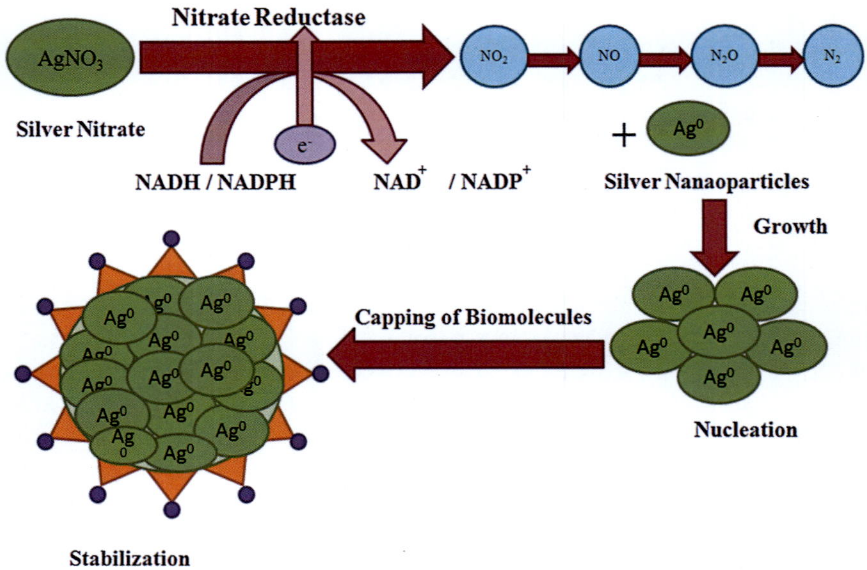


**Fig. 6.3** Schematic representation of intracellular MNP synthesis in bacteria stepwise

Intracellular Fabrication of MNPs

Intracellular formation of MNPs takes place through reductase enzymes. These reductases are able to detoxify heavy metals and reduce metal salts to MNPs which results in to less polydispersity (Sengani et al. 2017) (Fig. 6.3).

Intracellular MNP synthesis is similar to extracellular synthesis. Incubation is followed by biomass elimination through ultrasonication, washing, and centrifugation. These steps enable cell wall degradation and release of MNPs. The mixture is washed and centrifuged for MNPs collection (Singh et al. 2016).

#### 6.3.1.2 Mycosynthesis of MNPs Using Fungi and Yeast (Eukaryotic Synthesis)

Like bacteria, fungi also serve as promising route for biological synthesis of MNPs. Fungi have higher bioaccumulation capacity with simple downstream processing desirable for efficient and economical synthesis of MNPs. Three mechanisms have been put forward to describe fungal synthesis of MNPs, which includes fugal enzymes like nitrate reductase, electron shuttle quinones, or both (Alghuthaymi et al. 2015).

In *Penicillium* species and *Fusarium* reductase enzymes are required for the synthesis of MNPs, likewise in *Oxysporum* α-NADPH-dependent reductases and nitrate reductase play a crucial role in NP synthesis (Anil et al. 2007).

Yeast has the capacity to absorb, adsorb, and accumulate heavy metals and noxious chemicals from the surrounding habitat and performs self-defense through detoxification. These detoxification mechanisms consist of oxidation and reduction through enzymes, intracellular-sequestration, bio-absorption, bio-precipitation, and chelation (Apte et al. 2013). For example, *Yarrowia* and *Saccharomyces cerevisiae* show bioremediation of arsenic from industries (Zinjarde et al. 2014). Many yeast species have been explored for their potential to synthesize NPs (Table 6.3). Yeast can synthesize MNPs by both intracellular and extracellular approaches including membrane bound as well as cytosolic oxido-reductase and quinones like that of other fungi and bacteria (Boroumand et al. 2015).

#### 6.3.1.3 Virus-Mediated Sustainable Fabrication of MNPs

Virus is considered as connecting link between living and nonliving. Virus is composed of genetic material (DNA or RNA) enclosed by protein capsid (Selvakumar et al. 2014). Like other microbes, virus also shows bio-mineralization process for detoxification and metal absorption to cope up with harsh ecological conditions.

Virus shows a great potential toward interconnecting and assembling nanocomponents to develop organized assemblies of MNPs. Viruses serve as good platform for molecular assembly in nano-scale devices because of their monodispersity, size, and availability of chemical groups for modification. Thus, viruses can be used as scaffold for the development of nano-conjugates with noble MNPs. These organized assemblies can be used as engineering material to develop smart nano objects and in technologies like lithography. These nano-conjugates are also used in cancer treatment and drug delivery (Gahlawat and Choudhury 2019).

Disadvantages of virus-mediated synthesis of MNPs include requirement of host for protein expression, limited information on large-scale production and difficult handling. Recently bacteriophages and plant viruses are being utilized for MNP synthesis due to their easy production, no pathogenicity in animals, and chemical and structural stability (Table 6.4).

#### 6.3.1.4 Green Synthesis of MNPs Using Algal System (Eukaryotes)

Algae are polyphyletic in evolution which ranges from unicellular to multicellular autotrophic organisms. Algae have also been employed for bioremediation; e.g., *Stichococcus bacillaris* has the capacity to degrade synthetic polymer like silicone resins (Cappitelli and Sorlini 2008). Algae also have certain active biological compounds and secondary metabolite which act as stabilizing, capping, and reducing factors in the process of MNP synthesis (Fawcett et al. 2017). These algal-generated MNPs can be used as anticancer, antibacterial, antimicrobial, and anti-diabetic agents (Da Silva Ferreira et al. 2017). List of various algae exploited for MNP synthesis has been given in Table 6.5.

**Table 6.3** List of fungi and yeast for mycosynthesis of MNPs

| Fungi | Type of MNPs | Site | Applications | References |
|---|---|---|---|---|
| *Neurospora crassa* | Silver, gold, and bimetallic | Intra- and extracellular | | Singh et al. (2016) |
| *Verticillium* | Silver spherical | Intracellular | | Gahlawat and Choudhury (2019) |
| *Trichoderma viride* | Gold spherical | Extracellular | | |
| *Trichoderma harzianum* | Selenium spherical | Extracellular | | Hu et al. (2019) |
| *Trichoderma longibrachiatum* | Silver, hexagonal, triangular, and cuboid | Extracellular | | Omran et al. (2019) |
| *Trichoderma harzianum* | CdS spherical | Extracellular | | Khan and Lee (2020) |
| *Verticillium luteoalbum* | Gold spherical | Intracellular | | |
| *Alternaria alternate* | Gold spherical | Extracellular | Antimicrobial | |
| *Aspergillus fumigates* | Gold spherical | Intracellular | Antifungal | Khan and Lee (2020) |
| *Aspergillus terreus* | ZnO spherical | Intracellular | Antifungal | |
| *Aspergillus fumigates* | Silver spherical | Extracellular | Antifungal | |
| *Fusarium oxysporum* | Platinum hexagons, pentagons, circles, squares, | Extracellular | | Riddin et al. (2006) |
| *Cladosporium cladosporioides* | Silver spherical | Extracellular | | Balaji et al. (2009) |
| *Phoma* sp. | Silver spherical | Extracellular | | Khan and Lee (2020) |
| *Fusarium oxysporu* | $Fe_3O_4$ quasi-spherical | Extracellular | | |
| *Verticillium* sp. | $Fe_3O_4$ octahedral | Extracellular | | |
| *Penicillium diversum* | Silver spherical | Extracellular | Antimicrobial | Gahlawat and Choudhury (2019) |
| *Botrytis cinerea* | Gold spherical, hexagonal, pyramidal, triangular | Extracellular | | Castro et al. (2014) |
| *Nigrospora oryzae* | Gold cubic and spherical | | Anti-helminthic activity | Kar et al. (2014) |
| *Curvularia lunata* | Silver spherical | Extracellular | Antimicrobial | Ramalingmam et al. (2015) |

(continued)

**Table 6.3** (continued)

| Fungi | Type of MNPs | Site | Applications | References |
|---|---|---|---|---|
| *Metarhizium anisopliae* | Silver rods | Extracellular | Inhibitory effect on *Anopheles culicifacies* | Amerasan et al. (2016) |
| *Rhizopus oryzae* | Gold spherical and flower like | | | Kitching et al. (2016) |
| *Pleurotus ostreatus* | Gold spherical | Extracellular | Antimicrobial and anti cancer | El Domany et al. (2018) |
| *Colletotrichum sp.* | Aluminium oxide spherical | | Antimicrobial activity | Suryavanshi et al. (2017) |
| *Trichoderma asperellum* | CuO spherical | Extracellular | Antimicrobial activity | Saravanakumar et al. (2019) |
| *Cochliobolus geniculatus* | ZnO quasi-spherical | Extracellular | | Kadam et al. (2019) |
| *Trichoderma reesei* | Silver cubic | Extracellular | Antimicrobial | Gemishev et al. (2019) |
| *Fusarium solani* | Silver, copper and zinc | Extracellular | Biocide | El Sayed Manal and El Sayed Ashraf (2020) |
| Yeast species | | | | |
| *Yarrowia lipolytica* NCYC 789 | Silver spherical | Extracellular | Anti-bio-film | Singh et al. (2016) |
| Extremophilic yeast | Silver and gold | Extracellular | | Singh et al. (2016) |
| *Rhodosporidium diobovatum* | Lead | Intracellular | | Gahlawat and Choudhury (2019) |
| *Candida utilis* NCIM 3469 | Silver spherical | Extracellular | Antibacterial | Singh et al. (2016) |
| *Candida albicans* | CdS spherical | Extracellular | | Kumar et al. (2019) |
| *Pichia pastoris* | Silver spherical | Extracellular | Antimicrobial | Gahlawat and Choudhury (2019) |
| *Saccharomyces cerevisiae* | ZnS spherical | Intracellular | | Gahlawat and Choudhury (2019) |
| *Saccharomyces cerevisiae* | $Sb_2O_3$ spherical | Extracellular | | |
| *Saccharomyces cerevisiae* | SeS spherical | Intracellular | | Asghari-Paskiabi et al. (2019) |
| *Cryptococcus* | Silver spherical | Extracellular | Antimicrobial | Gahlawat and Choudhury (2019) |

**Table 6.4** Different viruses employed for MNP fabrication

| Virus | Type of NPs | Morphology | Site of synthesis | References |
|---|---|---|---|---|
| Tobacco mosaic virus (TMV) | $SiO_2$, $Fe_2O_3$,CdS, PbS | Film, prismatic disordered | On surface | Das et al. (2017), Gahlawat and Choudhury (2019) |
| Bacteriophage M13 | ZnS, CdS and $TiO_2$ | Film / wire | On surface | |
| TMV | Nickel | Rod | On surface | Khan and Lee (2020) |
| TMV | Nickel | Nano-wires | Central channel | |
| TMV | Zinc oxide | Rod | On surface | |
| Cowpea mosaic virus | Cobalt, nickel, Fe, palladium, Ni-Fe, and Co-Pt | Diverse shapes | On surface | |
| TMV | CoPt, $FePt_3$ | Nano-wires | Central channel | |
| Z1 peptide | ZnO | Nano-rod | On surface | |
| M13 virus | $Co_3O_4$ | Nano-wire | On surface | |
| Chilo iridescent virus | Gold | Nano-shell | On surface | |
| TMV | Palladium | Multi-walled carbon nanotubes | | |
| Hepatitis E virus | Nano-conjugates in treatment of cancer | 34 icosahedral | | Gahlawat and Choudhury (2019) |
| Cucumber mosaic virus | Nano-assemblies in treatment of cancer and delivery of drug | Icosahedral | | |
| Potato virus X | Specific doxorubicin delivery in treatment of cancer | Helical | | |
| TMV | Gold | Spherical | | |
| Potato virus X | Nano-conjugates for delivery of herceptin to treat breast cancer | Filamentous rod shaped | | |
| Red clover necrotic mosaic virus | As nano-carriers for specific doxorubicin delivery | Icosahedral | | |

**Table 6.5** List of algal members illustrated for synthesis of MNPs

| Algae | Type of MNPs | Morphology | Application | References |
|---|---|---|---|---|
| *Spirogyra varians* | Silver | Quasi-sphere | Antibacterial activity | Gahlawat and Choudhury (2019) |
| *Sargassum* | Palladium | Octahedral | Electro catalytic activity towards $H_2O_2$ | Singh et al. (2016) |
| *Chlorella vulgaris* | Gold | Spherical self assembled cores | Anti-pathogenic activity | Gahlawat and Choudhury (2019) |
| *Ulva lactuca* | Silver | Cubical | Anti-plasmodial | |
| *Porphyra vietnamensis* | Silver | Spherical | Antibacterial activity | |
| *Tetraselmis kochinensis* | Gold | Spherical | | |
| *Sargassum plagiophyllum* | Silver | Spherical | Antibacterial | |
| *Scenedesmus* sp. | Silver | Spherical crystalline | Antibacterial | |
| *Caulerpa racemosa* | Silver | Spherical | Antibacterial activity | |
| *Pithophora oedogonia* | Silver | Cubical, hexagonal | Antibacterial activity | |
| *Ecklonia cava* | Gold | Spherical | Anti-microbial | |
| *Caulerpa racemosa* | Silver | Distorted spherical | Degradation of methylene blue | |
| *Chlorella vulgaris* | Silver | Spherical | Antibacterial activity | Da Silva Ferreira et al. (2017) |
| *Sargassum tenerrimum* and *Turbinaria conoides* | Gold | Spherical | Reduction of dyes rhodamine B and sulforhodamine 101 | Gahlawat and Choudhury (2019) |
| *Cystoseira baccata* | Gold | Spherical | Anti-cancer | |
| *Sargassum muticum* | Zinc oxide | Hexagonal | Anti-angiogenesis and anti-apoptotic | |
| *Galaxaura elongata* | Gold | Hexagonal, rods, spherical and triangular | Activity against bacteria | |
| *Padina tetrastromatica* and *Turbinaria conoides* | Zinc oxide | Pentagonal, hexagonal, spherical and triangles | Activity against bacteria | |
| *Laminaria japonica* | Silver | Spherical/oval | Phyto-toxicity | Kim et al. (2018) |
| *Gelidium amansii* | Silver | Spherical | Antibacterial activity | Pugazhendhi et al. (2019) |
| *Polysiphonia* | Silver | Spherical | Activity against MCF-7 breast cancer cell lines | Mukherjee et al. (2014) |

## 6.3.2 *Fabrication of MNPs by Higher Eukaryotic System*

### 6.3.2.1 Plant-Mediated Formation of MNPs

Plants can adsorb, accumulate, and deteriorate heavy metal and their ions from ecological surroundings (Kulkarni and Muddapur 2014). Plants have been utilized in various diseases from ancient times (Ashraf et al. 2018).

Phyto-nanotechnology has generated diverse possibility for cost-effective, rapid, and eco-friendly sustainable bulk production of MNPs. MNPs generated from plants are more stable, biocompatible, and non-toxic. They have wide application in medical field (Noruzi 2015).

Different plant organs like leaf, stem, roots, flower, fruits, and their extracts can be exploited for plant-mediated MNP synthesis (Fig. 6.4). The accurate mechanism of MNP synthesis is under elucidation, but it can be considered that certain vitamins, proteins, amino acids, organic acid, polysaccharides, and secondary metabolites (alkaloids, flavonoids, polyphenols, terpenoids, heterocyclic compounds, etc.) perform crucial role in the reduction of metal salts and provide stability to MNPs (Duan et al. 2015). Different mechanisms are accountable for the synthesis and stability of MNPs in plants (Baker et al. 2013) for example in *Corallina officinalis* carbonyl group from proteins and hydroxyl group from polyphenols perform key role during synthesis gold NPs and provide additional stability (El-Kassas and El-Sheekh 2014). Plant-mediated production and stabilization of gold and silver NPs are assisted by the addition of bio-molecules in *Murraya koenigii* (Philip et al. 2011).

Likewise a purgative resin-named emodin and quinones are responsible for silver NP synthesis in xerophytes, whereas remirin, dietchequinone, and cyperoquinone are useful for the synthesis of MNPs in mesophytes. Eugenol in *Cinnamomum zeylanicum* required for gold and silver NP synthesis (Makarov et al. 2014) (Table 6.6).

Sustainable synthesis of MNPs using plants have certain advantages like no requirement of sophisticated lab facilities, one-step process, easy, harmless, less time taking in comparison to microbial fabrication of MNPs (Nasrollahzadeh and Sajadi 2015) (Fig. 6.5).

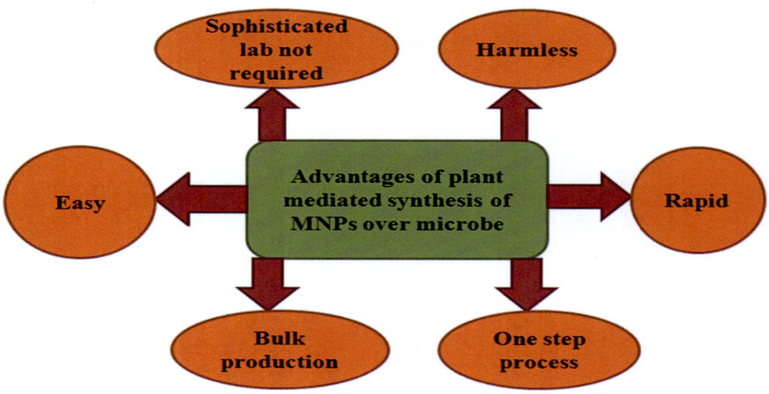


**Fig. 6.5** Advantages of phytogenic MNPs synthesis over microbes

**Table 6.6** List of higher plants for biogenic synthesis of MNPs

| Plant | Plant part | Type of NPs | Applications | References |
|---|---|---|---|---|
| *Ginkgo biloba* | Leaf | Copper spherical | Catalytic | Singh et al. (2016) |
| *Euphorbia prostrata* | Leaf | Ag and ($TiO_2$) spherical | Leishmanicidal | |
| *Nyctanthes arbortristis* | Flower | Silver | Cytotoxic and effective against bacteria | |
| *Gardenia jasminoides* | Leaf | Fe rock like | Effective against bacteria | |
| *Citrus medica* | Fruit | Copper | Antimicrobial | Singh et al. (2016) |
| *Lawsonia inermis* | Leaf | Fe hexagonal | Effective against bacteria | |
| *Artocarpus gomezianus* | Fruit | Zn spherical | Antioxidant, photo-catalytic and luminescent | |
| Orange and pineapple | Fruits | Ag spherical | | Singh et al. (2016) |
| *Anogeissus latifolia* | Gum powder | Silver spherical | Antibacterial | |
| *Catharanthus roseus* | Leaf | Palladium spherical | Dye break down | |
| *Panax ginseng* | Root | Au and Ag spherical | Antibacterial | |
| *Cymbopogon citratus* | Leaf | Au rod, hexagonal, spherical and triangular | Biocidal activity against mosquito | |
| Red ginseng | Root | Ag spherical | Activity against bacteria | |
| *Abutilon indicum* | Leaf | Ag spherical | Activity against bacteria | |
| *Cocos nucifera* | Leaf | Pb spherical | Activity against bacteria and photo-catalytic | |
| *Azadirachta indica* | Leaf | Silver | Larvicidal activity against mosquito | Poopathi et al. (2015) |
| *Nigella sativa* | Leaf | Silver spherical | Cytotoxic | Singh et al. (2016) |
| Banana | Peel | CdS | | |
| *Pinus densiflora* | Cones | Silver oval/ triangular shaped | Activity against microbes | |
| *Pistacia atlantica* | Seeds | Silver spherical | Activity against bacteria | |
| *Abutilon indicum* | Leaf | Copper | Antimicrobial | Ijaz et al. (2017) |
| *Origanum vulgare* | Leaf | $TiO_2$ | | Shaik et al. (2018) |
| *Thymus vulgaris* | Leaf and stem | ZnO | | Abolghasemi et al. (2019) |

(continued)

**Table 6.6** (continued)

| Plant | Plant part | Type of NPs | Applications | References |
|---|---|---|---|---|
| *Geranium graveolens* | Leaf | Ag | Antibacterial | Das et al. (2017) |
| *Aloe vera* | Leaf | Ag and Au | Antimicrobial | |
| *Carica papaya* | Fruit | Ag | Antibacterial | Jain et al. (2009) |
| *Rosa rugosa* | Leaf | Ag and Au | Antimicrobial | Dubey et al. (2010) |
| *Capsicum annuum* | Fruit | Ag | Antibacterial | Jha and Prasad (2011) |
| *Rosa damascena* | Petal | Ag | Antibacterial | Solgi (2012) |
| *Punica granatum* | Peel | | Antibacterial | Solgi (2012) |
| *Anogeissus latifolia* | Gum powder | Ag | | Kora et al. (2012) |
| *Crocus sativus* | Petal | Ag | Antibacterial | Solgi (2014) |
| *Phyllanthus emblica* | Fruit extract | Silver | Antimicrobial | Renuka et al. (2020) |
| *Clerodendrum infortunatum, Abutilon indicum,* and *Clerodendrum inerme* | Leaf | ZnO and Cu-doped ZnO NPs | | Khan et al. (2018) |
| *Convolvulus fruticosus* | Extract of flower | Gold spherical | Antibacterial | Ebrahimzadeh et al. (2020) |
| *Rheum ribes* | | Silver spherical | Antimicrobial and anticancer | Aygun et al. (2020) |
| *Elytraria acaulis* | Extract of leaf | Cuboid | Antibacterial | Rangayasami et al. (2020) |

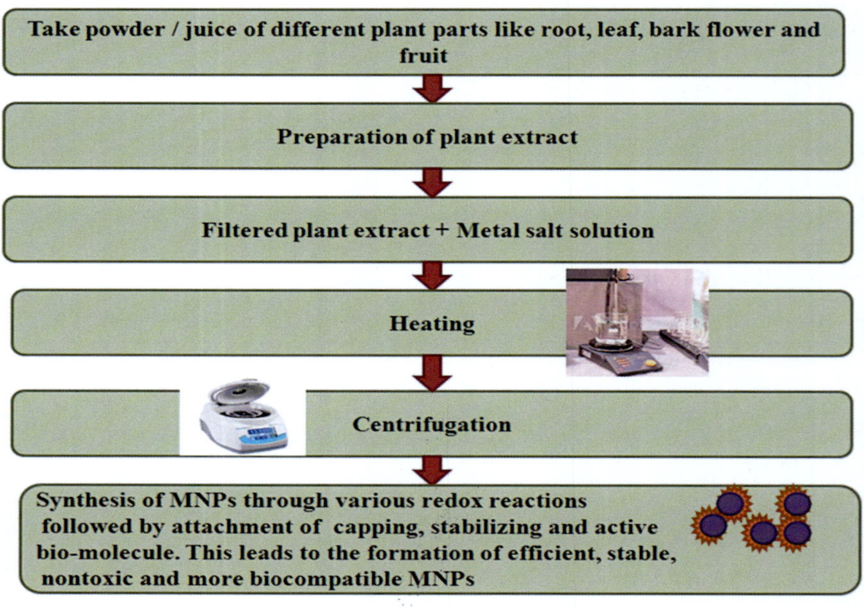


**Fig. 6.4** Step-wise illustration of process of MNP synthesis in plants

**Table 6.7** Human cell employed for metal nanoparticle synthesis

| Human cell line | Types of nanoparticle | References |
|---|---|---|
| HeLa | Gold intracellular | Singh et al. (2016) |
| SiHa | Gold intracellular | |
| HEK 293 | Gold intracellular | |
| SKNSH | Gold intracellular | |

#### 6.3.2.2 Human Cell Lines as Nano-Factories

Human cancer as well as non-cancerous cells has been found to produce MNPs in vitro conditions. Human cancer cells as SKNSH (human neuroblastoma), SiHa (malignant cells of cervical epithelium), and HeLa (malignant cells of cervical epithelium), non-cancer cells such as HEK-293 (non-malignant human embryonic kidney cells) have potential to synthesize gold NPs when incubated with 1 mM solution of tetrachloroaurate. MNPs are present in nucleus (smaller) and cytoplasm (larger) (Singh et al. 2016) (Table 6.7).

## 6.4 Limitations and Key Challenges for the Biogenic Formation of MNPs

Besides many benefits of biogenic MNPs, polydispersity and establishment of stable system for the production of MNPs with homogenous size and morphology remains a challenge (Singh et al. 2016). In the whole process, environmental, growth factors, and functional molecules play an important role. In *Ganoderma* sp. improvement in environmental conditions of temperature, pH, salt concentration, redox condition, irradiation, and aeration bring about the synthesis of 20 nm, biocompatible, and monodispersed gold NPs. In case of microorganisms rise in temperature favors the synthesis of MNPs because this elevation in temperature leads to the activation of most of the enzymes, required for the synthesis of MNPs. Elevation in temperature favors high yield of Ag NPs in *Magnolia* plant and size diminish from 50 nm at 25 °C to 16 nm at 95 °C (Singh et al. 2016).

The pH also plays a key role in MNP synthesis. Different pH ranges are required to induce NP synthesis in different microorganisms and any alteration in optimum pH can alter the physical properties (shape and size) of NPs. For example, MNP synthesis in *Escherichia coli* takes place at pH 10, in *Isaria fumoscrosea* at alkaline pH, in *Penicillium fellutanum* at pH 6, In *Fusarium acuminatum* at acidic pH. In plants, changes in pH leads to the changes in natural phyto-chemistry (charge distribution) of plants which bring about alteration in their binding properties and reduction of metal salts. Synthesis of gold NPs using *Avena sativa* extract, takes place at pH 3.0 and 4.0, because at pH 2.0 processes like NP aggregation have been observed. Thus it can be concluded that, at acidic pH, process like aggregation of NPs dominates over metal salts reduction (Singh et al. 2016). This can be explained with the fact that the functional groups which attach and nucleate metal ions are more readily available at pH 3.0 and 4.0 than at pH 2.0. At very low acidic pH, metal ions take part in nucleation reactions as a result of which agglomeration of the metals takes place. Synthesis of silver NPs employing tuber powder of *Curcuma longa* at alkaline pH requires many negatively charged functional groups which are responsible for the reduction of metal salts to metal ions. As a result of which efficient synthesis of NPs takes place. Likewise other environmental conditions as salt concentration, time period, and localization also affect MNP synthesis in different species of plants (Singh et al. 2016).

## 6.5 Advantages and Potential Employment of Biogenic MNPs

Biogenic synthesis of MNPs is rapid in comparison to physiochemical synthesis. Scientists have optimized the rapid process with bulk production from a variety of plant sources, for example, silver NPs can be synthesized within 2, 5, 45 min, 1, and 2 h from different plant extracts. Gold NPs can be synthesized within 3, 5, and 10 min using the extracts of diverse plant sources (Singh et al. 2016).

For biomedical applications of NPs, less cytotoxicity and biocompatibility are essential. MNPs which are synthesized through physiochemical methods become toxic due to the attachment of toxic byproducts released during the process and are less biocompatible which limits their application in biomedical field. But in contrast MNPs produced using biogenic routs are free from toxins, more biocompatible, and eco-friendly. These MNPs do not require any stabilizing agent further; many biomolecules present in living system serves as stabilizing and capping agent (Makarov et al. 2014). Biological MNPs adsorb many bio-molecules at their surface (selectively and progressively) when they are exposed to complex fluids of living system. This step plays a key role in the formation of corona around MNPs which interact with living systems. This corona provides additional stability, efficiency, and biocompatibility to biological MNPs. Medicinal plants are rich in metabolites with pharmacological activity. Corona of pharmacologically active metabolites provides additional stability and biocompatibility to the MNPs (Singh et al. 2016). Broad spectrum applications of biogenic MNP biomedical field have been summarized below (Fig. 6.6).

### *6.5.1 Activity Against Pathogenic Bacteria*

Many bacteria have developed resistance against one or more commercially available antibiotics (multi-drug resistance). Thus, there is a continuous demand of new antibiotics. Potential of biogenic MNPs have been investigated for better efficiency especially against MDR bacteria. To archive that MNPs with active biological molecules are being utilized against bacteria (MDR). These MNPs can easily penetrate inside the pathogenic bacteria and active biological molecules present around them increase their efficacy as antibacterial agents. Silver NPs developed through mycological route have been proved effective against gram-negative and gram-positive bacteria (Khan et al. 2018). Likewise, CuO, ZnO, and Cu-doped ZnO NPs developed using plants also show efficacy against bacterial strains (both gram-negative and -positive). Biologically synthesized NPs are found to have higher antimicrobial activity in

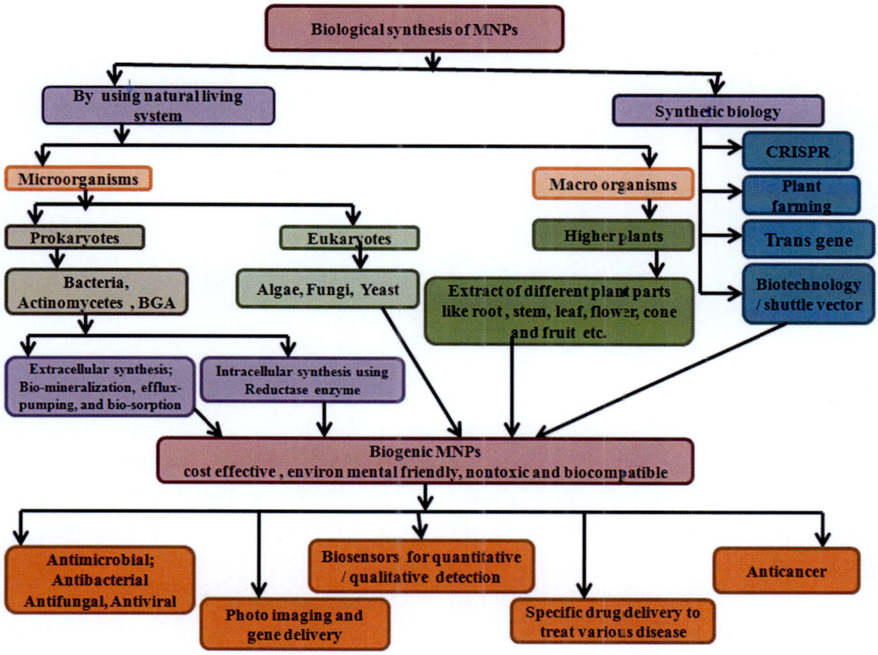


**Fig. 6.6** Biogenic green synthesis of MNPs and their biomedical application

comparison to chemically synthesized NPs, for example, Mukherjee et al. shown that 30 mM concentration of biologically synthesized MNPs have 96.67% antibacterial activity, whereas the same concentration of chemically formed MNPs did not have any significant activity. Biologically developed MNPs have higher biocidal activity against *Salmonella typhimurium* ATCC 14028, *Micrococcus luteus* ATCC 9341, and *B. subtilis* ATCC 6633 in comparison to chemically synthesized MNPs (Abeer Mohamad 2015). Various mechanisms have been put forward to define the antimicrobial activity of MNPs which includes alteration in protein and DNA, nutrient uptake, membrane potential, production of ROS (reactive oxygen species), protein oxidation or disruption of cell membrane which results in cell death (Khan et al. 2018).

## 6.5.2 *Activity Against Fungi*

Biogenic MNPs have also been exploited as potential antifungal agents for instance phytogenic NPs of ZnO and Cu-doped ZnO developed from *C. inerme, A. indicum*, and *C. infortunatum* were found effective against *T. harzianum*, *A. niger*, and *A. flavus*. CdS NPs developed through microbes have been found to have increased fungicidal activity against *A. niger* and *A. flavus* (Rajeshkumar et al. 2014). Likewise, biogenic NPs of Ag, Au, Ni, Pt, and Pd also have been reported to have antifungal properties. Mechanism of fungicidal activity is not cleared yet, but according to Reidy et al. (2013), it includes attachment of NPs on fungal cell surface followed by penetration inside the cell and interaction with phosphorus containing entities like DNA which leads to the inhibition of respiration and ultimately death of fungal cell (similar to antibacterial mode of action).

### 6.5.3 Biogenic MNPs as Potent Anticancer Agent

Several biogenic MNPs like Au, Ag, Ni, NiO, CuO, ZnO, MgO, and Cu-doped ZnO have been reported to have anticancer properties (Khan et al. 2018). Silver NPs synthesized from *Olax scandens* leaf have better biocompatibility for drug delivery, better efficiency as anticancer and better imaging facilitator in comparison to MNPs synthesized through physiochemical processes (Singh et al. 2016).

Biogenic MNPs show higher biocompatibility and anticancer activity in human endothelial cells of umbilical vein (HUVEC), B16 (mouse melanoma), A549 (human lung cancer), cardiomyoblast cell line of rat (H9C2), MCF7 (human breast cancer), and Chinese hamster ovary cells (CHO). Biological MNPs inside the living system show bright red fluorescence which can be utilized to monitor the location of drug within the cancer cells (Mukherjee et al. 2014). Gold NPs synthesized from extract of *Corallina officinalis* (red seaweed) show cytotoxic effect on human breast cancer cell line MCF7 (Singh et al. 2016).

Gold nano-conjugates developed using novel pro-angiogenic system found to boost blood vessels growth by redox signaling. Fabricated gold nano-bio-conjugates which were biosynthesized using leaf extract of *Olax scandens* can be used to treat breast (MCF-7), lung (A549), and colon (COLO 205) cancer cell lines (Mukherjee et al. 2014). These biologically synthesized gold and silver NPs are more biocompatible in case of ECV-304 cell and HUVEC cell line than physiochemically synthesized MNPs. It was found that biological MNPs have higher anti-cancerous activity on B16F10 and MCF-7 cell line when combined with drug like doxorubicin (Patra et al., 2015). MgO NPs synthesized using aqueous extract of *Sargassum wightii* were found effective against lung cancer cells (Pugazhendhi et al. 2019).

Furthermore silver and gold nanoparticles synthesized using extracts of different medicinal plants like *Butea monosperma* found more stable, biocompatible and effective against ECV-304, HUVEC, HNGC2, A549, B16F10, and MCF-7 cell lines (Singh et al. 2016).

Possible anticancer mechanism of MNPs involves the generation of reactive oxygen species (ROS) which damages the cellular components or apoptosis via caspase-dependent pathways and mitochondria-dependent pathways (Arokiyaraj et al. 2014).

### 6.5.4 Biogenic MNPs as Biosensor

Immobilization of bio-molecule-conjugated MNPs on the surface will result in the development of optical and electronic biosensor. Silver and gold NPs show plasmon resonance in visible spectrum (controlled through size variation of biogenic MNPs). Thus optical effects of MNPs can be regulated by association with different bio-molecules which leads to the quantitative and ion detection. For example, properties of gold NPs change significantly after agglomeration. This approach is exploited for bioassay labeling and tissue staining (Razavi et al. 2015). Agglomeration of MNPs leads to the spectral shift which is the key step required for the development of biosensors. Gold NPs functionalized with DNA have been employed for colorimetric detection of DNA triplex (Han et al. 2006).

### 6.5.5 *Biogenic MNPs as Vehicles for Drug Delivery*

For efficient treatment, it is required that effective dose of drug reach to the specific target within the scheduled time. Ability of MNPs (in conjugation with biomolecules) as drug delivery system has been explored (Ghosh et al. 2008). Non-toxic and non-immunogenic gold NPs can be used as carrier for specific delivery of drug. Aubin-Tam and Hamad-Schifferli (2008) have developed Au NPs and infrared light-based drug delivery system.

## 6.6 Conclusion, Outlook, and Future Aspects

Green synthesis of MNPs using biogenic routes has many advantages like cost-effective, controlled shape, size, stable synthesis, and rapid bulk production, biocompatible, non-toxic, and environmental friendly. Thus, it can be concluded that biological MNPs have higher potential as antimicrobial and anticancer agents. They also facilitate drug delivery to specific cells and are also used as biosensor, photo-imaging, thermal therapies, medical appliances, and cosmetic industry. However, there are certain limitations for efficient and successful synthesis of MNPs using biological systems. The yield of synthesized MNPs with respect to metal salt concentration needs to be elucidated and parameters which can conquer polydispersity and morphology of MNPs should be emphasized and optimized. Besides this, application of bio-films for enhanced and effective MNP synthesis has been recently recognized. Bio-films act as active growth mode with interesting features like capacity to overcome electrochemical reactions, catalyzing power and highly reducing matrix, which provides suitable conditions for efficient synthesis of MNPs. Bio-films have limited diffusion from outer environment which protect them from contamination. Another major gap of biogenic synthesis of MNPs is lack of much information about the mechanism and active bio-molecules involved in the synthesis and stabilization of MNPs. Furthermore development of scalable and reproducible process for commercial synthesis of MNPs with controlled size, morphology, and monodispersity is one of the major concerns. High yield of biogenic MNPs of monodispersity can be obtained by optimizing process parameters as temperature, pH, C:N ratio, and salt concentration. Apart from all these other parameters like biocompatibility, bioavailability, clearance and release kinetics of MNPs requires more emphasis and consideration for their biomedical application (in vivo). Further signaling processes involved in the synthesis of MNPs must be explored in details.

**Acknowledgments** I would like to thank Dr. Shyama Roy (Principal, J.D.W.C. Patna) for her immense support throughout.

**Conflicts of Interest** The authors declare that there is no conflict of interests.

Ethical Issues This chapter does not carry any studies with human participants or animals performed by any of the authors. All the articles reviewed for this manuscript have been properly cited.

## References

Abeer Mohamad, A. B. (2015). Comparison between biological and chemical synthesis of zinc oxide nanoparticles and its influence over some antibiotics. *International Journal of Pharma and Bio Science, 6*(3), 1357–1364.

Abolghasemi, R., Haghighi, M., Solgi, M., & Mobinikhaledi, A. (2019). Rapid synthesis of ZnO nanoparticles by waste thyme (*Thymus vulgaris* L.). *International Journal of Environmental Science and Technology, 16*, 6985–6990.

Ahmed, E., Kalathil, S., Shi, L., et al. (2018). Synthesis of ultra-small platinum, palladium and gold nanoparticles by *Shewanella loihica* PV-4 electrochemically active biofilms and their enhanced catalytic activities. *Journal of Saudi Chemical Society, 22*(8), 919–929.

Alghuthaymi, M. A., Almoammar, H., Mahindra, R., et al. (2015). Myconanoparticles: Synthesis and their role in phytopathogens management. *Biotechnology and Biotechnological Equipment, 29*, 221–236.

Amerasan, D., Nataraj, T., Murugan, K., et al. (2016). Myco-synthesis of silver nanoparticles using *Metarhizium anisopliae* against the rural malaria vector *Anopheles culicifacies* Giles (Diptera: Culicidae). *Journal of Pest Science, 89*, 249–256.

Anil Kumar, S., Abyaneh, M. K., Gosavi, S. W., et al. (2007). Nitrate reductase-mediated synthesis of silver nanoparticles from $AgNO_3$. *Biotechnology Letters, 29*, 439–445.

Apte, M., Sambre, D., Gaikawad, S., et al. (2013). Psychrotrophic yeast yarrowia lipolytica NCYC 789 mediates the synthesis of antimicrobial silver nanoparticles via cell-associated melanin. *AMB Express, 3*(1), 32.

Arokiyaraj, S., Arsu, M. V., Vincent, S., et al. (2014). Rapid green synthesis of silver nanoparticles from Chrysanthemum indicum L and its antibacterial and cytotoxic effects: An in vitro study. *International Journal of Nanomedicine, 9*, 379–388.

Asghari-Paskiabi, F., Imani, M., Rafii-Tabar, H., et al. (2019). Physicochemical properties, antifungal activity and cytotoxicity of selenium sulfide nanoparticles green synthesized by Saccharomyces cerevisiae. *Biochemical and Biophysical Research Communications, 516*(4), 1078–1084.

Ashraf, I., Zubair, M., Rizwan, K., et al. (2018). Chemical composition, antioxidant and antimicrobial potential of essential oils from different parts of Daphne mucronata Royle. *Chemistry Central Journal, 12*(1), 135.

Aubin-Tam, M.-E., & Hamad-Schifferli, K. (2008). Structure and function of nanoparticle–protein conjugates. *Biomedical Materials, 3*, 034001.

Aygun, A., Gulbağca, F., Nas, M. S., et al. (2020). Biological synthesis of silver nanoparticles using Rheum ribes and evaluation of their anticarcinogenic and antimicrobial potential: A novel approach in phytonanotechnology. *Journal of Pharmaceutical and Biomedical Analysis, 179*, 113012.

Azharuddin, M., Zhu, G. H., Das, D., et al. (2019). A repertoire of biomedical applications of noble metal nanoparticles. *Chemical Communications, 55*, 6964.

Baker, S., Rakshith, D., Kavitha, K. S., et al. (2013). Plants: Emerging as nanofactories towards facile route in synthesis of nanoparticles. *BioImpacts: BI, 3*(3), 111–117.

Balaji, D. S., Basavaraja, S., Deshpande, R., et al. (2009). Extracellular biosynthesis of functionalized silver nanoparticles by strains of Cladosporium cladosporioides fungus. *Colloids and Surfaces. B, Biointerfaces, 68*(1), 88–92.

Boroumand, M. A., Namvar, F., Moniri, M., et al. (2015). Nanoparticles biosynthesized by fungi and yeast: A review of their preparation, properties, and medical applications. *Molecules, 20*(9), 16540–16565.

Cappitelli, F., & Sorlini, C. (2008). Microorganisms attack synthetic polymers in items representing our cultural heritage. *Applied and Environmental Microbiology, 74*(3), 564–569.

Castro, M. E., Cottet, L., & Castillo, A. (2014). Biosynthesis of gold nanoparticles by extracellular molecules produced by the phytopathogenic fungus *Botrytis cinerea. Materials Letters, 115*, 42–44.

Da Silva Ferreira, V., ConzFerreira, M. E., Lima, L. M., et al. (2017). Green production of microalgae-based silver chloride nanoparticles with antimicrobial activity against pathogenic bacteria. *Enzyme and Microbial Technology, 97*, 114–121.

Das, R. K., Pachapur, V. L., Lonappan, L., et al. (2017). Biological synthesis of metallic nanoparticles: Plants, animals and microbial aspects. *Nanotechnology and Environmental Engineering, 2*, 18.

Deplanche K., Caldelari I., Mikheenko I. P., Sargent F., Macaskie L. E. (2010). Involvement of hydrogenases in the formation of highly catalytic Pd (0) nanoparticles by bioreduction of Pd (II) using Escherichia coli mutant strains. *Microbiology, 156*(9), 2630–40. https://doi.org/10.1099/mic.0.036681-0.

Divya, M., George, S. K., Saqib, H., & Joseph, S. (2019). Biogenic synthesis and effect of silver nanoparticles (AgNPs) to combat catheter-related urinary tract infections. *Biocatalysis and Agricultural Biotechnology, 18*, 101037.

Doria, G., Conde, J., Veigas, B., et al. (2012). Noble metal nanoparticles for biosensing applications. *Sensors, 12*(2), 1657–1687.

Duan, H., Wang, D., & Li, Y. (2015). Green chemistry for nanoparticle synthesis. *Chemical Society Reviews, 44*, 5778–5792.

Dubey, S. P., Lahtineb, M., & Sillanpaa, M. (2010). Green synthesis and characterization of silver and gold nanoparticles using leaf extract of *Rosa rugosa. Colloids and Surfaces, A: Physiochemical and Engineering Aspects, 364*, 34–41.

Dudek, N. K., Sun, C. L., & Burstein, D. (2017). Novel microbial diversity and functional potential in the marinemammal oral microbiome. *Current Biology, 27*(24), 3752–3762.

Ebrahimzadeh, M. A., Naghizadeh, A., Mohammadi-Aghdam, S., et al. (2020). Enhanced catalytic and antibacterial efficiency of biosynthesized Convolvulus fruticosus extract capped gold nanoparticles (CFE@AuNPs). *Journal of Photochemistry and Photobiology B: Biology, 209*, 111949.

El Domany, E. B., Essam, T. M., Ahmed, A. E., & Farghali, A. A. (2018). Biosynthesis physicochemical optimization of gold nanoparticles as anti-cancer and synergetic antimicrobial activity using Pleurotus ostreatus fungus. *Journal of Applied Pharmaceutical Science, 8*, 119–128.

El Sayed Manal, T., & El Sayed Ashraf, S. A. (2020). Biocidal activity of metal nanoparticles synthesized by Fusarium solani against multidrug-resistant bacteria and mycotoxigenic fungi. *Journal of Microbiology and Biotechnology, 30*(2), 226–236.

El-Kassas, H. Y., & El-Sheekh, M. M. (2014). Cytotoxic activity of biosynthesized gold nanoparticles with an extract of the red seaweed Corallina officinalis on the MCF-7 human breast cancer cell line. *Asian Pacific Journal of Cancer Prevention, 15*, 4311–4317.

Fawcett, D., Verduin, J. J., Shah, M., et al. (2017). A review of current research into the biogenic synthesis of metal and metal oxide nanoparticles via marine algae and seagrasses. *Journal of Nanoscience, 2017*, 1–15.

Gadd, G. M. (2007). Geomycology: Biogeochemical transformations of rocks, minerals, metals and radionuclides by fungi, bioweathering and bioremediation. *Mycological Research, 111*(1), 3–49.

Gahlawat, G., & Choudhury, A. R. (2019). A review on the biosynthesis of metal and metal salt nanoparticles by microbes. *RSC Advances, 9*, 12944.

Gemishev, O., Panayotova, M. I., Mintcheva, N., et al. (2019). A green approach for silver nanoparticles preparation by cell-free extract from Trichoderma reesei fungi and their characterization. *Materials Research Express, 6*(9), 095040.

Ghosh, P., Han, G., De, M., et al. (2008). Gold nanoparticles in delivery applications. *Advanced Drug Delivery Reviews, 60*, 1307–1315.

Han, M. S., Lytton-Jean, A. K., & Mirkin, C. A. (2006). A gold nanoparticle based approach for screening triplex DNA binders. *Journal of the American Chemical Society, 128*(15), 4954–4955.

Hu, D., Yu, S., Yu, D., Liu, N., et al. (2019). Biogenic Trichoderma harzianum-derived selenium nanoparticles with control functionalities originating from diverse recognition metabolites against phytopathogens and mycotoxins. *Food Control, 106*, 106748.

Husseiney, M. I., El-Aziz, M. A., Badr, Y., & Mahmoud, M. A. (2007). Biosynthesis of gold nanoparticles using Pseudomonas aeruginosa. *Spectrochimica Acta A, 67*, 1003–1006.

Ijaz, F., Shahid, S., Khan, S. A., et al. (2017). Green synthesis of copper oxide nanoparticles using Abutilon indicum leaf extract: Antimicrobial, antioxidant and photocatalytic dye degradation activities. *Tropical Journal of Pharmaceutical Research, 16*(4), 743–753.

Iravani, S. (2014). Bacteria in nanoparticle synthesis: Current status and future prospects. *International Scholarly Research Notices, 2014*, 1–18.

Jain, D., Kumar, D. H., Kachhwaha, S., & Kothari, S. L. (2009). Synthesis of plant-mediated silver nanoparticles using papaya fruit extract and evaluation of their antimicrobial activities. *Digest Nanomaterials and Biostructures, 4*(3), 557–563.

Jha A. K., & Prasad K. (2011). Green fruit of chili (Capsicum Annum L.) synthesizes nano silver. *Digest Journal of Nanomaterials and Biostructures, 6*(4), 1717–1723.

Jogee, P. S., Ingle, A. P., & Raj, M. (2017). Isolation and identification of toxigenic fungi from infected peanuts and efficacy of silver nanoparticles against them. *Food Control, 71*, 143.

Kadam, V. V., Ettiyappan, J. P., & Balakrishnan, R. M. (2019). Mechanistic insight into the endophytic fungus mediated synthesis of protein capped ZnO nanoparticles. *Materials Science and Engineering B, 243*, 214–221.

Kar, P. K., Murmu, S., Saha, S., et al. (2014). Anthelmintic efficacy of gold nanoparticles derived from a phytopathogenic fungus, *Nigrospora oryzae*. *PLoS One, 9*(1), 84693.

Karthik, L., Kumar, G., Kirthi, A. V., et al. (2014). *Streptomyces* sp. LK3 mediated synthesis of silver nanoparticles and its biomedical application. *Bioprocess and Biosystems Engineering, 37*, 261–267.

Khan, S. A., & Lee, C. S. (2020). Green biological synthesis of nanoparticles and their biomedical applications. In A. A. Inamuddin (Ed.), *Applications of nanotechnology for green synthesis, nanotechnology in the life sciences*. Cham: Springer.

Khan, S. A., Noreen, F., Kanwal, S., et al. (2018). Green synthesis of ZnO and Cu-doped ZnO nanoparticles from leaf extracts of Abutilon indicum, Clerodendrum infortunatum, Clerodendrum inerme and investigation of their biological and photocatalytic activities. *Material Science and Engineering C, 82*, 46–59.

Kim, K. J., Sung, W. S., Suh, B. K., et al. (2009). Antifungal activity and mode of action of silver nano-particles on *Candida albicans*. *Biometals, 22*, 235–242.

Kim, D. Y., Saratale, R. G., Shinde, S., et al. (2018). Green synthesis of silver nanoparticles using *Laminaria japonica* extract: Characterization and seedling growth assessment. *Journal of Cleaner Production, 172*, 2910–2918.

Kitching, M., Choudhary, P., Inguva, S., et al. (2016). Fungal surface protein mediated one-pot synthesis of stable and hemocompatible gold nanoparticles. *Enzyme and Microbial Technology, 95*, 76–84.

Kora, A. J., Beedu, S. R., & Jayaraman, A. (2012). Size-controlled green synthesis of silver nanoparticles mediated by gum ghatti (*Anogeissus latifolia*) and its biological activity. *Organic and Medicinal Chemistry Letters, 2*, 17–85.

Kralj, S., & Makovec, D. (2015). Magnetic assembly of superparamagnetic iron oxide nanoparticle clusters into nanochains and nanobundles. *ACS Nano, 9*(10), 9700–9707.

Kulkarni, N., & Muddapur, U. (2014). Biosynthesis of metal nanoparticles: A review. *Journal of Nanotechnology, 2014*, 1–8.

Kumar, Sowmya, B., & Geetha, R. (2019). Preparation of yeast mediated semiconductor nanoparticles by candida albicans and its bactericidal potential against Salmonella typhi and Staphylococcus aureus. *International Journal of Pharmaceutical Sciences and Research, 10*(2), 861–864.

Lin, J., Zhang, Y., Qian, J., & He, S. (2014). A nano-plasmonic chip for simultaneous sensing with dual-resonance surface-enhanced Raman scattering and localized surface plasmon resonance. *Laser and Photonics Review, 8*, 610–616.

Makarov, V. V., Love, A. J., Sinitsyna, O. V., et al. (2014). Green nanotechnologies: Synthesis of metal nanoparticles using plants. *Acta Naturae, 6*, 35–44.

Malam, Y., Lim, E. J., & Seifalian, A. M. (2011). Current trends in the application of nanoparticles in drug delivery. *Current Medicinal Chemistry, 18*, 1067–1078.

Mann, S. (1993). Molecular tectonics in biomineralization and biomimetic materials chemistry. *Nature, 365*, 499–505.

Mazloom, M. S., Hemmati-Sarapardeh, A., Husein, M. M., et al. (2020). Application of nanoparticles for asphaltenes adsorption and oxidation: A critical review of challenges and recent progress. *Fuel, 279*, 117763.

Mitra, S., Patra, P., Pradhan, S., et al. (2015). Detailed antifungal mode of action study: Understanding the insights into oxidative stress. *Journal of Colloid and Interface Science, 444*, 97.

Mukherjee, S., Chowdhury, D., Kotcherlakota, R., et al. (2014). Potential theranostics application of bio-synthesised silver nanoparticles (4-in-1 system). *Theranostics, 4*, 316–335.

Nasrollahzadeh, M., & Sajadi, M. S. (2015). Green synthesis of copper nano-particles using Ginkgo biloba L. leaf extract and their catalytic activity for the Huisgen [3+2] cycloaddition of azides and alkynes at room temperature. *Journal of Colloid and Interface Science, 457*, 141–147.

Noruzi, M. (2015). Biosynthesis of gold nanoparticles using plant extracts. *Bioprocess and Biosystems Engineering, 38*, 1–14.

Omran, B. A., Nassar, H. N., Younis, S. A., et al. (2019). Physiochemical properties of Trichoderma longibrachiatum DSMZ 16517-synthesized silver nanoparticles for the mitigation of halotolerant sulphate-reducing bacteria. *Journal of Applied Microbiology, 126*(1), 138–154.

Pal, G., Rai, P., & Pandey, A. (2019). Green synthesis of nanoparticles: A greener approach for a cleaner future. In *Green synthesis, characterization and applications of nanoparticles* (pp. 1–26). Amsterdam: Elsevier.

Patra, S., et al. (2015). Green synthesis, characterization of gold and silver nanoparticles and their potential application for cancer therapeutics. *Materials Science and Engineering C: Materials for Biological Applications, 53*, 298–309.

Patrick, T. S., Naphtaly, C., Moro Ouma Stephanus, M. O., et al. (2019). Application of nanoparticles in biofuels, An overview. *Fuel, 237*, 380–397.

Pestovsky, Y. S., & Martínez-Antonio, A. (2017). The use of nanoparticles and Nanoformulations in agriculture. *Journal of Nanoscience and Nanotechnology, 17*, 8699–8630.

Philip, D., Unni, C., Aromal, A., et al. (2011). Murraya Koenigii leaf-assisted rapid green synthesis of silver and gold nanoparticles. *Spectrochimica Acta. Part A, Molecular and Biomolecular Spectroscopy, 78*, 899–904.

Poopathi, S., De Britto, L. J., Praba, V. L., et al. (2015). Synthesis of silver nanoparticles from *Azadirachta indica*—A most effective method for mosquito control. *Environmental Science and Pollution Research, 22*, 2956–2963.

Pradhan, S., Roy, I., Lodh, G., et al. (2013). Entomotoxicity and biosafety assessment of PEGylated acephate nanoparticles: A biologically safe alternative to neurotoxic pesticides. *Journal of Environmental Science and Health, Part B, 48*, 559.

Pugazhendhi, A., Prabhu, R., Muruganantham, K., Shanmuganathan, R., & Natarajan, S. (2019). Anticancer, antimicrobial and photocatalytic activities of green synthesized magnesium oxide nanoparticles (MgONPs) using aqueous extract of Sargassum wightii. *Journal of Photochemistry and Photobiology, 190*, 86–97.

Rai, M., Ingle, A. P., Birla, S., et al. (2015). Strategic role of selected noble metal nanoparticles in medicine. *Critical Review of Microbiology*. Informa Healthcare USA early online, pp. 1–24.

Rajeshkumar, S., Ponnanikajamideen, M., Malarkodi, C., et al. (2014). Microbe-mediated synthesis of antimicrobial semiconductor nanoparticles by marine bacteria. *Journal of Nanostructure and Chemistry, 4*(2), 96.

Ramalingmam, P., Muthukrishnan, S., & Thangaraj, P. (2015). Biosynthesis of silver nanoparticles using an endophytic fungus, Curvularialunata and its antimicrobial potential. *Journal of Nanoscience and Nanoengineering, 1*, 241–247.

Rangayasami, S., Kannan, K., Joshi, S., & Subban, M. (2020). Bioengineered silver nanoparticles using Elytraria acaulis (L.f.) Lindau leaf extract and its biological applications. *Biocatalysis and Agricultural Biotechnology, 27*, 101690.

Ravishankar R. V. and Jamuna B. A. (2011). Nanoparticles and their potential application as antimicrobials Science against microbial pathogens, communicating current research and technological advance A, Méndez_Vilas (Ed) 197–209.

Razavi, M., Salahinejad, E., Fahmy, M., et al. (2015). Green chemical and biological synthesis of nanoparticles and their biomedical applications. In *Green processes for nanotechnology* (pp. 207–235). Cham: Springer.

Reidy, B., Haase, A., Luch, A., et al. (2013). Mechanisms of silver nanoparticle release, transformation and toxicity: A critical review of current knowledge and recommendations for future studies and applications. *Materials, 6*, 2295–2350.

Renuka, R., Renuka, D. K., Sivakami, M., et al. (2020). Biosynthesis of silver nanoparticles using *Phyllanthus emblica* fruit extract for antimicrobial application. *Biocatalysis and Agricultural Biotechnology, 24*, 101567.

Riddin, T. L., Gericke, M., & Whiteley, C. G. (2006). Analysis of the inter-and extracellular formation of platinum nanoparticles by Fusarium oxysporum f. sp. lycopersici using response surface methodology. *Nanotechnology, 17*(14), 3482.

Ruffini Castiglione, M., Giorgetti, L., Cremonini, R., et al. (2014). Impact of $TiO_2$ nanoparticles on *Vicia narbonensis* L.: Potential toxicity effects. *Protoplasma, 251*, 1471–1479.

Saravanakumar, K., Shanmugam, S., Varukattu, N. B., et al. (2019). Biosynthesis and characterization of copper oxide nanoparticles from indigenous fungi and its effect of photothermolysis on human lung carcinoma. *Journal of Photochemistry and Photobiology. B, 190*, 103–109.

Selvakumar, R., Seethalakshmi, N., Thavamani, P., et al. (2014). Recent advances in the synthesis of inorganic nano/microstructures using microbial biotemplates and their applications. *RSC Advances, 4*(94), 52156–52169.

Sengani, M., Grumezescu, A. M., & Rajeswari, V. D. (2017). Recent trends and methodologies in gold nanoparticle synthesis–a prospective review on drug delivery aspect. *Open Nano, 2*, 37–46.

Shaik, M. R., Khan, M., Kuniyil, M., et al. (2018). Plant-extract-assisted green synthesis of silver nanoparticles using Origanum vulgare L, extract and their microbicidal activities. *Sustainability, 10*, 913.

Singh, R., & Nalwa, H. S. (2011). Medical applications of nanoparticles in biological imaging, cell labeling, antimicrobial agents, and anticancer Nanodrugs. *Journal of Biomedical Nanotechnology, 7*(4), 489–503.

Singh, P., Kim, Y. J., Zhang, D., & Yang, D. C. (2016). Biological synthesis of nanoparticles from plants and microorganisms. *Trends in Biotechnology, 34*(7), 588–599.

Solgi, M. (2012). Application of nanotechnology and "Smart Packaging" in marketing and postharvest of cut flowers. In *First Nanotechnology and Its Application in Agriculture and Natural Resources Conference*, May 15–16, Karaj, Iran.

Solgi, M. (2014). Evaluation of plant-mediated silver nanoparticles synthesis and its application in postharvest physiology of cut flowers. *Physiology and Molecular Biology of Plants, 20*, 279–285.

Soni, N., & Prakash, S. (2015). Antimicrobial and mosquitocidal activity of microbial synthesized silver nanoparticles. *Journal of Parasitology Research, 114*(3), 1023–1030.

Suryavanshi, P., Pandit, R., Gade, A., et al. (2017). Colletotrichum sp.-mediated synthesis of sulphur and aluminium oxide nanoparticles and its in vitro activity against selected food-borne pathogens. *LWT - Food Science and Technology, 81*, 188–194.

Xiao, H., Xu, L., Xiao, Z., et al. (2019). Biological metabolism synthesis of metal oxides nanorods from bacteria as a biofactory toward high-performance lithium-ion battery anodes. *Small, 15*(38), e1902032.

Zhang, Y. J., Huang, R., Zhu, X. F., et al. (2012). Synthesis, properties, and optical applications of noble metal nanoparticle-biomolecule conjugates. *Materials Science, 57*, 238–246.

Zinjarde, S., Apte, M., Mohite, P., & Kumar, A. R. (2014). Yarrowia lipolytica and pollutants: Interactions and applications. *Biotechnology Advances, 32*(5), 920–933.

# Chapter 7
# Nanomaterial-Based Bio Scaffolds for Enhanced Biomedical Applications

**Bernadette K. Madathil, P. R. Anil Kumar, Vinoy Thomas, and Shiny Velayudhan**

## 7.1 Introduction

Tissue engineering and regenerative medicine aim to replace, repair, or regenerate damaged tissues and organs and involve the in vitro and in vivo use of scaffolds as substrates or support for cell growth and delivery. These scaffolds are fabricated using synthetic materials mainly polymers like polylactic acid (PLA), polyvinyl chloride (PVC), polylactic glycolic acid (PLGA), polycaprolactone (PCL), or natural biological materials like collagen, chitosan, alginate, fibronectin, etc. Alternatively, decellularized autologous, allogeneic, or xenogenic tissues have been used in tissue engineering as scaffolds (Costa et al. 2017). Bioscaffolds make use of biocompatible, biodegradable, and bioresorbable materials that serve as substrates onto which cells are seeded to form a three-dimensional (3D) structure that is comparable to the tissue in which it is intended to be implanted. In order to enhance

B. K. Madathil
Advanced Center for Tissue Engineering, Department of Biochemistry, University of Kerala, Thiruvananthapuram, Kerala, India

P. R. Anil Kumar
Department of Applied Biology, Biomedical Technology Wing, Sree Chitra Tirunal Institute for Medical Sciences and Technology, Thiruvananthapuram, Kerala, India

V. Thomas
Department of Materials Science and Engineering, Center for Nano-scale Materials & Biointegration (CNMB), University of Alabama at Birmingham (UAB), Birmingham, AL, USA

S. Velayudhan (✉)
Division of Dental Products, Department of Biomaterial Science and Technology, Biomedical Technology Wing, Sree Chitra Tirunal Institute for Medical Sciences and Technology, Thiruvananthapuram, Kerala, India
e-mail: shiny@sctimst.ac.in


K. Pal (ed.), *Bio-manufactured Nanomaterials*,
https://doi.org/10.1007/978-3-030-67223-2_7

degradability, control porosity, modulate mechanical properties, and affect the desired biological interaction with the host, bioscaffolds with nanostructures are designed to better emulate the nano-architecture of native tissues and organs.

Nanotechnology is a rapidly evolving field of research that deals with the synthesis of materials of 1–100 nm size. The physicochemical properties of materials vary from the micro- to the nanoscale. Nanomaterials can have any or all of their dimensions within the nanoscale range and are classified based on the number of dimensions that are not within the nanoscale. One-dimensional (1D) nanostructured materials like nanorods, nanofibers, nanowires, and nanotubes have one dimension of their structure that is not confined to the nanoscale. Two-dimensional (2D) nanostructured materials like nanocoatings, nanoflowers, nano-films, and nanolayers have two dimensions of their structure outside the nanoscale. Nanoparticles, however, have all dimensions of their structure within the nanoscale and are called zero-dimensional (0D) structures (Tripathi and Chung 2018). Three-dimensional (3D) nanomaterials are called bulk nanomaterials wherein typical hierarchical 3D nanoarchitectures with abundant porosity are formed using nanomaterials as building blocks (Arulman et al. 2018). Dispersions of nanoparticles, nano-sized bundles and multilayers of nanowires or nanotubes are some of the examples.

It has been documented that cells interact with their biological milieu in the nanoscale level and also produce nanostructured extracellular matrix (ECM) moieties. Hence, the biomimetic and physiochemical behaviors of nanomaterials play a pivotal role in directing cell–material interactions, cellular proliferation, migration, differentiation, and tissue regeneration. The proteins in the micro-environment adhere to the nanomaterial to form a protein corona that dictates the downstream biomolecular processes (Bhattacharya et al. 2016). Bioscaffolds have been fabricated from nanoparticles, nanofibers, nanotubes, nanocomposites, and hydrogels to better mimic the nanostructure of the ECM (Razavi 2017). There have been reports that modulating and modifying nanoparticle surface chemistry can effect changes in its nanostructural assembly, protein adsorption and downstream interactions at the biointerface (Gagner et al. 2012).

This chapter aims to review the different types of nanomaterial based bioscaffolds, their characterization, and the recent advances and milestones in the application of these bioscaffolds to biomedical application. The use of cutting-edge technologies to bioengineer nanomaterials that can induce desired biological effects will also be briefly discussed. Finally, concluding remarks on future perspectives have been summarized.

## 7.2 Basic Principles of Nanomaterial-Based Scaffolds

Scaffolds for tissue engineering must be developed keeping in mind the basic requirements of biocompatibility, biodegradability, and ability to elicit minimal immune reactions. A scaffold is expected to maintain a three-dimensional architecture, promote the growth of cells specific to the tissue it is intended to be implanted

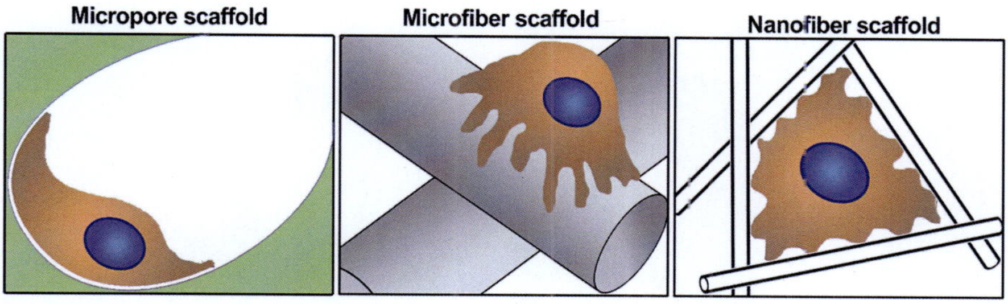


**Fig. 7.1** Schematics of cell morphology on a micropore scaffold, microfiber scaffold, and a nanofiber scaffold

in thus encouraging the development of functional neo-tissues (Kim and Mooney 1998). However, in the past most bioengineered scaffolds failed to recapitulate the dynamic and hierarchically organized nanoscaled extracellular environment (Dvir et al. 2011a).

In this scenario, the nanoscaled configuration of the ECM plays a key role in promoting and directing cellular interactions. It has been well documented that a scaffold with nanofibrous topography/architecture increases cell–material and cell–cell interactions due to the virtue of the vastly increased surface area to volume ratio (Fig. 7.1). Hence, it becomes imperative to engineer advanced biomaterials with nanostructured features to better emulate the nanoarchitecture of the ECM which would direct the formation of complex tissues/organs when compared to traditional bulk biomaterials.

## 7.3 Biomedical Nanomaterials for Bioscaffolds

In tissue engineering, biomedical nanomaterials form the basic building blocks from which scaffolds are fabricated. They are categorized as metallic nanomaterials, carbon nanotubes, nanoceramics, nanoparticles for biomedical imaging, and nanofibers, based on their structural nature and applications. Figure 7.2 depicts various biomedical applications of different nanomaterials. A brief insight on these nanomaterials is also given in the following sections.

### *7.3.1 Metallic Nanomaterials*

Metallic nanostructures such as copper, titanium, silver, gold, and oxides of iron and zinc find use in cancer diagnostics, therapeutics, and also tissue engineering of hard tissues. Metallic nanomaterials are the researcher's favorite as they have a plethora of interesting properties. Apart from ease of fabrication, surface chemistry of metallic nanoparticles allows for the attachment of biological moieties such as antibodies,

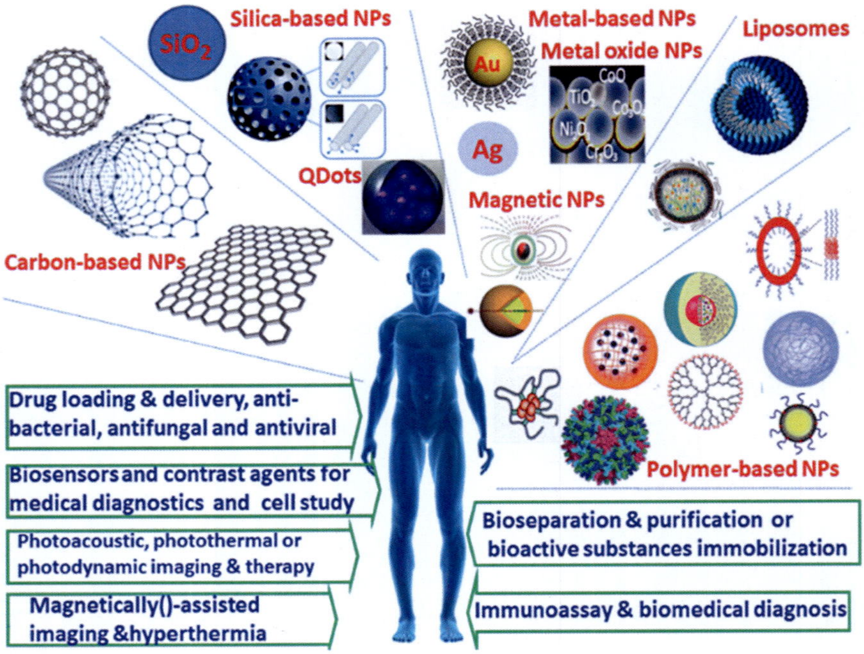


**Fig. 7.2** Various biomedical applications of different nanomaterials. (Adapted from Zhao et al. 2015)

peptides, nucleic acids, and polymer to the metal. The optically tunable properties of metallic nanomaterials enable the conversion of light or radio frequencies to heat, resulting in thermal ablation of the target tissues, allowing for simultaneous diagnosis and photothermal therapy in the patient (Lee and El-Sayed 2006; Jain et al. 2008; Day et al. 2009; Chen et al. 2010).

Nanostructured gold has been extensively used in medical applications owing to its low toxicity and conductivity. Alginate scaffolds reinforced with gold nanowires have demonstrated improved electrical conductivity and also enhanced tissue thickness and cellular alignment, enabling the cells to contract with electrical stimuli between adjacent cardiac cells (Dvir et al. 2011b). In a recent study, gold nanoparticle (Au-NPs)-integrated nanofibrous polymer scaffolds were fabricated and biological performance evaluated using various stem cells (Yadid et al. 2019). It was found that presence of gold nanoparticle facilitated improved stem cell proliferation and differentiation, thus making (Au-NPs) integrated polymer scaffolds a promising multimodal tool for tissue engineering and regenerative medicine.

Nanometals also find application in bone and cartilage regeneration. It has been reported that nanophase metals of Ti, $Ti_6Al_4V$ alloy, and CoCrMo alloy aid in improved osteoblast adhesion to the surface due to the increased surface particle boundaries in nanophase materials (Webster and Ejiofor 2004).

### 7.3.2 Carbon Nanotubes and Nanoceramics

Carbon nanotubes (CNTs) and single, double, or multi-wall structures are comprised of graphite sheet/sheets rolled into a cylinder. Due to its hexagonal structure and π electron conjugation, carbon nanotubes exhibit high mechanical strength, flexibility, and electrical conductivity (Iijima et al. 1992; Hummer et al. 2001). CNTs are the commonly used organic nanomaterials in tissue engineering and regenerative medicine owing to their excellent biocompatibility, superior mechanical properties, and exceptional chemical inertness (Veetil and Ye 2009; Dvir et al. 2011a). Coatings of CNT used over bone scaffolds not only improve their mechanical performance (Yu et al. 2000) but also assist in osteoblast adhesion (Nayak et al. 2010). CNTs exhibit excellent blood compatibility and are used in cardiopulmonary applications (Meng et al. 2005). CNTs have been used to design conductive substrates, especially in neural tissue engineering as they possess good mechanical and electrical properties (Mazzatenta et al. 2007). It has been reported that culturing neural stem cells (NSC) on CNT substrates not only increased the neural outgrowth but also promoted the expression of microtubule-associated protein 2 (MAP-2) (Dowell-Mesfin et al. 2004). Hence, CNTs are being used as culture substrates for enhancing the neuronal differentiation of NSCs (Hu et al. 2004; Kam et al. 2009). Apart from NSCs, CNT substrates have shown to elevate the differentiation of human embryonal stem cells (hESCs) toward neuronal lineages up to twofolds than when cultured on poly(L-ornithine) substrates (Chao et al. 2009).

Nanoceramics were first discovered in the early 1980s and are classified as inorganic, heat-resistant, nonmetallic solids that are less than 100 nm in diameter. Nanoceramics possess unique and often improved electric, magnetic, mechanical, and surface characteristics. By the virtue of their mechanical (strength, toughness) and surface characteristics (superplasticity, bioactivity), nanoceramics are used as coating materials for implants used in orthopedics and dentistry. The nanophase ceramics coatings enables modification of the surface properties to improve performance, reliability and biocompatibility and have been found to increase osteoblast adhesion when compared to non-nanophase materials in vitro (Webster et al. 1999). The enhanced adhesion of osteoblasts has been attributed to the grain size of the ceramic, suggesting the potential of nanoceramics to mimic the surface grain size of natural bone, facilitating protein interactions, and promoting osteoblast adhesion.

### 7.3.3 Nanoparticles in Medical Imaging

Biological investigations using engineered nanoparticles have increased exponentially in recent years. Biomedical imaging enables the real-time, non-invasive evaluation of the function of engineered tissues and monitoring of cell–cell and cell–ECM interactions at the molecular and organelle level (Nune et al. 2009). At the nanometer scale, materials depict unique physical, chemical, and optical properties and hence can be appropriately tuned for applications such as molecular imaging, diagnosis, or treatment of cancer (Lee and Chen 2009). Although molecular imaging applications using nanoparticles are being adopted into clinical applications, their shelf life, stability, and long-term toxicology continue to be challenges that have to be addressed.

### 7.3.4 Nanofibers

Nanofibers are fibers having diameters within the range of 1–100 nm. Nanofibers have a large surface area and high aspect ratio, which make them structurally very similar to the native cellular microenvironment. In addition, they also possess superior surface properties and the ability to fast-absorb biomolecules, which provides numerous binding sites to cellular receptors enabling stronger cell–matrix interactions (Stevens 2005). These attributes make them more efficient for tissue engineering applications as compared to their micro- and macro-scale counterparts (Yang et al. 2005). Furthermore, tailoring fiber dimension, orientation, and alignment can modulate the functional properties of the nanofiber scaffolds such as mechanical strength, morphology, porosity, structural integrity, and chemical functionality.

Nitti et al. reported that electrospun mats fabricated using rotating drum collector possessed significantly higher Young's modulus and the stress at break along the fiber direction rather than in the direction perpendicular to fibers, demonstrating the anisotropic mechanical behavior of these aligned fibers mats. This architecture is similar to native biological tendon and ligaments which exhibit greater strength along the fibers direction (Nitti et al. 2018).

Fiber orientation is also known to control cell orientation and tissue growth, influences cell proliferation, and is found to dictate functions of structure and organization of certain tissues (muscle fibers). Choi et al. fabricated PCL/collagen-based nanofibers and examined the ability of these electrospun nanofibers to guide morphogenesis of skeletal muscle cells and promote cellular organization. Their studies confirmed that uni-directionally oriented nanofibers were able to enhance muscle cell alignment and myotube formation when compared to randomly oriented nanofibers. Based on this study, it was concluded that the aligned nanofiber scaffolds seeded with skeletal muscle cells holds potential as implantable functional muscle tissues for patients with large muscle defects (Choi et al. 2008).

Nanofibrous structures are known to regulate the focal adhesions of cells seeded on the scaffolds by providing different nano-topographical cues for the cells. The distribution and organization of the cytoskeletal proteins and focal adhesions of endothelial cells on electrospun PCL/collagen scaffolds with different fiber diameters revealed that cells showed a better-developed cytoskeletal organization and improved focal adhesion when seeded on nanoscaled fibers (0.5μm) than on fibers with larger diameters. It was later established that when cells are seeded on scaffolds with nanosized fibers, they organize on the fibers by spreading and attaching to adsorbed proteins at multiple focal points (Elias et al. 2002; Xu et al. 2004).

## 7.4 Fabrication of Nanomaterial-Based Bioscaffolds

One of the most common methods adopted is to create nanomaterial-based bioscaffold which involves reinforcement of a bioscaffold with a nanomaterial. In this method, nanomaterials are synthesized by adopting either a "top-down" (large materials are broken down into small nano-sized particles) or a "bottom-up" (materials are grown by stacking molecules to form clusters and finally into nanomaterials) approach and subsequently incorporated into polymer scaffold/template to give rise to nanomaterial-based composite bioscaffolds. Nanomaterials are used as reinforcement for the enhancement of biological, electrical, and mechanical properties, for patterning of the substrate to facilitate the growth of various types of cells, and for molecular detection and biosensing. The polymer scaffolds for this purpose are fabricated by conventional methods of solvent casting, phase separation, compression molding, gas foaming, polymer etching, particle leaching, freeze drying, and electrospinning. In recent decades, additive manufacturing (AM) techniques such as rapid prototyping (RP), three-dimensional (3D) bioprinting and fused deposition modeling (FDM), and nanotechnology-enabled approaches such as self-assembly methods (Parisi et al. 2018), and various nanospinning approaches (Thomas et al. 2006) have also been employed for scaffold fabrication.

Besides nano-sized reinforcements, alternative methods focus on controlling the structure of scaffold at micro/nanoscale to emulate the size scale of extracellular matrix (ECM). In this approach, the focus is on the manipulation of the scaffolding techniques so as to engineer scaffold architecture and morphology very similar to the native ECM. Several fabrication techniques have been attempted to date; however, only three, i.e., electro-spinning, self-assembly, and phase separation, are being used in the fabrication of nanofibrous scaffolds for use in tissue engineering (Fig. 7.3). The following section discusses the fabrication of scaffolds that employs both the methods simultaneously to generate tissue engineering scaffolds.

### *7.4.1 Electrospinning*

Electrospinning (ES) is a process that is frequently used for the generation of nanofibers by virtue of its ease of technique, versatility, and cost-efficiency. The high surface area and porosity make electrospun nanofibers ideal scaffolds for tissue engineering (Thomas et al. 2006). The typical set up of ES is shown in Fig. 7.3. In ES, an electric field is used to create a charge on the surface of a solution that produces a force opposing its surface tension (Doshi and Reneker 1995). The electric field is increased till the solution is drawn from a tip to form Taylor cone and eventually a jet is shot out from the tip of the cone to a collector plate. The solvent evaporates, and continuous non-woven polymer fibers thus generated have diameters ranging from tens of nanometers to micrometers in diameter.

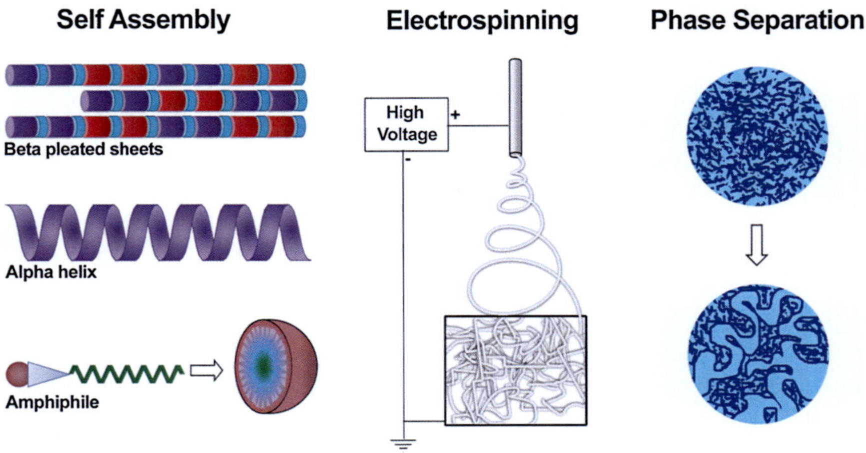


**Fig. 7.3** Schematic of the electrospinning, phase separation, and self-assembly process for the fabrication of scaffolds with a fibrous structure. (Adapted from Wade and Jason 2014)

The parameters such as polymer chain length, polymer concentration, polymer and solvent chemical properties, flow rate, applied voltage, and collector distance are optimized for different polymers (Thomas et al. 2011). Solution viscosity is an important parameter that directly affects the fiber size and morphology. Depending on the concentration of polymer dissolved in a solvent, the polymer solution formed can be classified as (a) dilute (polymer chains in solution rarely interact with each other); (b) semi-dilute I (polymer chains interact at a critical concentration); (c) semi-dilute II (polymer chains are entangled); and (d) concentrated (polymer chains entangle very tightly) (Sperling 2005). For the formation of smooth fibers, the polymer chains in the solution should entangle. The concentration at which entanglement takes place is called entanglement concentration (Ce). As a general rule, when the concentration of the polymer solution exceeds Ce, smooth and continuous fibers will be formed (Tiwari and Venkatraman 2012; Gupta et al. 2005). The molecular weight of polymer also plays a vital role in the solution viscosity. Generally, solutions containing low-molecular weight polymers give rise to beads rather than fibers, whereas fibers with larger diameters are formed with high-molecular weight polymer solutions (Koski et al. 2004; Wen et al. 2005).

The quality of the fiber produced depends also on the surface tension of the solvent used (Agarwal et al. 2008). While a solution with high surface tension destabilizes the polymer jet resulting in dispersion of droplets, instead of fibers (Rezaei et al. 2015), a lower surface tension leads to the formation of fibers at electric fields (Haghi and Akbari 2007). Apart from these the other factors such as electrical conductivity of the solvent (Li and Wang 2013; Mituppatham et al. 2004; Angammana and Jayaram 2011), polymer solution flow rate (Greiner and Wendorff 2007), and processing parameters (Sun et al. 2014; Reneker and Chun 1996; Thomas et al. 2011) also affect the fiber quality of the electrospun fiber (Table 7.1).

**Table 7.1** Electrospinning parameters affecting fiber morphology

| Parameter | Effect on fiber morphology |
|---|---|
| Increase in solution concentration | Increase in fiber diameter |
| Decrease in solution concentration | Electrospraying; bead formation |
| Increase in applied voltage | Decrease in fiber diameter; bead formation |
| Decrease in applied voltage | Fiber structure defect |
| Increase in flow rate | Increase in fiber diameter and porosity |
| Decrease in flow rate | Obstruction in capillary tip |
| Increase in temperature | Obstruction in capillary tip |
| Increase in humidity | Porous fibers |
| Increase in distance between capillary tip and collector | Process difficulties |

Over the last decade, many researchers have experimented with different ES parameters and have created electrospun scaffolds using varieties of synthetic (Xin et al. 2007; Liu et al. 2017; Qi et al. 2012; Rezabeigi and Demarquette 2019; Boland et al. 2004; Thomas et al. 2006) as well as natural (Jayakumar et al. 2010; Tao et al. 2020; Wakuda et al. 2018; Sell et al. 2009; Wang et al. 2019) polymers. These scaffolds were used as medical prosthesis, wound dressings, drug delivery matrixes, and for tissue engineering (Al-Enizi et al. 2018). A few of these are mentioned below.

Yang et al. (2005) prepared a highly aligned PLA nanofibrous scaffold using the ES method for neural tissue engineering applications. The researchers used a mix of solvent dichloromethane (DCM) and dimethylformamide (DMF) in the ratio 70:30 and varied the polymer concentration to optimize the diameter of the electrospun fibers. Duan et al. fabricated a nanofibrous membrane of PLGA and chitosan with varying compositions. The hydrophilicity and crosslinking of chitosan enhanced the biocompatibility and mechanical properties of the PLGA (Duan et al. 2007). Chilarski et al. prepared fibrous scaffolds using dibutyryl chitin (DBC) for wound healing applications and proved that these scaffolds encouraged healing of burn wounds (Chilarski et al. 2007). The following years witnessed many electrospun scaffolds being successfully used as wound dressings (Charernsriwilaiwat et al. 2012; Elzatahry et al. 2012; Li et al. 2012; Rnjak-Kovacina et al. 2012).

Drug-loaded nanofibers can be used (a) as implants (Nguyen et al. 2012), (b) for targeted drug delivery, and (c) as a support system for cellular differentiation and proliferation (Liang et al. 2007). Kenawy et al. (2002) reported the pioneering work on the use of drug-loaded electrospun matrix for the treatment of periodontal disease. Later, many investigators explored various drug delivery systems by (a) combining electrospinning of polymer and drug dissolved in the same solvent (Jannesari et al. 2011; Nguyen et al. 2012; Gunn and Zhang 2010; Ma et al. 2011; Yu et al. 2010), (b) using pH-responsive nanofibers (Demirci et al. 2014), (c) in situ cross-linking (Slemming-Adamsen et al. 2015), egg albumin nanofibers (Zahedi and Fallah-Darrehchi 2015), monolithic core-based core–shell nanofibers (Zupančič et al. 2016), cross-linked cellulose nanofibers (Fiorati et al. 2017), and hyperthermia nanofibers (Kim et al. 2013), with the aim of enhancing drug loading, cell attachment, and mass transfer performance.

Another important application of electrospun nanofibrous scaffolds are in tissue engineering, which is dealt in detail under Sect. 7.6.

### *7.4.2 Phase Separation Method*

Thermally induced phase separation (TIPS) is also a widely used technique for the fabrication of highly porous and interconnected scaffolds and is based on the change in temperature to effect a demixing of the homogenous polymer solution to create polymer-enriched and polymer-deficient phases. The architecture of the scaffold is modulated by varying parameters such as polymer concentration, solvent/non-solvent ratio, and temperature (Conoscenti et al. 2017). The TIPS process involves five steps: (a) polymer dissolution, (b) phase separation and gelation, (c) solvent extraction, (d) freezing, and (e) freeze-drying under vacuum (Smith et al. 2009). Li et al. used TIPS for the synthesis of electroactive nanofibrous scaffolds using polylactide and electroactive tetraniline-polylactide-tetraaniline. Variation in the diameter of the nanofibers of varying process parameters was also observed. These scaffolds were found to promote proliferation in C2C12 myoblasts (Li et al. 2014).

Thermally induced phase separation can be combined with other processing techniques to obtain 3D nano/bioscaffolds with well-defined porous structures/architecture. The most common techniques adopted for this is to combine TIPS with particle leaching (Zhang and Ma 2000). Aminolysis in combination with TIPS also allows for surface modification of the nanofibers. Chen et al. generated nanofibrous scaffolds from aminolyzed modified PLA, which exhibited enhanced attachment and proliferation of MC3T3-E1 subclone 14 cells (Chen et al. 2016).

Jack et al. (2009) used to modify TIPS for the fabrication and characterization of highly porous nano-sized hydroxyapatite (HA)/poly(hydroxyabutyrate-co-hydroxyvalerate) (PHBV) polymer composite scaffolds (Fig. 7.4). The researchers employed varying processing conditions and found that the pore architectures could be modulated by controlling the phase-separation parameters. They also found that HA particles were dispersed in the pore walls of the scaffolds, and the mechanical properties and the in vitro bioactivity of the scaffolds were improved (Jack et al. 2009).

### *7.4.3 Self-Assembly*

Self-assembly involves the spontaneous hierarchical organization into ordered supramolecular structures. It is mediated through hydrogen bonds, van der Waals forces, electrostatic, and hydrophobic interactions. Nanofibrous structures can be formed using the precepts of self-assembly under stringently regulated micro-environments. The concept of self-organization arises from the chemical

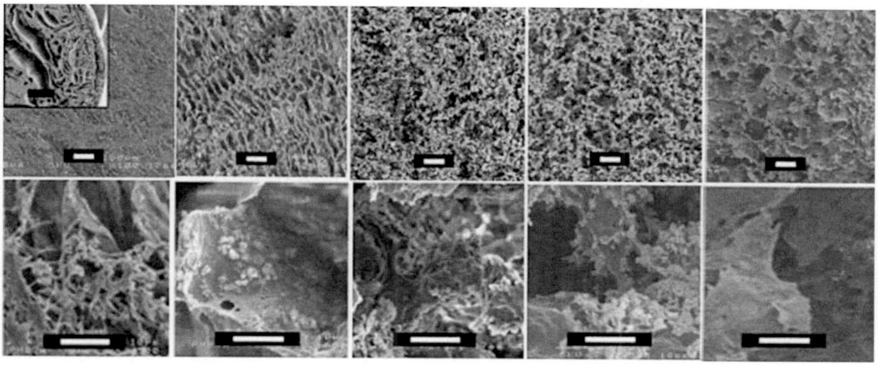

**Fig. 7.4** SEM images of HA/PHBV scaffolds produced from fabricated with dioxane under the different cooling regimes. The scale bars are 100μm (upper row) and 10μm (lower row). The inset shown is from a pure PHBV scaffold using regime I (scale bar = 1 mm) and shows the inhomogeneous nature of this sample

complementarity and structural compatibility of the atoms and molecules that form the building blocks in the fabrication process (Zhang 2003). While many biomolecules like lipids, sugars, nucleic acid, and proteins can self-assemble, it is the peptides and proteins that have been extensively researched. Self-assembled molecules can also be de novo synthesized and includes amphiphilic, cyclic, ionic complementary, and hybrid peptides. Amphiphilic peptides have a hydrophobic tail and a hydrophilic head which undergo self-assembly in aqueous solutions to form nanostructures like nanotubes, nanofibers, nanovesicles, nanorods, and micellar nanostructures (Qiu et al. 2018). These peptides can also be conjugated to alkyl groups which can modulate self-assembly. Ionic complementary peptides have a side group with charged side chains and hydrophobic chains as the other side group. In aqueous medium, the peptide self-assembles with the charged side on the outside and the hydrophobic side chains within the nanofiber. Environmental parameters mainly temperature, pH, ionic strength, and mechanical force modulate the self-assembly process (Sun et al. 2017). Supramolecular hydrogels and nanofibrous structure imbibed in water that is formed by the assembly of molecular building blocks are known as hydrogelators. These hydrogelators may be low molecular weight or polymeric. The low-molecular-weight hydrogelators assemble into structures like nanofibers and helices via non-covalent interactions mainly hydrogen bonding (H bonding), electrostatic interactions, hydrophobic interactions, and van der Waals interactions. Additionally they can also be crosslinked through metal–ligand coordination and host–guest recognition. Supramolecular hydrogels exhibit stimuli-responsive properties and exhibit gel–sol or sol–gel transitions in response to either physical, chemical, biological, or multi-stimuli (Hoque et al. 2019; Webber et al. 2016). These supramolecular hydrogels can be designed through the incorporation of an appropriate stimuli-responsive moiety into the hydrogelator scaffold (Shigemitsu and Hamachi 2017). Due to their biocompatibility and biodegradability supra-molecular hydrogels find numerous biomedical applications ranging from bioimaging to drug delivery and tissue engineering (Dong et al. 2015).

## 7.5 Characterization of Nanomaterial-Based Bioscaffolds

Nanomaterials exhibit different optical, magnetic, electrical, and mechanical properties when they are in their bulk state, this difference being attributed to their small size and sensitivity to external conditions. The physicochemical properties of nanomaterial depend on its size, shape, crystallinity and composition.

### *7.5.1 Structural and Morphological Characterization of Nanomaterials*

The size of the nanomaterial determines whether it falls in the nano or micro scale. Optical microscopes that use visible light allow for resolutions up to 0.2μm. Electron microscopes facilitate magnifications of the order of one million and resolution up to 0.1 nm (Yesilkir-Baydar et al. 2017). The particle size, shape, and aggregation characteristics of nanomaterials are thus routinely measured using electron microscopy like transmission electron microscope (TEM) and scanning electron microscope (SEM). The images are used for the measurement of nanomaterials and clusters. Some of the shapes include spherical, flat, conical, cylindrical, and irregular shapes. Through high-resolution transmission electron microscopy (HRTEM), the crystalloid structures at atomic scales can be analyzed. Here the electron interacts with the specimen, and the electron wave undergoes a phase change and interference as it passes through the imaging system of the microscope's (Yesilkir-Baydar et al. 2017). SEM also provides information on surface topology and morphology and can provide information on composition when coupled to energy dispersive X-ray spectroscopy (EDX).

Scanning probe microscopy techniques namely atomic force microscopy (AFM) and scanning tunneling microscopy (STM) have also been adopted for structural elucidation of nanomaterials. Here a probe is used to scan the surface of the material, and the data obtained is used to generate a real-time 3D surface topography with high spatial resolution. AFM makes use of a cantilever/tip assembly as the probe, and the motion of the probe is monitored via a laser beam that is reflected off the cantilever and tracked by a photosensitive diode. Changes in amplitude give information on the surface topography, while changes in phase differentiate between different types of materials. Therefore, noncontact AFM is used to study surface topography, adhesion forces, viscosity, and rigidity (Ando et al. 2008). STM is based on monitoring the tunneling current generated between a sharp metal probe and the sample surface when a small electric voltage is applied. As the distance between the tip and the sample is a few angstroms, electrons can tunnel across the gap from the tip to the surface to create a tunneling current that is used to generate the 3D image of the surface topography (Ghasempour and Narei 2018). A piezoelectric element attached to the tip controls facilitates a feedback loop system which keeps the distance between the tip and the surface of the specimen constant. AFM is used for characterizing non-conducting materials while STM is preferred for characterizing conductive materials like carbon nanotubes and graphene layers.

The particles in liquid phase are measured using dynamic light scattering (DLS) techniques such as photon correlation spectroscopy (Ealias and Saravanakumar 2017). Dynamic light scattering is also used to determine particle size distribution. Nanoparticles in suspension are in a continuous Brownian movement. Dynamic light scattering measures the light scattering as a function of time and calculates the hydrodynamic radius of the nanoparticle. It helps to delineate the colloidal stability of the nanoparticles and also evaluates particle–particle interaction, and the aggregation property of the nanoparticles as cluster size can be determined.

Nanomaterial size can also be analyzed using differential centrifugal sedimentation (DCS) where particle size is measured based on the sedimentation rate, which depends upon their size and density. DCS has a higher resolution than DLS, both the techniques can however be coupled to get more reliable data on particle size and aggregation. Nanoparticle tracking analysis (NTA) is a newer technique based on light scattering and Brownian movement that is used for measuring particle size and requires very low concentrations of the sample. The NTA software detects single particles by light scattering and tracks single nanoparticles that move under Brownian motion. The distance the particle moves in a given time interval is correlated to its hydrodynamic diameter and gives information on the nanoparticle size. Larger nanoparticles and heavy aggregates move with less speed, while smaller nanoparticles move faster. On compiling and processing this information from a significant number of particles, particle concentration and size distributions can be obtained (Laborda et al. 2016; Mourdikoudis et al. 2018). The size distribution of both mono and poly disperse samples can be obtained (Filipe et al. 2010). Nuclear magnetic resonance (NMR) is also used for measuring the hydrodynamic radius of metal nanoparticles. A schematic representation of the structural and morphological characterization of nanomaterials is given in Fig. 7.5.

### *7.5.2 Characterization of the Chemical Composition of Nanomaterials*

The purity of the nanomaterial is determined by the chemical or elemental composition which also provides information about the presence of some undesired elements that may act as contaminants. The elemental composition measurements and oxidation states are analyzed via X-ray photoelectron spectroscopy (XPS). It also analyses ligand exchange interactions and surface functionalization of nanoparticles (Mourdikoudis et al. 2018). Some techniques involve chemical digestion of the particles followed by wet chemical analysis such as inductively coupled plasma mass spectrometry (ICP-MS), atomic absorption spectroscopy (AAA), and ion chromatography. Fourier transmission infrared spectroscopy (FTIR) is used to examine the functional groups or metabolites present on the surface of

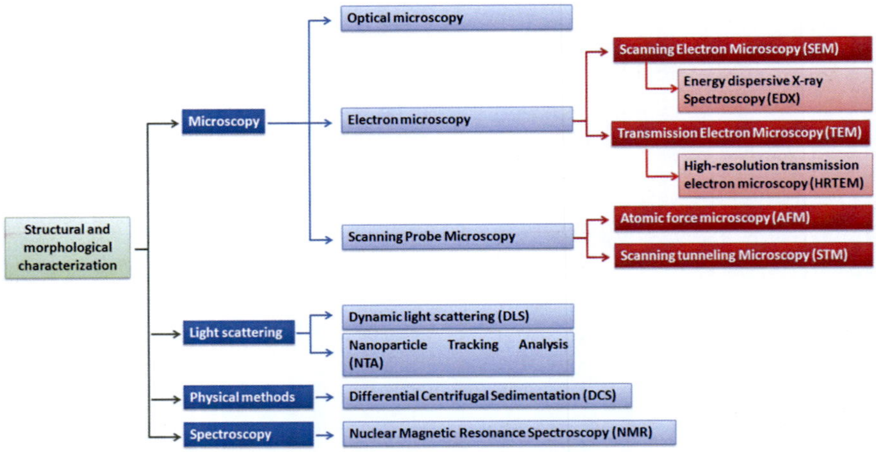


**Fig. 7.5** Schematic representation of the structural and morphological characterization of nanomaterials

nanoparticles, which might be responsible for reduction and stabilization of nanoparticles (Gour and Jain 2019). Energy dispersive spectroscopy (EDS) is used to detect the elemental composition of metals and is based on the separation of the X-rays of different elements into an energy spectrum which is characteristic of that element. The EDS is usually coupled to either SEM or TEM. The reduction of metal ions in their metals can be estimated using UV–visible spectroscopy.

The oxidation state and binding energy within the elemental nanoparticle can be characterized. The metal–oxygen bonds can be confirmed using FTIR and XPS. The oxidation state, the bond length and bond vibration can be determined by using, X-ray Absorption spectroscopy (XAS) and FTIR. XAS measures the X-ray absorption coefficient of a material as a function of energy. Each element has a characteristic set of absorption edges corresponding to different binding energies of its electrons. XAS can be either extended X-ray absorption fine structure (EXAFS) or X-ray absorption near edge structure (XANES). EXAFS identifies the chemical state of a species, interatomic distances even in very low concentrations while XANES probes the density of states of empty/partially filled electronic states by considering the excitation of an inner shell electron to those states that are permitted by dipole selection rules (Mourdikoudis et al. 2018).

The surface charge of a nanoparticle determines its interactions with the target. The surface charge is difficult to measure directly and so the zeta potential is calculated. A zeta potentiometer is used for the measurement of surface charges and dispersion stability of the nanoparticle in solution. The zeta potential depends on the concentration of suspension and composition of the solvent. Colloids with values of zeta potential in the range 20–30 mV or higher are considered as stable (Mourdikoudis et al. 2018).

Crystallography is the study of atoms and molecular arrangement in crystal solids and is analyzed by powder X-ray or electron diffraction to determine the

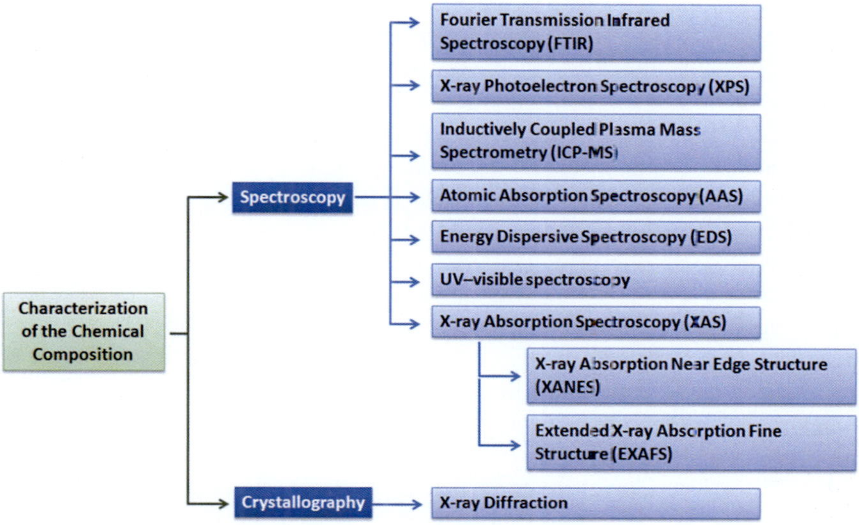


**Fig. 7.6** Characterization of the chemical composition of nanomaterials

structural arrangement. Data from X-ray diffractogram's gives information on the structure, phase type, lattice parameters, and crystalline grain size. It also helps delineate the mineral composition of crystalline nanoparticles. Percentage crystallinity and dimensions of each unit cell can also be calculated by XRD. Figure 7.6 depicts the schematics of characterization of the chemical composition of nanomaterials.

### *7.5.3 Characterization of the Mechanical Properties of Nanomaterial Bioscaffolds*

The stiffness and mechanical characteristics of the bioscaffold must resemble those of the native tissue in which it is intended to be implanted (Discher 2005). The stiffness of the substrate has been reported to affect cell migration, proliferation and differentiation (Petersen et al. 2012; Keogh et al. 2010). The rheological characteristics such as Young's modulus, tensile strength, compressive strength, and elastic response to deformation are important parameters that are needed to be evaluated. Elastic modulus measures strain in response to a given tensile or compressive stress along the plane of the applied force; flexural modulus measures the relationship between a bending stress and the resulting strain after a compressive stress applied perpendicularly; tensile strength is the maximum stress a material can withstand prior to its breaking and maximum strain is the ductility exhibited by the material before a fracture (Parisi et al. 2018). Owing to their fragility, the characterization of mechanical properties of soft scaffolds are also difficult. Classical methods of mechanical testing of biomaterials are challenging when they have to be adopted to the micro and nanometer scale. Micro/nanomechanical testing using modified techniques of contact atomic force microscopy (Tan and Lim 2006; Zhang et al. 2018) depth sensing indentation (Perepelkin et al. 2019) and ultrasound viscoelastography (Hong et al. 2018) has been reported. Nanoindentation is an emerging technique for the measurement of nanomechanical properties of biomaterials and scaffolds. Here a probe is brought in contact with a surface, pushed into the material and then retracted, recording load (P) and displacement (h) over time (t). The P–h–t data are analyzed to generate information on mechanical properties (Oyen 2011; Oyen and Cook 2009).

In the nanoindentation test, small loads and a small tip can be used along with an atomic force microscope (Barone et al. 2010). Contact AFM can measure forces at the nanometer level, force curves are generated from which mechanical data such as Youngs modulus can be calculated using different mathematical models. Garcia et al. provides an in-depth review on the use of AFM for nanomechanical mapping (Garcia 2020). The changes in mechanical properties of the scaffold from protein deposition or ECM deposition by the seeded cells have been observed (Sakai et al. 2005). Zhu et al. had employed a modified AFM nanoindentation technique to demonstrate an increase in the tensile strength of collagen–chitosan scaffolds after they were cultured with cells (Zhu et al. 2011). The use of nanoparticles as filler materials for the development of nanocomposites also enhances their mechanical properties (Thomas et al. 2007; Stanishevsky et al. 2008; Tyagi et al. 2009). Gaihre et al. had used nano-calcium zirconate as fillers for a chitosan scaffold to form nanocomposite structures. The authors had adopted the technique of peak force quantitative nanomechanical mapping (PFQNM) using an atomic force microscope (AFM) and observed that the Young's modulus of the composite material was greater than that of the chitosan scaffold alone (Gaihre and Jayasuriya 2018). Perepelkin et al. proposed the development of a the Borodich–Galanov (BG) method and its extended variant (eBG) that takes into account the adhesion effects of in-depth sensing indentation (Perepelkin et al. 2019). Ultrasound elastography is another technique that has been adopted for nanomechanical measurements. Here a mechanical excitation is used to stimulate the material, and the response of the material, i.e., resulting deformation or strain is monitored through ultrasound imaging. The information on the excitation and response of the material is used to obtain quantitative data on the stiffness of the material. Based on the source of excitation, different ultrasound elastography methods have been developed, e.g., strain elastography, acoustic radiation force elastography (Lin 2020). The technique of acoustic radiation force elastography has been adopted for many ultrasound elastography techniques such as vibroacoustography, shear wave dispersion ultrasound vibrometry, shear wave spectroscopy, monitored steady-state excitation and recovery, and multimode ultrasound viscoelastography (Lin 2020; Shigao et al. 2009; Deffieux et al. 2009; Hong et al. 2016). Hong et al. had used multimode ultrasound viscoelastography (MUVE) in the creep mode to study the viscoelastic properties of 3D agarose, collagen, and fibrin hydrogels. Ultrasound pulses were used to generate and image noninvasively deformations within soft hydrogels, at sub-millimeter resolution. Data at different

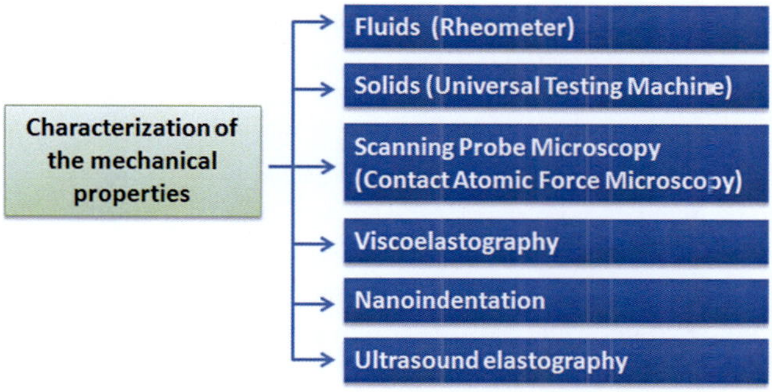


**Fig. 7.7** Characterization of the mechanical properties of nanomaterial bioscaffolds

time intervals were used to obtain quantitative parameters that characterized the elastic and viscoelastic properties of the material (Hong et al. 2018). The authors have proposed the technique to evaluate the development of native and engineered tissues with time. Mattei et al. adopted the epsilon dot method, used generally for bulk compression testing to nanoindentation studies. The nano-elipson dot method was used for the nanoscale testing of polydimethylsiloxane and gelatin. This technique has potential for the nanomechanical testing of soft scaffolds and biological tissues (Mattei et al. 2015). Porosity in the scaffold facilitates cell migration and vascularization into the scaffold and enhances the passage of nutrients to cells lying deep within the scaffold. However, it does reduce the compression strength and overall mechanical strength of the bioscaffold (Prasadh and Wong 2018). The mechanical characterization techniques is schematically given in Fig. 7.7.

### *7.5.4 Characterization of Magnetic Properties of Nanomaterials*

The International Standards Organization (ISO) has also laid down the standards and specification for characterization of magnetic nanoparticles (ISO/TS 19807-1:2019 2019). Magnetic properties of nanoparticles are studied as a function of magnetic field, temperature, and time. The techniques adopted for analysis are vibrating sample magnetometry (VSM), superconducting quantum interference device magnetometry (SQUID), and ferromagnetic resonance (FMR).

An efficient technique to measure magnetic behavior of a material is by using magnetometry techniques that employ direct current (dc). Here, the samples are moving within a dc field that is uniform and constant, thus producing a change in the magnetic field. The moment by force is measured using a vibrating-sample magnetometer (VSM). In this technique, a sample is placed in a copper coil with an applied field and then vibrated perpendicular to the field, and the magnetic moment is recorded by measuring a change in voltage (Sandler et al. 2019). SQUID is used to analyze properties such as the magnetization saturation (MS), the magnetization remanence (MR), and the blocking temperature (TB). The output voltage of a SQUID is recorded in the form of a flux profile and provides valuable information about the magnetic properties of a material. Hybrid VSM SQUID equipment has been developed and used (Sandler et al. 2019). Mössbauer spectroscopy is based on the recoil-free resonance fluorescence of $\gamma$-photons in matter with Mössbauer-active elements, such as Fe. This technique gives information on the oxidation state, symmetry and spin state and the magnetic ordering of the Fe atoms in a nanoparticle sample, thus identifying the magnetic phases in a sample (Fock et al. 2017). Electron paramagnetic resonance (EPR) provides spectroscopic data on unpaired electronic spin states at both the atomic (conventional EPR) and the nanoscale (ferromagnetic resonance-FMR) (Smirnov 2011). FMR is a spectroscopic technique that observes the magnetization of ferromagnetic materials. It probes the sample magnetization that results from the magnetic moments of dipolar-coupled but unpaired electrons. FMR spectra gives information on the average shape and size of particles, which are composed of ferromagnetic elements (Fe, Ni, Co) (Mourdikoudis et al. 2018).

## 7.6 Bioengineering of Nanomaterial Scaffolds to Modulate Biological Functions: Applications of Nanomaterial-Based Scaffolds

The bioscaffold plays a major role in modulating cell proliferation, differentiation, migration, and the host tissue response (Budhwani et al. 2016). Hence, surface modification of the scaffold is a way forward to engineer substrates that utilizes the properties of the biomaterial and the biological milieu to effect the desired physiological response. These surface modifications include topographical modification, protein adsorption, coating with minerals, incorporation of functional groups, and immobilization of biomacromolecules (Richbourg et al. 2019). Topographical modifications are physical changes induced in the sub-micron to nanometer range and are aimed at increasing the surface roughness, surface area, and hydrophilicity/hydrophobicity (Tucker et al. 2019) and also to act as conduits for guided regeneration of cells and tissues. Such modifications are made possible using various techniques like alkali acid etching, incorporation of nanofibers onto the surface of the scaffold, replica molding, heat embossing, and via the use of partial solvents such as toluene, acetone, and ethyl acetate (Limongi et al. 2016). Nanoscale alterations in topography of the scaffold effect cellular adhesion, motility, orientation, cell, surface antigen display, activation and modulation of various intracellular signaling pathways that regulate gene expression, and adsorption and conformation of integrin binding proteins (Curtis and Wilkinson 1999; Webster et al. 2001). Protein adsorption onto the scaffold produces chemical cues that aid in tuning the biological responses to the scaffold (Jose et al. 2010; Phipps et al. 2011). The type of protein that can be adsorbed rely on the properties of both the protein and the biomaterial.

It is influenced by external parameters such as solution pH, temperature, ion concentration, buffer composition of the protein solution, surface energy, polarity, charge, and morphology of the biomaterial. Many studies have reported on the ability of protein adsorption to increase cell adhesion, spreading proliferation, and even modulated differentiation (Rajabi et al. 2018; Noh et al. 2016). Mineral-based surface modification facilitates deposition of inorganic moieties such as calcium phosphate to scaffolds without exerting any changes to the bulk properties of the biomaterial and finds application mainly in bone tissue engineering. Mineral–ion deposition occurs by nucleation of inorganic crystalline particles from ion-rich solution onto the surface of the scaffold. Simulated body fluid (SBF) and its derivatives have been used to produce hydroxyapatite like nodules on scaffold surfaces (Richbourg et al. 2019). The presence of functional groups such as amines, carboxylic acids, thiols, hydroxyl, or carbonyl groups on the surface enables binding to bioactive molecules and cell surface proteins and also facilitates cross-linking with larger polypeptides, glycosaminoglycans, and other ECM moieties. Certain proteins have preferential binding to certain specific chemical moieties (Parisi et al. 2018). The methods used for functionalizing surfaces are plasma deposition, phototherapy, physical entrapment of small functional molecules, aminolysis, and hydrolysis. Most of these functionalized scaffolds are used for conjugation and immobilization of biomolecules. Short peptide sequences, bioactive polypeptides of the ECM proteins, enzymes, growth factors, and antibodies are either conjugated or entrapped to the bioscaffold (Katti et al. 2008). However, it must be ensured that immobilization does not affect any changes in the structure of the biomolecule, and the desired physiological response is attained. Research is currently aimed at binding many different biomolecules to the scaffold surface and to incorporate more than one surface modification in a single scaffold so as to better mimic the ECM and elicit the desired physiological response.

The application potential for nanomaterial bioscaffolds in tissue engineering is vast and spans many areas such as skin, cartilage, bone, tendons, cardiac, vascular, liver, ophthalmology, neural, pancreas, and drug delivery among others. This review however will outline the various nanomaterials that are being researched and developed in the field of bone, neural, and cardiac tissue engineering.

### *7.6.1 Bioscaffolds for Bone Tissue Engineering*

Bone is a highly complex structured hard tissue where organic and inorganic mineral phases play a pivotal role in its functional and physicochemical properties. When an injury crosses the critical size of a defect, the normal physiological healing may not lead to complete recovery and restoration of normal bone architecture. Such cases mandate the use of implants and hold scope for the use of cell-based therapies and tissue-engineered scaffolds. Since the architecture of the bone is complex spanning across micro- and nanoscales, nanomaterial-based scaffolds are being used to mimic the bone tissue ECM. Most of these bioscaffolds are nanocomposites fabricated via incorporation of nanoparticles, nanotubes, and nanofibers as reinforcements of polymer and ceramic scaffolds. Li et al. used an in vivo goat shank model to study the bone regeneration capacity of porous scaffolds of nano-hydroxyapatite (nHA), collagen, and poly-L-lactic acid (PLLA), reinforced by chitin fibers. New bone formation and complete repair of the defect were noted 15 weeks post-surgery (Li et al. 2006). Mi et al. had adopted the technique of electro-deposition to coat calcium phosphate onto the nanofibrous PLLA scaffolds. The nanofibrous polymer was generated using the techniques of TIPS and porogen-leaching. These scaffolds were seeded with bone marrow-derived mesenchymal stem cells and osteogenic differentiation was observed in an in vivo rat subcutaneous model (Mi et al. 2020). Rezk et al. developed a bifunctional composite coating composed of PCL and nHA loaded with simvastatin and deposited them on the magnesium AZ31 alloy via electrospinning technique. The coating was observed to reduce corrosion and promote adhesion of mouse pre-osteoblast cells in vitro (Rezk 2018). Stem cells can also be incorporated into the bioscaffolds. Carbon-based nanomaterial scaffolds like fullerenes, graphene oxides, and carbon nanotubes have shown promising results in the differentiation of mesenchymal stem cells into the osteogenic lineage (Kang et al. 2017). Glass nanotubes have also been reported to promote cell differentiation and osteogenesis. Xiao et al. had used nanotubes of bioactive glass with ordered mesoporous structure as scaffolds into which the small molecule simvastatin was loaded. These scaffolds could induce the mesenchymal stem cells to express osteogenic-related genes osteopontin, osteocalcin, collagen 1, and Runx2 (Xiao et al. 2019a, b). Copper can also be doped into the mesoporous bioactive glass to affect its antibacterial activity (Luo et al. 2020; Baino 2020). Growth factors like vascular endothelial growth factor (VEGF) and bone morphogenetic protein (BMP) have also been incorporated into the scaffold as control delivery systems that can promote bone tissue engineering (Li et al. 2015; Xiao et al. 2019a, b; Sharma et al. 2018).

The bottom-up approach for the synthesis of nanomaterials has also been adopted for developing bioscaffolds for bone tissue engineering. Hosseinkhani et al. demonstrated that a 3D network of nanofibers formed by the self-assembly of peptide amphiphile molecules increases the proliferation and differentiation of mesenchymal stem cells into the osteogenic lineage as noted by increased alkaline phosphatase activity and osteocalcin content (Hosseinkhani et al. 2006a, b). The group had also reported that a hybrid scaffold comprising of a hydrogel formed from the self-assembly of peptide-amphiphile (PA) with mesenchymal stem cell suspensions in media, and a collagen sponge reinforced with poly(glycolic acid) (PGA) fibers, could induce the osteogenic differentiation in a perfusion culture bioreactor. In vivo studies in the subcutis of rats demonstrated that these cells seeded hybrid scaffolds could subsequently also induce ectopic bone formation (Hosseinkhani et al. 2006c).

In order to mimic closely both the morphology and inorganic composition, fibroporous nanocomposite scaffolds have also been developed using electrospinning or by self-assembly. Nano-hydroxyapatite (nHA) incorporated electrospun fiber scaffolds of polymers with or without proteins or peptides have been reported (Thomas et al. 2006, 2007; Tyagi et al. 2009; Jose et al. 2009, 2010). Presence of nHA and collagen protein in these bone-mimetic composite scaffolds exhibited enhanced regeneration properties both in vitro and in vivo (Jose et al. 2010; Phipps et al. 2011). Fibrous nanocomposite scaffold have also been developed using electrospun nHA fibers and arginine-glycine-aspartic acid (RGD)-bearing peptide-amphiphile gel produced by self-assembly. In vitro studies using preosteoblastic cultures have shown enhanced expression of the osteogenic genes for bone sialoprotein and osteocalcin when cultured on the nanocomposite (Çakmak et al. 2013). Nanotopographical changes also affect the cellular response to the substrate. Dalby et al. had demonstrated that random circular nanostructures promote and direct osteoblast differentiation of MSCs and do not require an osteoinductive cell culture medium (Dalby et al. 2007).

### 7.6.2 *Bioscaffold for Cardiac Tissue Engineering*

Myocardial infarction is caused by the reduced supply of blood to the heart, and the damaged tissue forms a non-contractile scar and normally does not regenerate. Cardiac tissue engineering aims to replace the damage tissue with new ones. Cardiomyocytes are the primary cell type of the heart and the Purkinje fibers propagate electrical communication between adjacent cells (Kankala et al. 2018). Hence, tissue engineering strategies have incorporated the use of conductive nanomaterial bioscaffolds such as carbon nanotubes, carbon nanofibers, graphene oxide, metallic nanoparticles, and electroactive polymers (Ashtari et al. 2019) which promote the synchronous beating of seeded cardiomyocytes in vitro. The main approaches include the design of scaffolds, patches, and injectable materials (Amezcua et al. 2016). The unidirectional arrangement of the nanofibers results in an anisotropic scaffold which promotes cell alignment in one direction (Sapir et al. 2013). Martins et al. had developed nanocomposites of carbon nanofibers in a chitosan matrix that could support attachment and proliferation of rat neonatal heart cells. These cells expressed genes involved in cardiac muscle contraction and electrical coupling (Martins et al. 2014). Similar results have also been observed in other nanocomposite systems for myocardial tissue engineering such as carbon nanofibers with PLGA (Stout et al. 2011) and carbon nanofibers, self-assembled rosette nanotubes, and poly(2-hydroxyethyl methacrylate) (pHEMA) hydrogel (Meng et al. 2012). Carbon nanotubes (CNT) for cardiac tissue engineering have also been extensively researched. Martinelli et al. reported that polydimethyl siloxane (PDMS) substrate with deposited multi-walled CNT could promote cardiomyocyte growth and differentiation and improved the cells electrophysiological characteristics, intracellular calcium signaling, and enhanced the maturation of functional syncytia. These results suggest that CNTs reinforce the electrical coupling between cardiomyocytes (Martinelli et al. 2018). Nanocomposites of carbon nanotubes with various polymer matrices such as poly(glycerol sebacate)/gelatin, PVA/chitosan nanofibers, PCL mats, and fibers have also exhibited promising results as bioscaffolds for cardiac tissue engineering (Kharaziha et al. 2014; Liao et al. 2011; Wickham et al. 2014).

Simon-Yarza et al. generated a recombinant human neuregulin-1 beta isoform (Nrg)-incorporated PLGA nanofibrous scaffold. Nrg is a growth factor which promotes angiogenesis and cardiomyocyte replication. In vivo experiments proved that when the scaffold was implanted into a myocardial ischemic rat model, it could successfully integrated with the cardiac tissue, although inflammatory responses were noted (Simón-Yarza et al. 2015). Graphene and graphene oxide have also been advocated as potential nanomaterials for myocardial tissue engineering. Smith et al. developed nano-patterned hybrid scaffolds that were conductive using graphene and polyethylene glycol (PEG). These scaffolds enhanced the myofibrils and sarcomeric structures and also increased the electrical coupling of cardiac cells (Smith et al. 2017). Shin et al. developed a cell construct via layer-by-layer technology using alternate cell-seeded layers and graphene oxide-coated poly-L-lysine (PLL) thin nanofilm deposition. These constructs exhibited strong spontaneous beating behavior in vitro, demonstrating their potential as 3D cardiac tissue constructs (Shin et al. 2014). Bioscaffolds may be incorporated with metallic nanoparticles like gold and iron oxide to increase their conductivity. In a study by Fleischer et al., gold nanoparticles were encapsulated in the coiled electrospun PCL nanofibers. These scaffolds facilitated the cardiomyocytes to contract efficiently by propagating the electrical signals between the adjacent cells (Fleischer et al. 2014). You et al. incorporated gold nanoparticles into a thiol-HEMA/HEMA hybrid hydrogel to mimic the mechanical properties of native myocardial ECM (You et al. 2011). Gold nanoparticle-incorporated scaffolds have also been demonstrated to promote the differentiation of mesenchymal stem cells into the cardiomyogenic lineage (Ravichandran et al. 2014; Sridhar et al. 2015). Ishii et al. bioengineered multilayered sheets of adipose-derived stem cells labeled with magnetic (Magnetite $Fe_3O_4$) nanoparticle-containing liposomes (MCLs). The therapeutic potential of these sheets were evaluated in a mouse myocardial infarction model. The bioengineered cell sheets were transplanted onto the infarcted myocardium using a magnet prior to skin closure. The treated animals showed improved systolic function and thinning of the infarcted wall, and angiogenesis was observed, the latter may be due to an increased expression of VEGF and bFGF in ischemic hearts (Ishii et al. 2014). Cerium oxide-incorporated electrospun composite scaffolds were also introduced for angiogenesis (Xu et al. 2020) in tissue engineering applications.

The potential of electro active polymers like polypyrrole and polyaniline for cardiac tissue engineering has only been evaluated in the last decade (Nishizawa et al. 2007; Bidez et al. 2006). Wang et al. developed a cardiac tissue patch using conductive polypyrrole nanoparticles, gelatin methacrylate (GelMA), and poly-(ethylene glycol) diacrylate (PEGDA) that were cross-linked to a cryogel form with a dopamine crosslinker. Cardiomyocyte-loaded patch was found to improve the cardiac function in vivo in rat myocardial infarction (MI)-affected models. The results showed that the infarct size is reduced by 42.6%. Furthermore, the patch promoted synchronous contractions of the cardiomyocytes through the upregulation of α-actinin and CX-43 in these cells (Wang et al. 2016). Gelmi et al. had coated PLGA fibers with polypyrrole, and the electroactive scaffold showed positive results for differentiation of human iPSCs into the cardiac lineage under electromechanical stimulation (Gelmi et al. 2016). The blend of polyaniline with PLGA can be electrospun to form nanofibrous scaffolds with aligned fibers. In vitro cardiomyocytes adhered onto the scaffold and formed clusters which exhibited synchronous beating and expression of the gap junctional protein connexin 43 (Hsiao et al. 2013). Polyaniline electrospun nanofibers modified with hyperbranched poly-L-lysine dendrimers can also promote the differentiation of neonatal rat cardiac cells cultured on them following stimulation with an electric current (Fernandes et al. 2010). The potential of nanopatterned surface as substrates for cardiac tissue engineering has also been explored. Kim et al. cultured cardiac cells on polyethylene glycol hydrogel having grooved arrays of widths between 50 and 800 nm. Guided by the nanoscale mechanical cues, these tissue constructs in vitro exhibited anisotropic action potential propagation and produced the aligned contractility similar to that of native cardiac tissue (Kim et al. 2009). A detailed review of nanopatterning for cardiac tissue engineering has been given by Cristallini et al. (2020).

### *7.6.3 Bioscaffolds for Neural Tissue Engineering*

Over the last decade, much research has been undertaken in the field of nanomaterials for neural tissue engineering and regeneration. Tissue engineering approaches find applications in the treatment of injuries to the peripheral nervous system than for treatment of damages to the central nervous system as crossing of the blood–brain barrier remains a major hurdle. Some of the main approaches in neural tissue engineering and regeneration include (a) delivery of small-molecule drugs, biochemical analogs, and nucleic acid based, i.e., DNA/RNA vectors to direct stem cell and neural differentiation and (b) modulating the surface chemistry and nanotopography of substrates to tune neural stem cell morphology, proliferation, and differentiation (Shah et al. 2016). Graphene- and graphene oxide (GO)-based nanomaterials possess good flexibility, mechanical strength, and electrical conductivity which makes them excellent bioscaffolds for neural tissue engineering. Nanocomposite of PLLA nanofibers with graphene oxide were used to enhance surface roughness and have been found to promote neural growth (Zhang et al. 2016). Chiacchiaretta et al. had observed that astrocytes earlier exposed to graphene oxide were able to enhance the intrinsic excitability and density of GABA synapses in primary neurons that were cocultured with them (Chiacchiaretta et al. 2018). Akhavan et al. used graphene nanogrids as photocatalytic stimulators to promote neural stem cells growth into 2D neural networks (Akhavan and Ghaderi 2013). Park et al. had reported on improved adhesion of neuronal stem cells and their differentiation into neurons when cultured on graphene substrates (Park et al. 2011a). Alternatively graphene oxide can be used to coat electrospun poly (vinyl chloride) nanofibers. These fibers when used as scaffolds enhanced proliferation and differentiation of primary motor neurons (Feng et al. 2015). Carbon nanotubes (CNT) have been used in vitro as substrates for neuronal proliferation, differentiation, and stimulation. The surface electric charge of CNTs exerts a strong influence on their biological effects. It has been observed that more the positive surface charge, cells grew out more neurite branching and growth cones (Hu 2004). CNTs which have good electrical conductivity have been used as substrates for support and differentiation of stem cells into neural cells in vitro (Chao et al. 2009). Park et al. had also explored the effect of carbon nanotube coatings of various geometries on the growth and differentiation of human neural stem cells. They observed that the optimal nanotopography and patterns provided cues of selective laminin adsorption which resulted in selective adhesion and growth of neural stem cells. The CNT network patterns effected structural-polarization modulated neuronal differentiation (Park et al. 2011b). Electron spun polymer fibers can also be immobilized with collagen to better mimic the extracellular matrix. These nanofibers provide biological cues that were found to enhance attachment and viability of neural stem cells in vitro (Li et al. 2008). CNT-hydrogel nanocomposites with tunable porous structures have also been fabricated by 3D printing. Here amine functionalized multi-walled carbon nanotubes were incorporated into PEGDA polymer. The nanocomposite scaffold was observed to enhance proliferation and early differentiation of neural stem cell. Additional electrical stimulations were able to promote neurite outgrowth and could be used as a therapeutic in nerve regeneration (Lee et al. 2018). Electro-conductive polymers such as pyrrole have also been used for the fabrication of nanoscaffolds for neural tissue engineering. Poly pyrrole has been used to coat PLA electrospun nanofibers, and these conductive scaffolds could support attachment and migration of neural progenitor cells (Sudwilai et al. 2014). Similar observations were also reported by other groups (Stewart et al. 2015; Lee and Chen 2009). Nanoparticles have also been employed for promoting neural tissue regeneration although most studies employ in vitro models. These include metallic nanoparticles of gold, silver, iron oxide, zinc oxide, and titanium oxide. Being conductive, they can be used as nanocomposites for the electrical stimulation for the regeneration of nerve tissue (Arzaghi et al. 2020). Riggio et al. had functionalized magnetic nanoparticles with beta-NGF, and a magnetic field was used to align the neurite outgrowths in PC12 cells (Riggio et al. 2014). Yuan et al. had used nerve growth factor to functionalize superparamagnetic iron oxide–gold core–shell nanoparticle which could modulate neuronal activities under photo-stimulation using low intensity LED light. Higher intensities were found to enhance differentiation promoting more branching points and longer neurite length (Yuan et al. 2019). Neurotrophic growth factor therapy accelerates neurogenesis, angiogenesis, remyelination, synaptogenesis, and axonal outgrowth of the neuronal cells (Arzaghi et al. 2020). Xia et al. reported that Schwann cells proliferated better on aligned electrospun poly(methyl methacrylate) nanofibers than on the non-aligned fibers. These cells elongated in the direction of the aligned nanofibers and developed longer cell processes than when cultured on the non-aligned nanofibers. Co-cultures of Schwann cells and dorsal root ganglion neurons demonstrated that both the cells co-localized and elongated along the orientation of the aligned nanofibers and would facilitate the process of myelination (Xia et al. 2016). Hu et al. had reported that neural cells differentiated from adipose-derived mesenchymal stem cells and treated with FGF2miR-218 could be cultured onto three dimensionally aligned poly(3-hydroxybutyrate-co-3-hydroxyvalerate)

PHBV nanofibers. These nerve grafts were able to restore motor function in vivo in a rat with a 10 mm critical size sciatic nerve defect (Hu et al. 2017). Nanopatterned surfaces also find potential in neural tissue engineering. Cesca et al. had pioneered a fast-prototyping method using a single-step plasma treatment for generating nanopatterned PCL scaffolds. These substrates promoted the attachment and growth of primary hippocampal neurons. These cells formed a functional network that exhibited a dense pattern of synaptic contacts (Cesca et al. 2014).

## 7.7 Conclusions and Future Perspectives

Advanced applications of nanomaterials as bioscaffolds in tissue engineering necessitate that they mimic the architecture and physical properties of the natural biological tissue or organ. These bioscaffolds are designed and fabricated with biocompatible and biodegradable materials that facilitate nutrient transport and promote cellular growth and differentiation. Modifications of these bioscaffolds, especially via non-thermal cold plasma processing, help further knowledge of cell–material interaction and tissue development and increase their potential as advanced regenerative therapeutics. Albeit promising there are also numerous ongoing challenges, mainly those of controlling signals that are presented to the cells and long-term bioaccumulation and toxicity when dealing with nanoparticles like metals, graphene, and carbon nanotubes. Finally scaling up of production, while maintaining reproducibility and morphology of the nanomaterials, to meet commercial demands, is also mandated.

## References


Agarwal, S., Wendorff, J. H., & Greiner, A. (2008). Use of electrospinning technique for biomedical applications. *Polymer, 49*, 5603–5621.

Akhavan, O., & Ghaderi, E. (2013). Differentiation of human neural stem cells into neural networks on graphene nano-grids. *Journal of Materials Chemistry B, 1*(45), 6291.

Al-Enizi, A. M., Zagho, M. M., & Elzatahry, A. A. (2018). Polymer-based electrospun nano-fibers for biomedical applications. *Nano-materials (Basel), 8*(4), 259.

Amezcua, R., Shirolkar, A., Fraze, C., & Stout, D. (2016). Nano-materials for cardiac myocyte tissue engineering. *Nano-materials, 6*(7), 133.

Ando, T., Uchihashi, T., & Fukuma, T. (2008). High-speed atomic force microscopy for nano-visualization of dynamic biomolecular processes. *Progress in Surface Science, 83*(7–9), 337–437.

Angammana, C. J., & Jayaram, S. H. (2011). Analysis of the effects of solution conductivity on electrospinning process and fiber morphology. *IEEE Transactions on Industry Applications, 47*, 1109–1117.

Arulmani, S., Anandan, S., Ashokkumar, M. (2018). Chapter 1—Introduction to advanced nano-materials. In B. A. Bhanvase, V. B. Pawade, S. J. Dhoble, S. H. Sonawane, & M. Ashokkumar (Eds.), I*n micro and nano technologies. Nanomaterials for green energy* (pp. 1–53). Amsterdam: Elsevier. ISBN 9780128137314. https://doi.org/10.1016/B978-0-12-813731-4.00001-1.

Arzaghi, H., Adel, B., Jafari H., Askarian-Amiri, S., Shiralizadeh Dezfuli, A., Akbarzadeh, A., Pazoki-Toroudi, H. Nanomaterial integration into the scaffolding materials for nerve tissue engineering: a review. Rev Neurosci. 2020 Aug 10:/j/revneuro.ahead-of-print/revneuro-2020-0008/revneuro-2020-0008.xml. https://doi.org/10.1515/revneuro-2020-0008.

Ashtari, K., Nazari, H., Ko, H., Tebon, P., Akhshik, M., Akbari, M., & Khademhosseini, A. (2019). Electrically conductive nano-materials for cardiac tissue engineering. *Advanced Drug Delivery Reviews, 144*, 162–179.

Baino, F. (2020). Copper-doped ordered mesoporous bioactive glass: A promising multifunctional platform for bone tissue engineering. *Bioengineering, 7*(2), 45–55.

Barone, A. C., Salerno, M., Patra, N., Gastaldi, D., Bertarelli, E., Carnelli, D., & Vena, P. (2010). Calibration issues for nano-indentation experiments: Direct atomic force microscopy measurements and indirect methods. *Microscopy Research and Technique, 73*(10), 996–1004.

Bhattacharya, K., Mukherjee, S. P., Gallud, A., Burkert, S. C., Bistarelli, S., Bellucci, S., & Fadeel, B. (2016). Biological interactions of carbon-based nano-materials: From coronation to degradation. *Nano-medicine: Nano-technology, Biology and Medicine, 12*(2), 333–351.

Bidez, P. R., Li, S., MacDiarmid, A. G., Venancio, E. C., Wei, Y., & Lelkes, P. I. (2006). Polyaniline, an electroactive polymer, supports adhesion and proliferation of cardiac myoblasts. *Journal of Biomaterials Science, Polymer Edition, 17*(1–2), 199–212.

Boland, E. D., Telemeco, T. A., Simpson, D. G., Wnek, G. E., & Bowlin, G. L. (2004). Processing of polymer nano-fibers through electrospinning as drug delivery systems. *Journal of Biomedical Materials Research Part B, 71*, 144–152.

Budhwani, K., Wood, A. T., Sethu, P., & Thomas, V. (2016). Nanofiber and stem cell enabled biomimetic systems and regenerative medicine. *Journal of Nanoscience and Nanotechnology, 16*(9), 8923–8934.

Çakmak, S., Çakmak, A. S., & Gümüşderelioğlu, M. (2013). RGD-bearing peptide-amphiphile-hydroxyapatite nano-composite bone scaffold: An *in vitro* study. *Biomedical Materials, 8*(4), 045014.

Cesca, F., Limongi, T., Accardo, A., Rocchi, A., Orlando, M., Shalabaeva, V., & Benfenati, F. (2014). Fabrication of biocompatible free-standing nano-patterned films for primary neuronal cultures. *RSC Advances, 4*(86), 45696–45702.

Chao, T. I., Xiang, S. H., Chen, C. S., Chin, W. C., Nelson, A. J., Wang, C. C., & Lu, J. (2009). Carbon nano-tubes promote neuron differentiation from human embryonic stem cells. *Biochemical and Biophysical Research Communications, 384*, 426–430.

Charernsriwilaiwat, N., Opanasopit, P., Rojanarata, T., & Ngawhirunpat, T. (2012). Lysozyme-loaded, electrospun chitosan-based nano-fiber mats for wound healing. *International Journal of Pharmaceutics, 427*(2), 379–384.

Chen, H., Shao, L., Ming, T., Sun, Z., Zhao, C., Yang, B., & Wang, J. (2010). Understanding the photothermal conversion efficiency of gold nano-crystals. *Small, 6*, 2272–2280.

Chen, S., He, Z., Xu, G., & Xiao, X. (2016). Fabrication and characterization of modified nano-fibrous poly(L-lactic acid) scaffolds by thermally induced phase separation technique and aminolysis for promoting cyctocompatibility. *Journal of Biomaterials Science, Polymer Edition, 27*(10), 1058–1068.

Chiacchiaretta, M., Bramini, M., Rocchi, A., Armirotti, A., Giordano, E., Vázquez, E., & Benfenati, F. (2018). Graphene oxide upregulates the homeostatic functions of primary astrocytes and modulates astrocyte-to-neuron communication. *Nano Letters, 18*(9), 5827–5838.

Chilarski, A., Szosland, L., Krucińska, I., Kiekens, P., Błasińska, A., Schoukens, G., Cisło, R., & Szumilewicz, J. (2007). Novel dressing materials accelerating wound healing made from dibutyrylchitin. *Fibres & Textiles in Eastern Europe, 15*, 77–81.

Choi, J. S., Lee, S. J., Christ, G. J., Atala, A., & Yoo, J. J. (2008). The influence of electrospun aligned poly(epsilon-caprolactone)/collagen nano-fiber meshes on the formation of self-aligned skeletal muscle myotubes. *Biomaterials, 29*(19), 2899–2906.

Conoscenti, G., Carrubba, V. L., & Brucato, V. (2017). A versatile technique to produce porous polymeric scaffolds: The thermally induced phase separation (TIPS) method. *Archives in Chemical Research, 1*, 2.

Costa, A., Naranjo, J. D., Londono, R., & Badylak, S. F. (2017). Biologic scaffolds. *Cold Spring Harbor Perspectives in Medicine, 7*(9), 1–23.

Cristallini, C., Vitale, E., Giachino, C., & Rastaldo, R. (2020). Nano-engineering in cardiac regeneration: Looking back and going forward. *Nano-materials, 10*(8), 1587.
Curtis, A., & Wilkinson, C. (1999). New depths in cell behaviour: Reactions of cells to nano-topography. *Biochemical Society Symposium, 65*, 15–26.
Dalby, M. J., Gadegaard, N., Tare, R., Andar, A., Riehle, M. O., Herzyk, P., & Oreffo, R. O. C. (2007). The control of human mesenchymal cell differentiation using nano-scale symmetry and disorder. *Nature Materials, 6*(12), 997–1003.
Day, E. S., Morton, J. G., & West, J. L. (2009). Nano-particles for thermal cancer therapy. *Journal of Biomechanical Engineering, 131*, 074001.
Deffieux, T., Montaldo, G., Tanter, M., & Fink, M. (2009). Shear wave spectroscopy for in vivo quantification of human soft tissues visco-elasticity. *IEEE Transactions on Medical Imaging, 28*(3), 313–322.
Demirci, S., Celebioglu, A., Aytac, Z., & Uyar, T. (2014). pH-responsivenano-fibers with controlled drug release properties. *Polymer Chemistry, 5*, 2050–2056.
Discher, D. E. (2005). Tissue cells feel and respond to the stiffness of their substrate. *Science, 310*(5751), 1139–1143.
Dong, R., Pang, Y., Su, Y., & Zhu, X. (2015). Supramolecular hydrogels: Synthesis, properties and their biomedical applications. *Biomaterials Science, 3*(7), 937–954.
Doshi, J., & Reneker, D. H. (1995). Electrospinning process and applications of electrospun fibers. *Journal of Electrostatics, 35*, 151–160.
Dowell-Mesfin, N. M., Abdul-Karim, M. A., Turner, A. M. P., Schanz, S., Craighead, H. G., Roysam, B., Turner, J. N., & Shain, W. (2004). Topographically modified surfaces affectorientation and growth of hippocampal neurons. *Journal of Neural Engineering, 1*, 78–90.
Duan, B., Wu, L., Yuan, X., Hu, Z., Li, X., Zhang, Y., Yao, K., & Wang, M. (2007). Hybrid nano-fibrous membranes of PLGA/chitosan fabricated via an electrospinning array. *Journal of Biomedical Materials Research Part A, 83*, 868–878.
Dvir, T., Timko, B. P., Kohane, D. S., & Langer, R. (2011a). Nano-technological strategies for engineering complex tissues. *Nature Nanotechnology, 6*, 13–22.
Dvir, T., Timko, B. P., Brigham, M. D., Naik, S. R., Karajanagi, S. S., Levy, O., Jin, H., Parker, K. K., Langer, R., & Kohane, D. S. (2011b). Nano-wired three-dimensional cardiac patches. *Nature Nanotechnology, 6*, 720–725.
Ealias, A. M., & Saravanakumar, M. P. (2017). A review on the classification, characterisation, synthesis of nano-particles and their application. *IOP Conference Series: Materials Science and Engineering, 263*, 032019.
Elias, K. L., Price, R. L., & Webster, T. J. (2002). Enhanced functions of osteoblasts on nano-meter diameter carbon fibers. *Biomaterials, 23*(15), 3279–3287.
Elzatahry, A. A., Al-Enizi, A. M., Elsayed, E. A., Butorac, R. R., Al-Deyab, S. S., Wadaan, M. A. M., & Cowley, A. H. (2012). Nano-fiber composites containing N-heterocyclic carbene complexes with antimicrobial activity. *International Journal of Nanomedicine, 7*, 2829–2832.
Feng, Z. Q., Wang, T., Zhao, B., Li, J., & Jin, L. (2015). Soft graphene nano-fibers designed for the acceleration of nerve growth and development. *Advanced Materials, 27*(41), 6462–6468.
Fernandes, E. G. R., Zucolotto, V., & De Queiroz, A. A. A. (2010). Electrospinning of hyper-branched poly-L-lysine/polyaniline nano-fibers for application in cardiac tissue engineering. *Journal of Macromolecular Science, Part A, 47*(12), 1203–1207.
Filipe, V., Hawe, A., & Jiskoot, W. (2010). Critical evaluation of nano-particle tracking analysis (NTA) by nano-sight for the measurement of nano-particles and protein aggregates. *Pharmaceutical Research, 27*(5), 796–810.
Fiorati, A., Turco, G., Travan, A., Caneva, E., Pastori, N., Cametti, M., Punta, C., & Melone, L. (2017). Mechanical and drug release properties of sponges from cross-linked cellulose nano-fibers. *ChemPlusChem, 82*, 848–858.
Fleischer, S., Shevach, M., Feiner, R., & Dvir, T. (2014). Coiled fiber scaffolds embedded with gold nano-particles improve the performance of engineered cardiac tissues. *Nanoscale, 6*(16), 9410–9414.
Fock, J., Bogart, L. K., González-Alonso, D., Espeso, J. I., Hansen, M. F., Varón, M., & Pankhurst, Q. A. (2017). On the "centre of gravity" method for measuring the composition of magnetite/

maghemite mixtures, or the stoichiometry of magnetite-maghemite solid solutions, via 57Fe Mössbauer spectroscopy. *Journal of Physics D: Applied Physics, 50*(26), 265005.
Gagner, J. E., Shrivastava, S., Qian, X., Dordick, J. S., & Siegel, R. W. (2012). Engineering nanomaterials for biomedical applications requires understanding the nano-bio interface: A perspective. *The Journal of Physical Chemistry Letters, 3*(21), 3149–3158.
Gaihre, B., & Jayasuriya, A. C. (2018). Comparative investigation of porous nano-hydroxyapaptite/chitosan, nano-zirconia/chitosan and novel nano-calcium zirconate/chitosan composite scaffolds for their potential applications in bone regeneration. *Materials Science and Engineering: C, 91*, 330–339.
Garcia, R. (2020). Nano-mechanical mapping of soft materials with the atomic force microscope: Methods, theory and applications. *Chemical Society Reviews, 49*, 5850–5884.
Gelmi, A., Cieslar-Pobuda, A., de Muinck, E., Los, M., Rafat, M., & Jager, E. W. H. (2016). Direct mechanical stimulation of stem cells: A beating electromechanically active scaffold for cardiac tissue engineering. *Advanced Healthcare Materials, 5*(12), 1471–1480.
Ghasempour, R., & Narei, H. (2018). CNT basics and characteristics. In *Carbon nano-tube-reinforced polymers* (pp. 1–24). Amsterdam: Elsevier.
Gour, A., & Jain, N. K. (2019). Advances in green synthesis of nano-particles. *Artificial Cells, Nano-medicine, and Biotechnology, 47*(1), 844–851.
Greiner, A., & Wendorff, J. H. (2007). Electrospinning: A fascinating method for the preparation of ultrathin fibers. *Angewandte Chemie, International Edition, 46*, 5670–5703.
Gunn, J., & Zhang, M. (2010). Polyblend nano-fibers for biomedical applications: Perspectives and challenges. *Trends in Biotechnology, 28*(4), 189–197.
Gupta, P., Elkins, C., Long, T. E., & Wilkes, G. L. (2005). Electrospinning of linear homopolymers of poly(methyl methacrylate): Exploring relationships between fiber formation, viscosity, molecular weight and concentration in a good solvent. *Polymer, 46*, 4799–4810.
Haghi, A. K., & Akbari, M. (2007). Trends in electrospinning of natural nano-fibers. *Physica Status Solidia-Applications and Materials Science, 204*, 1830–1834.
Hong, X., Stegemann, J. P., & Deng, C. X. (2016). Microscale characterization of the viscoelastic properties of hydrogel biomaterials using dual-mode ultrasound elastography. *Biomaterials, 88*, 12–24.
Hong, X., Annamalai, R. T., Kemerer, T. S., Deng, C. X., & Stegemann, J. P. (2018). Multimode ultrasound viscoelastography for three-dimensional interrogation of microscale mechanical properties in heterogeneous biomaterials. *Biomaterials, 178*, 11–22.
Hoque, J., Sangaj, N., & Varghese, S. (2019). Stimuli-responsive supramolecular hydrogels and their applications in regenerative medicine. *Macromolecular Bioscience, 19*(1), e1800259.
Hosseinkhani, H., Hosseinkhani, M., & Kobayashi, H. (2006a). Proliferation and differentiation of mesenchymal stem cells using self-assembled peptide amphiphile nano-fibers. *Biomedical Materials, 1*(1), 8–15.
Hosseinkhani, H., Hosseinkhani, M., Tian, F., Kobayashi, H., & Tabata, Y. (2006b). Osteogenic differentiation of mesenchymal stem cells in self-assembled peptide-amphiphile nano-fibers. *Biomaterials, 27*(22), 4079–4086.
Hosseinkhani, H., Hosseinkhani, M., Tian, F., Kobayashi, H., & Tabata, Y. (2006c). Ectopic bone formation in collagen sponge self-assembled peptide–amphiphile nano-fibers hybrid scaffold in a perfusion culture bioreactor. *Biomaterials, 27*(29), 5089–5098.
Hsiao, C. W., Bai, M. Y., Chang, Y., Chung, M. F., Lee, T. Y., Wu, C. T., & Sung, H. W. (2013). Electrical coupling of isolated cardiomyocyte clusters grown on aligned conductive nano-fibrous meshes for their synchronized beating. *Biomaterials, 34*(4), 1063–1072.
Hu, H., Ni, Y., Montana, V., Haddon, R. C., & Parpura, V. (2004). Chemically functionalized carbon nano-tubes as substrates for neuronal growth. *Nano Letters, 4*, 507–511.
Hu, F., Zhang, X., Liu, H., Xu, P., Doulathunnisa Teng, G., & Xiao, Z. (2017). Neuronally differentiated adipose-derived stem cells and aligned PHBV nano-fiber nerve scaffolds promote sciatic nerve regeneration. *Biochemical and Biophysical Research Communications, 489*(2), 171–178.

Hummer, G., Rasaiah, J. C., & Noworyta, J. P. (2001). Water conduction through the hydrophobic channel of a carbon nano-tube. *Nature, 414*, 188–190.

Iijima, S., Ajayan, P. M., & Ichihashi, T. (1992). Growth model for carbon nano-tubes. *Physical Review Letters, 69*, 3100–3103.

Ishii, M., Shibata, R., Shimizu, Y., Yamamoto, T., Kondo, K., Inoue, Y., & Murohara, T. (2014). Multilayered adipose-derived regenerative cell sheets created by a novel magnetite tissue engineering method for myocardial infarction. *International Journal of Cardiology, 175*(3), 545–553.

ISO/TS 19807-1:2019. (2019). Nano-technologies–magnetic nano-materials—Part 1: Specification of characteristics and measurements for magnetic nano-suspensions.

Jack, K., Velayudhan, S., Lukmann, P., Trau, M., Lisbeth, G., & Cooper-White, J. (2009). Fabrication of nano-composite biodegradable polymer scaffolds by thermally-induced phase separation. *Acta Biomaterialia, 5*, 2657–2667.

Jain, P. K., Huang, X., El-Sayed, I. H., & El-Sayed, M. A. (2008). Noble metals on the nano-scale: Optical and photothermal properties and some applications in imaging, sensing, biology, and medicine. *Accounts of Chemical Research, 41*, 1578–1586.

Jannesari, M., Varshosaz, J., Morshed, M., & Zamani, M. (2011). Composite poly(vinyl alcohol)/poly(vinyl acetate) electrospun nano-fibrous mats as a novel wound dressing matrix for controlled release of drugs. *International Journal of Nanomedicine, 6*, 993–1003.

Jayakumar, R., Prabaharan, M., Nair, S. V., & Tamura, H. (2010). Novel chitin and chitosan nanofibers in biomedical applications. *Biotechnology Advances, 28*(1), 142–150.

Jose, M. V., Thomas, V., Johnson, K., Dean, D. R., & Nyairo, E. (2009). Aligned nanocomposite scaffolds based on PLGA/HA for bone tissue engineering. *Acta Biomaterialia, 5*, 305–315.

Jose, M. V., Thomas, V., Xu, Y., Bellis, S., Nyairo, E., & Dean, D. R. (2010). Multifunctional aligned nanofibrous nanocomposite scaffolds of PLGA/Collagen/HA for bone tissue engineering. *Macromolecular Bioscience, 10*, 433–444.

Kam, N. W., Jan, E., & Kotov, N. A. (2009). Electrical stimulation of neural stem cells mediated by humanized carbon nano-tube composite made with extracellular matrix protein. *Nano Letters, 9*, 273–278.

Kang, E. S., Kim, D. S., Suhito, I. R., Choo, S. S., Kim, S. J., Song, I., & Kim, T. H. (2017). Guiding osteogenesis of mesenchymal stem cells using carbon-based nano-materials. *Nano-Convergence, 4*(1), 2.

Kankala, R. K., Zhu, K., Sun, X. N., Liu, C. G., Wang, S. B., & Chen, A. Z. (2018). Cardiac tissue engineering on the nano-scale. *ACS Biomaterials Science & Engineering, 4*(3), 800–818.

Katti, D., Vasita, R., & Shanmugam, K. (2008). Improved biomaterials for tissue engineering applications: Surface modification of polymers. *Current Topics in Medicinal Chemistry, 8*(4), 341–353.

Kenawy, el-R., Bowlin, G. L., Mansfield, K., Layman, J., Simpson, D. G., Sanders, E. H., Wnek, G. E. (2002). Release of tetracycline hydrochloride from electrospun poly(ethylene-covinylacetate), poly(lactic acid), and a blend. J Control Release. 2002 May 17;81(1–2):57-64. https://doi.org/10.1016/s0168-3659(02)00041-x.

Keogh, M. B., O'Brien, F. J., & Daly, J. S. (2010). Substrate stiffness and contractile behaviour modulate the functional maturation of osteoblasts on a collagen–GAG scaffold. *Acta Biomaterialia, 6*(11), 4305–4313.

Kharaziha, M., Shin, S. R., Nikkhah, M., Topkaya, S. N., Masoumi, N., Annabi, N., & Khademhosseini, A. (2014). Tough and flexible CNT–polymeric hybrid scaffolds for engineering cardiac constructs. *Biomaterials, 35*(26), 7346–7354.

Kim, B. S., & Mooney, D. J. (1998). Development of biocompatible synthetic extracellular matrices for tissue engineering. *Trends in Biotechnology, 16*, 224–230.

Kim, D. H., Lipke, E. A., Kim, P., Cheong, R., Thompson, S., Delannoy, M., & Levchenko, A. (2009). Nano-scale cues regulate the structure and function of macroscopic cardiac tissue constructs. *Proceedings of the National Academy of Sciences, 107*(2), 565–570.

Kim, Y. J., Ebara, M., & Aoyagi, T. (2013). A smart hyperthermia nano-fiber with switchable drug release for inducing cancer apoptosis. *Advanced Functional Materials, 23*, 5753–5761.

Koski, A., Yim, K., & Shivkumar, S. (2004). Effect of molecular weight on fibrous PVA produced by electrospinning. *Materials Letters, 58*, 493–497.

Laborda, F., Bolea, E., Cepriá, G., Gómez, M. T., Jiménez, M. S., Pérez-Arantegui, J., & Castillo, J. R. (2016). Detection, characterization and quantification of inorganic engineered nanomaterials: A review of techniques and methodological approaches for the analysis of complex samples. *Analytica Chimica Acta, 904*, 10–32.

Lee, S., & Chen, X. Y. (2009). Dual-modality probes for in vivo molecular imaging. *Molecular Imaging, 8*(2), 87–100.

Lee, K. S., & El-Sayed, M. A. (2006). Gold and silver nano-particles in sensing and imaging: Sensitivity of plasmon response to size, shape, and metal composition. *The Journal of Physical Chemistry B, 110*, 19220–19225.

Lee, J. Y., Bashur, C. A., Goldstein, A. S., & Schmidt, C. E. (2009). Polypyrrole-coated electrospun PLGA nano-fibers for neural tissue applications. *Biomaterials, 30*(26), 4325–4335.

Lee, S. J., Zhu, W., Nowicki, M., Lee, G., Heo, D. N., Kim, J., & Zhang, L. G. (2018). 3D printing nano-conductive multi-walled carbon nano-tube scaffolds for nerve regeneration. *Journal of Neural Engineering, 15*(1), 016018.

Li, Z., & Wang, C. (2013). Effects of working parameters on electrospinning. In *One-dimensional nano-structures* (pp. 15–28). Berlin: Springer.

Li, X., Feng, Q., Liu, X., Dong, W., & Cui, F. (2006). Collagen-based implants reinforced by chitin fibres in a goat shank bone defect model. *Biomaterials, 27*(9), 1917–1923.

Li, W., Guo, Y., Wang, H., Shi, D., Liang, C., Ye, Z., & Gong, J. (2008). Electrospun nanofibers immobilized with collagen for neural stem cells culture. *Journal of Materials Science: Materials in Medicine, 19*(2), 847–854.

Li, Y., Chen, F., Nie, J., & Yang, D. (2012). Electrospun poly(lactic acid)/chitosan core–shell structure nano-fibers from homogeneous solution. *Carbohydrate Polymers, 90*, 1445–1451.

Li, L., Ge, J., Wang, L., Guo, B., & Ma, P. X. (2014). Electroactive nano-fibrous biomimetic scaffolds by thermally induced phase separation. *Journal of Materials Chemistry B, 2*(36), 6119.

Li, L., Zhou, G., Wang, Y., Yang, G., Ding, S., & Zhou, S. (2015). Controlled dual delivery of BMP-2 and dexamethasone by nano-particle-embedded electrospun nano-fibers for the efficient repair of critical-sized rat calvarial defect. *Biomaterials, 37*, 218–229.

Liang, D., Hsiao, B. S., & Chu, B. (2007). Functional electrospun nano-fibrous scaffolds for biomedical applications. *Advanced Drug Delivery Reviews, 59*, 1392–1412.

Liao, H., Qi, R., Shen, M., Cao, X., Guo, R., Zhang, Y., & Shi, X. (2011). Improved cellular response on multiwalled carbon nano-tube-incorporated electrospun polyvinyl alcohol/chitosan nano-fibrous scaffolds. *Colloids and Surfaces B: Biointerfaces, 84*(2), 528–535.

Limongi, T., Tirinato, L., Pagliari, F., Giugni, A., Allione, M., Perozziello, G., & Di Fabrizio, E. (2016). Fabrication and applications of micro/nano-structured devices for tissue engineering. *Nano-Micro Letters, 9*(1), 1.

Lin, C. Y. (2020). Ramp-creep ultrasound viscoelastography for measuring viscoelastic parameters of materials. *Materials, 13*(16), 3593.

Liu, X., Baldursdottir, S. G., Aho, J., Qu, H., Christensen, L. P., Rantanen, J., & Yang, M. (2017). Electrospinnability of poly lactic-co-glycolic acid (PLGA): The role of solvent type and solvent composition. *Pharmaceutical Research, 34*(4), 738–749.

Luo, H., Xiao, J., Peng, M., Zhang, Q., Yang, Z., Si, H., & Wan, Y. (2020). One-pot synthesis of copper-doped mesoporous bioglass towards multifunctional 3D nano-fibrous scaffolds for bone regeneration. *Journal of Non-Crystalline Solids, 532*, 119856.

Ma, G., Liu, Y., Peng, C., Fang, D., He, B., & Nie, J. (2011). Paclitaxel loaded electrospun porous nano-fibers as mat potential application for chemotherapy against prostate cancer. *Carbohydrate Polymers, 86*, 505–512.

Martins, A. M., Eng, G., Caridade, S. G., Mano, J. F., Reis, R. L., & Vunjak-Novakovic, G. (2014). Electrically conductive chitosan/carbon scaffolds for cardiac tissue engineering. *Biomacromolecules, 15*(2), 635–643.

Martinelli, V., Bosi, S., Peña, B., Baj, G., Long, C. S., Sbaizero, O., … Mestroni, L. (2018). 3D Carbon Nanotube-Based Composites for Cardiac Tissue Engineering. ACS *Applied Bio Materials*. https://doi.org/10.1021/acsabm.8b00440.

Mattei, G., Gruca, G., Rijnveld, N., & Ahluwalia, A. (2015). The nano-epsilon dot method for strain rate viscoelastic characterisation of soft biomaterials by spherical nano-indentation. *Journal of the Mechanical Behavior of Biomedical Materials, 50*, 150–159.

Mazzatenta, A., Giugliano, M., Campidelli, S., Gambazzi, L., Businaro, L., Markram, H., Prato, M., & Ballerini, L. (2007). Interfacing neurons with carbon nano-tubes: Electrical signal transfer and synaptic stimulation in cultured brain circuits. *The Journal of Neuroscience, 27*, 6931–6936.

Meng, J., Kong, H., Xu, H. Y., Song, L., Wang, C. Y., & Xie, S. S. (2005). Improving the blood compatibility of polyurethane using carbon nano-tubes as fillers and its implications to cardio-vascular surgery. *Journal of Biomedical Materials Research. Part A, 74*, 208–214.

Meng, X., Stout, D. A., Sun, L., Beingessner, R. L., Fenniri, H., & Webster, T. J. (2012). Novel injectable biomimetic hydrogels with carbon nano-fibers and self-assembled rosette nano-tubes for myocardial applications. *Journal of Biomedical Materials Research. Part A, 101A*(4), 1095–1102.

Mi, X., Gupte, M. J., Zhang, Z., Swanson, W. B., McCauley, L. K., & Ma, P. X. (2020). Three-dimensional electrodeposition of calcium phosphates on porous nano-fibrous scaffolds and their controlled release of calcium for bone regeneration. *ACS Applied Materials & Interfaces, 2*(29), 32503–32513.

Mituppatham, C., Nithitanakul, M., & Supaphol, P. (2004). Ultrafine electrospun polyamide-6 fibers: Effect of solution conditions on morphology and average fiber diameter. *Macromolecular Chemistry and Physics, 205*, 2327–2338.

Mourdikoudis, S., Pallares, R. M., & Thanh, N. T. K. (2018). Characterization techniques for nano-particles: Comparison and complementarity upon studying nano-particle properties. *Nanoscale, 10*(27), 12871–12934.

Nayak, T. R., Jian, L., Phua, L. C., Ho, H. K., Ren, Y., & Pastorin, G. (2010). Thin films of functionalized multiwalled carbon nano-tubes as suitable scaffold materials for stem cells proliferation and bone formation. *ACS Nano, 4*, 7717–7725.

Nguyen, T. T., Ghosh, C., Hwang, S. G., Chanunpanich, N., & Park, J. S. (2012). Porous core/sheath composite nano-fibers fabricated by coaxial electrospinning as a potential mat for drug release system. *International Journal of Pharmaceutics, 439*, 296–306.

Nishizawa, M., Nozaki, H., Kaji, H., Kitazume, T., Kobayashi, N., Ishibashi, T., & Abe, T. (2007). Electrodeposition of anchored polypyrrole film on microelectrodes and stimulation of cultured cardiac myocytes. *Biomaterials, 28*(8), 1480–1485.

Nitti, P., Gallo, N., Natta, L., Scalera, F., Palazzo, B., Sannino, A., & Gervaso, F. (2018). Influence of nano-fiber orientation on morphological and mechanical properties of electrospun chitosan mats. *Journal of Healthcare Engineering, 2018*, 1–12.

Noh, Y. K., Du, P., Kim, I. G., Ko, J., Kim, S. W., & Park, K. (2016). Polymer mesh scaffold combined with cell-derived ECM for osteogenesis of human mesenchymal stem cells. *Biomaterials Research, 20*, 6, Open access.

Nune, S. K., Gunda, P., Thallapally, P. K., Lin, Y. Y., Forrest, M. L., & Berkland, C. J. (2009). Nano-particles for biomedical imaging. *Expert Opinion on Drug Delivery, 6*(11), 1175–1194.

Oyen, M. L. (2011). Nano-indentation of biological and biomimetic materials. *Experimental Techniques, 37*(1), 73–87.

Oyen, M. L., & Cook, R. F. (2009). A practical guide for analysis of nano-indentation data. *Journal of the Mechanical Behavior of Biomedical Materials, 2*(4), 396–407.

Parisi, L., Toffoli, A., Ghiacci, G., & Macaluso, G. (2018). Tailoring the interface of biomaterials to design effective scaffolds. *Journal of Functional Biomaterials, 9*(3), 50.

Park, S. Y., Choi, D. S., Jin, H. J., Park, J., Byun, K. E., Lee, K. B., & Hong, S. (2011a). Polarization-controlled differentiation of human neural stem cells using synergistic cues from the patterns of carbon nano-tube monolayer coating. *ACS Nano, 5*(6), 4704–4711.

Park, S. Y., Park, J., Sim, S. H., Sung, M. G., Kim, K. S., Hong, B. H., & Hong, S. (2011b). Enhanced differentiation of human neural stem cells into neurons on graphene. *Advanced Materials, 23*(36), H263–H267.

Perepelkin, N. V., Borodich, F. M., Kovalev, A. E., & Gorb, S. N. (2019). Depth-sensing indentation as a micro- and nano-mechanical approach to characterisation of mechanical properties of soft, biological, and biomimetic materials. *Nano-materials, 10*(1), 15.

Petersen, A., Joly, P., Bergmann, C., Korus, G., & Duda, G. N. (2012). The impact of substrate stiffness and mechanical loading on fibroblast-induced scaffold remodeling. *Tissue Engineering Part A, 18*(17–18), 1804–1817.

Phipps, M. C., Clem, W. C., Catledge, S. A., Xu, Y., Hennessy, K. M., Thomas, V., Jablonsky, M. J., Chowdhury, S., Stanishevsky, A., Vohra, Y. K., & Bellis, S. L. (2011). Mesenchymal stem cells response to bone-mimetic elctrospun matrices composed of polycaprolactone, collagen I and nanoparticulate hydroxyapatite. *PLoS One, 6*(2), e16813.

Prasadh, S., & Wong, R. C. W. (2018). Unraveling the mechanical strength of biomaterials used as a bone scaffold in oral and maxillofacial defects. *Oral Science International, 15*(2), 48–55.

Qi, R., Cao, X., Shen, M., Guo, R., Yu, J., & Shi, X. (2012). Biocompatibility of electrospun halloysite nano-tube-doped poly(lactic-co-glycolic acid) composite nano-fibers. *Journal of Biomaterials Science. Polymer Edition, 23*(1–4), 299–313.

Qiu, F., Chen, Y., Tang, C., & Zhao, X. (2018). Amphiphilic peptides as novel nano-materials: Design, self-assembly and application. *International Journal of Nano-medicine, 13*, 5003–5022.

Rajabi, M., Firouzi, M., Hassannejad, Z., Haririan, I., & Zahedi, P. (2017). Fabrication and characterization of electrospun laminin-functionalized silk fibroin/poly(ethylene oxide) nano-fibrous scaffolds for peripheral nerve regeneration. *Journal of Biomedical Materials Research Part B: Applied Biomaterials, 106*(4), 1595–1604.

Rajabi M., Firouzi M., Hassannejad Z., Haririan I., Zahedi P. (2018). Fabrication and characterization of electrospun lamininfunctionalized silk fibroin/poly(ethylene oxide) nanofibrous scaffolds for peripheral nerve regeneration. J Biomed Mater Res B Appl Biomater. May;*106*(4):1595–1604. https://doi.org/10.1002/jbm.b.33968.

Ravichandran, R., Sridhar, R., Venugopal, J. R., Sundarrajan, S., Mukherjee, S., & Ramakrishna, S. (2014). Gold nano-particle loaded hybrid nano-fibers for cardiogenic differentiation of stem cells for infarcted myocardium regeneration. *Macromolecular Bioscience, 14*(4), 515–525.

Razavi, M. (2017). Introduction. In *Nano-biomaterials science, development and evaluation* (pp. 1–5). Amsterdam: Elsevier.

Reneker, D. H., & Chun, I. (1996). Nano-metre diameter fibres of polymer, produced by electrospinning. *Nanotechnology, 7*, 216–223.

Rezabeigi, E., & Demarquette, N. R. (2019). Ultraporous membranes electrospun from nonsolvent-induced phase-separated ternary systems. *Macromolecular Rapid Communications, 40*(9), e1800880.

Rezaei, A., Nasirpour, A., & Fathi, M. (2015). Application of cellulosic nano-fibers in food science using electrospinning and its potential risk comprehensive. *Reviews in Food Science and Food Safety, 14*, 269–284.

Richbourg, N. R., Peppas, N. A., & Sikavitsas, V. I. (2019). Tuning the biomimetic behavior of scaffolds for regenerative medicine through surface modifications. *Journal of Tissue Engineering and Regenerative Medicine, 13*(8), 1275–1293.

Riggio, C., Calatayud, M. P., Giannaccini, M., Sanz, B., Torres, T. E., Fernández-Pacheco, R., & Raffa, V. (2014). The orientation of the neuronal growth process can be directed via magnetic nano-particles under an applied magnetic field. *Nano-medicine: Nano-technology, Biology and Medicine, 10*(7), 1549–1558.

Rnjak-Kovacina, J., Wise, S. G., Li, Z., Maitz, P. K. M., Young, C. J., Wang, Y., & Weiss, A. S. (2012). Electrospun synthetic human elastin: Collagen composite scaffolds for dermal tissue engineering. *Acta Biomaterialia, 8*, 3714–3722.

Sakai, S., Masuhara, H., Yamada, Y., Ono, T., Ijima, H., & Kawakami, K. (2005). Transition of mechanical property of porous alginate scaffold with cells during culture period. *Journal of Bioscience and Bioengineering, 100*(1), 127–129.

Sandler, S., Fellows, B. D., & Mefford, O. T. (2019). Best practices for characterization of magnetic nano-particles for biomedical applications. *Analytical Chemistry, 91*(22), 14159–14169.

Sapir, Y., Polyak, B., & Cohen, S. (2013). Nanomaterials for cardiac tissue engineering. In *Nanomaterials in tissue engineering* (pp. 244–275). Cambridge: Woodhead Publishing.

Sell, S. A., McClure, M. J., Garg, K., Wolfe, P. S., & Bowlin, G. L. (2009). Electrospinning of collagen/biopolymers for regenerative medicine and cardiovascular tissue engineering. *Advanced Drug Delivery Reviews, 61*(12), 1007–1019.

Shah, S., Solanki, A., & Lee, K. B. (2016). Nano-technology-based approaches for guiding neural regeneration. *Accounts of Chemical Research, 49*(1), 17–26.

Sharma, S., Sapkota, D., Xue, Y., Rajthala, S., Yassin, M. A., Finne-Wistrand, A., & Mustafa, K. (2018). Delivery of VEGFA in bone marrow stromal cells seeded in copolymer scaffold enhances angiogenesis, but is inadequate for osteogenesis as compared with the dual delivery of VEGFA and BMP2 in a subcutaneous mouse model. *Stem Cell Research & Therapy, 9*(1), 23.

Shigao, C., Urban, M., Pislaru, C., Kinnick, R., Zheng, Y., Yao, A., & Greenleaf, J. (2009). Shearwave dispersion ultrasound vibrometry (SDUV) for measuring tissue elasticity and viscosity. *IEEE Transactions on Ultrasonics, Ferroelectrics, and Frequency Control, 56*(1), 55–62.

Shigemitsu, H., & Hamachi, I. (2017). Design strategies of stimuli-responsive supramolecular hydrogels relying on structural analyses and cell-mimicking approaches. *Accounts of Chemical Research, 50*(4), 740–750.

Shin, S. R., Aghaei-Ghareh-Bolagh, B., Gao, X., Nikkhah, M., Jung, S. M., Dolatshahi-Pirouz, A., & Khademhosseini, A. (2014). Layer-by-layer assembly of 3D tissue constructs with functionalized graphene. *Advanced Functional Materials, 24*(39), 6136–6144.

Simón-Yarza, T., Rossi, A., Heffels, K. H., Prósper, F., Groll, J., & Blanco-Prieto, M. J. (2015). Polymeric electrospun scaffolds: Neuregulin encapsulation and biocompatibility studies in a model of myocardial ischemia. *Tissue Engineering Part A, 21*(9–10), 1654–1661.

Slemming-Adamsen, P., Song, J., Dong, M., Besenbacher, F., & Chen, M. (2015). In situ cross-linked pnipam/gelatin nano-fibers for thermo-responsive drug release. *Macromolecular Materials and Engineering, 300*, 1226–1231.

Smirnov, A. (2011). EPR studies of nano-materials. In S. K. Misra (Ed.), *Multifrequency electron paramagnetic resonance* (pp. 825–843). Hoboken: Wiley.

Smith, I. O., Liu, X. H., Smith, L. A., & Ma, P. X. (2009). Nano-structured polymer scaffolds for tissue engineering and regenerative medicine. *Wiley Interdisciplinary Reviews: Nano-medicine and Nano-biotechnology, 1*(2), 226–236.

Smith, A. S. T., Yoo, H., Yi, H., Ahn, E. H., Lee, J. H., Shao, G., & Kim, D. H. (2017). Micro- and nano-patterned conductive graphene–PEG hybrid scaffolds for cardiac tissue engineering. *Chemical Communications, 53*(53), 7412–7415.

Sperling, L. H. (2005). *Introduction to physical polymer science*. New York: Wiley.

Sridhar, S., Venugopal, J. R., Sridhar, R., & Ramakrishna, S. (2015). Cardiogenic differentiation of mesenchymal stem cells with gold nano-particle loaded functionalized nano-fibers. *Colloids and Surfaces B: Biointerfaces, 134*, 346–354.

Stanishevsky, A., Chowdhury, S., Chinoda, P., & Thomas, V. (2008). Hydroxyapatite nanoparticle loaded collagen fiber composites: Microarchitecture and nanoindentation study. *Journal of Biomedical Materials Research Part A, 86*(4), 873–882.

Stevens, M. M. (2005). Exploring and engineering the cell surface interface. *Science, 310*(5751), 1135–1138.

Stewart, E., Kobayashi, N. R., Higgins, M. J., Quigley, A. F., Jamal, S., Moulton, S. E., & Crook, J. M. (2015). Electrical stimulation using conductive polymer polypyrrole promotes differentiation of human neural stem cells: A biocompatible platform for translational neural tissue engineering. *Tissue Engineering Part C: Methods, 21*(4), 385–393.

Stout, D. A., Basu, B., & Webster, T. J. (2011). Poly(lactic–co-glycolic acid): Carbon nano-fiber composites for myocardial tissue engineering applications. *Acta Biomaterialia, 7*(8), 3101–3112.

Sudwilai, T., Ng, J. J., Boonkrai, C., Israsena, N., Chuangchote, S., & Supaphol, P. (2014). Polypyrrole-coated electrospun poly(lactic acid) fibrous scaffold: Effects of coating on electrical conductivity and neural cell growth. *Journal of Biomaterials Science, Polymer Edition, 25*(12), 1240–1252.

Sun, B., Long, Y. Z., Zhang, H. D., Li, M. M., Duvail, J. L., Jiang, X. Y., & Yinf, H. L. (2014). Advances in three-dimensional nano-fibrous macrostructures via electrospinning. *Progress in Polymer Science, 39*, 862–890.

Sun, L., Zheng, C., & Webster, T. (2017). Self-assembled peptide nano-materials for biomedical applications: Promises and pitfalls. *International Journal of Nano-medicine, 12*, 73–86.

Tan, E. P. S., & Lim, C. T. (2006). Mechanical characterization of a single nano-fiber. In *Nanomechanics of materials and structures* (pp. 121–137). Dordrecht: Springer.

Tao, F., Cheng, Y., Shi, X., Zheng, H., Du, Y., Xiang, W., & Deng, H. (2020). Applications of chitin and chitosan nano-fibers in bone regenerative engineering. *Carbohydrate Polymers, 230*, 115658.

Thomas, V., Dean, R., & Vohra, Y. K. (2006). Nanostructured biomaterials for regenerative medicine. *Current Nanoscience, 2*, 155–117.

Thomas, V., Dean, D. R., Jose, M. V., Mathew, B., Chowdhury, S., & Vohra, Y. K. (2007). Nanostructured biocomposite scaffolds based on collagen coelectrospun with nanohydroxyapatite. *Biomacromolecules, 8*(2), 631–637.

Thomas, V., Donahoe, T., Dean, D. R., Nyairo, E., & Vohra, Y. K. (2011). Electrospun tubular conduits based on bioresorbable biosyn®: Mechano-morphological characterization. *Acta Biomaterialia, 7*, 2070–2079.

Tiwari, S. K., & Venkatraman, S. S. (2012). Importance of viscosity parameters in electro-spinning: Of monolithic and core-shell fibers. *Materials Science and Engineering C, 32*, 1037–1042.

Tripathi, R. M., & Chung, S. J. (2018). Biogenic nano-materials: Synthesis, characterization, growth mechanism, and biomedical applications. *Journal of Microbiological Methods, 157*, 65–80.

Tucker, B. S., Vijayan, V. M., Vohra, Y. K., & Thomas, V. (2019). Novel magneto-plasma processing for enhanced modification of electrospun biomaterials. *Materials Letters, 250*, 96–98.

Tyagi, P., Catledge, S., Stanishevsky, A., Thomas, V., & Vohra, Y. K. (2009). Nanomechanical properties of electrospun polycaprolactone/nanohydroxyapatite fibers. *Journal of Nanoscience and Nanotechnology, 9*, 4839–4845.

Veetil, J. V., & Ye, K. (2009). Tailored carbon nano-tubes for tissue engineering applications. *Biotechnology Progress, 25*, 709–721.

Wade, R. J., & Jason, A. B. (2014). Advances in nanofibrous scaffolds for biomedical applications: From electrospinning to self-assembly. *Nano Today, 9*, 722–742.

Wakuda, Y., Nishimoto, S., Suye, S. I., & Fujita, S. (2018). Native collagen hydrogel nano-fibres with anisotropic structure using core-shell electrospinning. *Scientific Reports, 8*(1), 6248.

Wang, L., Jiang, J., Hua, W., Darabi, A., Song, X., Song, C., & Qiu, X. (2016). Mussel-inspired conductive cryogel as cardiac tissue patch to repair myocardial infarction by migration of conductive nano-particles. *Advanced Functional Materials, 26*(24), 4293–4305.

Wang, Y., Guo, Z., Qian, Y., Zhang, Z., Lyu, L., Wang, Y., & Ye, F. (2019). Study on the electrospinning of gelatin/pullulan composite nano-fibers. *Polymers (Basel), 11*(9), 1424.

Webber, M. J., Appel, E. A., Meijer, E. W., & Langer, R. (2016). Supramolecular biomaterials. *Nature Materials, 15*(1), 13–26.

Webster, T. J., & Ejiofor, J. U. (2004). Increased osteoblast adhesion on nano-phase metals: Ti, Ti6Al4V, and CoCrMo. *Biomaterials, 25*, 4731–4739.

Webster, T. J., Siegel, R. W., & Bizios, R. (1999). Osteoblast adhesion on nano-phase ceramics. *Biomaterials, 20*, 1221–1227.

Webster, T. J., Schadler, L. S., Siegel, R. W., & BiziosR. (2001). Mechanisms of enhanced osteoblast adhesion on nano-phase alumina involve vitronectin. *Tissue Engineering, 7*(3), 291–301.

Wen, X. T., Fan, H. S., Tan, Y. F., Cao, H., Li, H., Bi, C., & Zhang, X. D. (2005). Preparation of electrospun PLA nano-fiber scaffold and the evaluation in vitro. *Key Engineering Materials, 288–289*, 139–142.

Wickham, A. M., Islam, M. M., Mondal, D., Phopase, J., Sadhu, V., Tamás, É., & Griffith, M. (2014). Polycaprolactone-thiophene-conjugated carbon nano-tube meshes as scaffolds for cardiac progenitor cells. *Journal of Biomedical Materials Research Part B: Applied Biomaterials, 102*(7), 1553–1561.

Xia, H., Sun, X., Liu, D., Zhou, Y., & Zhong, D. (2016). Oriented growth of rat Schwann cells on aligned electrospun poly(methyl methacrylate) nano-fibers. *Journal of the Neurological Sciences, 369*, 88–95.

Xiao, J., Luo, H., Ao, H., Huang, Y., Yao, F., Zhang, Q., & Wan, Y. (2019a). A rhBMP-2-loaded three-dimensional mesoporous bioactive glass nano-tubular scaffold prepared from bacterial cellulose. *Colloids and Surfaces A: Physicochemical and Engineering Aspects, 581*, 123838.

Xiao, J., Wan, Y., Yang, Z., Huang, Y., Zhu, Y., Yao, F., & Luo, H. (2019b). Simvastatin-loaded nano-tubular mesoporous bioactive glass scaffolds for bone tissue engineering. *Microporous and Mesoporous Materials, 109*, 570.

Xin, X., Hussain, M., & Mao, J. J. (2007). Continuing differentiation of human mesenchymal stem cells and induced chondrogenic and osteogenic lineages in electrospun PLGA nano-fiber scaffold. *Biomaterials, 28*, 316–325.

Xu, C., Inai, R., Kotaki, M., & Ramakrishna, S. (2004). Electrospun nano-fiber fabrication as synthetic extracellular matrix and its potential for vascular tissue engineering. *Tissue Engineering, 10*(7–8), 1160–1168.

Xu, Z., Xu, Y., Basuthakur, P., Patra, C. R., Ramakrishna, S., Liu, Y., Thomas, V., & Nanda, H. S. (2020). Fibro-porous PLLA/gelatin composite membrane doped with cerium oxide nanoparticles as bioactive scaffolds for future angiogenesis. *Journal of Materials Chemistry B, 8*(39), 9110–9120.

Yadid, M., Feiner, R., & Dvir, T. (2019). Gold nano-particle-integrated scaffolds for tissue engineering and regenerative medicine. *Nano Letters, 19*(4), 2198–2206.

Yang, F., Murugan, R., Wang, S., & Ramakrishna, S. (2005). Electrospinning of nano-/micro scale poly-(l-lactic acid) aligned fibers and their potential in neural tissue engineering. *Biomaterials, 26*, 2603–2610.

Yesilkir-Baydar, S., Oztel, O. N., Cakir-Koc, R., & Candayan, A. (2017). Evaluation techniques. In *Nano-biomaterials science, development and evaluation* (pp. 211–232). Amsterdam: Elsevier.

You, J. O., Rafat, M., Ye, G. J. C., & Auguste, D. T. (2011). Nano-engineering the heart: Conductive scaffolds enhance connexin 43 expression. *Nano Letters, 11*(9), 3643–3648.

Yu, M. F., Files, B. S., Arepalli, S., & Ruoff, R. S. (2000). Tensile loading of ropes of single wall carbon nano-tubes and their mechanical properties. *Physical Review Letters 84*, 5552–5555.

Yu, D. G., Branford-White, C., Li, L., Wu, X. M., & Zhu, L. M. (2010). The compatibility of acyclovir with polyacrylonitrile in the electrospun drug-loaded nano-fibers. *Journal of Applied Polymer Science, 117*, 1509–1515.

Yuan, M., Wang, Y., & Qin, Y. X. (2019). Engineered nano-medicine for neuroregeneration: Light emitting diode-mediated superparamagnetic iron oxide-gold core-shell nano-particles functionalized by nerve growth factor. *Nano-medicine: Nano-technology, Biology and Medicine, 102*, 052.

Zahedi, P., & Fallah-Darrehchi, M. (2015). Electrospun egg albumin-PVA nano-fibers containing tetracycline hydrochloride: Morphological, drug release, antibacterial, thermal and mechanical properties. *Fibers and Polymers, 16*, 2184–2192.

Zhang, S. (2003). Fabrication of novel biomaterials through molecular self-assembly. *Nature Biotechnology 200, 321*(10), 1171–1178.

Zhang, R. Y., & Ma, P. X. (2000). Synthetic nano-fibrillar extracellular matrices with predesigned macro-porous architectures. *Journal of Biomedical Materials Research, 52*(2), 430–438.

Zhang, K., Zheng, H., Liang, S., & Gao, C. (2016). Aligned PLLA nano-fibrous scaffolds coated with graphene oxide for promoting neural cell growth. *Acta Biomaterialia, 37*, 131–142.

Zhang, H., Zhang, Q., zhe Ruan, L., Duan, J., Wan, M., & Insana, M. F. (2018). Modeling ramp-hold indentation measurements based on Kelvin–Voigt fractional derivative model. *Measurement Science and Technology, 29*(3), 035701.

Zhao, F., Yao, D., Guo, R., Deng, L., Dong, A., & Zhang, J. (2015). Composites of polymer hydrogels and nanoparticulate systems for biomedical and pharmaceutical applications. *Nanomaterials, 5*, 2054–2130.

Zhu, Y., Dong, Z., Wejinya, U. C., Jin, S., & Ye, K. (2011). Determination of mechanical properties of soft tissue scaffolds by atomic force microscopy nano-indentation. *Journal of Biomechanics, 44*(13), 2356–2361.

Zupanˇciˇc, Š., Sinha, R. S., Sinha, R. S., Kristl, J., & Yarin, A. L. (2016). Controlled release of ciprofloxacin from core–shell nano-fibers with monolithic or blended core. *Molecular Pharmaceutics, 13*, 1393–1404.

# Chapter 8
# Highly Toxic Nanomaterials for Cancer Treatment

**Mazhar Salim Al Zoubi, Alaa A. A. Aljabali, and Kaushik Pal**

## 8.1 Research Background and Discoveries of Toxic Nanomaterials

Nanotechnology is a new science that is invading broad sectors in our lives, including industrial, environmental, and health applications. This new era of biotechnology applications must be evaluated from time to time to reveal the impact and efficiency of our health. Promising outcomes have been reported in many in vitro and in vivo studies (Wang et al. 2017a; Aljabali et al. 2018a, b, 2019, 2020; Al-Trad et al. 2019; Alomari et al. 2020). The most common diseases, such as cancers and diabetes, are eager to find safe and effective treatments, including natural products and nanotechnology approached to treat or deliver the drug (Bisht and Rayamajhi 2016; Wang et al. 2017a, b; Farooq et al. 2018; Liu et al. 2018; Yang et al. 2018; Arvanag et al. 2019; Peng et al. 2019; Agarwal et al. 2020). Although contradictory outcomes regarding nanoparticles efficacy in the treatment of chronic diseases such as cancer, the toxicity of these nanomaterial treatments need to be evaluated in vivo. Two decades ago, there was an uncertainty report by the Royal Society and Royal Academy of Engineering (Pidgeon et al. 2004; Maynard 2014; Bowman 2017).

M. S. Al Zoubi (✉)
Faculty of Medicine, Department of Basic Medical Sciences, Yarmouk University, Irbid, Jordan
e-mail: mszoubi@yu.edu.jo

A. A. A. Aljabali
Faculty of Pharmacy, Department of Pharmaceutics and Pharmaceutical Technology, Yarmouk University, Irbid, Jordan

K. Pal
Laboratório de Biopolímeros e Sensores, Instituto de Macromoléculas, Universidade Federal do Rio de Janeiro (LABIOS/IMA/UFRJ)–Cidade Universitária, Rio de Janeiro, Brazil

Wuhan University, Wuchang Dist., Hubei, China


K. Pal (ed.), *Bio-manufactured Nanomaterials*,
https://doi.org/10.1007/978-3-030-67223-2_8

Then, different countries established different plans to assess the clinical manifestations and toxicity issues of the nanoscale materials (Handy and Shaw 2007). The wide range of nanomaterial sizes (1–100 nm) and types are making these assessments not easy going. This—not surprisingly—resulted in a vast number of studies on individual nanoparticles in particular applications. In this chapter, we will discuss significant issues and concerns about nanomaterials toxicity in cancer treatments. Therefore, this chapter will be subsection according to the application routes of the nanomaterials.

## 8.2 Advantages of Nanoparticle Formulation: Drug Delivery Treatment

Encapsulation of targeted drug systems have been proposed and designed to overcome the adverse effect of specific cancer therapies such as chemotherapies. This strategic option will provide a particular targeting of cancer sites (Wang et al. 2016). Generally, there are two primary mechanisms of targeted drug delivery, which are known as (a) passive targeting and (b) active targeting. In the former mechanism, the nanoparticles rely on the high permeability of certain tumors due to the lack of lymphatic drainage, which will generate a space of more than 10 nm, such as the normal tissues, up to 150 nm (Bazak et al. 2014). Therefore, the larger size nanoparticles have an advantageous step over the small size molecules of chemotherapies that can pass the 10 nm space of all tissues causing systematic toxicity. Different studies demonstrated lower systemic cytotoxicity of encapsulated cancer chemotherapies (Drummond et al. 1999; Heo et al. 2012; Tang et al. 2012; Mundra et al. 2015).

On the other hand, active targeting therapy is characterized by taking advantage of the overexpression of specific receptors on the cancer cell membrane (Fig. 8.1). This approach involves the generation of different ligand-binding site mechanism.

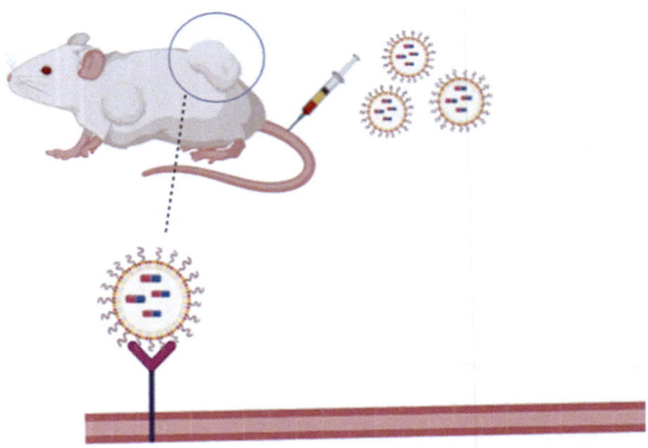

**Fig. 8.1** A representative diagram showing the possibility of using nanoparticles as a drug delivery system in the treatment of xenografted tumors

For instance, it can be a conjugation of antibodies, aptamers, carbohydrates, and peptide molecules (Byrne et al. 2008; Cho et al. 2008). In light of this specific targeting of cancer cells, there were some promising outcomes when the researchers targeted stem cells that are involved in the initiation and metastasis of cancer. In addition, these studies reported lower systemic cytotoxicity of the active targeting therapies (Rao et al. 2015).

As a new approach, the endogenous and exogenous stimuli-responsive nanoparticle systems have been proposed to be used in the treatment of cancers depending on endogenous factors of cancer microenvironment such as hypoxia, ATP, and acidic conditions or exogenous factors such as using physical properties of heat, light, and MRI (Raza et al. 2019). For instance, hyperthermia has been reported to be beneficial for the drug release at the target site of tumor growth (Chen et al. 2011; Wang et al. 2012a) In addition to the previous approaches of nanoparticles in cancer treatment, the "combination therapy" is considered as a practical option. This involves the delivery of more than one drug through the nanoparticle courier systems.

## 8.3 Toxicity of Nanoparticles

Lately, the use of nanoparticles in cancer therapy has received much attention. Research shows that different nanoparticles of metal oxide induce cytotoxic effects in cancerous cells and not in normal cells. In certain instances, nanoparticles alone and in comparison, to the other two treatments, such as photocatalytic therapy or anticancer drugs, have been shown to use such activity against cancer. Zinc oxide nanoparticles were demonstrated impartial of such an activity or loaded with a drug for cancer, such as doxorubicin. Cobalt oxide, iron oxide, and copper oxide are also nanoparticles that show cytotoxic effects on cancer cells. The antitumor mechanism might work with, among other possibilities, reactive oxygen species (ROS) or apoptosis and necrosis (Fig. 8.10) (Vinardell and Mitjans 2015). Due to the differences in the delivery system using nanoparticle approaches, there will be different clinical manifestations of these options.

### *8.3.1 Liposomal Nanoparticles*

Liposomal anticancer drugs are one of the few FDA-approved therapies in cancer (Doxil or Caelyx and DaunoXome). The European approved liposomal formula is non-pegylated doxorubicin (Myocet) (Haley and Frenkel 2008). Some studies have investigated the cytotoxic effect of liposomal nanoparticles based on different properties. For instance, positively charged lipid nanoparticles (+NP) showed hepatotoxicity compared to the negatively charged particles. Kedmi et al. showed significant overexpression of specific cytokines such as IF-$\gamma$ and TNF-$\alpha$ by leukocytes through the induction of TLR-4 (Kedmi et al. 2010). Other studies showed an induction of the reactive oxygen species (ROS) production and lower the viability of cells by different concentrations of cationic liposomal nanoparticles (Takano et al. 2001; Bae et al. 2009; Soenen et al. 2009). These ROS activation processes are related to macrophage activation induced by nanoparticle treatment.

### *8.3.2 Gold Nanoparticles (Au NPs)*

Gold nanoparticles are one of the proposed options for the treatment of cancer. Therefore, many studies have been conducted to evaluate the toxicity of Au NPs in vitro and in vivo. Some studies showed variable cellular uptake of gold nanoparticles due to the differences in the size of the prepared nanoparticles. It has been shown that 2–6 nm Au NPs can be localized in nuclear and cytoplasmic sites of the cell compared to the 15 nm particles which have been found in the cytosolic location only (Huang et al. 2012). Interestingly, a study showed that Au NPs could be taken by receptor-mediated endocytosis processes in breast cancer cell lines (Zhang et al. 2013). In an in vivo study, the results showed an accumulation of gold nanoparticles (4–10 nm) in different organs, including the kidney, liver, brain, and spleen (Hillyer and Albrecht 2001).

Therefore, safety concerns are limiting the evaluation of the toxic effect of Au NPs in humans. However, a phase I trial showed the localization of 27 nm TNF-α-Au NPs in the breast cancer tumor tissues but not in the parenchymal cells. In contrast, normal liver cells revealed the presence of those particles in liver tumor cases (Libutti et al. 2010). It has been shown that more than 90% of gold nanoparticles circulate in the blood for more than a week, and most of it accumulates in the liver, the perseverance of Au NPs is attributed to the large surface area to mass ratio (Oberdörster et al. 2005).

Different studies reported the low toxicity of gold nanoparticles. For example, gold nanoparticles sized 4, 12, and 18 nm showed no effect on the mortality of leukocytes in vitro (Connor et al. 2005). In another study, the 30 nm Au NPs showed concentration-dependent hemolysis of RBC after incubation with no aggregation of platelets or ROS formation (Love et al. 2012). In an in vivo study, the Au NPs did not show signs of toxicity within 24 h of injection (Glazer et al. 2011). In another in vivo study, the embryos of zebrafish did not show malformation or death after exposure to gold nanoparticles compared to the silver nanoparticles (Asharani et al. 2011). In addition, Shayne et al. evaluated the toxicity of gold nanoshells (150 nm) in mice and dogs. The study showed no signs of toxicity in the injected animals. However, the cleared nanoparticles were mainly accumulated in the reticuloendothelial system (liver and spleen), where small concentrations were detected in the kidney and adrenal glands with no signs of accumulation in other tissues. On the other hand, dogs showed a slight variation in blood chemistry and the formation of black pigments in Kupffer cells in the liver and spleen tissue (Gad et al. 2012). In vitro experiment demonstrated stable gold nanoparticles with a green synthesis approach (Fig. 8.2) (Asiya et al. 2020).

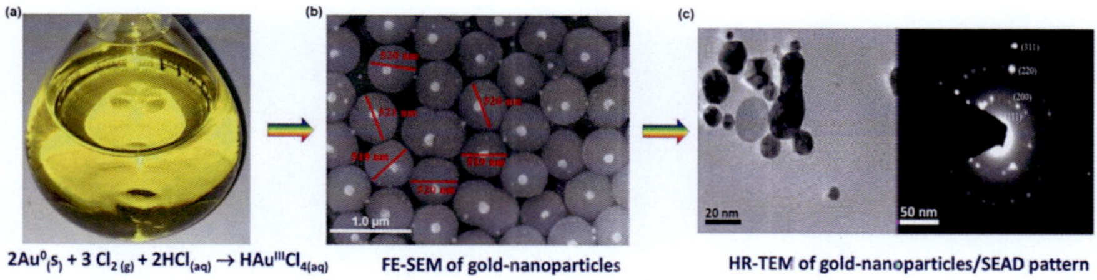


**Fig. 8.2** Illustration of green synthesized gold nanoparticles demonstrating the FE-SEM and HR-TEM (Asiya et al. 2020) [Copyright @ 2020, Elsevier]

### *8.3.3 Silver Nanoparticles (Ag NPs)*

A group of studies have been conducted to evaluate the toxicity of silver nanoparticles in different systems, including environmental, in vitro, and in vivo (Braydich-Stolle et al. 2005; Santoro et al. 2007; Ahamed et al. 2008; Carlson et al. 2008; Navarro et al. 2008; Sereemaspun et al. 2008; Arora et al. 2009; Choi et al. 2009; Griffitt et al. 2009; Kawata et al. 2009; Kvitek et al. 2009; Roh et al. 2009). For instance, an in vivo study reported a concentration-dependent malformation and mortality in zebrafish embryos due to exposure to silver nanoparticles (Ag NPs) sized 5–35 nm. Additionally, Ag NPs and platinum nanoparticles caused a delay in hatching and a reduction in the heart rate (Asharani et al. 2011). Treatment of mouse cells by two types of Ag NPs revealed the DNA damage effect in vitro. In particular, the polysaccharide-coated Ag nanoparticles showed more severe DNA damage compared to uncoated Ag NPs. However, both the types have shown upregulation of p53, Rad51, and p-H2AX proteins. Moreover, there was an induction of cell apoptosis. The difference in the cell response was attributed to the particle surface chemistry, which makes the coated Ag NPs more distributed in organelles (Ahamed et al. 2008). In another study, the researchers reported size-dependent ROS and inflammatory induction by smaller Ag NPs (15 nm) in lung macrophages (Carlson et al. 2008).

On the other hand, Arora et al. reported a safe interaction between the liver cells and fibroblast with Ag NPs and induction of anti-oxidative mechanism; however, they mentioned the presence of dark pigments in the mitochondria (Arora et al. 2009). In a model of a unicellular organism (*Paramecium caudatum*), the results showed no toxicity of Ag NPs below 25 mg/mL, while the Ag ion was found to be toxic at 0.4 mg/mL (Kvitek et al. 2009). Moreover, an in vitro study using the A549 lung cell line showed higher toxicity of Ag NP with a higher fraction of silver ions (Beer et al. 2012). These results are supported by another study when the researchers reported the more toxic effect of prolonged storage Ag NPs. The dissolution of Ag ions was found to be the cause of toxicity (Kittler et al. 2010). In another study, the Ag NPs exhibited toxicity when added to *Chlamydomonas reinhardtii* alga (Navarro et al. 2008). Thirugnanasambandan et al. showed the formation of silver nanoparticles using FE-SEM and HR-TEM (Fig. 8.3) (Thirugnanasambandan et al. 2018).

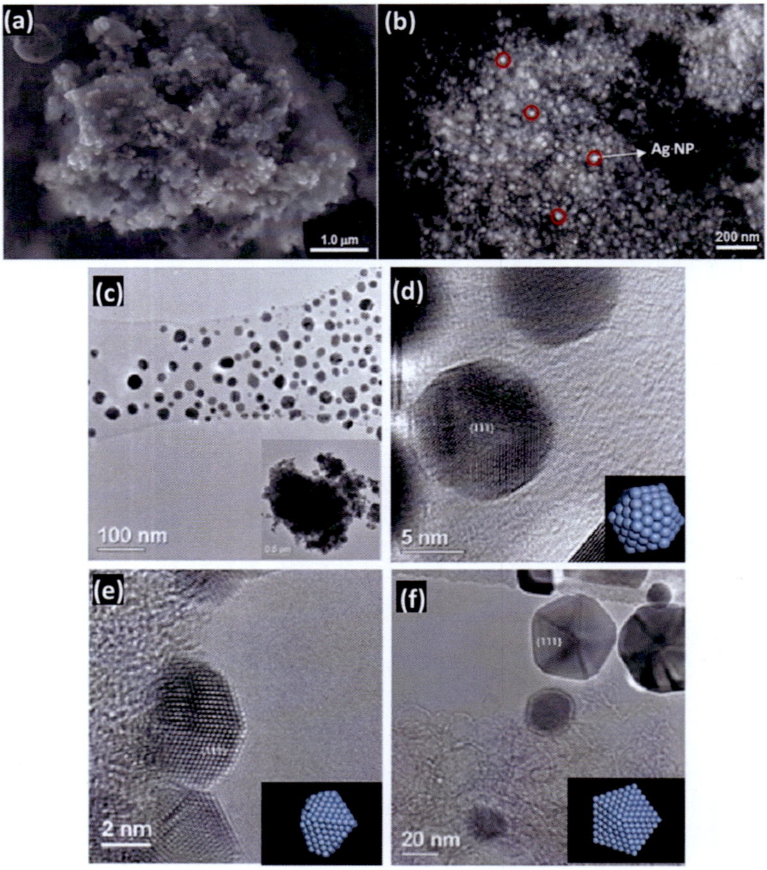


**Fig. 8.3** Morphological illustration of silver nanoparticles as appeared by FE-SEM and HR-TEM. (**a**) FE-SEM at 10µm scale, (**b**) FE-SEM at 200µm scale, (**c**) HR-TEM of uniform silver nanoparticles (**d–f**) the stabilized silver nanoparticles with PVP/S at different scales (Thirugnanasambandan et al. 2018) [Copyright @ 2018, Elsevier]

## *8.3.4 Copper Oxide Nanoparticles (CuO NPs)*

Green synthesize CuO NPs showed significant cytotoxicity against the human cancer cell line (A549) and breast cancer cell line (MCF7) (Sankar et al. 2014; Sivaraj et al. 2014). The antitumor cytotoxic effect has been attributed to the induction of apoptosis and ROS (Wang et al. 2012b). In an in vivo study, the researchers demonstrated a reduction in the tumor volume and metastasis with rapid clearance of CuO NPs within 7 days after Venus injection (Wang et al. 2013). In an in vitro study, copper nanoparticles (CuO NPs) showed high toxicity using human cell lines. In particular, DNA damage and ROS induction have been reported in the human lung epithelial cell line (A549) after exposure to CuO NPs. In the same study, ZnO showed a negative effect on cell viability and DNA damage, while $TiO_2$ caused DNA

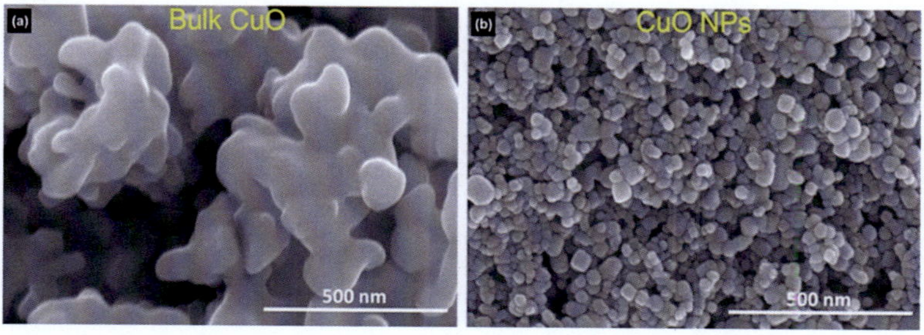


**Fig. 8.4** SEM image of CuO nanoparticles showing the difference between bulk CuO (left) and CuO nanoparticles that are formed by different colloidal conditions (Bai et al. 2012) [Copyright @ 2012, Elsevier]

damage only. Another study showed that the high doses of CuO NPs displayed induction of severe acute inflammatory signs in the lung of rats (Yokohira et al. 2009).

The toxic effect mechanism of CuO NPs is attributed to the induction of oxidative stress as a primary mechanism (Wang et al. 2012c). It has been shown that the exposure of cells to CuO NPs induce a reduction in antioxidant enzymes (catalase and glutathione reductase) as well as higher activity of glutathione peroxidase (Fahmy and Cormier 2009). These findings are supporting the previous report about the effect of waterborne Cu on the apoptosis and necrosis of the branchial chloride cells (Li et al. 1998). Furthermore, a group of studies showed that CuO NPs have toxic effects on animals in aqueous environments such as fish, mussels, and Alae (Pelgrom et al. 1995; Griffitt et al. 2007; Grosell et al. 2007; Aruoja et al. 2009; Gomes et al. 2012). Shi et al. showed that CuO NPs reduce the chlorophyll content of the duckweed much more than the ionic form of Cu (Shi et al. 2011). As well, bacterial cells showed induction of ROS formation and DNA damage due to the ionic release of the Cu from the CuO NPs (Bondarenko et al. 2012). It has been shown that different preparation conditions will affect the size and features of copper oxide nanoparticles. For instance, the concentrations of ascorbic acid and the stirring rate will affect the uniform structure of the CuO nanoparticles as shown in Fig. 8.4 (Bai et al. 2012).

### 8.3.5 *Zinc Oxide Nanoparticles (ZnO NP)*

Zinc oxide nanoparticles (ZnO NPs) have shown efficient toxicity against certain cancer cell lines like T98G and low toxic effects against human HEK normal cells (Wahab et al. 2013b). The antitumor activity of ZnO NPs has been attributed to the genotoxicity and apoptosis activation. Such findings were verified by a melanoma cell study that showed the dose-dependent induction of ROS and apoptosis pathways (Wahab et al. 2013a). Another in vitro study showed the efficacy of low concentration of ZnO NPs against HepG2 and MCF7 cancer cell lines. The results of these studies demonstrated a dose-dependent effect of ZnO NPs (Wahab et al. 2014). In addition, ZnO NP treatment with UV irradiation in an in vitro study has shown substantial apoptosis of squamous cell carcinoma cell lines (HNSCC) of the head and neck (Hackenberg et al. 2010).

Despite the difference in metal components between CuO NPs and ZnO NPs, both the nanomaterials showed similar toxic effects (Franklin et al. 2007; Mortimer et al. 2008, 2010; Chang et al. 2012). Human skin is the most important route for exposure to ZnO NPs because of its common use in sunscreen products. It has been shown that ZnO NPs have high skin adsorption with no signs of penetration in Franz-type diffusion cells (Cross et al. 2007). On the other hand, the oral route of ZnO NPs found histopathological changes in the ZnO NP-treated animals in a dose-related manner (Wang et al. 2008). Consistently, another study showed significant damage in the heart, liver, lung, and kidney with normal spleen and brain (Zheng et al. 2009). In an in vitro study, the small size ZnO NPs at low concentration exhibited high protein adsorption and cytotoxicity (Horie et al. 2009). It has been shown that ZnO NP toxicity is time- and concentration-dependent (Sharma et al. 2009). For instance, the concentration of ZnO NPs more than 15 ppm was reported to be lethal for most human cells as well as rodent cells (Brunner et al. 2006). The cytotoxicity of ZnO NPs was attributed to the induction of oxidative stress mechanisms. The human colon cancer cell line (LoVo) showed cytotoxicity through the induction of oxidative stress response (De Berardis et al. 2010). In another in vivo study, *C. elegans* showed an inhibition of the reproduction and growth when treated with ZnO NPs (Wang et al. 2009).

Furthermore, environmental studies showed that smaller ZnO NPs are more toxic than bulk formulas. For instance, it has been demonstrated that Zinc ions and ZnO NPs are inhibiting the growth of plant root and seed germination (Xiong et al. 2011). In another study, the researchers reported inhibition of microalga growth by ZnO NPs (Franklin et al. 2007). Zebrafish experiments showed lethal results when the concentration of the ZnO NPs reached 2000μg/L (Xiong et al. 2011). However, the $LC_{50}$ of CuO NPs was found to be higher than the ZnO NPs (Blinova et al. 2010). At the microorganism level, the smaller ZnO NPs showed higher inhibition in the bacterial growth like *E. coli* (Yamamoto 2001; Heinlaan et al. 2008; Jiang et al. 2009). Moreover, ZnO NPs showed significant inhibition in yeast (*Saccharomyces cerevisiae*) growth (Kasemets et al. 2009). The antibacterial activity of ZnO NPs was attributed to the increased membrane permeability and membrane disorganization (Brayner et al. 2006). Figure 8.5 shows an example of ZnO NPs that have synthesized in ethylene glycol (EG) and diethylene glycol as shown by SEM and TEM (Chieng and Loo 2012).

### 8.3.6 Silica Nanoparticles ($SiO_2$ NPs)

Silicon oxide nanoparticles have been known as acceptable carriers for pancreatic cancer treatments like paclitaxel (Meng et al. 2015). This is attributable to the pores of $SiO_2$ NPs. The in vivo study showed no toxic effect of camptothecin-loaded

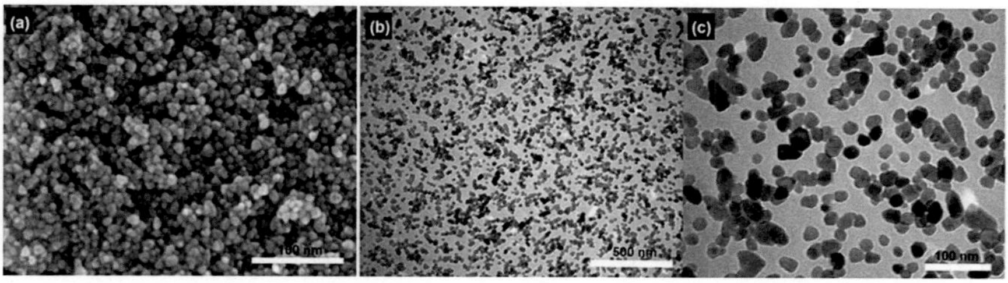


**Fig. 8.5** Images of ZnO nanoparticles. (**a**) SEM image of ZnO NP synthesized in EG, (**b**) TEM image of ZnO NP synthesized in EG, (**c**) TEM image of ZnO NP synthesized in DEG (Chieng and Loo 2012) [Copyright @ 2012, Elsevier]

particles with proper elimination (Lu et al. 2010). Coating strategies have also been suggested to reduce the aggregation of $SiO_2$ NPs, which can enhance the elimination capacity (Bagwe et al. 2006). In vitro study demonstrated a significant reduction in the viability of lung cancer cell line (human bronchoalveolar carcinoma-derived cells) in a dose- and time-dependent manner after exposure to silica nanoparticles ($SiO_2$) (15 and 46 nm). In particular, signs of oxidative stress due to $SiO_2$ nanoparticles exposure were found by increasing ROS combined with glutathione reduction. In addition, membrane damage was associated with lipid peroxidation that is indicated by increased production of malondialdehyde and lactate dehydrogenase (Lin et al. 2006a, b). Figure 8.6 shows the SEM images of $SiO_2$ NPs that have been prepared in different solutions of surfactants with different chain lengths (Singh et al. 2011).

### *8.3.7 Platinum Nanoparticles (Pt NPs)*

Platinum nanoparticles have been used for more than two decades in different medical applications, such as imaging contrasts and anticancer therapeutics. For instance, a study of platinum nanoparticles (Pt NPs) showed an inhibition in the growth of the xenografted A549 cells (lung cancer cell line) in vivo. The inhibition of cancer growth showed a dose-dependent effect (Yogesh et al. 2016). In another study, the researchers reported a higher specificity of platinum folate nanoparticles to cancerous tissues compared to the conventional Pt NPs (Mironava et al. 2013). Moreover, Pt NPs has been shown non-toxic and effective anticancer activity as a photothermal targeting treatment (Samadi et al. 2018). However, the toxicity of Pt NPs (3–10 nm) has been evaluated in different in vitro and in vivo models. Pt NPs have been found to induce a delay hatching of zebrafish; moreover, the researchers demonstrated a reduction in the heart rate as the concentration increases (Asharani et al. 2011). In another approach using green microalgae species *Pseudokirchneriella subcapitata* and *Chlamydomonas reinhardtii*, the results showed that *P. subcapitata* was more sensitive to Pt NPs may be due to the presence of polysaccharide-rich cell wall in this algal species (Sørensen et al. 2016). In another experiment, cell analysis showed that Pt NPs are not inducing cytotoxicity or oxidative stress in human endothelial

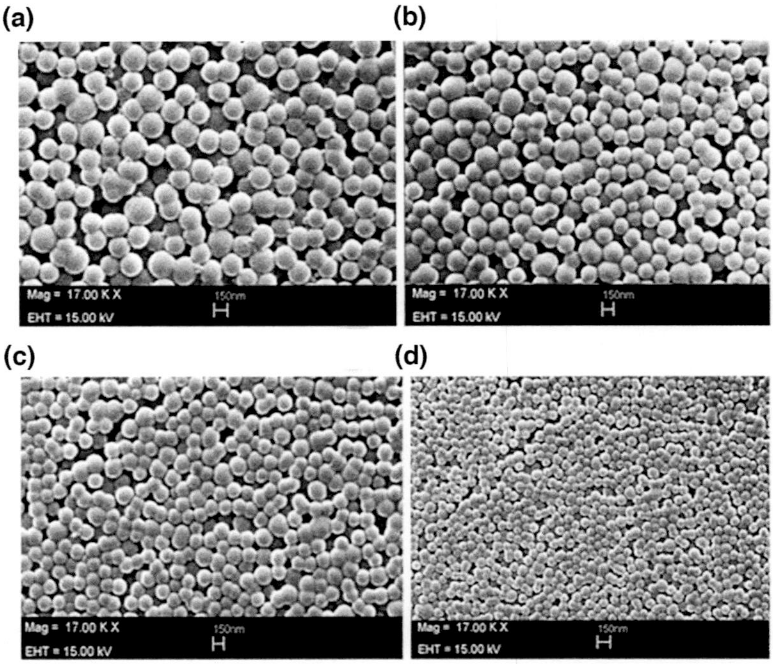


**Fig. 8.6** SEM images of $SiO_2$ nanoparticles prepared (**a**) without surfactant, (**b**) span 20, (**c**) span 40, (**d**) span 60 (Singh et al. 2011) [Copyright @ 2011, SAGE Journal, L. P. Singh et al.]

and lung epithelial. Besides, in vivo, results showed that the respiratory tract exposures to the particles would cause a retain of Pt NPs in the lung tissues and association with lung tissue inflammation (Elder et al. 2007). In an in vitro study, the capping of Pt NPs with folic acid displayed a significant reduction in the viability of the MCF7 breast cancer cell line in a concentration-dependent manner (Teow and Valiyaveettil 2010). In in vitro and in vivo experiments by Brown et al., the researchers showed no significant reduction in the viability of three cell lines (4T1-tdTomato-luc, HepG2, and NIH/3T3) after exposure to Pt NPs. Additionally, intravenous injection of Pt NPs using BALB/c mice, the blood chemistry analyses showed an average level of liver function test and renal function test with no change in body weight or behavioral changes. Histological examination of the liver, spleen, kidneys, and heart showed no structural and morphological changes; however, the spleen exhibited minor changes in tissue architecture. Interestingly, the biodistribution experiment for short-term and long-term injection showed an accumulation of Pt NP in liver and spleen specifically with normal level of serum biochemical parameters of liver function test (Brown et al. 2018). Platinum nanoparticles were prepared in different shapes according to the type of chemical and physical template. For instance, liposomal templates are used to generate dendrite and cage three-dimensional Pt nanostructures (Fig. 8.7) (Peng and Yang 2009).

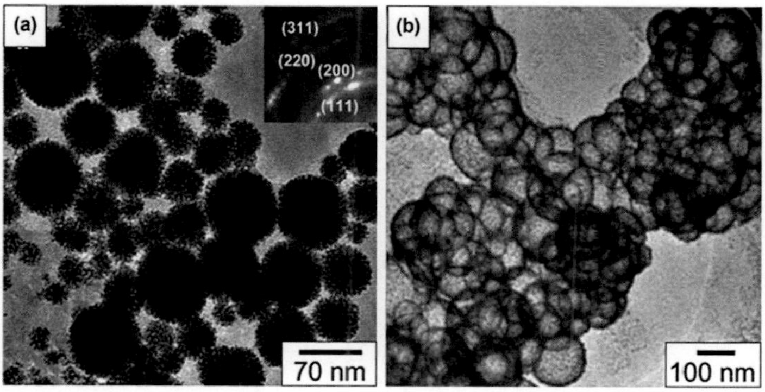


**Fig. 8.7** Illustration of TEM images of Pt nanoparticles showing different shapes using liposome templates. (**a**) Dendrite, (**b**) cage (Peng and Yang 2009) [Copyright @ 2019, Nanomaterials]

### *8.3.8 Cerium Oxide Nanoparticles ($CeO_2$ NPs)*

$CeO_2$ nanoparticles have been proposed as a potential therapeutic option for cancer (Celardo et al. 2011). Therefore, the cytotoxicity of $CeO_2$ was investigated in a few studies. For example, irradiation with nanoparticles of $CeO_2$ revealed unusual properties in cancer cells but not in regular cells by enhancing oxidative stress and apoptosis (Tarnuzzer et al. 2005; Colon et al. 2010). These findings are supported by the results reporting low toxicity of $CeO_2$ NP against normal hTERT-HPNE cells and significant toxicity against pancreatic cancer cell lines (Wason et al. 2013). In vivo study showed antioxidant and low toxicity of $CeO_2$NP in Wistar rats (Khorrami et al. 2019). In an in vitro study, the $CeO_2$ nanoparticles were assessed for its cytotoxicity and oxidative stress effect on lung cancer cell lines (human bronchoalveolar carcinoma-derived cell line (A549)). The production of oxidative stress markers was induced significantly. As well, cell viability exhibited a significant reduction in a dose- and time-dependent manner after exposure to $CeO_2$ NP (Lin et al. 2006a, b). In another study, human lung epithelial cells (BEAS-2B) were exposed to different sizes of $CeO_2$ NP to evaluate the cytotoxicity and oxidative stress effect. The results showed a significant reduction in the viability of exposed cells. Besides, the ROS markers showed significant induction by overexpressing catalase, glutathione *S*-transferase, and reduction of glutathione. The induction of ROS was associated with the activation of the apoptosis mechanism through caspase-3.

Interestingly, $CeO_2$ NP showed cytoplasmic penetration capability by localization in the perinuclear region, which may suggest the ability of $CeO_2$ NP to interact with biological molecules and cause adverse effects (Park et al. 2008). In addition, in a toxicological assay using *Daphnia magna* plankton, the results showed an extremely toxic effect of $CeO_2$ NP at low concentrations (García et al. 2011). The contradictory results of the $CeO_2$ NP products have been attributed to many reasons, including particle size, surface chemistry, cell type, animal model, concentration, exposure route, and preparation method (Gagnon and Fromm 2015). However, most

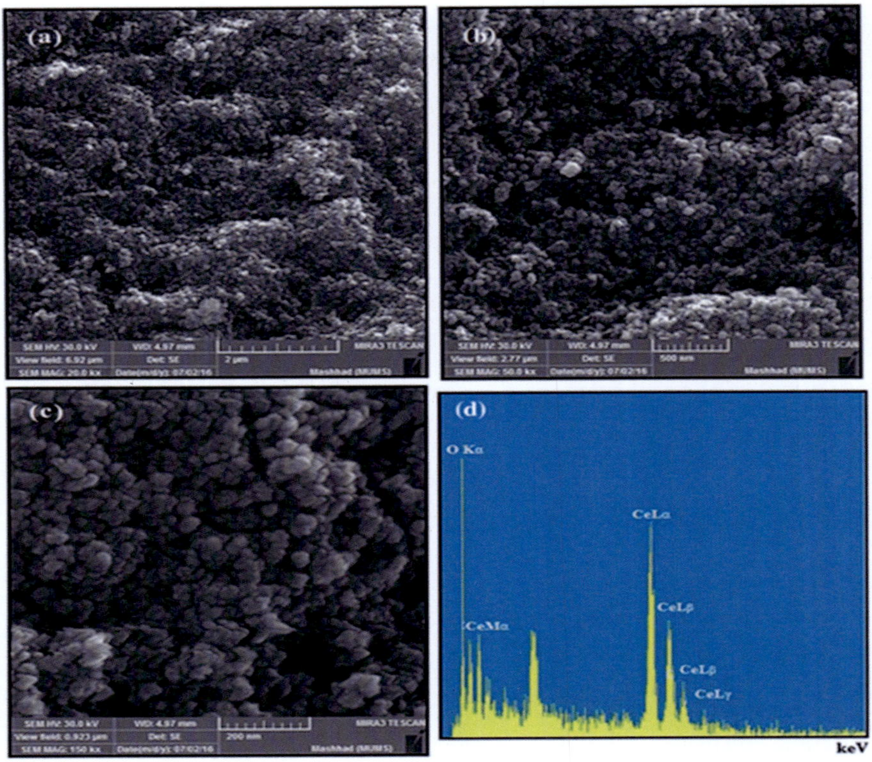


**Fig. 8.8** Illustration of the synthesized $CeO_2$ NP. (**a–c**) FE-SEM images of $CeO_2$ NPs at different magnifications, (**d**) energy-dispersive X-ray (Miri et al. 2020) [Copyright @ 2019, Nanomaterials]

of the tumor studies demonstrated a promising efficacy of $CeO_2$ NP due to the selective toxicity against cancer tissues. For instance, $CeO_2$ NP prevented invasiveness and increased apoptosis in different types of cancers, including melanoma and skin cancer, with a protective effect on normal stromal cells (Alili et al. 2011; De Marzi et al. 2013; Sack et al. 2014). Additionally, the genotoxicity of $CeO_2$ NP has been described as a size-dependent property. In particular, 5 nm particle size showed no DNA damage in cancer cells, while 16–22 nm size particles induced DNA damage in cancer cell lines (Alili et al. 2013). Figure 8.8 shows the FE-SEM image of synthesized $CeO_2$ NPs showing the uniform of spherical nanoparticles with a size of 15–20 nm (Miri et al. 2020).

### *8.3.9 Titanium Oxide Nanoparticles ($TiO_2$ NPs)*

Titanium oxide nanoparticles ($TiO_2$ NPs) have been reported as a safe and non-toxic therapeutic agent in photothermal therapy of cancer (Ou et al. 2016). Another in vitro study showed that titanium oxide nanoparticles have cytoplasmic and membranous incorporation in different types of cancer cell lines such as T-24, HeLa, and

**Fig. 8.9** SEM image of $TiO_2$ nanoparticles (Hu et al. 2014) [Copyright @ 2020, Wiley Online Library]

U937 cells (Thevenot et al. 2008). $TiO_2$ showed long time availability in the body with no signs of toxicity (Vinardell and Mitjans 2015). Using *Daphnia magna* plankton, the results showed a non-toxic effect of titanium NPs (García et al. 2011). Despite the non-toxic impact of $TiO_2$ nanoparticles, the major challenge that faces $TiO_2$ nanoparticles uses in the photodynamic therapy is limited penetration of the triggering UV-visible lights (Cui et al. 2013). Figure 8.9 shows the SEM image of the $TiO_2$ nanoparticles (Hu et al. 2014).

### 8.3.10 Iron Oxide Nanoparticles ($Fe_3O_4$ NPs)

Hyperthermia and ROS induction are effective strategies in targeting tumor cells through the administration of nanoparticles, such as iron oxide nanoparticles ($Fe_3O_4$ NPs) (Laurent et al. 2011). Therefore, $Fe_3O_4$ NPs have been approved for clinical use in the EU as a medical device in the treatment of glioma, glioblastoma, and prostate cancer through the hyperthermia approach (van Landeghem et al. 2009; Silva et al. 2011). In vitro study showed the ability of $Fe_3O_4$ NPs to induce oxidative stress and apoptosis and reduction in the viability of the MCF7 cell line. In addition, a genotoxic effect has been demonstrated in $Fe_3O_4$ NP-treated cell lines (Alarifi et al. 2014). Another study showed selective induction of the ROS, autophagy, and mitochondrial damage in the lung cancer cell line (A549) treated with $Fe_3O_4$ NPs (Khan et al. 2012). These findings are supported by another study when the researchers showed a selective anticancer effect of $Fe_3O_4$ NPs which has been attributed to the induction of ROS through the p53 pathway. The result of that study showed the selective killing of HepG2 (a human hepatocellular carcinoma) and A549 (a human lung adenocarcinoma) but not the normal cell line (IMR-90) (Ahamed et al. 2013). However, there is a big concern regarding the use of $Fe_3O_4$ NPs and their toxicological impact (Cho et al. 2010; Soenen and De Cuyper 2010). $Fe_3O_4$ NP toxicity is associated with the application of those particles in the medical field. For instance,

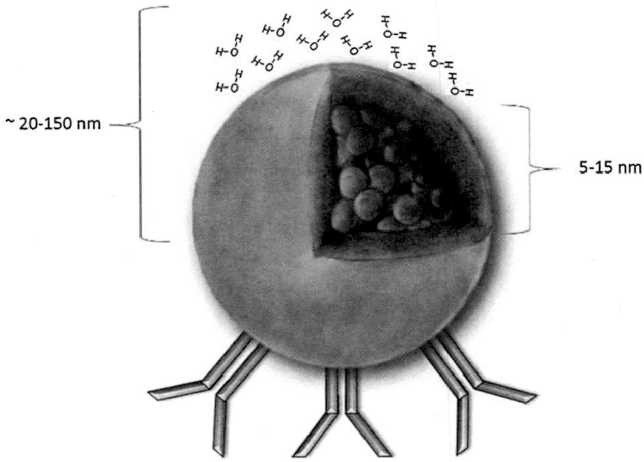


**Fig. 8.10** Illustration scheme of superparamagnetic iron oxide nanoparticle showing the core with a radius (5–15 nm) and the hydrodynamic with a radius of 20–150 nm, and a possible conjugation of antibodies (Dulińska-Litewka et al. 2019) [Copyright @ 2019, Materials]

$Fe_3O_4$ NPs are used in many medical approaches such as MRI, drug delivery, hyperthermia induction, cell labeling, and gene therapy. Therefore, the toxicity effect is related to the application of usage and type of $Fe_3O_4$ NPs (Liu et al. 2013). The generation of ROS is the most common toxic effect of $Fe_3O_4$ NPs in MRI applications (Shubayev et al. 2009; Kim et al. 2012; Schrand et al. 2012). These ROS are very aggressive genotoxic products that can lead to DNA damage as well as other biological molecules harms (Soenen and De Cuyper 2010). Eventually, there will be activation of specific pathways that may lead to cell apoptosis (Könczöl et al. 2011; Murray et al. 2013). Many in vitro studies using different cell types showed a cytotoxic effect of magnetic $Fe_3O_4$ NPs (Berry et al. 2003; Kim et al. 2006; Müller et al. 2007, 2008; Martin et al. 2008; Pawelczyk et al. 2008; Mahmoudi et al. 2009). Consequently, the development of magnetic iron oxide nanoparticles needs more investigation about the biocompatibility and lower toxicity effect in biomedical applications (Liu et al. 2013). Figure 8.10 represents an illustration scheme of iron oxide nanoparticles, showing the core and hydrodynamic radiuses with a possible antibody conjugation (Dulińska-Litewka et al. 2019).

## 8.4 Molecular Mechanisms of Nanoparticle Toxicity

Several factors, such as size, charge, metal composition, dose, dissolution, and route of exposure of nanoparticles, have been associated with toxicity and body clearance. For instance, the positive surface charge showed better renal excretion due to the stimulation of hydrodynamic diameters when interacting with serum proteins. In other words, particles with hydrodynamic diameters of less than 6 mm display improved renal clearance. Additionally, the surface coating of nanoparticles with

PEG or smaller molecules such as cysteine or thiolated polyaminocarboxylate enhanced the renal clearance of nanoparticles by neutralization of the charge (Choi et al. 2007; Alric et al. 2013). Smaller ZnO NPs and CuO NPs showed higher toxicity effects compared to the bulk NPs. For instance, smaller nanoparticles were found to cause thrombocyte and granulocyte activation, and induction of inflammation and hemolysis of human blood in vitro (Mayer et al. 2009). Generally, the smaller sized nanoparticles have a larger specific surface area to size ratio, which may increase the NP reactivity with biological molecules and may facilitate the tissue distribution to different organs (De Jong et al. 2008). The reported accumulation of nanoparticles raises the warning for the need for more studies to evaluate the retaining of administrated particles. The liver and spleen showed a collection of nanoparticles as an indication of low elimination and clearance. Gold and iron-containing nanoparticles showed more decomposition, clearance, and elimination; on the other hand, inert and hydrophobic nanomaterials showed higher ate of accumulation and less clearance (Brown et al. 2018).

However, size, shape, and surface charge effects are also particle-type dependent. In particular, a systematic study that compared a group of metal nanoparticles concluded that: (a) Atomic number of the building element is proportionally associated with the cytotoxicity of NPs; (b) Cell viability is related to the particle surface charge, availability of the binding site, and the dissolution of NPs (Huang et al. 2017). Many studies showed an induction of the oxidative stress response in the metal NP-treated cells. Naturally, oxidative stress species are very active products that can attack different biological molecules including nucleic acids, proteins, and proteins. Eventually, oxidative stress products will cause severe cell damages through cell apoptosis and cell cycle arrest (Fig. 8.11). Interestingly, intracellular calcium ion ($Ca^{++}$) release has been associated with NP-induced oxidative stress which leads to different cell signaling cascades, suggesting the involvement of

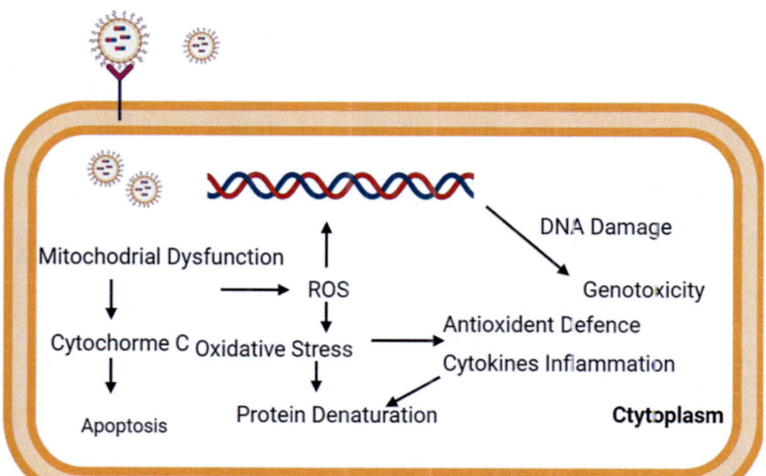


**Fig. 8.11** A representative diagram showing the intracellular cytotoxic effect of nanoparticles treatments (*ROS* reactive oxygen species)

diverse pathways in the cell injuries and death (Huang et al. 2010; Tang et al. 2013). It has been shown that the mitochondrial potential membrane was decreased after exposing lung cell lines to ZnO NPs or $TiO_2$ NPs (Lai et al. 2015; Wang et al. 2015). Additionally, different metal nanoparticles showed protein degradation in vitro (Chang et al. 2012; Saptarshi et al. 2013). Moreover, cell cycle arrest has been reported in vitro after exposure to metal nanoparticles such as $TiO_2$, ZnO, $Fe_3O_4$, CuO, and NiO (Wu et al. 2010; Kai et al. 2011; Lai et al. 2015; Wang et al. 2015; Periasamy et al. 2016). Nevertheless, a group of studies showed that the cell cycle arrest mechanism is cell-type- and nanoparticle-type-dependent (Huang et al. 2017).

## 8.5 Conclusions, Outlook, and Future Aspects

Nanomaterials have been used in different strategies and formulation to treat different types of cancers through in vitro and in vivo approaches. These approaches included direct action of nanoparticles on the target tumor tissue and specific drug delivery systems through encapsulation mechanisms to reduce the cytotoxicity of chemotherapies or by facilitating the hypothermic action. Despite the outcomes of different researches that demonstrated the antitumor effects of different types of nanotechnology approaches, some limitations have to be considered in future studies before the final approval for the medical use of those nanomaterials. These limitations include the heterogeneity of the in vitro studies, like the use of different cell lines in different research. In addition, the differences in the size, structural properties, and methods of preparation are making the decision more difficult to reach the right conclusion. Furthermore, the diverse inorganic nanoparticles formulation has been extensively explored to efficiently deliver to the target site for its nanoscale drug delivery functions. Moreover, the interaction of nanoparticles with the serum proteins (protein corona) raises the important need for more unified approaches and experimental models to have consistent results for the safe use and efficient clearance and bio-distribution of nanoparticles in the medical application. Further evidence and facts regarding the systematic distribution, biocompatibility, and cytotoxicity in vivo studies are essential for proceeding to clinical trials. Collectively, a huge effort and collaboration using in vivo approaches have to be conducted at an international scale to move to the next step for the use of nanotechnology in clinical trials. We may predict that medical nanotechnology is an efficient tool for repelling cancer as well as being kept into others potential applications that may curtail environmental pollution, energy consumption, greenhouse gas emission, and help inhibit syndromes.

**Acknowledgments** We are very thankful to Mr. Adel Rababah for his help in fine art production.

Ethical Clearance We declare that no animals or human subjects were used in this article.

**Conflict of Interest** All authors declare there is no conflict of interest for this springer book chapter publication.

# References

Agarwal, H., Kumar, S. V., & Rajeshkumar, S. (2020). Antidiabetic effect of silver nanoparticles synthesized using lemongrass (Cymbopogon Citratus) through conventional heating and microwave irradiation approach. *Journal of Microbiology, Biotechnology and Food Sciences, 9*(6), 371–376.

Ahamed, M., Karns, M., Goodson, M., Rowe, J., Hussain, S. M., Schlager, J. J., & Hong, Y. (2008). DNA damage response to different surface chemistry of silver nanoparticles in mammalian cells. *Toxicology and Applied Pharmacology, 233*(3), 404–410.

Ahamed, M., Alhadlaq, H. A., Khan, M. M., & Akhtar, M. J. (2013). Selective killing of cancer cells by iron oxide nanoparticles mediated through reactive oxygen species via p53 pathway. *Journal of Nanoparticle Research, 15*(1), 1225.

Alarifi, S., Ali, D., Alkahtani, S., & Alhader, M. (2014). Iron oxide nanoparticles induce oxidative stress, DNA damage, and caspase activation in the human breast cancer cell line. *Biological Trace Element Research, 159*(1–3), 416–424.

Alili, L., Sack, M., Karakoti, A. S., Teuber, S., Puschmann, K., Hirst, S. M., Reilly, C. M., Zanger, K., Stahl, W., & Das, S. (2011). Combined cytotoxic and anti-invasive properties of redox-active nanoparticles in tumor–stroma interactions. *Biomaterials, 32*(11), 2918–2929.

Alili, L., Sack, M., von Montfort, C., Giri, S., Das, S., Carroll, K. S., Zanger, K., Seal, S., & Brenneisen, P. (2013). Downregulation of tumor growth and invasion by redox-active nanoparticles. *Antioxidants & Redox Signaling, 19*(8), 765–778.

Aljabali, A. A., Akkam, Y., Al Zoubi, M. S., Al-Batayneh, K. M., Al-Trad, B., Abo Alrob, O., Alkilany, A. M., Benamara, M., & Evans, D. J. (2018a). Synthesis of gold nanoparticles using leaf extract of Ziziphus zizyphus and their antimicrobial activity. *Nanomaterials, 8*(3), 174.

Aljabali, A. A., Hussein, E., Aljumaili, O., Al Zoubi, M., Altrad, B., Albatayneh, K., & Abd Al-razaq, M. A. (2018b). Rapid magnetic nanobiosensor for the detection of Serratia marcescen. In *IOP Conference Series: Materials Science and Engineering*.

Aljabali, A. A., Al Zoubi, M. S., Al-Batanyeh, K. M., Al-Radaideh, A., Obeid, M. A., Al Sharabi, A., Alshaer, W., AbuFares, B., Al-Zanati, T., & Tambuwala, M. M. (2019). Gold-coated plant virus as computed tomography imaging contrast agent. *Beilstein Journal of Nanotechnology, 10*(1), 1983–1993.

Aljabali, A. A., Al Zoubi, M. S., Alzoubi, L., Al-Batanyeh, K. M., Obeid, M. A., & Tambwala, M. M. (2020). *Chemical engineering of protein cages and nanoparticles for pharmaceutical applications. Nanofabrication for smart nanosensor applications* (pp. 415–433). Amsterdam: Elsevier.

Alomari, G., Al-Trad, B., Hamdan, S., Aljabali, A., Al-Zoubi, M., Bataineh, N., Qar, J., & Tambuwala, M. M. (2020). Gold nanoparticles attenuate albuminuria by inhibiting podocyte injury in a rat model of diabetic nephropathy. *Drug Delivery and Translational Research, 10*(1), 216–226.

Alric, C., Miladi, I., Kryza, D., Taleb, J., Lux, F., Bazzi, R., Billotey, C., Janier, M., Perriat, P., & Roux, S. (2013). The biodistribution of gold nanoparticles designed for renal clearance. *Nanoscale, 5*(13), 5930–5939.

Al-Trad, B., Aljabali, A., Al Zoubi, M., Shehab, M., & Omari, S. (2019). Effect of gold nanoparticles treatment on the testosterone-induced benign prostatic hyperplasia in rats. *International Journal of Nanomedicine, 14*, 3145.

Arora, S., Jain, J., Rajwade, J., & Paknikar, K. (2009). Interactions of silver nanoparticles with primary mouse fibroblasts and liver cells. *Toxicology and Applied Pharmacology, 236*(3), 310–318.

Aruoja, V., Dubourguier, H.-C., Kasemets, K., & Kahru, A. (2009). Toxicity of nanoparticles of CuO, ZnO and TiO2 to microalgae Pseudokirchneriella subcapitata. *Science of the Total Environment, 407*(4), 1461–1468.

Arvanag, F. M., Bayrami, A., Habibi-Yangjeh, A., & Pouran, S. R. (2019). A comprehensive study on antidiabetic and antibacterial activities of ZnO nanoparticles biosynthesized using Silybum marianum L seed extract. *Materials Science and Engineering: C, 97*, 397–405.

Asharani, P., Lianwu, Y., Gong, Z., & Valiyaveettil, S. (2011). Comparison of the toxicity of silver, gold and platinum nanoparticles in developing zebrafish embryos. *Nanotoxicology, 5*(1), 43–54.

Asiya, S., Pal, K., Kralj, S., El-Sayyad, G., de Souza, F., & Narayanan, T. (2020). Sustainable preparation of gold nanoparticles via green chemistry approach for biogenic applications. *Materials Today Chemistry, 17*, 100327.

Bae, Y. S., Lee, J. H., Choi, S. H., Kim, S., Almazan, F., Witztum, J. L., & Miller, Y. I. (2009). Macrophages generate reactive oxygen species in response to minimally oxidized low-density lipoprotein: Toll-like receptor 4– and spleen tyrosine kinase–dependent activation of NADPH oxidase 2. *Circulation Research, 104*(2), 210–218.

Bagwe, R. P., Hilliard, L. R., & Tan, W. (2006). Surface modification of silica nanoparticles to reduce aggregation and nonspecific binding. *Langmuir, 22*(9), 4357–4362.

Bai, Y., Yang, T., Gu, Q., Cheng, G., & Zheng, R. (2012). Shape control mechanism of cuprous oxide nanoparticles in aqueous colloidal solutions. *Powder Technology, 227*, 35–42.

Bazak, R., Houri, M., El Achy, S., Hussein, W., & Refaat, T. (2014). Passive targeting of nanoparticles to cancer: A comprehensive review of the literature. *Molecular and Clinical Oncology, 2*(6), 904–908.

Beer, C., Foldbjerg, R., Hayashi, Y., Sutherland, D. S., & Autrup, H. (2012). Toxicity of silver nanoparticles—Nanoparticle or silver ion? *Toxicology Letters, 208*(3), 286–292.

Berry, C. C., Wells, S., Charles, S., & Curtis, A. S. (2003). Dextran and albumin derivatised iron oxide nanoparticles: Influence on fibroblasts in vitro. *Biomaterials, 24*(25), 4551–4557.

Bisht, G., & Rayamajhi, S. (2016). ZnO nanoparticles: A promising anticancer agent. *Nanobiomedicine, 3*, 3–9.

Blinova, I., Ivask, A., Heinlaan, M., Mortimer, M., & Kahru, A. (2010). Ecotoxicity of nanoparticles of CuO and ZnO in natural water. *Environmental Pollution, 158*(1), 41–47.

Bondarenko, O., Ivask, A., Käkinen, A., & Kahru, A. (2012). Sub-toxic effects of CuO nanoparticles on bacteria: Kinetics, role of Cu ions and possible mechanisms of action. *Environmental Pollution, 169*, 81–89.

Bowman, D. M. (2017). More than a decade on: Mapping today's regulatory and policy landscapes following the publication of nanoscience and nanotechnologies: Opportunities and uncertainties. *NanoEthics, 11*(2), 169–186.

Braydich-Stolle, L., Hussain, S., Schlager, J. J., & Hofmann, M.-C. (2005). In vitro cytotoxicity of nanoparticles in mammalian germline stem cells. *Toxicological Sciences, 88*(2), 412–419.

Brayner, R., Ferrari-Iliou, R., Brivois, N., Djediat, S., Benedetti, M. F., & Fiévet, F. (2006). Toxicological impact studies based on Escherichia coli bacteria in ultrafine ZnO nanoparticles colloidal medium. *Nano Letters, 6*(4), 866–870.

Brown, A. L., Kai, M. P., DuRoss, A. N., Sahay, G., & Sun, C. (2018). Biodistribution and toxicity of micellar platinum nanoparticles in mice via intravenous administration. *Nanomaterials, 8*(6), 410.

Brunner, T. J., Wick, P., Manser, P., Spohn, P., Grass, R. N., Limbach, L. K., Bruinink, A., & Stark, W. J. (2006). In vitro cytotoxicity of oxide nanoparticles: Comparison to asbestos, silica, and the effect of particle solubility. *Environmental Science & Technology, 40*(14), 4374–4381.

Byrne, J. D., Betancourt, T., & Brannon-Peppas, L. (2008). Active targeting schemes for nanoparticle systems in cancer therapeutics. *Advanced Drug Delivery Reviews, 60*(15), 1615–1626.

Carlson, C., Hussain, S. M., Schrand, A. M., Braydich-Stolle, L. K., Hess, K. L., Jones, R. L., & Schlager, J. J. (2008). Unique cellular interaction of silver nanoparticles: Size-dependent generation of reactive oxygen species. *The Journal of Physical Chemistry B, 112*(43), 13608–13619.

Celardo, I., Traversa, E., & Ghibelli, L. (2011). Cerium oxide nanoparticles: A promise for applications in therapy. *Journal of Experimental Therapeutics & Oncology, 9*(1), 47–51.

Chang, Y.-N., Zhang, M., Xia, L., Zhang, J., & Xing, G. (2012). The toxic effects and mechanisms of CuO and ZnO nanoparticles. *Materials, 5*(12), 2850–2871.

Chen, J., Liu, M., Chen, C., Gong, H., & Gao, C. (2011). Synthesis and characterization of silica nanoparticles with well-defined thermoresponsive PNIPAM via a combination of RAFT and click chemistry. *ACS Applied Materials & Interfaces, 3*(8), 3215–3223.

Chieng, B. W., & Loo, Y. Y. (2012). Synthesis of ZnO nanoparticles by modified polyol method. *Materials Letters, 73*, 78–82.

Cho, K., Wang, X., Nie, S., & Shin, D. M. (2008). Therapeutic nanoparticles for drug delivery in cancer. *Clinical Cancer Research, 14*(5), 1310–1316.

Cho, E. C., Glaus, C., Chen, J., Welch, M. J., & Xia, Y. (2010). Inorganic nanoparticle-based contrast agents for molecular imaging. *Trends in Molecular Medicine, 16*(12), 561–573.

Choi, H. S., Liu, W., Misra, P., Tanaka, E., Zimmer, J. P., Ipe, B. I., Bawendi, M. G., & Frangioni, J. V. (2007). Renal clearance of quantum dots. *Nature Biotechnology, 25*(10), 1165–1170.

Choi, O., Clevenger, T. E., Deng, B., Surampalli, R. Y., Ross, L., Jr., & Hu, Z. (2009). Role of sulfide and ligand strength in controlling nanosilver toxicity. *Water Research, 43*(7), 1879–1886.

Colon, J., Hsieh, N., Ferguson, A., Kupelian, P., Seal, S., Jenkins, D. W., & Baker, C. H. (2010). Cerium oxide nanoparticles protect gastrointestinal epithelium from radiation-induced damage by reduction of reactive oxygen species and upregulation of superoxide dismutase 2. *Nanomedicine: Nanotechnology, Biology and Medicine, 6*(5), 698–705.

Connor, E. E., Mwamuka, J., Gole, A., Murphy, C. J., & Wyatt, M. D. (2005). Gold nanoparticles are taken up by human cells but do not cause acute cytotoxicity. *Small, 1*(3), 325–327.

Cross, S. E., Innes, B., Roberts, M. S., Tsuzuki, T., Robertson, T. A., & McCormick, P. (2007). Human skin penetration of sunscreen nanoparticles: In-vitro assessment of a novel micronized zinc oxide formulation. *Skin Pharmacology and Physiology, 20*(3), 148–154.

Cui, S., Yin, D., Chen, Y., Di, Y., Chen, H., Ma, Y., Achilefu, S., & Gu, Y. (2013). In vivo targeted deep-tissue photodynamic therapy based on near-infrared light triggered upconversion nanoconstruct. *ACS Nano, 7*(1), 676–688.

De Berardis, B., Civitelli, G., Condello, M., Lista, P., Pozzi, R., Arancia, G., & Meschini, S. (2010). Exposure to ZnO nanoparticles induces oxidative stress and cytotoxicity in human colon carcinoma cells. *Toxicology and Applied Pharmacology, 246*(3), 116–127.

De Jong, W. H., Hagens, W. I., Krystek, P., Burger, M. C., Sips, A. J., & Geertsma, R. E. (2008). Particle size-dependent organ distribution of gold nanoparticles after intravenous administration. *Biomaterials, 29*(12), 1912–1919.

De Marzi, L., Monaco, A., De Lapuente, J., Ramos, D., Borras, M., Di Gioacchino, M., Santucci, S., & Poma, A. (2013). Cytotoxicity and genotoxicity of ceria nanoparticles on different cell lines in vitro. *International Journal of Molecular Sciences, 14*(2), 3065–3077.

Drummond, D. C., Meyer, O., Hong, K., Kirpotin, D. B., & Papahadjopoulos, D. (1999). Optimizing liposomes for delivery of chemotherapeutic agents to solid tumors. *Pharmacological Reviews, 51*(4), 691–744.

Dulińska-Litewka, J., Łazarczyk, A., Hałubiec, P., Szafrański, O., Karnas, K., & Karewicz, A. (2019). Superparamagnetic iron oxide nanoparticles—Current and prospective medical applications. *Materials, 12*(4), 617.

Elder, A., Yang, H., Gwiazda, R., Teng, X., Thurston, S., He, H., & Oberdörster, G. (2007). Testing nanomaterials of unknown toxicity: An example based on platinum nanoparticles of different shapes. *Advanced Materials, 19*(20), 3124–3129.

Fahmy, B., & Cormier, S. A. (2009). Copper oxide nanoparticles induce oxidative stress and cytotoxicity in airway epithelial cells. *Toxicology In Vitro, 23*(7), 1365–1371.

Farooq, M. U., Novosad, V., Rozhkova, E. A., Wali, H., Ali, A., Fateh, A. A., Neogi, P. B., Neogi, A., & Wang, Z. (2018). Gold nanoparticles-enabled efficient dual delivery of anticancer therapeutics to HeLa cells. *Scientific Reports, 8*(1), 1–12.

Franklin, N. M., Rogers, N. J., Apte, S. C., Batley, G. E., Gadd, G. E., & Casey, P. S. (2007). Comparative toxicity of nanoparticulate ZnO, bulk ZnO, and ZnCl2 to a freshwater microalga (Pseudokirchneriella subcapitata): The importance of particle solubility. *Environmental Science & Technology, 41*(24), 8484–8490.

Gad, S. C., Sharp, K. L., Montgomery, C., Payne, J. D., & Goodrich, G. P. (2012). Evaluation of the toxicity of intravenous delivery of auroshell particles (gold–silica nanoshells). *International Journal of Toxicology, 31*(6), 584–594.

Gagnon, J., & Fromm, K. M. (2015). Toxicity and protective effects of cerium oxide nanoparticles (nanoceria) depending on their preparation method, particle size, cell type, and exposure route. *European Journal of Inorganic Chemistry, 2015*(27), 4510–4517.

García, A., Espinosa, R., Delgado, L., Casals, E., González, E., Puntes, V., Barata, C., Font, X., & Sánchez, A. (2011). Acute toxicity of cerium oxide, titanium oxide and iron oxide nanoparticles using standardized tests. *Desalination, 269*(1–3), 136–141.

Glazer, E. S., Zhu, C., Hamir, A. N., Borne, A., Thompson, C. S., & Curley, S. A. (2011). Biodistribution and acute toxicity of naked gold nanoparticles in a rabbit hepatic tumor model. *Nanotoxicology, 5*(4), 459–468.

Gomes, T., Pereira, C. G., Cardoso, C., Pinheiro, J. P., Cancio, I., & Bebianno, M. J. (2012). Accumulation and toxicity of copper oxide nanoparticles in the digestive gland of Mytilus galloprovincialis. *Aquatic Toxicology, 118*, 72–79.

Griffitt, R. J., Weil, R., Hyndman, K. A., Denslow, N. D., Powers, K., Taylor, D., & Barber, D. S. (2007). Exposure to copper nanoparticles causes gill injury and acute lethality in zebrafish (Danio rerio). *Environmental Science & Technology, 41*(23), 8178–8186.

Griffitt, R. J., Hyndman, K., Denslow, N. D., & Barber, D. S. (2009). Comparison of molecular and histological changes in zebrafish gills exposed to metallic nanoparticles. *Toxicological Sciences, 107*(2), 404–415.

Grosell, M., Blanchard, J., Brix, K., & Gerdes, R. (2007). Physiology is pivotal for interactions between salinity and acute copper toxicity to fish and invertebrates. *Aquatic Toxicology, 84*(2), 162–172.

Hackenberg, S., Scherzed, A., Kessler, M., Froelich, K., Ginzkey, C., Koehler, C., Burghartz, M., Hagen, R., & Kleinsasser, N. (2010). Zinc oxide nanoparticles induce photocatalytic cell death in human head and neck squamous cell carcinoma cell lines in vitro. *International Journal of Oncology, 37*(6), 1583–1590.

Haley, B., & Frenkel, E. (2008). *Nanoparticles for drug delivery in cancer treatment. Urologic oncology: Seminars and original investigations*. Amsterdam: Elsevier.

Handy, R. D., & Shaw, B. J. (2007). Toxic effects of nanoparticles and nanomaterials: Implications for public health, risk assessment and the public perception of nanotechnology. *Health, Risk & Society, 9*(2), 125–144.

Heinlaan, M., Ivask, A., Blinova, I., Dubourguier, H.-C., & Kahru, A. (2008). Toxicity of nanosized and bulk ZnO, CuO and TiO2 to bacteria Vibrio fischeri and crustaceans Daphnia magna and Thamnocephalus platyurus. *Chemosphere, 71*(7), 1308–1316.

Heo, D. N., Yang, D. H., Moon, H.-J., Lee, J. B., Bae, M. S., Lee, S. C., Lee, W. J., Sun, I.-C., & Kwon, I. K. (2012). Gold nanoparticles surface-functionalized with paclitaxel drug and biotin receptor as theranostic agents for cancer therapy. *Biomaterials, 33*(3), 856–866.

Hillyer, J. F., & Albrecht, R. M. (2001). Gastrointestinal persorption and tissue distribution of differently sized colloidal gold nanoparticles. *Journal of Pharmaceutical Sciences, 90*(12), 1927–1936.

Horie, M., Nishio, K., Fujita, K., Endoh, S., Miyauchi, A., Saito, Y., Iwahashi, H., Yamamoto, K., Murayama, H., & Nakano, H. (2009). Protein adsorption of ultrafine metal oxide and its influence on cytotoxicity toward cultured cells. *Chemical Research in Toxicology, 22*(3), 543–553.

Hu, Y.-C., Dai, C.-L., & Hsu, C.-C. (2014). Titanium dioxide nanoparticle humidity microsensors integrated with circuitry on-a-chip. *Sensors, 14*(3), 4177–4188.

Huang, C.-C., Aronstam, R. S., Chen, D.-R., & Huang, Y.-W. (2010). Oxidative stress, calcium homeostasis, and altered gene expression in human lung epithelial cells exposed to ZnO nanoparticles. *Toxicology In Vitro, 24*(1), 45–55.

Huang, K., Ma, H., Liu, J., Huo, S., Kumar, A., Wei, T., Zhang, X., Jin, S., Gan, Y., & Wang, P. C. (2012). Size-dependent localization and penetration of ultrasmall gold nanoparticles in cancer cells, multicellular spheroids, and tumors in vivo. *ACS Nano, 6*(5), 4483–4493.

Huang, Y.-W., Cambre, M., & Lee, H.-J. (2017). The toxicity of nanoparticles depends on multiple molecular and physicochemical mechanisms. *International Journal of Molecular Sciences, 18*(12), 2702.

Jiang, W., Mashayekhi, H., & Xing, B. (2009). Bacterial toxicity comparison between nano- and micro-scaled oxide particles. *Environmental Pollution, 157*(5), 1619–1625.

Kai, W., Xiaojun, X., Ximing, P., Zhenqing, H., & Qiqing, Z. (2011). Cytotoxic effects and the mechanism of three types of magnetic nanoparticles on human hepatoma BEL-7402 cells. *Nanoscale Research Letters, 6*(1), 480.

Kasemets, K., Ivask, A., Dubourguier, H.-C., & Kahru, A. (2009). Toxicity of nanoparticles of ZnO, CuO and TiO2 to yeast Saccharomyces cerevisiae. *Toxicology In Vitro, 23*(6), 1116–1122.

Kawata, K., Osawa, M., & Okabe, S. (2009). In vitro toxicity of silver nanoparticles at noncytotoxic doses to HepG2 human hepatoma cells. *Environmental Science & Technology, 43*(15), 6046–6051.

Kedmi, R., Ben-Arie, N., & Peer, D. (2010). The systemic toxicity of positively charged lipid nanoparticles and the role of Toll-like receptor 4 in immune activation. *Biomaterials, 31*(26), 6867–6875.

Khan, M. I., Mohammad, A., Patil, G., Naqvi, S., Chauhan, L., & Ahmad, I. (2012). Induction of ROS, mitochondrial damage and autophagy in lung epithelial cancer cells by iron oxide nanoparticles. *Biomaterials, 33*(5), 1477–1488.

Khorrami, M. B., Sadeghnia, H. R., Pasdar, A., Ghayour-Mobarhan, M., Riahi-Zanjani, B., Hashemzadeh, A., Zare, M., & Darroudi, M. (2019). Antioxidant and toxicity studies of biosynthesized cerium oxide nanoparticles in rats. *International Journal of Nanomedicine, 14*, 2915.

Kim, J.-S., Yoon, T.-J., Yu, K.-N., Noh, M. S., Woo, M., Kim, B.-G., Lee, K.-H., Sohn, B.-H., Park, S.-B., & Lee, J.-K. (2006). Cellular uptake of magnetic nanoparticle is mediated through energy-dependent endocytosis in A549 cells. *Journal of Veterinary Science, 7*(4), 321–326.

Kim, J.-E., Shin, J.-Y., & Cho, M.-H. (2012). Magnetic nanoparticles: An update of application for drug delivery and possible toxic effects. *Archives of Toxicology, 86*(5), 685–700.

Kittler, S., Greulich, C., Diendorf, J., Koller, M., & Epple, M. (2010). Toxicity of silver nanoparticles increases during storage because of slow dissolution under release of silver ions. *Chemistry of Materials, 22*(16), 4548–4554.

Könczöl, M., Ebeling, S., Goldenberg, E., Treude, F., Gminski, R., Gieré, R., Grobéty, B., Rothen-Rutishauser, B., Merfort, I., & Mersch-Sundermann, V. (2011). Cytotoxicity and Genotoxicity of Size-Fractionated Iron Oxide (Magnetite) in A549 Human Lung Epithelial Cells: Role of ROS, JNK, and NF-κB. *Chemical Research in Toxicology, 24*, 1460–1475.

Kvitek, L., Vanickova, M., Panacek, A., Soukupova, J., Dittrich, M., Valentova, E., Prucek, R., Bancirova, M., Milde, D., & Zboril, R. (2009). Initial study on the toxicity of silver nanoparticles (NPs) against Paramecium caudatum. *The Journal of Physical Chemistry C, 113*(11), 4296–4300.

Lai, X., Wei, Y., Zhao, H., Chen, S., Bu, X., Lu, F., Qu, D., Yao, L., Zheng, J., & Zhang, J. (2015). The effect of Fe2O3 and ZnO nanoparticles on cytotoxicity and glucose metabolism in lung epithelial cells. *Journal of Applied Toxicology, 35*(6), 651–664.

Laurent, S., Dutz, S., Häfeli, U. O., & Mahmoudi, M. (2011). Magnetic fluid hyperthermia: Focus on superparamagnetic iron oxide nanoparticles. *Advances in Colloid and Interface Science, 166*(1–2), 8–23.

Li, J., Quabius, E., Bonga, S. W., Flik, G., & Lock, R. (1998). Effects of water-borne copper on branchial chloride cells and Na+/K+-ATPase activities in Mozambique tilapia (Oreochromis mossambicus). *Aquatic Toxicology, 43*(1), 1–11.

Libutti, S. K., Paciotti, G. F., Byrnes, A. A., Alexander, H. R., Gannon, W. E., Walker, M., Seidel, G. D., Yuldasheva, N., & Tamarkin, L. (2010). Phase I and pharmacokinetic studies of CYT-6091, a novel PEGylated colloidal gold-rhTNF nanomedicine. *Clinical Cancer Research, 16*(24), 6139–6149.

Lin, W., Huang, Y.-W., Zhou, X.-D., & Ma, Y. (2006a). In vitro toxicity of silica nanoparticles in human lung cancer cells. *Toxicology and Applied Pharmacology, 217*(3), 252–259.

Lin, W., Huang, Y.-W., Zhou, X.-D., & Ma, Y. (2006b). Toxicity of cerium oxide nanoparticles in human lung cancer cells. *International Journal of Toxicology, 25*(6), 451–457.

Liu, G., Gao, J., Ai, H., & Chen, X. (2013). Applications and potential toxicity of magnetic iron oxide nanoparticles. *Small, 9*(9–10), 1533–1545.

Liu, Y., Zeng, S., Liu, Y., Wu, W., Shen, Y., Zhang, L., Li, C., Chen, H., Liu, A., & Shen, L. (2018). Synthesis and antidiabetic activity of selenium nanoparticles in the presence of polysaccharides from Catathelasma ventricosum. *International Journal of Biological Macromolecules, 114*, 632–639.

Love, S. A., Thompson, J. W., & Haynes, C. L. (2012). Development of screening assays for nanoparticle toxicity assessment in human blood: Preliminary studies with charged Au nanoparticles. *Nanomedicine, 7*(9), 1355–1364.

Lu, J., Liong, M., Li, Z., Zink, J. I., & Tamanoi, F. (2010). Biocompatibility, biodistribution, and drug-delivery efficiency of mesoporous silica nanoparticles for cancer therapy in animals. *Small, 6*(16), 1794–1805.

Mahmoudi, M., Simchi, A., Imani, M., Milani, A. S., & Stroeve, P. (2009). An in vitro study of bare and poly (ethylene glycol)-co-fumarate-coated superparamagnetic iron oxide nanoparticles: A new toxicity identification procedure. *Nanotechnology, 20*(22), 225104.

Martin, A. L., Bernas, L. M., Rutt, B. K., Foster, P. J., & Gillies, E. R. (2008). Enhanced cell uptake of superparamagnetic iron oxide nanoparticles functionalized with dendritic guanidines. *Bioconjugate Chemistry, 19*(12), 2375–2384.

Mayer, A., Vadon, M., Rinner, B., Novak, A., Wintersteiger, R., & Fröhlich, E. (2009). The role of nanoparticle size in hemocompatibility. *Toxicology, 258*(2–3), 139–147.

Maynard, A. D. (2014). A decade of uncertainty. *Nature Nanotechnology, 9*(3), 159–160.

Meng, H., Wang, M., Liu, H., Liu, X., Situ, A., Wu, B., Ji, Z., Chang, C. H., & Nel, A. E. (2015). Use of a lipid-coated mesoporous silica nanoparticle platform for synergistic gemcitabine and paclitaxel delivery to human pancreatic cancer in mice. *ACS Nano, 9*(4), 3540–3557.

Miri, A., Darroudi, M., & Sarani, M. (2020). Biosynthesis of cerium oxide nanoparticles and its cytotoxicity survey against colon cancer cell line. *Applied Organometallic Chemistry, 34*(1), e5308.

Mironava, T., Simon, M., Rafailovich, M. H., & Rigas, B. (2013). Platinum folate nanoparticles toxicity: Cancer vs. normal cells. *Toxicology In Vitro, 27*(2), 882–889.

Mortimer, M., Kasemets, K., Heinlaan, M., Kurvet, I., & Kahru, A. (2008). High throughput kinetic Vibrio fischeri bioluminescence inhibition assay for study of toxic effects of nanoparticles. *Toxicology In Vitro, 22*(5), 1412–1417.

Mortimer, M., Kasemets, K., & Kahru, A. (2010). Toxicity of ZnO and CuO nanoparticles to ciliated protozoa Tetrahymena thermophila. *Toxicology, 269*(2–3), 182–189.

Müller, K., Skepper, J. N., Posfai, M., Trivedi, R., Howarth, S., Corot, C., Lancelot, E., Thompson, P. W., Brown, A. P., & Gillard, J. H. (2007). Effect of ultrasmall superparamagnetic iron oxide nanoparticles (Ferumoxtran-10) on human monocyte-macrophages in vitro. *Biomaterials, 28*(9), 1629–1642.

Müller, K., Skepper, J. N., Tang, T. Y., Graves, M. J., Patterson, A. J., Corot, C., Lancelot, E., Thompson, P. W., Brown, A. P., & Gillard, J. H. (2008). Atorvastatin and uptake of ultrasmall superparamagnetic iron oxide nanoparticles (Ferumoxtran-10) in human monocyte–macrophages: Implications for magnetic resonance imaging. *Biomaterials, 29*(17), 2656–2662.

Mundra, V., Li, W., & Mahato, R. I. (2015). Nanoparticle-mediated drug delivery for treating melanoma. *Nanomedicine, 10*(16), 2613–2633.

Murray, A. R., Kisin, E., Inman, A., Young, S.-H., Muhammed, M., Burks, T., Uheida, A., Tkach, A., Waltz, M., & Castranova, V. (2013). Oxidative stress and dermal toxicity of iron oxide nanoparticles in vitro. *Cell Biochemistry and Biophysics, 67*(2), 461–476.

Navarro, E., Piccapietra, F., Wagner, B., Marconi, F., Kaegi, R., Odzak, N., Sigg, L., & Behra, R. (2008). Toxicity of silver nanoparticles to Chlamydomonas reinhardtii. *Environmental Science & Technology, 42*(23), 8959–8964.

Oberdörster, G., Oberdörster, E., & Oberdörster, J. (2005). Nanotoxicology: An emerging discipline evolving from studies of ultrafine particles. *Environmental Health Perspectives, 113*(7), 823–839.

Ou, G., Li, Z., Li, D., Cheng, L., Liu, Z., & Wu, H. (2016). Photothermal therapy by using titanium oxide nanoparticles. *Nano Research, 9*(5), 1236–1243.

Park, E.-J., Choi, J., Park, Y.-K., & Park, K. (2008). Oxidative stress induced by cerium oxide nanoparticles in cultured BEAS-2B cells. *Toxicology, 245*(1–2), 90–100.

Pawelczyk, E., Arbab, A. S., Chaudhry, A., Balakumaran, A., Robey, P. G., & Frank, J. A. (2008). In vitro model of bromodeoxyuridine or iron oxide nanoparticle uptake by activated macrophages from labeled stem cells: Implications for cellular therapy. *Stem Cells, 26*(5), 1366–1375.

Pelgrom, S., Lock, R., Balm, P., & Bonga, S. W. (1995). Integrated physiological response of tilapia, Oreochromis mossambicus, to sublethal copper exposure. *Aquatic Toxicology, 32*(4), 303–320.

Peng, Z., & Yang, H. (2009). Designer platinum nanoparticles: Control of shape, composition in alloy, nanostructure and electrocatalytic property. *Nano Today, 4*(2), 143–164.

Peng, C., Xu, J., Yu, M., Ning, X., Huang, Y., Du, B., Hernandez, E., Kapur, P., Hsieh, J. T., & Zheng, J. (2019). Tuning the in vivo transport of anticancer drugs using renal-clearable gold nanoparticles. *Angewandte Chemie, 131*(25), 8567–8571.

Periasamy, V. S., Athinarayanan, J., Alhazmi, M., Alatiah, K. A., & Alshatwi, A. A. (2016). Fe3O4 nanoparticle redox system modulation via cell-cycle progression and gene expression in human mesenchymal stem cells. *Environmental Toxicology, 31*(8), 901–912.

Pidgeon, N., Porritt, J., Ryan, J., Seaton, A., Tendler, S., Welland, M., & Whatmore, R. (2004). *Nanoscience and nanotechnologies: Opportunities and uncertainties*. London: The Royal Society, The Royal Academy of Engineering.

Rao, W., Wang, H., Han, J., Zhao, S., Dumbleton, J., Agarwal, P., Zhang, W., Zhao, G., Yu, J., & Zynger, D. L. (2015). Chitosan-decorated doxorubicin-encapsulated nanoparticle targets and eliminates tumor reinitiating cancer stem-like cells. *ACS Nano, 9*(6), 5725–5740.

Raza, A., Rasheed, T., Nabeel, F., Hayat, U., Bilal, M., & Iqbal, H. (2019). Endogenous and exogenous stimuli-responsive drug delivery systems for programmed site-specific release. *Molecules, 24*(6), 1117.

Roh, J.-y., Sim, S. J., Yi, J., Park, K., Chung, K. H., Ryu, D.-y., & Choi, J. (2009). Ecotoxicity of silver nanoparticles on the soil nematode Caenorhabditis elegans using functional ecotoxicogenomics. *Environmental Science & Technology, 43*(10), 3933–3940.

Sack, M., Alili, L., Karaman, E., Das, S., Gupta, A., Seal, S., & Brenneisen, P. (2014). Combination of conventional chemotherapeutics with redox-active cerium oxide nanoparticles—A novel aspect in cancer therapy. *Molecular Cancer Therapeutics, 13*(7), 1740–1749.

Samadi, A., Klingberg, H., Jauffred, L., Kjær, A., Bendix, P. M., & Oddershede, L. B. (2018). Platinum nanoparticles: A non-toxic, effective and thermally stable alternative plasmonic material for cancer therapy and bioengineering. *Nanoscale, 10*(19), 9097–9107.

Sankar, R., Maheswari, R., Karthik, S., Shivashangari, K. S., & Ravikumar, V. (2014). Anticancer activity of Ficus religiosa engineered copper oxide nanoparticles. *Materials Science and Engineering: C, 44*, 234–239.

Santoro, C. M., Duchsherer, N. L., & Grainger, D. W. (2007). Minimal in vitro antimicrobial efficacy and ocular cell toxicity from silver nanoparticles. *NanoBiotechnology, 3*(2), 55–65.

Saptarshi, S. R., Duschl, A., & Lopata, A. L. (2013). Interaction of nanoparticles with proteins: Relation to bio-reactivity of the nanoparticle. *Journal of Nanobiotechnology, 11*(1), 26.

Schrand, A. M., Dai, L., Schlager, J. J., & Hussain, S. M. (2012). *Toxicity testing of nanomaterials. New technologies for toxicity testing* (pp. 58–75). Berlin: Springer.

Sereemaspun, A., Hongpiticharoen, P., Rojanathanes, R., Maneewattanapinyo, P., Ekgasit, S., & Warisnoicharoen, W. (2008). Inhibition of human cytochrome P450 enzymes by metallic nanoparticles: A preliminary to nanogenomics. *International Journal of Pharmacology, 4*(6), 492–495.

Sharma, V., Shukla, R. K., Saxena, N., Parmar, D., Das, M., & Dhawan, A. (2009). DNA damaging potential of zinc oxide nanoparticles in human epidermal cells. *Toxicology Letters, 185*(3), 211–218.

Shi, J., Abid, A. D., Kennedy, I. M., Hristova, K. R., & Silk, W. K. (2011). To duckweeds (Landoltia punctata), nanoparticulate copper oxide is more inhibitory than the soluble copper in the bulk solution. *Environmental Pollution, 159*(5), 1277–1282.

Shubayev, V. I., Pisanic, T. R., II, & Jin, S. (2009). Magnetic nanoparticles for theragnostics. *Advanced Drug Delivery Reviews, 61*(6), 467–477.

Silva, A. C., Oliveira, T. R., Mamani, J. B., Malheiros, S. M., Malavolta, L., Pavon, L. F., Sibov, T. T., Amaro, E., Jr., Tannús, A., & Vidoto, E. L. (2011). Application of hyperthermia induced by superparamagnetic iron oxide nanoparticles in glioma treatment. *International Journal of Nanomedicine, 6*, 591.

Singh, L., Agarwal, S., Bhattacharyya, S., Sharma, U., & Ahalawat, S. (2011). Preparation of silica nanoparticles and its beneficial role in cementitious materials. *Nanomaterials and Nanotechnology, 1*, 9.

Sivaraj, R., Rahman, P. K., Rajiv, P., Narendhran, S., & Venckatesh, R. (2014). Biosynthesis and characterization of Acalypha indica mediated copper oxide nanoparticles and evaluation of its antimicrobial and anticancer activity. *Spectrochimica Acta Part A: Molecular and Biomolecular Spectroscopy, 129*, 255–258.

Soenen, S. J., & De Cuyper, M. (2010). Assessing iron oxide nanoparticle toxicity in vitro: Current status and future prospects. *Nanomedicine, 5*(8), 1261–1275.

Soenen, S. J., Illyes, E., Vercauteren, D., Braeckmans, K., Majer, Z., De Smedt, S. C., & De Cuyper, M. (2009). The role of nanoparticle concentration-dependent induction of cellular stress in the internalization of non-toxic cationic magnetoliposomes. *Biomaterials, 30*(36), 6803–6813.

Sørensen, S. N., Engelbrekt, C., Lützhøft, H.-C. H., Jiménez-Lamana, J., Noori, J. S., Alatraktchi, F. A., Delgado, C. G., Slaveykova, V. I., & Baun, A. (2016). A multimethod approach for investigating algal toxicity of platinum nanoparticles. *Environmental Science & Technology, 50*(19), 10635–10643.

Takano, S., Aramaki, Y., & Tsuchiya, S. (2001). Lipoxygenase may be involved in cationic liposome-induced macrophage apoptosis. *Biochemical and Biophysical Research Communications, 288*(1), 116–120.

Tang, F., Li, L., & Chen, D. (2012). Mesoporous silica nanoparticles: Synthesis, biocompatibility and drug delivery. *Advanced Materials, 24*(12), 1504–1534.

Tang, T.-H., Chang, C.-T., Wang, H.-J., Erickson, J. D., Reichard, R. A., Martin, A. G., Shannon, E. K., Martin, A. L., Huang, Y.-W., & Aronstam, R. S. (2013). Oxidative stress disruption of receptor-mediated calcium signaling mechanisms. *Journal of Biomedical Science, 20*(1), 48.

Tarnuzzer, R. W., Colon, J., Patil, S., & Seal, S. (2005). Vacancy engineered ceria nanostructures for protection from radiation-induced cellular damage. *Nano Letters, 5*(12), 2573–2577.

Teow, Y., & Valiyaveettil, S. (2010). Active targeting of cancer cells using folic acid-conjugated platinum nanoparticles. *Nanoscale, 2*(12), 2607–2613.

Thevenot, P., Cho, J., Wavhal, D., Timmons, R. B., & Tang, L. (2008). Surface chemistry influences cancer killing effect of TiO2 nanoparticles. *Nanomedicine: Nanotechnology, Biology and Medicine, 4*(3), 226–236.

Thirugnanasambandan, T., Pal, K., Sidhu, A., Abd Elkodous, M., Prasath, H., Kulasekarapandian, K., Ayeshamariam, A., & Jeevanandam, J. (2018). Aggrandize efficiency of ultra-thin silicon solar cell via topical clustering of silver nanoparticles. *Nano-Structures & Nano-Objects, 16*, 224–233.

van Landeghem, F. K., Maier-Hauff, K., Jordan, A., Hoffmann, K.-T., Gneveckow, U., Scholz, R., Thiesen, B., Brück, W., & von Deimling, A. (2009). Post-mortem studies in glioblastoma patients treated with thermotherapy using magnetic nanoparticles. *Biomaterials, 30*(1), 52–57.

Vinardell, M. P., & Mitjans, M. (2015). Antitumor activities of metal oxide nanoparticles. *Nanomaterials, 5*(2), 1004–1021.

Wahab, R., Dwivedi, S., Umar, A., Singh, S., Hwang, I., Shin, H.-S., Musarrat, J., Al-Khedhairy, A. A., & Kim, Y.-S. (2013a). ZnO nanoparticles induce oxidative stress in Cloudman S91 melanoma cancer cells. *Journal of Biomedical Nanotechnology, 9*(3), 441–449.

Wahab, R., Kaushik, N. K., Kaushik, N., Choi, E. H., Umar, A., Dwivedi, S., Musarrat, J., & Al-Khedhairy, A. A. (2013b). ZnO nanoparticles induces cell death in malignant human T98G gliomas, KB and non-malignant HEK cells. *Journal of Biomedical Nanotechnology, 9*(7), 1181–1189.

Wahab, R., Siddiqui, M. A., Saquib, Q., Dwivedi, S., Ahmad, J., Musarrat, J., Al-Khedhairy, A. A., & Shin, H.-S. (2014). ZnO nanoparticles induced oxidative stress and apoptosis in HepG2 and MCF-7 cancer cells and their antibacterial activity. *Colloids and Surfaces B: Biointerfaces, 117*, 267–276.

Wang, B., Feng, W., Wang, M., Wang, T., Gu, Y., Zhu, M., Ouyang, H., Shi, J., Zhang, F., & Zhao, Y. (2008). Acute toxicological impact of nano-and submicro-scaled zinc oxide powder on healthy adult mice. *Journal of Nanoparticle Research, 10*(2), 263–276.

Wang, H., Wick, R. L., & Xing, B. (2009). Toxicity of nanoparticulate and bulk ZnO, Al2O3 and TiO2 to the nematode Caenorhabditis elegans. *Environmental Pollution, 157*(4), 1171–1177.

Wang, L., Dong, J., Ouyang, W., Wang, X., & Tang, J. (2012a). Anticancer effect and feasibility study of hyperthermia treatment of pancreatic cancer using magnetic nanoparticles. *Oncology Reports, 27*(3), 719–726.

Wang, Y., Zi, X.-Y., Su, J., Zhang, H.-X., Zhang, X.-R., Zhu, H.-Y., Li, J.-X., Yin, M., Yang, F., & Hu, Y.-P. (2012b). Cuprous oxide nanoparticles selectively induce apoptosis of tumor cells. *International Journal of Nanomedicine, 7*, 2641.

Wang, Z., Li, N., Zhao, J., White, J. C., Qu, P., & Xing, B. (2012c). CuO nanoparticle interaction with human epithelial cells: Cellular uptake, location, export, and genotoxicity. *Chemical Research in Toxicology, 25*(7), 1512–1521.

Wang, Y., Yang, F., Zhang, H., Zi, X., Pan, X., Chen, F., Luo, W., Li, J., Zhu, H., & Hu, Y. (2013). Cuprous oxide nanoparticles inhibit the growth and metastasis of melanoma by targeting mitochondria. *Cell Death & Disease, 4*(8), e783–e783.

Wang, Y., Cui, H., Zhou, J., Li, F., Wang, J., Chen, M., & Liu, Q. (2015). Cytotoxicity, DNA damage, and apoptosis induced by titanium dioxide nanoparticles in human non-small cell lung cancer A549 cells. *Environmental Science and Pollution Research, 22*(7), 5519–5530.

Wang, H., Yu, J., Lu, X., & He, X. (2016). Nanoparticle systems reduce systemic toxicity in cancer treatment. *Nanomedicine, 11*(2), 103–106.

Wang, Y., Deng, Y., Luo, H., Zhu, A., Ke, H., Yang, H., & Chen, H. (2017a). Light-responsive nanoparticles for highly efficient cytoplasmic delivery of anticancer agents. *ACS Nano, 11*(12), 12134–12144.

Wang, Y., Xie, Y., Li, J., Peng, Z.-H., Sheinin, Y., Zhou, J., & Oupický, D. (2017b). Tumor-penetrating nanoparticles for enhanced anticancer activity of combined photodynamic and hypoxia-activated therapy. *ACS Nano, 11*(2), 2227–2238.

Wason, M. S., Colon, J., Das, S., Seal, S., Turkson, J., Zhao, J., & Baker, C. H. (2013). Sensitization of pancreatic cancer cells to radiation by cerium oxide nanoparticle-induced ROS production. *Nanomedicine: Nanotechnology, Biology and Medicine, 9*(4), 558–569.

Wu, J., Sun, J., & Xue, Y. (2010). Involvement of JNK and P53 activation in G2/M cell cycle arrest and apoptosis induced by titanium dioxide nanoparticles in neuron cells. *Toxicology Letters, 199*(3), 269–276.

Xiong, D., Fang, T., Yu, L., Sima, X., & Zhu, W. (2011). Effects of nano-scale TiO2, ZnO and their bulk counterparts on zebrafish: Acute toxicity, oxidative stress and oxidative damage. *Science of the Total Environment, 409*(8), 1444–1452.

Yamamoto, O. (2001). Influence of particle size on the antibacterial activity of zinc oxide. *International Journal of Inorganic Materials, 3*(7), 643–646.

Yang, R., Xu, J., Xu, L., Sun, X., Chen, Q., Zhao, Y., Peng, R., & Liu, Z. (2018). Cancer cell membrane-coated adjuvant nanoparticles with mannose modification for effective anticancer vaccination. *ACS Nano, 12*(6), 5121–5129.

Yogesh, B., Vineeta, B., Rammesh, N., & Saili, P. (2016). Biosynthesized platinum nanoparticles inhibit the proliferation of human lung-cancer cells in vitro and delay the growth of a human lung-tumor xenograft in vivo:-In vitro and in vivo anticancer activity of bio-Pt NPs. *Journal of Pharmacopuncture, 19*(2), 114.

Yokohira, M., Hashimoto, N., Yamakawa, K., Suzuki, S., Saoo, K., Kuno, T., & Imaida, K. (2009). Lung carcinogenic bioassay of CuO and TiO2 nanoparticles with intratracheal instillation using F344 male rats. *Journal of Toxicologic Pathology, 22*(1), 71–78.

Zhang, W., Ji, Y., Wu, X., & Xu, H. (2013). Trafficking of gold nanorods in breast cancer cells: Uptake, lysosome maturation, and elimination. *ACS Applied Materials & Interfaces, 5*(19), 9856–9865.

Zheng, Y., Li, R., & Wang, Y. (2009). In vitro and in vivo biocompatibility studies of ZnO nanoparticles. *International Journal of Modern Physics B, 23*(6–7), 1566–1571.

# Chapter 9
# Applications of Nanomaterials in Tissue Engineering and Regenerative Medicine

**Rabia Aziz**

## 9.1 Overview of Nanomaterials: Tissue Manufacturing and Regenerative Medicine

Tissue engineering is a rapidly growing domain which involves and which is used to repair, replace, and/or create artificial cells, tissues and organs by utilizing the amalgamation of biological cells and biomaterials (Saha 2013). To regenerate, damaged or diseased cells, tissues, and organs, regenerative medicine provide remarkable insights by the combination of tissue engineering and life science principles (Maoa and Mooney 2015).

Nanotechnology is playing a promising role in the success of tissue engineering and regenerative medicine. Nanotechnology has several applications involving creation of nanofibers, nanostructured scaffolds, and nanopatterns in tissue engineering and regenerative medicine (Saha 2013).

Moreover, utilization of nanofabrication methods has numerous advantages in tissue engineering (Fig. 9.1). The formation of nanopatterns, nanofibers, as well as controlled release nanoparticles by using nanotechnology introduces many applications in tissue engineering including imitating local tissues as biomaterials to be built is of the size of nanometer, for example, cardiovascular tissue, bone marrow, extracellular liquids, and so on (Chung et al. 2007).

This chapter aims to throw light on the applications of nanomaterials in tissue engineering and regenerative medicine, highlighting the most promising and widely used nanomaterials used for the purpose.

---

R. Aziz (✉)
Digital Health and Bioinformatic Engineering, University of Salerno,
Fisciano, SA, Italy
e-mail: r.aziz@studenti.unisa.it



K. Pal (ed.), *Bio-manufactured Nanomaterials*,
https://doi.org/10.1007/978-3-030-67223-2_9

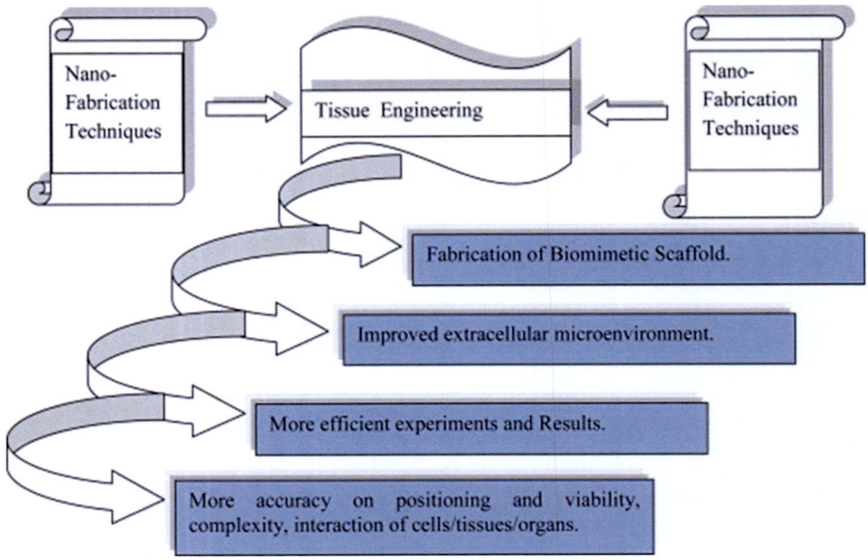


**Fig. 9.1** Schematic representation of benefits of using micro- and nanofabrication for tissue engineering (Saha 2013)

## 9.2 Application of Nanoparticles in Gene Delivery

Gene delivery is a promising technology for the explicit therapy of various gene-related maladies going from hemophilia, cancer, neurodegenerative diseases, hypercholesterolemia, autoimmune disorder to cancer (Choi et al. 2014). To prevent or cure the advancement of the relevant disease, this technique involves the introduction of genes to the destination cells or tissues by the modification of endogenous gene expression (Rapti et al. 2011; Ando et al. 2014).

With the extraordinary advancement of nanotechnology and bioscience, gene therapy depicts a tremendous aptitude in clinical implementation for several severe incurable human diseases (Ibraheem et al. 2014) (Fig. 9.2).

CALLA-01, a focused nanoparticle framework dependent on cyclodextrins, has been produced for the first in-human stage 1 clinical trial (Davis et al. 2010). Another biodegradable and biocompatible polycation is chitosan, which would be filled as a favorable transporter for gene therapy (Lee et al. 2014). One of the widely investigated nanoparticles for gene delivery are lipid-based nanoparticles having greater biocompatibility as well as close similarity with lipidic membranes that encourage their entrance into the cells (Hadinoto et al. 2013; Chitkara et al. 2015). Several other biocompatible nanoparticles pulled in incredible considerations as gene delivery systems, such as low molecular weight polyethylenimine (Dong et al. 2011), poly(β-amino ester)s (Deng et al. 2014), disulfide cross-connected polymers (Tai et al. 2015), polyamidoamine (Xu et al. 2014),and polyphosphoesters (Xu et al. 2015).

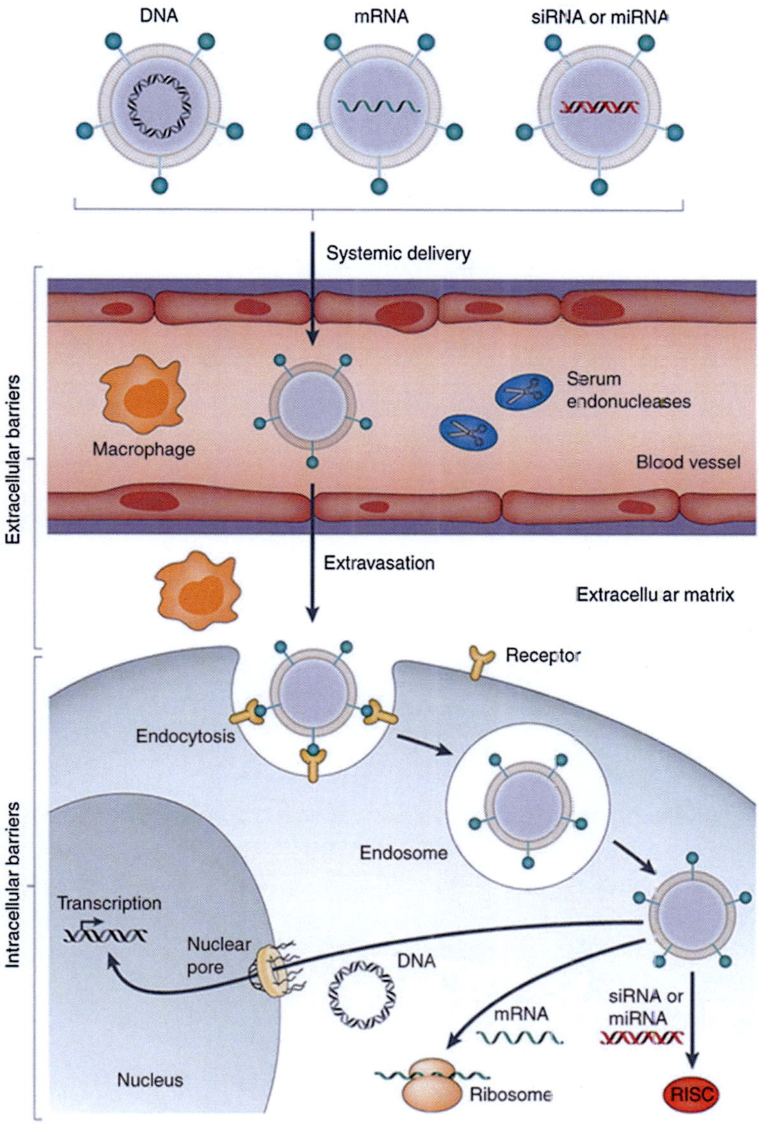


**Fig. 9.2** Schematic representation of gene therapy based on nanoparticles (Chen et al. 2016)

## 9.3 Transfection Agents Due to Nanoparticles

The process of introduction of nucleic acids into the living cells via non-viral means is termed as transfection (Neuhaus et al. 2016). Nowadays, nanoparticles are an effective alternative to insert non-viral DNA to the eukaryotic cells. Nanoparticles help DNA to efficiently link with proteins, ligands and lipids present in the cells by breaking the endosomal barriers and crossing the membranes in an effective manner (Dzięgiel 2016).

The crossing of membranes and breaking endosomal barriers can be done easily by the use of nanoparticles because despite having smaller size, they provide larger surfaced for adhesion as well as have high stability than other particles (Barkalina et al. 2014). There are many kinds of nanoparticles with different properties and traits and are proved as efficient transfection agents are described below.

### *9.3.1 Mesoporous Silica*

The nanoparticles of mesoporous silica are formulated in the form of structures like that of honeycomb. Different channels are present that enables the encapsulation of molecules and their delivery within the cells. They can also work in combination with micelles, magnetic nanoparticles, as well as polymers for the creation of delivery platforms within cells with high stability and biocompatibility (Slowing et al. 2008). We can also obtain nanoparticles of mesoporous silica with different morphology and pore sizes. These properties influence the procedure of loading the molecules to the pores (Wang et al. 2015).

### *9.3.2 Polymers*

Polyamidoamine (PAMAM), chitosan, poly-L-lactic acid (PLA), poly-L-lactide-coglycolide (PLGA), gelatine, and many other polymers are used as transfection agents in regenerative medicine. They can form diverse shapes of nanoparticles, such as dendrimers (Barkalina et al. 2014). In order to attain the efficient and accurately targeted delivery platforms, they can link with other functional groups by either using their natural shapes or mixed (i.e., natural and synthetic polymers) (Nitta and Numata 2013).

### *9.3.3 Lipids*

Another important type of biomimetic molecules are the lipid nanoparticles. Mono- and bi-layered structures are formed by phospholipid nanoparticles while nanospheres are formed by solid lipid nanoparticles. Polymers or surfactants are used to stabilize the lipid cores of such structures (Barkalina et al. 2014). Negatively charged nucleic acids can bind with cationic lipids by ionic reactions. Nanosphere is hydrophobic from inside which provides better water solubility and encapsulation of substations conveyed by them (Carmona-Ribeiro 2010). Still, nanocarriers made with polymers are more stable than this kind (Wang et al. 2015).

### 9.3.4 Carbon-Based Nanoparticles

Carbon nanotubes (CNT) and graphene oxide are the successfully used forms of carbon-based nanoparticles as transfection agents. In order to overlook the agglomeration and precipitation, these are used with water solvents as they are not dispersible in water. Functional groups are added along with carbon nanotubes to reduce the cytotoxicity caused by CNTs (Zanin et al. 2014).

### 9.3.5 Metals

Due to their less reactivity, nanoparticles of noble metals are of great interest as transfection agents (Austin et al. 2014). Gold nanoparticles are also successfully used for gene and drug delivery to the cells as their inner core is inert in nature. This allows molecules to bind with them by either covalent or non-covalent conjugation (Ghosh et al. 2008). Moreover, silver nanoparticles can also be used because of their anti-bacterial properties, but they can also produce toxicity (Austin et al. 2014).

## 9.4 Cell Patterning Via Nanoparticles

Cell patterning is one of the most significant areas of tissue engineering. Fabrication of artificial tissues as a replacement of damaged tissues can be done by different conventional methods and provide promising results, e.g., microcontact printing (Dike et al. 1999) and lithography (Bhatia et al. 1997). These methods provide precise results but are time consuming and expensive. Therefore, several physical methods are introduced for cell patterning involving inkjet printing (Xu et al. 2006) and cell spraying (Nahmias et al. 2005a, b). These methods also possess several shortcomings, for instance, cells got damaged due to high temperature or pressure in inkjet printing as well as deposition of cells is much slow in laser-guided direct writing (Nahmias et al. 2005a, b).

In order to reduce the drawbacks of the aforementioned methods, novel approaches are designed for cell patterning which involves the use of nanoparticles. Mostly, magnetite nanoparticles such as magnetite cationic liposomes (MCLs) are used along with magnetic force for the fabrication of cell pattens on the unspecialized surfaces and provide high resolution (Ino et al. 2007).

## 9.5 Elastin-Based Nanoparticles for Targeted Gene Treatment

Several genetic diseases can be cured by the gene therapy that aims to transport a specific gene of interest to inactivate, replace or correct the faulty gene. In viral gene delivery methods, viral vectors are used for the delivery of DNA into the host cells, usually, viruses, for instance, lentivirus (Escors and Breckpot 2010), adeno-associated viruses (Kotterman and Schaffer 2014), adeno viruses (Wold and Toth 2013), and herpes simplex viruses (Manservigi et al. 2010).

Nowadays, Elastin-like polypeptides (ELPs) are becoming popular as vectors for the delivery of drugs and genes because of genetic encodability and phase transition. These are basically protein-based polymers and used as new gene carriers. The property of self-assembly of nanostructures of ELPs into nanoparticles makes them a good choice as the viral gene delivery vectors (Monfort and Koria 2017).

## 9.6 Stem Cell Therapy by Nanoparticles

The treatment of several incurable and degenerative diseases has been made possible due to stem cell therapy Hasan et al. (2018). Nanotechnology is playing a promising role in the success of stem cell therapy due to their incredible properties. Several kinds of nanoparticles are used in stem cell therapy, which are described below.

### *9.6.1 Metal Nanoparticles*

Metal nanoparticles have accumulated great interest to be used in stem cell therapy. Nanoparticles of several metals including gold, silver, and some metal oxides can be used in the process. With regard to foundational microorganism treatment, specialists mean to follow transplanted cells stacked with AuNPs. An ongoing report effectively complexed 40 nm AuNPs along with the two ligands, including rhodamine B isothiocyanate (RITC) and poly-L-lysine (PLL), to rise nanoparticle uptake by human mesenchymal stem cells (hMSC). AuNP uptake did not restrain cell differentiation, and marked human mesenchymal stem cells demonstrated solid constriction, or perceivability, during an in vitro miniature CT imaging (Kim et al. 2016).

Silver nanoparticles possess anti-bacterial properties and can give promising results, but they also create neurodegenerative gene expression and inflammatory responses (Huang et al. 2015). Nanoparticles of cerium oxide (CeO), iron oxide ($Fe_3O_4$), and zinc oxide (ZnO) are used and due to their magnetic properties gave good results in human stem cells. For instance, modified nanoparticles of superparamagnetic iron oxide (SPIO) are used in human neural stem cells (hNSC) without hindering proliferation and viability of the cells (Yuan et al. 2018) and to decrease nitrosative stress and reactive oxygen species, ceramic and zinc oxides are used (Dowding et al. 2014).

### 9.6.2 Silica Nanoparticles

Silica nanoparticles are inert and transparent in nature, that is why can be linked to the different kinds of fluorescent probes. In Drosophila, silica nanoparticles penetrated to the neurons without causing cytotoxic effects in vivo and made them an exciting target for the treatment of neurodegeneration (Qian et al. 2008).

### 9.6.3 Polymeric Nanoparticles

Due to their flexible physical properties, rate of degradation in vivo and synthesis techniques, polymeric nanoparticles are of great importance in the stem cell therapy. Nanoparticles of poly lactic acid (PLA), poly aspartic acid, poly D,L-lactic-*co*-glycolic acid (PLGA), poly butylcyanoacrylate (PBCA), and poly glycolic acid (PGA) are commonly used and are promising materials for the treatment of neurodegeneration (Bhatt et al. 2017).

## 9.7 Nanofabrication Technology in Tissue Engineering

Nanofabrication refers to the formation of artifacts that can be measured in a nanoscale (Harvey and Ghantasala 2006). Nanofabrication can be divided in two basic approaches:

1. Top-down approach
2. Bottom-up approach

### 9.7.1 Top-Down Approach

The conventional top-down methodology includes cultivating cells into fully measured porous scaffolds to form tissue constructs. This methodology has numerous impediments such like slow vascularization, limitations of diffusion, lower density of cell, and heterogenous distribution of cells (Tiruvannamalai-Annamalai et al. 2014).

### 9.7.2 Bottom-Up Approach

The bottom-up or modular approach involves the engineering of complex tissues and organs from the microscale modules. This approach eliminates all the shortcomings of the conventional approach (Tiruvannamalai-Annamalai et al. 2014) (Fig. 9.3).

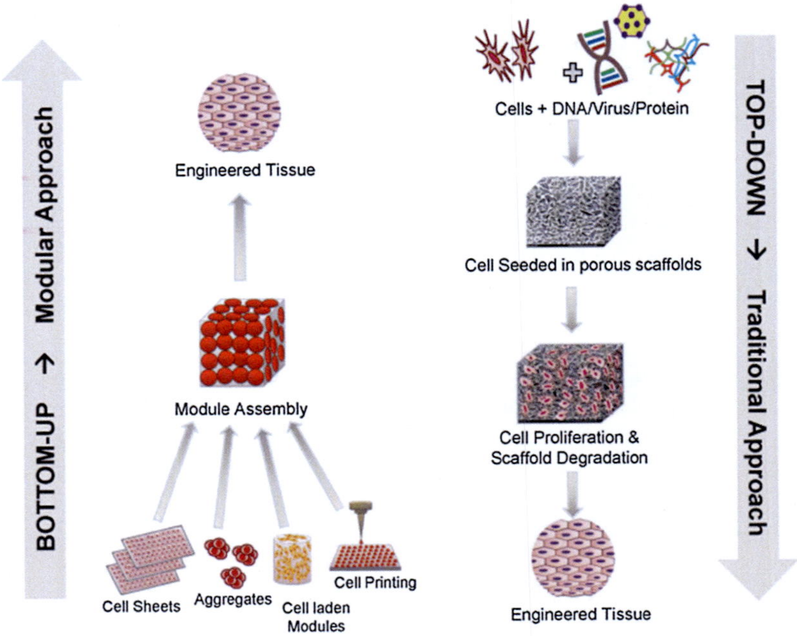


**Fig. 9.3** Comparison of top-down and bottom-up approaches for tissue engineering (Tiruvannamalai-Annamalai et al. 2014)

## 9.8 Nanofibrous Scaffolds in Tissue Engineering

Nanofibrous scaffolds are basically extracellular matrices that are artificially designed to deliver natural environment for the formation of tissues. Due to their greater surface to volume ratio, nanofibrous scaffolds efficiently promote cell proliferation, differentiation, and adhesion (Gupta et al. 2014).

Nanofibrous scaffolds have architectural features like that of extracellular matrix which has a complex 3D network for proliferation, growth, and differentiation of cell and consists of nanofiber-based cellular matrix. Nanofibrous scaffolds containing nanofibers have greater similarity with several extracellular matrix molecules, for instance, matrix proteins involving fibronectin, laminin, and collagen (5–500 nm in size) as well as proteoglycans including hyaluronic acid (450–1000 nm in size) (Lelkesa 2005).

## 9.9 Scaffolds of Nanostructured Materials for Replacing Damaged Organs

The demand of scaffolds designed from nanostructured materials is increasing tremendously because of their ability to mimic native tissues by configuring their geometry and optimizing biomaterials. Nowadays, the request for replacement of organs and regeneration of tissues is surging because of growing number of cases associated with tissue damage and organ failure as there is a scarcity of organs for the transplantation (Gupta et al. 2014).

### 9.9.1 *Neural Tissue Generation*

The key standard of neural tissue designing is to give a positive situation involving biomimetic scaffolds and cells in vitro, and further to stimulate the capability of body to practically recuperate beforehand irrevocable tissues instead of straightforwardly to embed the artificial tissues (Place et al. 2009). Nerve regeneration approaches involve the use of natural polymers (chitin, chitosan, alginate gelatin, collagen), synthetic biodegradable polymers (PLGA, poly L-lactic acid, poly ε-caprolactone), conducting polymers (polyaniline, polypyrrole) and synthetic no-degradable polymers (silicone). An ideal nerve channel must be flexible, biocompatible, thin, compliant, neuro-conductive, biodegradable, porous, and neuro-inductive (Verreck et al. 2005). Even though the above-mentioned biomaterials fulfill most of the aforementioned criteria, still they possess few drawbacks that have to solved in order to meet neuro regeneration applications (Haile et al. 2007).

To overcome those drawbacks and improve the properties of nerve scaffolds, researchers have incorporated the use of different techniques including electrospinning, polymer blending, and introducing nerve growth factors in the scaffolds (Sua 2005). Table 9.1 depicts the summary of techniques and biomaterials used to enhance nerve regeneration.

### 9.9.2 *Cardiovascular Attempts*

The rate of cardiovascular diseases (such as heart failure and myocardial infarction) is increasing day by day. The only solution to these diseases was heart transplantation which was not possible for every patient due to the scarcity of donors. Cardiovascular tissue engineering has revolutionized this concept by introducing injectable gels, artificial implantable blood vessels, cardiac patches, etc. created from biodegradable polymers (Curtis and Russell 2010; Kai et al. 2011). Biodegradable polymers are divided in to two principal classes, i.e., synthetic and natural polymers.

#### 9.9.2.1 Natural Biodegradable Polymers

Natural polymers are referred to as polymers obtained from nature (Vroman and Tighzert 2009). Natural polymers involve fibrin, collagen, gelatin, alginate, Matrigel, and chitosan. Natural biodegradable polymers possess several merits

**Table 9.1** Summary of techniques and biomaterials to enhance nerve regeneration (Subramanian et al. 2009)

| Biomaterials | Technique of fabrication | Enhanced properties | References |
|---|---|---|---|
| Star-poly(ethylene glycol) | Integration of polysaccharide (e.g., heparin) | Adjustable mechanical and physical properties for the adoption of definite tissue requirements | Freudenberg et al. (2009) |
| Chitosan | Altered with (γ-glycidoxypropyltri methoxysilane) | Mechanical strength | Amado et al. (2008) |
| Poly(β-hydroxybutarate) | Sheets infused with the molecules of extracellular matrix | Proliferation and adhesion of cell | Novikova et al. (2008) |
| Poly(ε-caprolactone) | Thermal fiber bonding and electrospinning | Mechanical strength | Lee et al. (2008) |
| Poly(ε-caprolactone) | Alignment of fibers by electrospinning | Contact control | Chew et al. (2008) |
| Poly(sialic acid) | Modification of hydrogel by adsorbed poly-L-ornithine, poly-L-lysine, collagen, or laminin | Mechanical compatibility and cell adhesion | Haile et al. (2008) |
| Poly(lactic-*co*-glycolic-acid) | Improved immersion precipitation method | Hydrophilicity and selective permeability | Oh et al. (2008) |
| Poly(D,L-lactide-*co*-ε-caprolactone) | PPy nanoparticle composite and PPy coating substrate | Electrical signal for the assembly of cell functions | Zhang et al. (2007) |
| Chitosan | Thermo-responsive chitosan-based hydrogel functionalized with polylysine | Surface characteristic (charge density, wettability), mechanical compatibility, injectable scaffold | Crompton et al. (2007) |
| Poly(ε-caprolactone) | Electrospinning (polymer blending using collagen) | Biological characteristic (adhesion of Schwann cell, differentiation, and migration) | Schnell et al. (2007) |
| Collagen | Crosslinked hydrogel using YIGSR peptide improved dendrimers | Biological property (enhance the corneal epithelial cell growth as well as outgrowth of neurite) | Duan et al. (2007) |
| Poly(glycerol-sebacate) | Replica molding | Flexibility, robust contact, control response, micropatterned reactants, surface degradable | Bettinger et al. (2006) |
| Poly(lactic-*co*-glycolic acid) | Microbraiding process | Porosity, flexibility | Burdick et al. (2006) |
| Poly(D,L-lactide-*co*-glycolide) | Low pressure injection molding | Lengthwise aligned channels, porosity, imitates the geometry of innate nerves | Bini et al. (2004) |
| Poly(2-hydroxyethyl methacrylate) | Fiber templating process | Physical properties like soft tissues, oriented scaffold | Sundback et al. (2003) |
| Poly(2-hydroxyethyl methacrylate) | Liquid-liquid centrifugal casting | Mechanical features like spinal cord | Flynn et al. (2003) |

involving ample accessibility, biodegradability, as well as renewability, whereas its demerits include inadequate electrical conductivity, fast degradation, weak mechanical properties, and immunological reaction (Dhandayuthapani 2011).

#### 9.9.2.2 Synthetic Biodegradable Polymers

Synthetic polymers refer to the man-made polymers (Toong et al. 2020). Synthetic polymers include poly(glycolic acid), poly(lactic acid), poly(lactic-*co*-glycolic acid), polycaprolactone, polyurethanes, and poly(ethylene glycol). Synthetic biodegradable polymers also possess some merits involving controlled structure, stable mechanical properties, flexibility, as well as no immunological concerns, whereas some of its demerits are low biocompatibility and absence of cell attachment (BaoLin and Ma 2014).

In order to improve the drawbacks of both natural and synthetic polymers, researchers have made novel natural/synthetic composites (combination of both natural and synthetic polymers). In this manner, properties of composites have improved to the greater extent. These include PLA/chitosan, $TiO_2$-PEG/chitosan, and gelatin/PCL/graphene. Natural/synthetic composites possess high biocompatibility, strong mechanical strength, increased electrical conductivity, and improved biological properties (Zhuab 2018; Chen et al. 2019).

## 9.10 Conclusion, Outlook, and Future Aspects

Nanotechnology is the promising tool for the advancement of tissue engineering and regenerative medicine. The amalgamation of nanotechnology along with tissue engineering and regenerative medicine has introduced new insights for the regeneration and repair of damaged or diseased cells, tissues, and organs. In future, it can be predicted that nanotechnology would be helpful in the formation of complex artificial organs such as heart.

**Conflict of Interest** There are no conflicts of interest.
Ethical Issues There are no concerned ethical issues.

## References

Amado, S., Simões, M. J., da Silva, P. A. S. A., Luís, A. L., Shirosaki, Y., Lopes, M. A., Santos, J. D., Fregnan, F., Gambarotta, G., Raimondo, S., Fornaro, M., Veloso, A. P., Varejão, A. S. P., Maurício, A. C., & Geuna, S. (2008). Use of hybrid chitosan membranes and N1E-115 cells for promoting nerve regeneration in an axonotmesis rat model. *Biomaterials, 29*(33), 4409–4419. https://doi.org/10.1016/j.biomaterials.2008.07.043.

Ando, M., Takahashi, Y., Yamashita, T., Fujimoto, M., Nishikawa, M., Watanabe, Y., & Takakura, Y. (2014). Prevention of adverse events of interferon $\gamma$ gene therapy by gene delivery of interferon $\gamma$-heparin-binding domain fusion protein in mice. *Molecular Therapy, Methods and Clinical Developments, 1*, 14023. https://doi.org/10.1038/mtm.2014.23.

Austin, L. A., Mackey, M. A., Dreaden, E. C., & El-Sayed, M. A. (2014). The optical, photothermal, and facile surface chemical properties of gold and silver nanoparticles in biodiagnostics, therapy, and drug delivery. *Archives of Toxicology, 88*(7), 1391–1417. https://doi.org/10.1007/s00204-014-1245-3.

BaoLin, G., & Ma, P. X. (2014). Synthetic biodegradable functional polymers for tissue engineering: A brief review. *Science China Chemistry, 57*(4), 490–500. https://doi.org/10.1007/s11426-014-5086-y.

Barkalina, N., Charalambous, C., Jones, C., & Coward, K. (2014). Nanotechnology in reproductive medicine: Emerging applications of nanomaterials. *Nanomedicine, 10*(5), 921–938. https://doi.org/10.1016/j.nano.2014.01.001.

Bettinger, C. J., Orrick, B., Misra, A., Langer, R., & Borenstein, J. T. (2006). Microfabrication of poly (glycerol-sebacate) for contact guidance applications. *Biomaterials, 27*(12), 2558–2565.

Bhatia, S. N., Yarmush, M. L., & Toner, M. (1997). Controlling cell interactions by micropatterning in co-cultures: Hepatocytes and 3T3 fibroblasts. *Journal of Biomedical Materials Research, 34*(2), 189–199. https://doi.org/10.1002/(sici)1097-4636(199702)34:2<189::aid-jbm8>3.0.co;2-m.

Bhatt, P. C., Verma, A., Al-Abbasi, F. A., Anwar, F., Kumar, V., & Panda, B. P. (2017). Development of surface-engineered PLGA nanoparticulate-delivery system of Tet1-conjugated nattokinase enzyme for inhibition of A$\beta$40 plaques in Alzheimer's disease. *International Journal of Nanomedicine, 12*, 8749–8768. Retrieved from https://www.scopus.com/record/display.uri?eid=2-s2.0-85038227580&origin=inward.

Bini, T. B., Gao, S., Xu, X., Wang, S., Ramakrishna, S., & Leong, K. W. (2004). Peripheral nerve regeneration by microbraided poly(L-lactide-co-glycolide) biodegradable polymer fibers. *Journal of Biomedical Materials Research A, 68*(2), 286–295.

Burdick, J. A., Ward, M., Liang, E., Young, M. J., & Langer, R. (2006). Stimulation of neurite outgrowth by neurotrophins delivered from degradable hydrogels. *Biomaterials, 27*(3), 452–459.

Carmona-Ribeiro, A. M. (2010). Biomimetic nanoparticles: Preparation, characterization and biomedical applications. *International Journal of Biomedicine, 5*, 249–259. https://doi.org/10.2147/ijn.s9035.

Chen, J., Guo, Z., Tian, H., & Chen, X. (2016). Production and clinical development of nanoparticles for gene delivery. *Molecular Therapy-Methods and Clinical Development, 3*, 16023. https://doi.org/10.1038/mtm.2016.23.

Chen, X., Feng, B., Zhu, D.-Q., Chen, Y.-W., Ji, W., Ji, T.-J., & Li, F. (2019). Characteristics and toxicity assessment of electrospun gelatin/PCL nanofibrous scaffold loaded with graphene in vitro and in vivo. *International Journal of Nanomedicine, 14*, 3669–3678. https://doi.org/10.2147/IJN.S204971.

Chew, S. Y., Mi, R., Hok, A., & Leong, K. (2008). The effect of the alignment of electrospun fibrous scaffolds on Schwann cell maturation. *Biomaterials, 29*(6), 653–661. https://doi.org/10.1016/j.biomaterials.2007.10.025.

Chitkara, D., Mittal, A., & Mahato, R. I. (2015). miRNAs in pancreatic cancer: Therapeutic potential, delivery challenges and strategies. *Advanced Drug Delivery Reviews, 18*, 34–52. https://doi.org/10.1016/j.addr.2014.09.006.

Choi, Y. S., Lee, M. Y., David, A. E., & Park, Y. S. (2014). Nanoparticles for gene delivery: Therapeutic and toxic effects. *Molecular & Cellular Toxicology, 10*, 1–8. Retrieved from https://link.springer.com/article/10.1007/s13273-014-0001-3.

Chung, B. G., Kang, L., & Khademhosseini, A. (2007). Micro- and nanoscale technologies for tissue engineering and drug discovery applications. *Expert Opinion on Drug Discovery, 2*(12), 1653–1668. https://doi.org/10.1517/17460441.2.12.1653.

Crompton, K. E., Goud, J. D., Bellamkonda, R. V., Gengenbach, T. R., Finkelstein, D. I., Horne, M. K., & Forsythe, J. S. (2007). Polylysine-functionalised thermoresponsive chitosan hydrogel for neural tissue engineering. *Biomaterials, 28*(3), 441–449.

Curtis, M. W., & Russell, B. (2010). Cardiac tissue engineering. *Journal of Cardiovascular Nursing, 24*(2), 87–92. https://doi.org/10.1097/01.JCN.0000343562.06614.49.

Davis, M. E., Zuckerman, J. E., Choi, C. H. J., Seligson, D., Tolcher, A., Alabi, C. A., Yen, Y., Heidel, J. D., & Ribas, A. (2010). Evidence of RNAi in humans from systemically administered siRNA via targeted nanoparticles. *Nature, 464*(7291), 1067–1070. Retrieved from https://www.nature.com/articles/nature08956.

Deng, X., Zheng, N., Song, Z., Yin, L., & Cheng, J. (2014). Trigger-responsive, fast-degradable poly(β-amino ester)s for enhanced DNA unpackaging and reduced toxicity. *Biomaterials, 35*(18), 5006–5015. https://doi.org/10.1016/j.biomaterials.2014.03.005.

Dhandayuthapani, B. (2011). Polymeric biomaterials for tissue engineering applications. *International Journal of Polymer Science, 2011*, 184623. https://doi.org/10.1155/2011/290602.

Dike, L. E., Chen, C. S., Mrksich, M., Tien, J., Whitesides, G. M., & Ingber, D. E. (1999). Geometric control of switching between growth, apoptosis, and differentiation during angiogenesis using micropatterned substrates. *In Vitro Cellular & Developmental Biology. Animal, 35*, 441–448. Retrieved from https://link.springer.com/article/10.1007/s11626-999-0050-4.

Dong, X., Tian, H., Chen, L., Chen, J., & Chen, X. (2011). Biodegradable mPEG-b-P(MCC-g-OEI) copolymers for efficient gene delivery. *Journal of Controlled Release, 152*(1), 135–142. Retrieved from https://www.sciencedirect.com/science/article/pii/S0168365911001611.

Dowding, J. M., Song, W., Bossy, K., Karakoti, A., Kumar, A., Kim, A., Bossy, B., Seal, S., Ellisman, M. H., Perkins, G., Self, W. T., & Bossy-Wetzel, E. (2014). Cerium oxide nanoparticles protect against Aβ-induced mitochondrial fragmentation and neuronal cell death. *Cell Death and Differentiation, 21*(10), 1622–1632. Retrieved from https //www.nature.com/articles/cdd201472.

Duan, X., McLaughlin, C., Griffith, M., & Sheardown, H. (2007). Biofunctionalization of collagen for improved biological response: Scaffolds for corneal tissue engineering. *Biomaterials, 28*(1), 78–88.

Dzięgiel, N. (2016). Nanoparticles as a tool for transfection and transgenesis—A review. *Annals of Animal Science, 16*(1), 53–64. https://doi.org/10.1515/aoas-2015-0077.

Escors, D., & Breckpot, K. (2010). Lentiviral vectors in gene therapy: Their current status and future potential. *Archivum Immunologiae et therapiae expermentalis, 58*(2), 107–119. https://doi.org/10.1007/s00005-010-0063-4.

Flynn, L., Dalton, P. D., & Shoichet, M. S. (2003). Fiber templating of poly(2-hydroxyethyl methacrylate) for neural tissue engineering. *Biomaterials, 24*, 4265–4272.

Freudenberg, U., Hermann, A., Welzel, P. B., Stirl, K., Schwarz, S. C., Grimmer, M., Zieris, A., Panyanuwat, W., Zschoche, S., Meinhold, D., Storch, A., & Werner, C. (2009). A star-PEG–heparin hydrogel platform to aid cell replacement therapies for neurodegenerative diseases. *Biomaterials, 30*(28), 5049–5060. https://doi.org/10.1016/j.biomaterials.2009.06.002.

Ghosh, P., Han, G., De, M., Kim, C. K., & Rotello, V. M. (2008). Gold nanoparticles in delivery applications. *Advanced Drug Dellivery Reviews, 60*(11), 1307–1315. https://doi.org/10.1016/j.addr.2008.03.016.

Gupta, K. C., Haider, A., Choi, Y.-r., & Kang, I.-K. (2014). Nanofibrous scaffolds in biomedical applications. *Biomaterials Research, 18*(2), 27–38. Retrieved from https //www.ncbi.nlm.nih.gov/pmc/articles/PMC4549138/#:~:text=Nanofibrous%20scaffolds%20are%20artificial%20extracellular,high%20surface%20to%20volume%20ratio.

Hadinoto, K., Sundaresan, A., & Cheow, W. S. (2013). Lipid–polymer hybrid nanoparticles as a new generation therapeutic delivery platform: A review. *European Journal of Pharmaceutics and Biopharmaceutics, 85*(3), 427–443. Retrieved from https://www.sciencedirect.com/science/article/pii/S0939641113002464.

Haile, Y., Haastert, K., Cesnulevicius, K., Stummeyer, K., Timmer, M., Berski, S., Dräger, G., Gerardy-Schahn, R., & Grothe, C. (2007). Culturing of glial and neuronal cells on polysialic acid. *Biomaterials, 28*(6), 1163–1173. https://doi.org/10.1016/j biomaterials.2006.10.030.

Haile, Y., Berski, S., Dräger, G., Nobre, A., Stummeyer, K., Gerardy-Schahn, R., & Grothe, C. (2008). The effect of modified polysialic acid based hydrogels on the adhesion and viability of primary neurons and glial cells. *Biomaterials, 29*(12), 1880–1891. https://doi.org/10.1016/j.biomaterials.2007.12.030.

Harvey, E., & Ghantasala, M. (2006). Nanofabrication. In E. Harvery & M. Ghantasala (Eds.), *Nanostructure control of materials* (pp. 303–330). Amsterdam: Elsevier. https://doi.org/10.1533/9781845691189.303.

Hasan, A., Morshed, M., Memic, A., Hassan, S., Webster, T. J., & Marei, H. E.-S. (2018). Nanoparticles in tissue engineering: Applications,challenges and prospects. *International Journal of Nanomedicine, 13*, 5637–5655. Retrieved from https://www.dovepress.com/nanoparticles-in-tissue-engineering-applications-challenges-and-prospe-peer-reviewed-article-IJN.

Hosseini, A. Y. (2019). Chapter 29—Applications of nanotechnology for regenerative medicine; healing tissues at the nanoscale. In *Principles of regenerative medicine* (3rd ed., pp. 485–504). Amsterdam: Elsevier. Retrieved from https://www.sciencedirect.com/science/article/pii/B9780128098806000291.

Huang, C.-L., Hsiao, I.-L., Lin, H.-C., Wang, C.-F., Huang, Y.-J., & Chuang, C.-Y. (2015). Silver nanoparticles affect on gene expression of inflammatory and neurodegenerative responses in mouse brain neural cells. *Environmental Research, 136*, 253–263. https://doi.org/10.1016/j.envres.2014.11.006.

Ibraheem, D., Elaissari, A., & Fessi, H. (2014). Gene therapy and DNA delivery systems. *International Journal of Pharmaceutics, 459*(1–2), 70–83. https://doi.org/10.1016/j.ijpharm.2013.11.041.

Ino, K., Ito, A., & Honda, H. (2007). Cell patterning using magnetite nanoparticles and magnetic force. *Biotechnology and Bioengineering, 97*(5), 1309–1317. https://doi.org/10.1002/bit.21322.

Kai, D., Prabhakaran, M. P., Jin, G., & Ramakrishna, S. (2011). Polypyrrole-contained electrospun conductive nanofibrous membranes for cardiac tissue engineering. *Journal of Biomedical Materials Research, Part A, 1*(99), 376–385. https://doi.org/10.1002/jbm.a.33200.

Kim, T., Lee, N., Arifin, D. R., Shats, I., Janowski, M., Walczak, P., Hyeon, T., & Bulte, J. W. M. (2016). In vivo micro-CT imaging of human mesenchymal stem cells labeled with gold-poly-l-lysine nanocomplexes. *Advanced Functional Materials, 27*(3), 1604213. https://doi.org/10.1002/adfm.201604213.

Kotterman, M. A., & Schaffer, D. V. (2014). Engineering adeno-associated viruses for clinical gene therapy. *Nature Review Genetics, 15*(7), 445–451. Retrieved from https://www.nature.com/articles/nrg3742.

Lee, J., Cuddihy, M. J., & Kotov, N. A. (2008). Three-dimensional cell culture matrices: State of the art. *Tissue Engineering, 14*(1), 61–86. https://doi.org/10.1089/teb.2007.0150.

Lee, S. J., Lee, A., Hwang, S. R., Park, J.-S., Jang, J., Huh, M. S., Jo, D.-G., Yoon, S.-Y., Byun, Y., Kim, S. H., Kwon, I. C., Youn, I., & Kim, K. (2014). TNF-α gene silencing using polymerized siRNA/thiolated glycol chitosan nanoparticles for rheumatoid arthritis. *Molecular Therapy: The Journal of the American Society of Gene Therapy, 22*(2), 397–408. https://doi.org/10.1038/mt.2013.245.

Lelkesa, M. L. (2005, October). Electrospun protein fibers as matrices for tissue engineering. Biomaterials, 26(30), 5999–6008. https://doi.org/10.1016/j.biomaterials.2005.03.030.

Manservigi, R., Argnani, R., & Marconi, P. (2010). HSV recombinant vectors for gene therapy. *Open Virol Journal, 4*, 123–156. https://doi.org/10.2174/1874357901004010123.

Maoa, A. S., & Mooney, D. J. (2015). Regenerative medicine: Current therapies and future directions. *Proceedings of the National Academy of Sciences of the United States of America, 112*(47), 14452–14459. https://doi.org/10.1073/pnas.1508520112.

Monfort, D. A., & Koria, P. (2017). Recombinant elastin-based nanoparticles for targeted gene therapy. *Gene Therapy, 24*(10), 610–620. Retrieved from https://www.nature.com/articles/gt201754.

Nahmias, Y., Arneja, A., Tower, T. T., Renn, M. J., & Odde, D. J. (2005a). Cell patterning on biological gels via cell spraying through a mask. *Tissue Engineering, 11*(5–6), 701–708. https://doi.org/10.1089/ten.2005.11.701.

Nahmias, Y., Arneja, A., Tower, T. T., Renn, M. J., & Odde, D. J. (2005b). Laser-guided direct writing for three-dimensional tissue engineering. *Biotechnology and Bioengineering, 92*(2), 129–136. https://doi.org/10.1002/bit.20585.

Neuhaus, B., Tosun, B., Rotan, O., Frede, A., Westendorf, A. M., & Epple, M. (2016). Nanoparticles as transfection agents: A comprehensive study with ten different cell lines. *RSC Advances, 22*, 18102–18112. Retrieved from https://pubs.rsc.org/en/content/articlelanding/2016/ra/c5ra25333k#!divAbstract.

Nitta, S. K., & Numata, K. (2013). Biopolymer-based nanoparticles for drug/gene delivery and tissue. *International Journal of Molecular Sciences, 14*(1), 1629–1654. https://doi.org/10.3390/ijms14011629.

Novikova, L. N., Pettersson, J., Brohlin, M., Wiberg, M., & Novikov, L. N. (2008). Biodegradable poly-beta-hydroxybutyrate scaffold seeded with Schwann cells to promote spinal cord repair. *Biomaterials, 29*(9), 1198–1206. https://doi.org/10.1016/j.biomaterials.2007.11.033.

Oh, S. H., Kim, J. H., Song, K. S., Jeon, B. H., Yoon, J. H., Seo, T. B., Namgung, U., Lee, I. W., & Lee, J. H. (2008). Peripheral nerve regeneration within an asymmetrically porous PLGA/Pluronic F127 nerve guide conduit. *Biomaerials, 29*(11), 1601–1609.

Place, E. S., Evans, N. D., & Stevens, M. M. (2009). Complexity in biomaterials for tissue engineering. *Nature Materials, 8*(6), 457–470. https://doi.org/10.1038/nmat2441.

Qian, J., Li, X., Wei, M., Gao, X., Xu, Z., & He, S. (2008). Bio-molecule-conjugated fluorescent organically modified silica nanoparticles as optical probes for cancer cell imaging. *Optics Express, 16*(24), 19568–19578. Retrieved from https://www.scopus.com/record/display.uri?eid=2-s2.0-56749170938&origin=inward.

Rapti, K., Chaanine, A. H., & Hajjar, R. J. (2011). Targeted gene therapy for the treatment of heart failure. *Canadian Journal of Cardiology, 27*(3), 265–283. https://doi.org/10.1016/j.cjca.2011.02.005.

Saha, J. K. (2013). Nanotechnology for tissue engineering: Need, techniques and applications. *Journal of Pharmacy Research, 7*(2), 200–204. Retrieved from https://www.sciencedirect.com/science/article/pii/S097469431300100X.

Schnell, E., Klinkhammer, K., Balzer, S., Brook, G., Klee, D., Dalton, P., & Mey, J. (2007). Guidance of glial cell migration and axonal growth on electrospun nanofibers of poly-epsilon-caprolactone and a collagen/poly-epsilon-caprolactone blend. *Biomaterials, 28*(19), 3012–3025.

Slowing, I. I., Vivero-Escoto, J. L., Wu, C.-W., & Lin, V. S.-Y. (2008). Mesoporous silica nanoparticles as controlled release drug delivery and gene transfection carriers. *Advanced Drug Delivery Reviews, 60*(11), 1278–1288. https://doi.org/10.1016/j.addr.2008.03.012.

Sua, P.-R. C.-H.-H.-Y. (2005, November). Release characteristics and bioactivity of gelatin-tricalcium phosphate membranes covalently immobilized with nerve growth factors. *Biomaterials, 26*(33), 6579–87. https://doi.org/10.1016/j.biomaterials.2005.03.037.

Subramanian, A., Krishnan, U. M., & Sethuraman, S. (2009). Development of biomaterial scaffold for nerve tissue engineering: Biomaterial mediated neural regeneration. *Journal of Biomedical Science, 16*(1), 108. Retrieved from https://jbiomedsci.biomedcentral.com/articles/10.1186/1423-0127-16-108#ref-CR62.

Sundback, C., Hadlock, T., Cheney, M., & Vacanti, J. (2003). Manufacture of porous polymer nerve conduits by a novel low-pressure injection molding process. *Biomaterials, 24*(5), 819–830.

Tai, Z., Wang, X., Tian, J., Gao, Y., Zhang, L., Yao, C., Wu, X., Zhang, W., Zhu, Q., & Gao, S. (2015). Biodegradable stearylated peptide with internal disulfide bonds for efficient delivery of sirna in vitro and in vivo. *Biomacromolecules, 16*(4), 1119–1130. Retrieved from https://pubs.acs.org/doi/full/10.1021/bm501777a.

Tiruvannamalai-Annamalai, R., Armant, D. R., & Matthew, H. W. T. (2014). A glycosaminoglycan based, modular tissue scaffold system for rapid assembly of perfusable, high cell density, engineered tissues. *PLoS One, 9*(1), e84287. https://doi.org/10.1371/journal.pone.0084287.

Toong, D. W. Y., Toh, H. W., Ng, J. C. K., Wong, P. E. H., Leo, H. L., Venkatraman, S., Tan, L. P., Ang, H. Y., & Huang, Y. (2020). Bioresorbable polymeric scaffold in cardiovascular applications. *International Journal of Molecular Sciences, 21*(10), 3444. Retrieved from https://webcache.googleusercontent.com/search?q=cache:PG6Z3laFggwJ:https://www.mdpi.com/1422-0067/21/10/3444/pdf+&cd=2&hl=en&ct=clnk&gl=it.

Verreck, G., Chun, I., Li, Y., Kataria, R., Zhang, Q., Rosenblatt, J., Decorte, A., Heymans, K., Adriaensen, J., Bruining, M., Van Remoortere, M., Borghys, H., Meert, T., Peeters, J., & Brewstera, M. E. (2005). Preparation and physicochemical characterization of biodegradable nerve guides containing the nerve growth agent sabeluzole. *Biomaterials, 26*(11), 1307–1315. https://doi.org/10.1016/j.biomaterials.2004.04.040.

Vroman, I., & Tighzert, L. (2009). Biodegradable polymers. *Materials, 2*(2), 307–344. https://doi.org/10.3390/ma2020307.

Wang, Y., Zhao, Q., Han, N., Bai, L., Li, J., Liu, J., Che, E., Hu, L., Zhang, Q., Jiang, T., & Wang, S. (2015). Mesoporous silica nanoparticles in drug delivery and biomedical applications. *Nanomedicine: Nanotechnology, Biology and Medicine, 11*(2), 313–327. https://doi.org/10.1016/j.nano.2014.09.014.

Wold, W. S. M., & Toth, K. (2013). Adenovirus vectors for gene therapy, vaccination and cancer gene therapy. *Current Gene Therapy, 13*(6), 421–433. Retrieved from https://www.ncbi.nlm.nih.gov/pmc/articles/PMC4507798/.

Xu, T., Gregory, C. A., Molnar, P., Cui, X., Jalota, S., Bhaduri, S. B., & Bolanda, T. (2006). Viability and electrophysiology of neural cell structures generated by the inkjet printing method. *Biomaterials, 27*(19), 3580–3588. https://doi.org/10.1016/j.biomaterials.2006.01.048.

Xu, X., Jian, Y., Li, Y., Zhang, X., Tu, Z., & Gu, Z. (2014). Bio-inspired supramolecular hybrid dendrimers self-assembled from low-generation peptide dendrons for highly efficient gene delivery and biological tracking. *ACS Nano, 8*(9), 9255–9264. Retrieved from https://pubs.acs.org/doi/10.1021/nn503118f.

Xu, W., Wang, Y., Li, S., Ke, Z., Yan, Y., Li, S., Xing, Z., Wang, C., Zeng, F., Liu, R., & Deng, F. (2015). Efficient gene and siRNA delivery with cationic polyphosphoramide with amino moieties in the main chain. *RSC Advances, 5*, 50425–50432. Retrieved from https://pubs.rsc.org/en/content/articlelanding/2015/ra/c5ra02721g#!divAbstract.

Yuan, M., Wang, Y., & Qin, Y.-X. (2018). Promoting neuroregeneration by applying dynamic magnetic fields to a novel nanomedicine: Superparamagnetic iron oxide (SPIO)-gold nanoparticles bounded with nerve growth factor (NGF). *Nanomedicine: Nanotechnology, Biology and Medicine, 14*(4), 1337–1347. https://doi.org/10.1016/j.nano.2018.03.004.

Zanin, H., Hollanda, L. M., Ceragioli, H. J., Ferreira, M. S., Machado, D., Lancellotti, M., Catharino, R. R., Baranauskas, V., & Lobo, A. O. (2014). Carbon nanoparticles for gene transfection in eukaryotic cell lines. *Material Science and Engineering C: Materials for Biological Applications, 39*, 359–370. https://doi.org/10.1016/j.msec.2014.03.016.

Zhang, Z., Rouabhia, M., Wang, Z., Roberge, C., Shi, G., Roche, P., Li, J., & Dao, L. (2007). Electrically conductive biodegradable polymer composite for nerve regeneration: Electricity-stimulated neurite outgrowth and axon regeneration. *Artificial Organs, 31*(1), 13–22.

Zhuab, N. L. (2018, October 1). Fabrication of engineered nanoparticles on biological macromolecular (PEGylated chitosan) composite for bio-active hydrogel system in cardiac repair applications. *International Journal of Biological Macromolecules, 117*, 553–58. https://doi.org/10.1016/j.ijbiomac.2018.04.196.

# Chapter 10
# Nanomaterials for the Management of Multidisciplinary Dental Sciences and Applications

**C. Pushpalatha, S. V. Sowmya, Neha Jain, Shruthi Nagaraja, Shrikanth, Dominic Augustine, Vanishri C. Haragannavar, Elizabeth Abbu Varghese, and Chhaya Kumar**

## 10.1 Introduction

The term nanodentistry is defined as "the science and technology of diagnosis, treatment and prevention of oral diseases, relieving pain, preserving and improving dental health using nanostructured material." The recent advancements in nanodentistry and innovations in dental field have eliminated the drawbacks associated with conventional methods. They have successfully maintained and achieved perfect oral health with enhanced physical properties and antimicrobial potential. The nanotechnology in dentistry promises in bringing a substantial transition in dental field and driving dental materials industry to substantial growth. The rising interest in the applications of nanotechnology in dentistry is leading to the development of newer technologies, dental materials, and therapies with significant success (Abou Neel et al. 2015). This technology has varied applications in dental field in the areas of

C. Pushpalatha (✉) · E. A. Varghese · C. Kumar
Faculty of Dental Sciences, Department of Pedodontics and Preventive Dentistry, M.S. Ramaiah University of Applied Sciences, Bangalore, India
e-mail: pushpalathag.pd.ds@msruas.ac.in

S. V. Sowmya · D. Augustine · V. C. Haragannavar
Faculty of Dental Sciences, Department of Oral Pathology and Microbiology, M.S. Ramaiah University of Applied Sciences, Bangalore, India

N. Jain
KLE Society's Institute of Dental Sciences, Bangalore, India

S. Nagaraja
Faculty of Dental Sciences, Department of Conservative Dentistry and Endodontics, M.S. Ramaiah University of Applied Sciences, Bangalore, India

Shrikanth
Mazumdar Shaw Medical Centre, Narayana Hrudayalaya Multispeciality Hospital, Bangalore, India


K. Pal (ed.), *Bio-manufactured Nanomaterials*,
https://doi.org/10.1007/978-3-030-67223-2_10

development, prevention, diagnosis, and treatment. It can help in the development of dental materials, prevention of dental caries and periodontal diseases, diagnosis of oral cancer, treatment of dentin hypersensitivity, endodontic diseases, and tooth tissue engineering technology. A thorough understanding of the nanotechnology discipline is of paramount importance to employ these materials in routine dental practice. Nanoparticles possess antimicrobial properties; their incorporation prevents biofilm formation on composite restorations, thereby minimizing microleakage and secondary caries formation. The bond strength between restorative materials and dentin is enhanced with the use of nanoparticles. White spot lesions can be prevented using nanoparticle incorporated adhesive systems in orthodontic treatments. Although in–vitro studies show that nanoparticles can be effectively used with dental biomaterials; however, in–vivo long-term results are necessary for clinical application (Abou Neel et al. 2015). The chapter highlights the recent advances of the use of nanoparticles in oral diagnostics, prevention, and therapeutic modalities that could revolutionize the clinical practices of dentistry.

## 10.2 Diagnostic Nanodentistry

Nanotechnology has been used for some decades in dentistry in the fields of atomic force microscopy (AFM), imaging contrast enhancers, quantum dots (QD), nanoscale cantilevers, single-wall carbon nanotubes, nanoelectromechanical systems, nanotexturing, optical nanobiosensor, biobarcode assay, and biochips (Fig. 10.1).

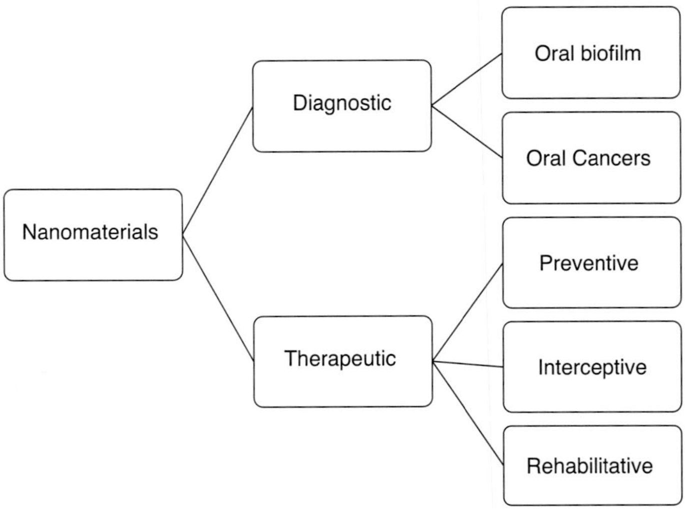


**Fig. 10.1** Application of nanoparticles in dentistry

### 10.2.1 Oral Biofilms

Biofilm-dependent oral diseases such as dental caries can be prevented and treated by an in-depth understanding of nanomechanics and its pathogenesis including bacterial adhesion and colonization (Chen et al. 2014). AFM is a recent nanotechnological tool that can measure the adhesion of bacteria to each other or to substrates like teeth and implants. The property of AFM that facilitates this process is its ability to interact and image live cells without disrupting the bacterial morphology (Abe et al. 2012). Increased sensitivity of a real-time scanning of live bacterial cells is obtained using nanomechanical biosensors and AFM cantilever. AFM also helps in obtaining information on elasticity of a cell and its membrane properties. Oral biofilm comprises of different adherent bacterial species in polysaccharide and protein matrix. Understanding the adhesion mechanism of bacteria to each other and to the substrate helps in understanding the mechanism of biofilm formation and pathogenesis of diseases (Zhang et al. 2011). Biofilm and bacterial infection-based ailments such as periodontal diseases and dental caries can also be managed using nanotechnology. Bacterial labeling using QDs was first employed by Kloepfer et al. (2003). Chalmers et al. (2007) established a single-cell resolution of oral biofilms by using a luminescent probe based QDs.

### 10.2.2 Oral Cancer

Structural imaging techniques such as "Computed tomography (CT), molecular resonance imaging (MRI), and ultrasound (US)" identify anatomical patterns, tumor size, location, and spread based on endogenous contrast (Van Acker et al. 2014). These imaging techniques may not be able to differentiate benign and malignant primary and metastatic tumors less than 5 mm. Therefore, nanotechnology-based CT imaging was developed by Hainfeld et al. (2011) using gold nanoparticles (GNPs) that exhibited high contrast due to strong absorption of X-rays. Originally GNPs were used to enhance in–vivo vascular contrast in CT imaging (Van Acker et al. 2014).

Nanotechnology also has a noteworthy part in tumor detection. They behave as an inorganic fluorophore when excited at a specific wavelength. They possess dimensional similarities with nucleic acids and proteins. QDs can also be functionalized by coupling with molecules ("transferrin, immunoglobulin G (IgG), biotin, streptavidin, avidin, nucleic acids, peptides, serotonin, adenine, adenine monophosphate, and wheat germ agglutinin") that bind to targets ("immunoglobulin G, antigens, glycoproteins, nucleic acid sequences, and receptors") forming luminescent probes. This luminescence finds application for in–vivo biological species labeling.

Bhirde (2010) demonstrated that a combination of QDs and carbon nanotube-based drug delivery targeted and killed HN13 cells. Functionalized, solubilized single-wall carbon nanotubes (SWNTs) are the other nanoparticles that are used for the diagnosis and treatment by combining with peptides, proteins, genes, and DNA. The diagnosis of oral cancer can also be obtained using mini scale beads that can effectively penetrate and destroy cancer cells. Bharadwaj and Medhi (2020) and Dogra et al. (2020) suggested that nanoparticles can be used as a drug delivery system for cancer therapy because of its flexible and dynamic nature.

## 10.3 Nanoparticles Used for Clinical Diagnostic Purposes

See Fig. 10.2.

### *10.3.1 Gold Nanoparticles*

Gold nanoparticles amplify and produce strong Raman spectra, so that it can be used in surface-enhanced Raman spectroscopy imaging which allows for deep tissue penetration. It can be synthesized by chemical, physical, and biological methods. Biological synthesis is a new attractive method being explored for developing biosensors, photoimaging, immunological assays, targeted drug delivery systems, and photothermal and photodynamic therapies. Various types of gold nanoparticles like gold nanorods, nanocages, nanoshells, nanostars, nanocubes, and nanospheres are used. However, cost-effective systems and long-term cell and immune responses need to be evaluated (Khlebtsov et al. 2013).

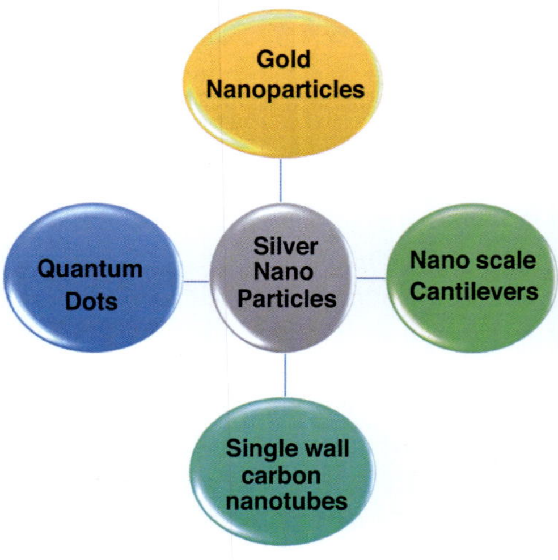


**Fig. 10.2** Common nanomaterials for clinical diagnostic purposes

### 10.3.2 Quantum Dots

They are semiconductor nanocrystals with broad absorption spectra, fluorescent probes used for ultra-sensitive, high-resolution 3D cellular imaging, DNA detection, and labeling biological species in–vivo. It is photostable and resistant to photo bleach and provides high imaging contrast, enhanced stability and brightness. Hence, it is used for in–vitro and in–vivo real-time imaging (Walling et al. 2009).

### 10.3.3 Silver Nanoparticles

Silver nanoparticles induce cell death and sensitize cancer cells. It causes changes in cell shape and lowers the cell metabolism and viability, causing mitochondrial and DNA damage (Liao et al. 2019). It is used to detect serum p53 in head and neck cancer, used in dual-imaging/therapy to detect, and destroy specific targeted cancer cells (Zhou et al. 2011).

### 10.3.4 Nanoscale Cantilevers

Cantilevers are tiny beams, the surface of which can be modified to impart specificity. It forms a part of a larger diagnostic device. When the cancer molecules bind to the cantilever, there are nanomechanical deflections and changes in its shape, providing quick and selective cancer cell diagnosis (Alharbi and Al-sheikh 2014).

### 10.3.5 Single-Wall Carbon Nanotubes (SWNTs)

Functionalized, solubilized SWNTs are used for peptide, proteins, gene, and DNA transport across cell membranes. It has enhanced biocompatibility and low cytotoxicity and can be used as a carrier for both the treatment and diagnosis of cancers (Tan et al. 2014).

## 10.4 Common Nanotechnologies for Clinical Diagnostic Purposes

See Fig. 10.3.

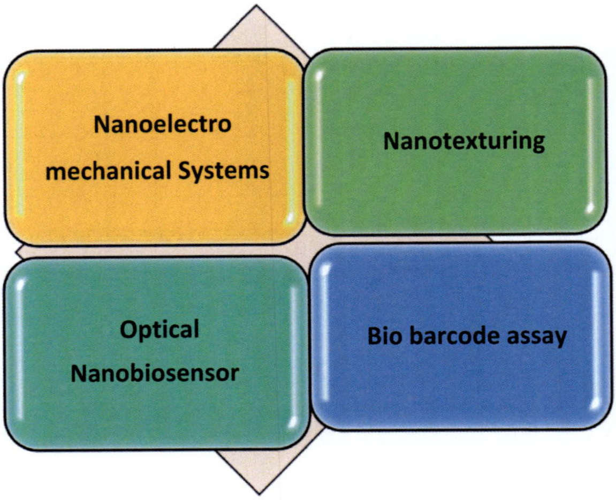


**Fig. 10.3** The most common nanotechnologies for clinical diagnostic purposes

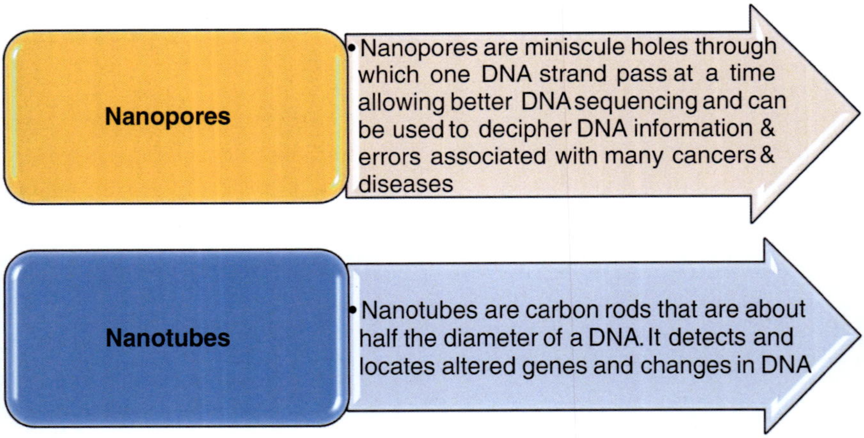


**Fig. 10.4** Classification of NEMS

## *10.4.1 Nanoelectromechanical Systems (NEMS)*

It is a noninvasive technology that converts biochemical signals to electrical signals which can monitor disease progression and treatment outcome. They have high specificity and sensitivity for enhanced single-molecule-level analyte detection. However, they are difficult to handle and non-biodegradable. This system is used for orthodontic tooth movement and to expand maxilla (Subramani et al. 2019).

They are further classified into two: nanopores and nanotubes (Fig. 10.4).

### 10.4.2 Nanotexturing

It provides physicochemical nanoscale modification of surfaces. Based on properties like size and selective capture, the molecular weights of proteins can be accurately analyzed. The electrical property of cells and chemical and physical properties of cells can be monitored through the use of nanowires. Nanotextured titanium oxide layer showed a promising result in reducing the bacteria surface colonization on orthopedic and dental implants (Ferraris et al. 2019).

### 10.4.3 Optical Nanobiosensor

It is a fiberoptic tool that allows for real-time and minimally invasive analysis of intracellular components. Components involved in cellular energy and cell death such as cytochrome can be analyzed. Here low concentrates of analyses are required; it is cost effective and allows rapid analysis and the chip can be reusable. Clinical tools based on optical nanobiosensor helps in monitoring of therapeutic drug, thus reducing the complications (Garzón et al. 2019).

### 10.4.4 Bio-Barcode Assay

Newer diagnostic tool can detect proteins, nucleic acid, and DNA targets using monoclonal antibody or complementary oligonucleotide. It has high sensitivity, shorter detection time, simple to use, and low cost. Additionally, another device, integrated microfluidic platform was developed to detect tumor necrosis factor-α and interleukin-6 in the picomolar range to detect and monitor periodontal disease. This assay is considered to be safe and do not denature on binding with target molecules (Wang et al. 2019a, b, c).

### 10.4.5 Biochips and Salivary Biomarkers

Small devices of less than a few millimeters on which microarrays are arranged constitute biochips that permit many high throughput diagnostic tests to be performed. Segal and Wong (2008) have identified mRNAs in saliva of individuals who are apparently normal and those having underlying pathologic conditions. This salivary transcriptome can be used for research and diagnostic purposes with the help of microelectromechanical systems and NEMS biosensors that have high sensitivity and specificity. Later, the oral fluid nanosensor that was hand-held, automated, integrated system was developed by the research consortium for the rapid detection of multiple salivary protein and nucleic acid targets. Kaczor-Urbanowicz et al. (2017) has recommended the use of this sensor for detection of multiple salivary biomarkers for oral cancer.

Profiling is another emerging diagnostic tool for detecting miRNA expression levels that are altered in oral squamous cell carcinomas. Profiling methods include quantitative PCR, microarray hybridization, and next-generation sequencing. The detection of miRNAs in saliva which is at very low concentration (nanogram range) forms the non-invasive diagnostic method. "Applied Biosystems Stem-loop RT based TaqMan MicroRNA assay" is a gold-standard and quantifies miRNA with a large dynamic range with high sensitivity and specificity. Cytology-on-a-chip diagnostic technique provides high sensitivity and specificity in the rapid detection of premalignant and malignant cells as pioneered by Weigum et al. (2010).

Bansal et al. (2014) have devised an electronic microchip-assay to detect C-reactive protein (CRP), an inflammatory biomarker in saliva that differentiates healthy periodontium from gingivitis. CRP is also a systemic inflammatory biomarker that is elevated in myocardial infarction, atherosclerosis, and arthritis. Studies have shown that chronic periodontitis patients with bone loss exhibit higher expression of tumor necrosis factor-α (TNF-α) and interleukin-6 (IL-6). TNF-α is also increased in bleeding on probing sites, pocket depth, and higher clinical attachment levels.

Lab on a chip—oral fluid nanosensor test or lab on a chip-based device combines self-assembled monolayers, microfluidics, bionanotechnology, and cyclic enzymatic amplification. It has high specificity and sensitivity when used as a point of care device. A novel, lab-on-a chip system, electronic microchip assay was developed to detect CRP in human saliva for early detection and monitoring of chronic periodontal diseases (Christodoulides et al. 2005).

There is a constant quest for developing new diagnostic tools for dental diseases by researchers. However, the use of nanotechnology-based diagnostic approaches in clinical trials is limited in the literature. Currently, nanotechnology has been applied in the field of periodontology and oral cancer. There is a need to explore the applications of nanotechnology in the diagnosis of other dental pathologies.

## 10.5 Preventive Nanodentistry

Nanotechnology provides novel strategies in the field of preventive dentistry. It helps in the prevention of bacterial biofilms or remineralization of submicrometer-sized dental caries.

### *10.5.1 Prevention of Oral Biofilm Formation*

Dental plaque or a biofilm is formed by bacterial colonization onto the pellicle through adhesin receptor interactions. Plaque maturation involves bacterial co-aggregation, quorum sensing, and metabolic interactions in the presence of dietary sugars. The prevention of biofilm formation and management of dental caries is challenging in dentistry. Zirconium oxide nanomaterials have antimicrobial action against gram-negative bacteria (Precious Ayanwale and Reyes-López 2019; Khan et al. 2020; Reyes-López et al. 2020). Ferumoxytol has been used by Liu et al. (2018) for the disruption of biofilm with the release of hydrogen peroxide. Recently oral biofilm was eliminated by combing iron-based nanozymes and hydrogen peroxide as new strategy (Wang et al. 2020).

There is evidence demonstrating that casein phosphopeptide-amorphous calcium phosphate (CPP–ACP) aggregates decrease microbial adherence and prevent biofilm formation. It is noted that non-aggregated and clustered hydroxyapatite nanocrystallite particles interfere with the binding of microorganisms onto the tooth surface and are also helpful in biofilm management (Dashper et al. 2016). Dextran-coated iron oxide nanoparticles are used as an antibiofilm agent for the disruption of biofilm without any adverse effects to the host tissues and oral microbiota (Naha et al. 2019).

### *10.5.2 Prevention of Dental Caries and Periodontal Diseases*

Nano-toothbrush has been developed by adding colloidal nanogold or nanosilver particles in the bristles of the toothbrush. It not only aids to improve the mechanical removal of the adherent oral biofilm on the tooth but also has an antibacterial effect. Nanorobotic dentifrices can be delivered through toothpaste or a mouthwash. It can clean all tooth surfaces, supragingival, and subgingival areas (Kasimoglu et al. 2020). It can reach areas which are inaccessible to brush and floss and metabolize the trapped organic matter and render the area plaque and calculus free. Toothpastes and mouthwashes containing nanocalcium fluoride, calcium carbonate nanoparticles, and 3% sodium tri-metaphosphate have shown to reduce hypersensitivity, promote remineralization, and have improved fluoride release. Manikandan et al. (2017) showed that silver oxide nanoparticles formed using *Ficus benghalensis* prop root extract in toothpaste exerted a strong antibacterial effect on oral microbes like *Streptococcus mutans* and *Lactobacilli* species (Manikandan et al. 2017). Ebadifar et al. (2017) proved that toothpastes containing nanohydroxyapatite (HA) crystals when used on worn teeth increased the microhardness of the tooth enamel.

## 10.6 Therapeutic Nanodentistry

### *10.6.1 Remineralization of Submicrometric-Sized Tooth Defects*

Fluoride is an effective remineralizing agent used for the prevention of mineral loss. Remineralization of initial enamel lesions is enhanced by CPP–ACP that may be released by chewing gum. Studies have shown that CPP–ACP sugar-free gum significantly reduces proximal caries in children (Morgan et al. 2008). A combination of CPP–ACP with fluoride has shown to increase the remineralization (Attiguppe et al. 2019).

### *10.6.2 Biomimetic Synthesis of Enamel and Repair of Microcavities*

Biomimetic carbonate hydroxyapatite nanoparticles that mimic the size of natural dentinal hydroxyapatite (20 nm) or enamel apatite (100 nm) have been incorporated into toothpastes or mouth-rinsing solutions to repair micrometer-sized tooth surface defects in–vitro (Roveri et al. 2008). An in–vitro study has demonstrated a 20-day use of nano-sized "amorphous calcium carbonate particles" is effective as an remineralizing agent against artificial white-spot enamel lesions (Ma et al. 2019). Biomimetic hydroxyapatite toothpaste containing micro-structured hydroxyapatite nanoparticles helps in producing biologic enamel hydroxyapatite with both structure and morphology. The disadvantage of this paste is that it is highly acidic (pH 3.5) and contains high hydrogen peroxide concentration (Bossù et al. 2019).

## 10.7 Interceptive Nanodentistry

### *10.7.1 Nanotechnology in Endodontics*

Bacterial biofilms are the major cause of endodontic pathogenesis. Despite the technological advancements in cleaning and shaping root canal systems, the root canal treatment failure rate has not declined. One of the attributing factors is ineffective disinfection of root canals. This has reduced the treatment strategies. Applications of nanotechnology in endodontics include: as biofilm eradication, root canal irrigants, intracanal medicaments, obturating materials, photodynamic therapy, perforation repair, and apical seal. Nanoparticles are being incorporated into root canal sealers to improve their antibacterial properties, penetration into dentinal tubules, and enhanced adhesion to the root surface.

Chitosan nanoparticles and zinc oxide nanoparticles were evaluated for *E. faecalis* biofilm disruption. These nanoparticles eliminated biofilms and showed a time- and dose-dependent action. Neutralizing effect of tissue inhibitors inhibit the effect of chitosan nanoparticles. Zinc oxide nanoparticles have been used to enhance the antibacterial effect of sealers. Incorporation of chitosan and zinc oxide nanoparticles into calcium hydroxide-based sealers exhibited significant antibiofilm activity against *E. faecalis* strain (Nair et al. 2018). Chitosan nanoparticles with chlorhexidine showed effective antibacterial action when used in membrane barriers placed in periapical surgeries (Barreras et al. 2016).

Bioactive glass (BAG) has an alkaline pH, potent remineralizing, and antibacterial action. BAG nanoparticles are added into sealers to reduce void formation and improve bonding to root canal walls. Polyisoprene and polycaprolactone incorporated with BAG nanoparticles can be used in monobloc obturation. Silver nanoparticles, because of the aforementioned properties, have been tested against *E. faecalis* biofilms and have shown complete elimination of the biofilm and superior penetration into the dentinal tubules. Silver nanoparticle (AgNP) gel (0.02%) was evaluated as an intracanal medicament and showed significant *E. faecalis* biofilm disruption (Balto et al. 2020). Two major drawbacks of AgNP include tooth discoloration and cytotoxic effects on mammalian cells. Nano-catalysis was used to eliminate biofilm disruption from dentinal tubules. Quaternized polyethyleneimine nanoparticles (QPEI-NPs) get adsorbed on bacterial cell membrane and penetrate into it, block exchange of essential ions, destabilize the cell membrane, and lead to cell lysis. Barros et al. (2014) concluded that addition of QPEI-NPs to AH Plus and Pulp Canal sealer resulted in increased surface charge and wettability but the antibacterial action was lost after 7 days after setting of sealers. However, KeslerShvero et al. (2013)showed that addition of QPEI-NPs to AH Plus sealer increased the antibacterial activity. Polymeric resin-based root canal sealer with bioactive nanoparticles exhibits both antibacterial and remineralization properties (Baras et al. 2020).

Functionalization alters the surface charge, composition, and structure without changing the original properties of the bulk material. There will be an inorganic or a polymeric core material which is decorated with different reactive molecules, ligands, and peptides. Photosensitizer functionalized nanoparticles have a unique advantage of increased physical and chemical reactivity. These photosensitizers can be supplemented, encapsulated or bound, and loaded with nanoparticles, or nanoparticles can themselves be the photosensitizers (Ravindran et al. 2013). These nanoparticles increase the antimicrobial photodynamic therapy effect as there is an increased concentration of photosensitizer per mass, increased production of reactive oxygen species, reduced antimicrobial resistance, enhanced bacterial targeting and better stability of photosensitizer after conjugation, and there is a controlled activity on photoactivation. Methylene blue–loaded poly(lactic-*co*-glycolic) acid nanoparticles exhibited a higher planktonic and biofilm phototoxicity against *E. faecalis*. Rose Bengal photosensitizer-bound polystyrene beads showed antimicrobial activity but had a slower release of reactive oxygen species. Rose Bengal photosensitizer functionalized chitosan nanoparticles showed significant improvement of antimicrobial effect on dental biofilms (Shrestha et al. 2014).

Mineral trioxide aggregate (MTA) is the most popular material used in pulp capping, pulpotomy, apexification, root end filling, perforation repair, and regenerative endodontics. Its advantages include superior biocompatibility and improved sealing ability. It can undergo setting in moist environment and has odontogenic capabilities. However, there are a few drawbacks including long setting time, difficult to handle, causes tooth discoloration, acidic environment decrease its hardness, and increase its porosity. Saghiri et al. (2012) developed nanomodified white MTA, and they claimed that it has improved resistance to acidic environments, increased strength, lower porosity, faster setting, improved micro hardness, smoother surface and finish, better adhesion that improved effect on regeneration potential of tissues. Nano bismuth oxide (10%) particles have been added to MTA which significantly improved the microhardness and compressive strength (Saghiri et al. 2015).

Bioaggregate is developed by Innovative BioCaramix Inc. (Vancouver, BC, Canada) is composed of nano-sized tricalcium silicate, tantalum oxide, calcium phosphate, and silicon dioxide. It causes remineralization by deposition of hydroxyapatite. The inventors claim that it shows an early calcium release when compared to MTA Angelus. It has enhanced biocompatibility, improved sealing ability, more acid resistance, higher fracture resistance, improved induction of osteoblasts, and improved mineralization abilities than MTA (Yan et al. 2010).

### 10.7.2 Nanomaterials in Restorative Dentistry

Nanocomposites have larger surface area available for enhanced bonding and smaller size allows for better molecular scale interactions. Smaller particle size leads to better translucency, smooth finish, superior wear resistance, and improved esthetics. Antimicrobial and remineralizing nanoparticles when added to composites will lead to decrease in secondary caries and reduced failure of restorations.

Bioactive multifunctional composite made of nanoamorphous calcium phosphate, silver nanoparticles, 2-methacryloyloxyethyl phosphoryl choline dimethylamino hexadecyl methacrylate used to restore root caries showed superior release of calcium and phosphorous and aided in remineralization and hardness of the root dentin. Antibacterial nanocomposite made of silver-loaded poly-cation functionalized nano-diamonds has antibacterial action against *Streptococcus mutans,* greater flexural strength, Vickers hardness, minimal cytotoxicity, and modulus of elasticity (Wang et al. 2019a, b, c). Zhang et al. (2018) tested calcium-doped mesoporous silica nanoparticle-based dental resin composites which had greater mechanical properties, remineralization, and antibacterial properties. Cheng et al. (2012) developed nanocomposites containing nano calcium phosphate and silver nanoparticles. This composite showed both antibacterial and remineralizing actions without altering the mechanical properties. Addition of 10% zinc oxide to composite had lesser antibacterial action when compared with silver nanoparticle. Addition of 1% quaternary ammonium polyethylenimine nanoparticles to composite resin exerted an immediate and prolonged antibacterial action against *Streptococcus mutans* without compromising its mechanical properties.

Dental composites containing nano calcium fluoride showed high fluoride release, superior remineralizing, and anti-caries properties without change in wear resistance and mechanical properties. Smart composites containing nano CPP-ACP show a twofold increase in flexural strength and increased acid buffering action and modulus of elasticity. Chitlac composite (lactose-modified chitosan) with silver nanoparticles showed superior biofilm disruption and prevented the formation of mature biofilm.

### 10.7.3 Nanomaterials for Stabilization of Dentin and Adhesive–Dentin Resin Interface

Dental caries and acid etching for composite restorations involve acid attacks on dentin which demineralizes and weakens the collagen present in dentin. Demineralized collagen is more prone to enzymatic degradation. These exposed collagen fibrils can be better protected using nanofillers of zirconia, hydroxyapatite, silica, reactive nanogels, bioactive calcium, and silicate.

Fluid nanoprecursors like polymer-induced precursor show intrafibrillar and interfibrillar remineralization. Use of self-assembling nano layers containing calcium or hydroxyapatite enhances the bonding at the resin tooth interface. Azad et al. (2018) added 0.2–0.5 wt% silinated silica nanoparticles to dental adhesives which showed to have enhanced bonding strength and physical and mechanical properties of the material. The combination of 2-methacryloxylethyldodecyl methyl ammonium bromide and nano amorphous calcium phosphate in orthodontic adhesive eradicates biofilm and remineralizes white spot lesions. Self-healing adhesives containing microcapsules of dimethylaminohexadecyl methacrylate and nanoparticles of amorphous calcium phosphate (NACP) showed significant crack healing property and decreased the biofilm formation (Yue et al. 2018). Al-Qarni et al. (2018) designed an adhesive which can release calcium and phosphate ions and repels proteins made of NACP; the protein repellent action prevents biofilm formation. Liang et al. (2018) investigated addition of poly(amidoamine) and NACP as adhesive showed remineralization and increased hardness of demineralized dentin. Dutra-Correa et al. (2018) added silver nanoparticles to the primer of three-step adhesive and found that it had potent antimicrobial activity and enhanced microtensile bond strength. Chlorhexidine-loaded poly(lactic-*co*-glycolic acid) nanoparticles were delivered into the dentinal tubules. These nanoparticles could infiltrate the dentinal tubules up to 10μm, which could improve the adhesive penetration.

### 10.7.4 Nano Glass Ionomer Cement (Nanoionomer)

Use of nanoparticle-sized powder of fluoroalumino silicate improved and decreased the setting time. Addition of nanohydroxyapatite and nanofluorapatite significantly enhanced the compressive, tensile, and flexural strength of glass ionomer cement. By substituting polyacrylic acid with *N*-vinylpyrrolidone copolymers and adding nanohydroxyapatite and fluoroapatite, the mechanical properties can be enhanced. The addition of nanohydroxyapatite and nanozirconium oxide (4% by volume) significantly enhanced the mechanical properties, but addition of nanozirconium oxide led to increased crack formation. Incorporation of nano-chitosan particles to glass ionomer cement improved the flexural strength. Equia system, containing nanosilica fillers (15% wt) in the matrix, shows improved wear resistance, optical, esthetic, and mechanical properties. Nanofillers were added to resin-modified glass ionomer cements, but no significant difference was seen when compared with conventional resin-modified glass ionomer cement.

## 10.8 Nanobiomaterials in Tissue Engineering

Tissue engineering includes a triad of stem cells, scaffolds, and biological signals/growth factors to develop biological substitutes that can restore, repair, maintain, or improve tissue functions. Nanotechnology in tissue engineering can be used for delivering bioactive molecules, growth factors, cell identification, cell targeting, and production of scaffolds. There is extensive research going on in the field of nanoengineered three dimensional matrix and scaffolds as these nanostructured materials have enhanced cellular interactions allowing better spreading, differentiation, and proliferation of stem cells. Biological signals-controlled delivery of drugs can also be added to functionalize these scaffolds. Nano-scaffolds can be used to deliver drugs, peptides, genetic material, or bioactive molecules in a localized and delivered in a spatiotemporal fashion. They also provide improved osteointegration, osteoconduction, and osteoinduction. Important features of a scaffold include adequate porosity, optimal mechanical strength, easy to handle the pores should allow easy diffusion of cells, growth factors, nutrients, toxic byproducts across it.

Nanohydroxyapatite and nano-calcium phosphate which is a bioactive nanoscaffolds have shown superior biocompatibility and bioactivity, and they promote remineralization and when stimulated cause increased protein absorption and cell response. Bioactive nanocarriers like nanotubes, nanospheres, and nanofibers are coated or encapsulated with specific drugs or signaling molecules. This allows slow and sustained release of the drugs or signaling molecules. The most common method for fabrication of these nano-scaffolds is electrospinning. Self-assembling peptides containing large amounts of Arg-Gly-Asp(RGD) sequences have shown to induce proliferation and differentiation of ameloblasts producing enamel nodules. Nanofibrous scaffolds infiltrated with bioactive glass, nanohydroxyapatite showed enhanced proliferation, differentiation, and mineralization of damaged and diseased dentin. Incorporation of magnesium phosphate into nanofibrous scaffolds when ectopically implanted in nude mice produced higher levels of extracellular matrix and hard tissue deposition. Self-assembling peptides containing RGD sequences have shown enhanced expression of odontoblast like phenotype and mineral deposition. Methacrylate based silica nanoparticles formed neodentinal cellular patterns. The bioactive glass nanocarriers have shown increased alkaline phosphatase activity, Runx2 and OCN expression, and increased osteoblastic differentiation and mineralization (Wang et al. 2019a, b, c). The bioactive nanoparticles are used for gene delivery and demonstrated effective transfection to mesenchymal stem cells and superior bone regeneration. The calcium-deficient hydroxyapatite nanocarriers with tetracycline demonstrated antibacterial activity against periodontal pathogens and improved proliferation of periodontal ligament cells. Addition of chitosan nanoparticles and silver nanoparticles in nanopoly(lactic-*co*-glycolic acid) scaffold was investigated for periodontal tissue regeneration. It was found that a ratio of 3:7 of nanopoly(lactic-*co*-glycolic acid)/chitosan and 50μg/mL silver particles was the optimal proportion. Nano-beta-tricalcium phosphate embedded in a collagen scaffold implanted with mesenchymal stem cells showed improved bone regeneration.

Hydroxyapatite nanowires in modified polylactic acid membrane promoted bone regeneration in rat mandible defect model. Nanofiber-reinforced hydroxyapatite-gelatin scaffold showed enhanced bone formation in critical sized alveolar defects in rabbit model.

## 10.9 Local Drug Delivery

The treatment strategies for periodontitis involve scaling and root planning followed by local drug delivery. These treatments are successful in the short term, but long-term success requires multiple follow-ups and patient compliance. Hence, use of targeted, sustained drug delivery using nanoparticles would be optimal. Poly(D,L-lactide-*co*-glycolide) and poly(D,L-lactide) nanoparticles and cellulose acetate phthalate mixed with triclosan were found to reduce periodontal inflammation. Niosomes are chemically stable, biodegradable, and biocompatible and have long shelf life and non-ionic vesicles. They were developed for controlled drug delivery and have been used to treat periodontitis and recurrent oral aphthous ulcers. Abbaraju et al. (2018) used mesoporous silica nanoparticles and J9 bacterial peptides to stimulate the production of molecules CD40 and CD86, J9 antibody production in–vivo. This could be used alternate to a vaccine to provoke an immune response. Poly(D,L-lactide-*co*-glycolide) acid–chitosan nanoparticles combined with lovastatin and tetracycline used as a drug delivery system was helpful in periodontal regenerative therapy (Lee et al. 2016). Lin et al. (2018) performed an in–vivo study to modulate the progression of periodontitis using PLGA–chitosan nanospheres with metronidazole and *N*-phenacyl thiazolium bromide. Inhibition of alveolar bone loss and reduced infiltration of inflammatory cell was seen. Chitosan nanoparticles were shown to be effective against periodontal pathogens and had a modulatory effect on pro-inflammatory mediators produced by gingival fibroblasts. Cisplatin in the matrix of polyethylene glycol-poly(glutamic acid) with antitumor nanoparticles is shown to be effective against oral squamous cell carcinoma with significantly lesser renal toxicity.

## 10.10 Modifications in Denture Bases, Liners, and Silicon Obturators

Adding 0.4% titanium dioxide nanoparticles into polymethylmethacrylate denture base improved its mechanical and antibacterial properties especially against *Candida albicans*. The biological behavior of polymethylmethacrylate modified with 7% wt nano zirconium oxide.

This modification improved hardness, fracture toughness, and flexural strength while Gad and Abualsaud (2019) showed that repair of denture base using polymethyl methacrylate with nano-zirconium oxide improved the transverse strength. The addition of hydrophobic nano-silica and pre-polymer to denture base acrylic resin showed an improvement in flexural strength.

## 10.11 Luting Cements

Nanoparticle-modified luting cements showed an improvement in bond strength, an increase in modulus of elasticity, and a reduction in polymerization shrinkage. Zinc oxide and magnesium oxide nanoparticles were added to zinc polycarboxylate cement to improve its compressive and tensile strengths while addition of nano-hydroxyapatite or fluorapatite to glass ionomer cements led to an enhanced tensile and compressive strength with good biaxial flexural strength.

## 10.12 Ceramics

Magalhaes et al. (2018) added titanium oxide nanotubes to yttria-stabilized tetragonal zirconia particles which enhanced the structural reliability while nano-ceramic-based CAD/CAM blocks by 3M Company showed enhanced esthetics, superior mechanical properties, fracture toughness, and excellent durability.

## 10.13 Impression Materials

Nanofillers were added to vinyl poly siloxane impression materials to produce less voids, superior accuracy, showed improved resistance to distortion, better flow, high tear resistance, and rapid snap set producing superior impressions (Pandey et al. 2019).

## 10.14 Implants

Various nanoparticle coatings have been used to improve the osseointegration and mechanical properties of titanium-based implants. The biofunctionalized coating of silver oxide nanoparticles with titanium oxide showed good antibacterial property and enhanced osseointegration of implants with minimal cytotoxic response. To improve the biological compatibility, Yang et al. (2018) attempted an extra-corporeal magnetic field coating of dental implants with reduced concentration of PLGA (Ag-$Fe_3O_4$). Higher biocidal ability of nanocopper when coated over titanium oxide against *Staphylococcus aureus* and *E. coli* were seen. The nanotitania coat roughens and optimizes the implant surfaces, hence reducing peri-implantitis. The nanosilver coating improves the antibacterial action against *Staphylococcus aureus* and *E. coli* with minimal cytotoxicity to fibroblast cells. The silver conjugated nano-chitosan coating suppressed the biofilm formation and improved corrosion resistance.

The implant surface with poly(lactic-*co*-glycolic acid)/sliver/zinc oxide nanorods showed sustained and improved antibacterial effect with minimal cytotoxicity. The gold nanoparticles improve the osteoconduction potential. The nanozinc and nano-hydoxyapatite coating on titanium oxide had superior antibacterial effect. The chitosan with hydroxyapatite and titanium oxide surface showed that this coating had better cell viability, adhesion, and differentiation potential. Super paramagnetic iron oxide nanoparticles promote osteoblastic activities and are efficient in treating biofilms formed on dental implants. The combination of chitosan and titanium oxide composite is effective shows antimicrobial activity against bacteria and fungi (Anaya-Esparza et al. 2020).

## 10.15 Future Outlook and Emerging Trends

### *10.15.1 Nanorobotics*

Nanorobots are devices in nanometers, when moved from bench to bed, would work in the nanoscale to generate cellular and molecular interactions. They can be used to treat hypersensitive dentin, to clean the dental tissues, induce analgesia, restore tooth defects, and realign soft tissues around the teeth (Kasimoglu et al. 2020).

### *10.15.2 2 Nanoanesthesia*

Nanoanesthesia makes use of nanotechnology or nanorobots that instill a suspension of a local anesthetic or analgesic drug in the tissues. These nanorobots move toward the pulp and are controlled by the dentist via a pre-programmed nanocomputer, temperature difference, and chemical gradient of the tooth structure. Once nanorobots have established control over the nerve impulses, the tooth is rendered numb, and the dental procedure can be performed. After the procedure, the nanorobots are swallowed and sensation is recovered. These nanoanalgesics have no needles, thereby reducing patient anxiety, improved precision, lowered toxic effects, rapid controlled, and reversible analgesia (Kasimoglu et al. 2020; Gupta 2018).

### *10.15.3 Nanosolutions and Nanoencapsulation*

Nanosolutions can be used as bonding agents as they result in dispersible and unique nanoparticles. Pathogens can be targeted by employing the nanoparticles in the form of nano-emulsified lipid droplets. Nanocapsules in targeted release systems are under trial for inclusion in vaccines and antibiotics.

### 10.15.4 Dentine Tubule Blocking to Alleviate Hypersensitivity

Nanorobots have been used in hypersensitive teeth that occlude dentinal tubules, thus providing quick and permanent cure. The tubules of hypersensitive teeth have twice the diameter and eight times the surface density of those in non-sensitive teeth (Kasimoglu et al. 2020; Gupta 2018).

### 10.15.5 Nanorobotic Dentifrices and Nanoneedles

All subgingival surfaces can be covered by nanorobotic dentifrices that are delivered through tooth paste or as a mouth wash. The pathogenic bacteria entrapped in plaque are destroyed by the dentifrobots introduced into the oral cavity (Kasimoglu et al. 2020; Gupta 2018). Suture needles with the inclusion of nanosized stainless steel crystals have been developed. There is a future scope of use of nano tweezers for cell surgery.

### 10.15.6 Orthodontic Therapy

Tooth movements such as uprighting, rotating, vertical repositioning, and tissue repair have been initiated using orthodontic robots that permit painless procedures. Recently, investigation on the use of a new stainless steel wire using nanotechnology that fulfills physical properties has been sought after (Kasimoglu et al. 2020; Gupta 2018).

### 10.15.7 Bone Replacement Materials

Development of bone by nanotechnology is targeted toward natural structure formation that is suitable for orthopedic and dental applications. The loose structure of nanocrystals with nanopores adsorbs protein, due to the presence of silica molecules. The treatment of bone defects can be instituted using hydroxyapatite nanoparticles.

### 10.15.8 Zinc Oxide Nanoparticles

Zinc oxide is the most common filler used in dentistry. Zinc oxide has a strong affinity to lipids and binds strongly to the bacterial cell membranes and altering its permeability. It also creates reactive oxygen species which leads to cell death. The antibacterial effect of nano-zinc oxide is dose dependent. Zinc oxide nanoparticles have a cytotoxic effect on human gingival fibroblasts. Nevertheless, this can be reduced by coating zinc oxide nanoparticles with chitosan. The nano-zinc oxide particles are observed to have anti-biofilm activities.

### *10.15.9 Titanium Dioxide Nanoparticles*

When exposed to UVA radiation or near UV rays, titanium oxide nanoparticles undergo photocatalysis and release reactive oxygen species causing cell death. Cytotoxic studies show that titanium oxide nanoparticles have a dose-dependent proinflammatory effect on human gingival fibroblasts.

### *10.15.10 Curcumin Nanoparticles*

Maghsoudi et al. (2017) analyzed and showed the antimicrobial properties of nano-curcumin against *Streptococcus mutans*. Nano-curcumin has anticancer, antioxidant, anti-inflammatory, and antimicrobial properties. The antimicrobial activity is linked with cell wall leakage due to the formation of transmembranous pores and disruption of cell wall. However, cytotoxicity of nanocurcumin has to be evaluated.

### *10.15.11 Copper Nanoparticles*

Copper nanoparticles adhere to the amine and carboxylic groups of microorganism's surfaces leading to modification of cell membrane, formation of reactive oxygen species, disrupts the protein, and DNA synthesis. González et al. (2016) developed copper nanoparticles in a chitosan shell and showed that it had improved antibiofilm action.

### *10.15.12 Nano-Quaternary Ammonium Compounds*

They are highly positively charged ions that bind to the negatively charged bacterial cell membranes, altering the cell permeability, and causing cell lysis. Quaternary ammonium compound (1%) nanoparticles have an antibacterial effect against *Streptococcus mutans* and *Lactobacillus* species.

## 10.16 Conclusions

Nanotechnology is set to revolutionize clinical dental practice and has shown various potential benefits to patient outcome. Nanomechanics such as micro-scale robots allow dental practitioners to perform therapeutic and restorative procedures at molecular and cellular levels with safety norms in place. Improvements will be seen in oral health by optimal utilization of these opportunities offered by nanotechnology. Similar to other technologies, nanotechnology carries a significant potential for misuse and abuse on a scale, scope never seen before if not properly controlled and directed. The current nanomaterials available have satisfactory safety norms, improved qualities and promising prospective. But with advancements happening over time with newer developments, according to the economic and technical resources, and human needs will determine which of the newer applications are most beneficial. Nanotechnology is bound to bring massive changes in the fields of medicine and dentistry by transforming the entire outlook of how a clinician treats patients in the future. There appears to be many potential benefits to patient outcome from nanotechnology use in dentistry with increased performance of the dental materials. Once nanomechanics are available, there will be a revolution in the practicing methods at a clinical set-up. Nanotechnology helps to perform therapeutic and restorative procedures at the molecular and cellular levels with safety norms in place. The current nanomaterials available are through green nanotechnology having satisfactory safety norms, improved qualities, and promising prospective.

**Acknowledgments** I sincerely thank all the co-authors those who have contributed in this chapter for successful completion of this chapter. All the authors are thankful for the editors to give an opportunity to contribute in this book as well as the reviewers for their valuable comments. The authors are thankful to Ramaiah University of Applied sciences for its library facility without which the book chapter would not possible.

Ethical Clearance In this chapter the data is compiled from the available literature, so the ethical clearance may not be essential. The data is put together according to the guidelines of the Springer Nature publishing.

**Conflicts of Interest** The authors in this chapter express no conflict of interest related to financial or any other personal considerations to that may impact the objectivity of this chapter.

## References

Abbaraju, P. L., Jambhrunkar, M., Yang, Y., Liu, Y., Lu, Y., & Yu, C. (2018). Asymmetric mesoporous silica nanoparticles as potent and safe immunoadjuvants provoke high immune responses. *Chemical Communications, 54*(16), 2020–2023.

Abe, Y., Skali-Lami, S., Block, J. C., & Francius, G. (2012). Cohesiveness and hydrodynamic properties of young drinking water biofilms. *Water Research, 46*(4), 1155–1166.

Abou Neel, E. A., Bozec, L., Perez, R. A., Kim, H. W., & Knowles, J. C. (2015). Nanotechnology in dentistry: Prevention, diagnosis, and therapy. *International Journal of Nanomedicine, 10*, 6371.

Alharbi, K. K., & Al-sheikh, Y. A. (2014). Role and implications of nanodiagnostics in the changing trends of clinical diagnosis. *Saudi journal of Biological Sciences, 21*(2), 109–117.

al-Qarni, F. D., Tay, F., Weir, M. D., Melo, M. A., Sun, J., Oates, T. W., Xie, X., & Xu, H. H. (2018). Protein-repelling adhesive resin containing calcium phosphate nanoparticles with repeated ion-recharge and re-releases. *Journal of Dentistry, 78*, 91–99.

Anaya-Esparza, L. M., Ruvalcaba-Gómez, J. M., Maytorena-Verdugo, C. I., González-Silva, N., Romero-Toledo, R., Aguilera-Aguirre, S., & Pérez-Larios, A. (2020). Chitosan-TiO2: A versatile hybrid composite. *Materials, 13*(4), 811.

Attiguppe, P., Malik, N., Ballal, S., & Naik, S. V. (2019). CPP–ACP and fluoride: A synergism to combat caries. *International Journal of Clinical Pediatric Dentistry, 12*(2), 120.

Azad, E., Atai, M., Zandi, M., Shokrollahi, P., & Solhi, L. (2018). Structure–properties relationships in dental adhesives: Effect of initiator, matrix monomer structure, and nano-filler incorporation. *Dental Materials, 34*(9), 1263–1270.

Balto, H., Bukhary, S., Al-Omran, O., BaHammam, A. and Al-Mutairi, B. (2020). Combined Effect of a Mixture of Silver Nanoparticles and Calcium Hydroxide against Enterococcus faecalis Biofilm. *Journal of Endodontics, 46*(11), 1689–1694.

Bansal, T., Pandey, A., Deepa, D., & Asthana, A. K. (2014). C-reactive protein (CRP) and its association with periodontal disease: A brief review. *Journal of Clinical and Diagnostic Research: JCDR, 8*(7), ZE21.

Baras, B. H., Melo, M. A. S., Thumbigere-Math, V., Tay, F. R., Fouad, A. F., Oates, T. W., Weir, M. D., Cheng, L., & Xu, H. H. (2020). Novel bioactive and therapeutic root canal sealers with antibacterial and remineralization properties. *Materials, 13*(5), 1096.

Barreras, U. S., Méndez, F. T., Martínez, R. E. M., Valencia, C. S., Rodríguez, P. R. M., & Rodríguez, J. P. L. (2016). Chitosan nanoparticles enhance the antibacterial activity of chlorhexidine in collagen membranes used for periapical guided tissue regeneration. *Materials Science and Engineering C, 58*, 1182–1187.

Barros, J., Silva, M. G., Rôças, I. N., Gonçalves, L. S., Alves, F. F., Lopes, M. A., Pina-Vaz, I., & Siqueira, J. F., Jr. (2014). Antibiofilm effects of endodontic sealers containing quaternary ammonium polyethylenimine nanoparticles. *Journal of Endodontics, 40*(8), 1167–1171.

Bharadwaj, R., & Medhi, S. (2020). Oral squamous cell carcinoma and the cutting edge of nanotechnology. *Multidisciplinary Cancer Investigation, 4*(2), 36–45.

Bhirde, A. A. (2010). *Targeted drug delivery using carbon nanotubes for cancer therapeutics*. Storrs: University of Connecticut.

Bossù, M., Saccucci, M., Salucci, A., Di Giorgio, G., Bruni, E., Uccelletti, D., Sarto, M. S., Familiari, G., Relucenti, M., & Polimeni, A. (2019). Enamel remineralization and repair results of biomimetic hydroxyapatite toothpaste on deciduous teeth: An effective option to fluoride toothpaste. *Journal of Nanobiotechnology, 17*(1), 17.

Chalmers, N. I., Palmer, R. J., Du-Thumm, L., Sullivan, R., Shi, W. & Kolenbrander, P. E. (2007). Use of quantum dot luminescent probes to achieve single-cell resolution of human oral bacteria in biofilms. *Applied and Environmental Microbiology, 73*(2), 630–636.

Chen, C., Weir, M. D., Cheng, L., Lin, N. J., Lin-Gibson, S., Chow, L. C., Zhou, X., & Xu, H. H. (2014). Antibacterial activity and ion release of bonding agent containing amorphous calcium phosphate nanoparticles. *Dental Materials, 30*(8), 891–901.

Cheng, L., Weir, M. D., Xu, H. H., Antonucci, J. M., Kraigsley, A. M., Lin, N. J., Lin-Gibson, S., & Zhou, X. (2012). Antibacterial amorphous calcium phosphate nanocomposites with a quaternary ammonium dimethacrylate and silver nanoparticles. *Dental Materials, 28*(5), 561–572.

Christodoulides, N., Mohanty, S., Miller, C. S., Langub, M. C., Floriano, P. N., Dharshan, P., Ali, M. F., Bernard, B., Romanovicz, D., Anslyn, E., & Fox, P. C. (2005). Application of microchip assay system for the measurement of C-reactive protein in human saliva. *Lab on a Chip, 5*(3), 261–269.

Dashper, S. G., Catmull, D. V., Liu, S. W., Myroforidis, H., Zalizniak, I., Palamara, J. E., Huq, N. L., & Reynolds, E. C. (2016). Casein phosphopeptide-amorphous calcium phosphate reduces streptococcus mutans biofilm development on glass ionomer cement and disrupts established biofilms. *PLoS One, 11*(9), e0162322.

Dogra, S., Gupta, A., Goyal, V., Chib, A. S., & Kataria, V. (2020). Recent trends, therapeutic applications, and future trends of nanomaterials in dentistry. In *Nanomaterials in diagnostic tools and devices* (pp. 257–292). Amsterdam: Elsevier.

Dutra-Correa, M., Leite, A. A., de Cara, S. P., Diniz, I. M., Marques, M. M., Suffredini, I. B., Fernandes, M. S., Toma, S. H., Araki, K., & Medeiros, I. S. (2018). Antibacterial effects and cytotoxicity of an adhesive containing low concentration of silver nanoparticles. *Journal of Dentistry, 77*, 66–71.

Ebadifar, A., Nomani, M., & Fatemi, S. A. (2017). Effect of nano-hydroxyapatite toothpaste on microhardnessofartificial carious lesions created on extracted teeth. *Journal of Dental Research, Dental Clinics, Dental Prospects, 11*(1), 14.

Ferraris, S., Cochis, A., Cazzola, M., Tortello, M., Scalia, A., Spriano, S., & Rimondini, L. (2019). Cytocompatible and anti-bacterial adhesion nanotextured titanium oxide layer on titanium surfaces for dental and orthopedic implants. *Frontiers in Bioengineering and Biotechnology, 7*, 103.

Gad, M. M., & Abualsaud, R. (2019). Behavior of PMMA denture base materials containing titanium dioxide nanoparticles: A literature review. *International Journal of Biomaterials, 2019*, 6190610.

Garzón, V., Pinacho, D. G., Bustos, R. H., Garzón, G., & Bustamante, S. (2019). Optical biosensors for therapeutic drug monitoring. *Biosensors, 9*(4), 132.

González, J., Covarrubias, C., Cádiz, M., Corral, C., Cuadra, F., Fuentevilla, I., & Bittner, M. (2016). Design of antimicrobial release systems based on chitosan and copper nanoparticles for localized periodontal therapy. *Journal of Dentistry & Oral Disorders, 2*(7), 1035.

Gupta, J. (2018). Nanotechnology and its application in dentistry-a review. *Nanotechnology. Research Review International Journal of Multidisciplinary, 3*, 647–650.

Hainfeld, J. F., O'Connor, M. J., Dilmanian, F. A., Slatkin, D. N., Adams, D. J., & Smilowitz, H. M. (2011). Micro-CT enables microlocalisation and quantification of Her2-targeted gold nanoparticles within tumour regions. *The British Journal of Radiology, 84*(1002), 526–533.

Kaczor-Urbanowicz, K. E., Martin Carreras-Presas, C., Aro, K., Tu, M., Garcia-Godoy, F., & Wong, D. T. (2017). Saliva diagnostics—Current views and directions. *Experimental Biology and Medicine, 242*(5), 459–472.

Kasimoglu, Y., Tabakcilar, D., Guclu, Z. A., Yamamoto-Nemoto, S., Tuna, E. B., Ozen, B., Tuzuner, T., & Ince, G. (2020). Nanomaterials and nanorobotics in dentistry: A review. *Journal of Dentistry Indonesia, 27*(2), 77–84.

KeslerShvero, D., Abramovitz, I., Zaltsman, N., Perez Davidi, M., Weiss, E. I., & Beyth, N. (2013). Towards antibacterial endodontic sealers using quaternary ammonium nanoparticles. *International Endodontic Journal, 46*(8), 747–754.

Khan, M., Shaik, M. R., Khan, S. T., Adil, S. F., Kuniyil, M., Khan, M., Al-Warthan, A. A., Siddiqui, M. R. H., & Nawaz Tahir, M. (2020). Enhanced antimicrobial activity of biofunctionalized zirconia nanoparticles. *ACS Omega, 5*(4), 1987–1996.

Khlebtsov, B. N., Khanadeev, V. A., Panfilova, E. V., Pylaev, T. E., Bibikova, O. A., Staroverov, S. A., Bogatyrev, V. A., Dykman, L. A., & Khlebtsov, N. G. (2013). New types of nanomaterials: Powders of gold nanospheres, nanorods, nanostars, and gold-silver nanocages. *Nanotechnologies in Russia, 8*(3–4), 209–219.

Kloepfer, J. A., Mielke, R. E., Wong, M. S., Nealson, K. H., Stucky, G., & Nadeau, J. L. (2003). Quantum dots as strain-and metabolism-specific microbiological labels. *Applied and Environmental Microbiology, 69*(7), 4205–4213.

Lee, B. S., Lee, C. C., Wang, Y. P., Chen, H. J., Lai, C. H., Hsieh, W. L., & Chen, Y. W. (2016). Controlled-release of tetracycline and lovastatin by poly (D, L-lactide-co-glycolide acid)-chitosan nanoparticles enhances periodontal regeneration in dogs. *International Journal of Nanomedicine, 11*, 285.

Liang, K., Xiao, S., Weir, M. D., Bao, C., Liu, H., Cheng, L., Zhou, X., Li, J., & Xu, H. H. (2018). Poly (amido amine) dendrimer and dental adhesive with calcium phosphate nanoparticles remineralized dentin in lactic acid. *Journal of Biomedical Materials Research Part B: Applied Biomaterials, 106*(6), 2414–2424.

Liao, C., Li, Y., & Tjong, S. C. (2019). Bactericidal and cytotoxic properties of silver nanoparticles. *International Journal of Molecular Sciences, 20*(2), 449.

Lin, J. H., Feng, F., Yu, M. C., Wang, C. H., & Chang, P. C. (2018). Modulation of periodontitis progression using pH-responsive nanosphere encapsulating metronidazole or N-phenacylthialzolium bromide. *Journal of Periodontal Research, 53*(1), 22–28.

Liu, Y., Naha, P. C., Hwang, G., Kim, D., Huang, Y., Simon-Soro, A., Jung, H. I., Ren, Z., Li, Y., Gubara, S., & Alawi, F. (2018). Topical ferumoxytol nanoparticles disrupt biofilms and prevent tooth decay in vivo via intrinsic catalytic activity. *Nature Communications, 9*(1), 1–12.

Ma, X., Lin, X., Zhong, T., & Xie, F. (2019). Evaluation of the efficacy of casein phosphopeptide-amorphous calcium phosphate on remineralization of white spot lesions in vitro and clinical research: A systematic review and meta-analysis. *BMC Oral Health, 19*(1), 1–11.

Magalhaes, A. P. R., Fortulan, C. A., Lisboa-Filho, P. N., Ramos-Tonello, C. M., Gomes, O. P., Cesar, P. F., Fukushima, K. A., Mondelli, R. F. L., & Borges, A. F. S. (2018). Effects of Y-TZP blank manufacturing control and addition of TiO2 nanotubes on structural reliability of dental materials. *Ceramics International, 44*(3), 2959–2967.

Maghsoudi, A., Yazdian, F., Shahmoradi, S., Ghaderi, L., Hemati, M., & Amoabediny, G. (2017). Curcumin-loaded polysaccharide nanoparticles: Optimization and anticariogenic activity against Streptococcus mutans. *Materials Science and Engineering: C, 75*, 1259–1267.

Manikandan, V., Velmurugan, P., Park, J. H., Chang, W. S., Park, Y. J., Jayanthi, P., Cho, M., & Oh, B. T. (2017). Green synthesis of silver oxide nanoparticles and its antibacterial activity against dental pathogens. *3 Biotech, 7*(1), 72.

Morgan, M. V., Adams, G. G., Bailey, D. L., Tsao, C. E., Fischman, S. L., & Reynolds, E. C. (2008). The anticariogenic effect of sugar-free gum containing CPP-ACP nanocomplexes on approximal caries determined using digital bitewing radiography. *Caries Research, 42*(3), 171–184.

Naha, P. C., Liu, Y., Hwang, G., Huang, Y., Gubara, S., Jonnakuti, V., Simon-Soro, A., Kim, D., Gao, L., Koo, H., & Cormode, D. P. (2019). Dextran-coated iron oxide nanoparticles as biomimetic catalysts for localized and pH-activated biofilm disruption. *ACS Nano, 13*(5), 4960–4971.

Nair, N., James, B., Devadathan, A., Johny, M. K., Mathew, J., & Jacob, J. (2018). Comparative evaluation of antibiofilm efficacy of chitosan nanoparticle-and zinc oxide nanoparticle-incorporated calcium hydroxide-based sealer: An in vitro study. *Contemporary Clinical Dentistry, 9*(3), 434.

Pandey, P., Mantri, S., Bhasin, A., & Deogade, S. C. (2019). Mechanical properties of a new vinyl polyether silicone in comparison to vinyl polysiloxane and polyether elastomeric impression materials. *Contemporary Clinical Dentistry, 10*(2), 203.

Precious Ayanwale, A., & Reyes-López, S. Y. (2019). ZrO2–ZnO nanoparticles as antibacterial agents. *ACS Omega, 4*(21), 19216–19224.

Ravindran, A., Chandran, P., & Khan, S. S. (2013). Biofunctionalized silver nanoparticles: Advances and prospects. *Colloids and Surfaces B: Biointerfaces, 105*, 342–352.

Reyes-López, S. Y., Cristobal, E., Francisco, L., Ayanwale, A. P., & Ruiz-Baltazar, A. D. J. (2020). *Bactericidal activity study of ZrO2-Ag2O nanoparticles*. Sao Paulo: Instituto de Ciencias Biomédicas.

Roveri, N., Battistella, E., Foltran, I., Foresti, E., Iafisco, M., Lelli, M., Palazzo, B. And Rimondini, L (2008) Synthetic biomimetic carbonate-hydroxyapatite nanocrystals for enamel remineralization. In Advanced materials research (Vol. 47, pp. 821-824). Trans Tech Publications Ltd Bäch .

Saghiri, M. A., Asgar, K., Lotfi, M., & Garcia-Godoy, F. (2012). Nanomodification of mineral trioxide aggregate for enhanced physiochemical properties. *International Endodontic Journal, 45*(11), 979–988.

Saghiri, M. A., Gutmann, J. L., Orangi, J., Asatourian, A., & Sheibani, N. (2015). Radiopacifier particle size impacts the physical properties of tricalcium silicate–based cements. *Journal of Endodontics, 41*(2), 225–230.

Segal, A., & Wong, D. T. (2008). Salivary diagnostics: Enhancing disease detection and making medicine better. *European Journal of Dental Education, 12*(Suppl 1), 22.

Shrestha, A., Hamblin, M. R., & Kishen, A. (2014). Photoactivated rose bengal functionalized chitosan nanoparticles produce antibacterial/biofilm activity and stabilize dentin-collagen. *Nanomedicine: Nanotechnology, Biology and Medicine, 10*(3), 491–501.

Subramani, K., Subbiah, U., & Huja, S. (2019). Nanotechnology in orthodontics—1: The past, present, and a perspective of the future. In *Nanobiomaterials in clinical dentistry* (pp. 279–298). Amsterdam: Elsevier.

Tan, J. M., Arulselvan, P., Fakurazi, S., Ithnin, H., & Hussein, M. Z. (2014). A review on characterizations and biocompatibility of functionalized carbon nanotubes in drug delivery design. *Journal of Nanomaterials, 2014*, 917024.

Van Acker, H., Van Dijck, P., & Coenye, T. (2014). Molecular mechanisms of antimicrobial tolerance and resistance in bacterial and fungal biofilms. *Trends in Microbiology, 22*(6), 326–333.

Walling, M. A., Novak, J. A., & Shepard, J. R. (2009). Quantum dots for live cell and in vivo imaging. *International Journal of Molecular Sciences, 10*(2), 441–491.

Wang, Y., Jin, M., Chen, G., Cui, X., Zhang, Y., Li, M., Liao, Y., Zhang, X., Qin, G., Yan, F., & Abd El-Aty, A. M. (2019a). Bio-barcode detection technology and its research applications: A review. *Journal of Advanced Research, 20*, 23–32.

Wang, L., Cao, W., Wang, X., Li, P., Zhou, J., Zhang, G., Li, X., & Xing, X. (2019b). Biodegradable silver-loaded polycation modified nanodiamonds/polyurethane scaffold with improved antibacterial and mechanical properties for cartilage tissue repairing. *Journal of Materials Science: Materials in Medicine, 30*(4), 41.

Wang, W., Liu, Y., Yang, C., Qi, X., Li, S., Liu, C., & Li, X. (2019c). Mesoporous bioactive glass combined with graphene oxide scaffolds for bone repair. *International Journal of Biological Sciences, 15*(10), 2156.

Wang, Y., Shen, X., Ma, S., Guo, Q., Zhang, W., Cheng, L., Ding, L., Xu, Z., Jiang, J., & Gao, L. (2020). Oral biofilm elimination by combining iron-based nanozymes and hydrogen peroxide-producing bacteria. *Biomaterials Science, 8*(9), 2447–2458.

Weigum, S. E., Floriano, P. N., Redding, S. W., Yeh, C. K., Westbrook, S. D., McGuff, H. S., Lin, A., Miller, F. R., Villarreal, F., Rowan, S. D., & Vigneswaran, N. (2010). Nano-bio-chip sensor platform for examination of oral exfoliative cytology. *Cancer Prevention Research, 3*(4), 518–528.

Yan, P., Yuan, Z., Jiang, H., Peng, B., & Bian, Z. (2010). Effect of bioaggregate on differentiation of human periodontal ligament fibroblasts. *International Endodontic Journal, 43*(12), 1116–1121.

Yang, Y., Ren, S., Zhang, X., Yu, Y., Liu, C., Yang, J., & Miao, L. (2018). Safety and efficacy of PLGA (Ag-Fe3O4)-coated dental implants in inhibiting bacteria adherence and osteogenic inducement under a magnetic field. *International Journal of Nanomedicine, 13*, 3751.

Yue, S., Wu, J., Zhang, Q., Zhang, K., Weir, M. D., Imazato, S., Bai, Y., & Xu, H. H. (2018). Novel dental adhesive resin with crack self-healing, antimicrobial and remineralization properties. *Journal of Dentistry, 75*, 48–57.

Zhang, T., Chao, Y., Shih, K., Li, X. Y., & Fang, H. H. (2011). Quantification of the lateral detachment force for bacterial cells using atomic force microscope and centrifugation. *Ultramicroscopy, 111*(2), 131–139.

Zhang, Y., Huang, C., & Chang, J. (2018). Ca-doped mesoporousSiO2/dental resin composites with enhanced mechanical properties, bioactivity and antibacterial properties. *Journal of Materials Chemistry, 6*(3), 477–486.

Zhou, W., Ma, Y., Yang, H., Ding, Y., & Luo, X. (2011). A label-free biosensor based on silver nanoparticles array for clinical detection of serum p53 in head and neck squamous cell carcinoma. *International Journal of Nanomedicine, 6*, 381.

# Chapter 11
# Biosafety and Toxicity of Nanomaterials for the Management of Drug and Gene Delivery

**V. Shwetha, N. Shruthi, Pushpalatha, R. Deveswaran, J. Anbu, and K. Vineeth Kumar**

## 11.1 Introduction

Norio Taniguchi from Tokyo Science University first defined nanotechnology in his 1976 paper (Moothedath et al. 2019; Brakmane et al. 2012; Kovvuru et al. 2012). Nanotechnology is an atomic- and molecular-scale analysis of micro-manufacturing matter requiring physical and chemical modifications to create nano-sized materials (Brakmane et al. 2012). Nanotechnology enables that the materials are synthesized and characterized at nanoscale ranging from 1 to 100 $nm^2$ with distinct morphology, shape and size (Brakmane et al. 2012). According to the number of the dimensions,

V. Shwetha (✉)
Faculty of Dental Sciences, Department of Oral Medicine and Radiology, Ramaiah University of Applied Sciences, Bangalore, Karnataka, India
e-mail: shwetha.or.ds@msruas.ac.in

N. Shruthi
Faculty of Dental Sciences, Department of Conservative and Endodontics, Ramaiah University of Applied Sciences, Bangalore, Karnataka, India

Pushpalatha
Faculty of Dental Sciences, Department of Pediatric Dentistry, Ramaiah University of Applied Sciences, Bangalore, Karnataka, India

R. Deveswaran
Pharmaceutics, Faculty of Pharmacy, Ramaiah University of Applied Sciences, Bangalore, Karnataka, India

J. Anbu
Faculty of Pharmacy, Department of Pharmacology, Ramaiah University of Applied Sciences, Bangalore, Karnataka, India

K. Vineeth Kumar
Faculty of Dental Sciences, Department of Oral and Maxillofacial Surgery, Ramaiah University of Applied Sciences, Bangalore, Karnataka, India



K. Pal (ed.), *Bio-manufactured Nanomaterials*,
https://doi.org/10.1007/978-3-030-67223-2_11

nanomaterials can be classified as zero dimensional, one dimensional, and two dimensional. In zero dimensional, the electrons are limited in all the directions and confined without allowing to move within the system, e.g. quantum dots. In one dimensional, the electron can move in unilateral direction within the system, e.g. nanotubes and nanorods. In two dimensional, the electron can move in two directions within the system, e.g. nanofilm and nanosheets (Tiwari et al. 2012; Mukherjee et al. 2014).

Nanomedicine involves the development of particles, structures and analytical techniques at the nanoscale that has various applications in medical diagnosis, imaging and targeted drug delivery system as shown in Fig. 11.1 (Mukherjee et al. 2014). When drugs with low solubility are taken orally, it will lower their bioavailability and cell permeability and result in undesirable adverse effects. Such drawbacks can be addressed with drug delivery systems that use nanotechnology (Patra et al. 2018). When a particle is converted to nanoscale size, it obtains unique properties like higher surface area per mass unit in comparison to macroscale particle (Mukherjee et al. 2014). It also induces predominance of quantum effects associated with a smaller size to adsorb and hold other compounds (Mukherjee et al. 2014). As a result, their electrical, optical and magnetic properties are altered. These properties make these molecules suitable for applications in drug and gene delivery (Mukherjee et al. 2014).

The use of nanoparticles in nanomedicine has revolutionized the delivery of drugs, permitting therapeutic agents to target specific levels of organs, tissues and cells selectively and minimizing healthy tissue exposure to drugs. Though normal modes of drug delivery systems are extensively used because of their efficiency in releasing drugs in a regulated manner, they have inherent limitations of influencing the tissues in general than affecting individual cells (Saidi et al. 2018). The purpose of nano drug delivery systems is to address the limitations of conventional macro drug delivery systems like reduced plasma half-life, dimensional instability and possible immune reactions (Saidi et al. 2018).

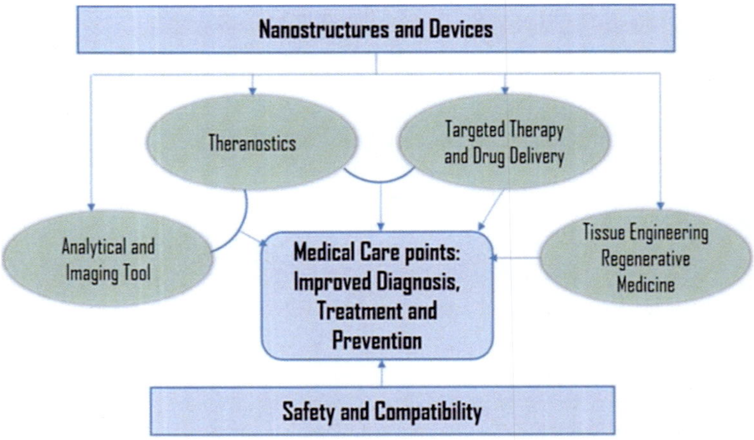


**Fig. 11.1** Varied applications of nanotechnology in diagnostic, therapeutic and regenerative medicine. (Picture adapted from Patra et al. 2018)

Nanoparticles have a distinct advantage over traditional drug delivery system due to their small scale and are favoured for medicines with low solubility and absorption. Nanoparticles can encapsulate the drugs and promote targeted drug delivery with sustained-release, as shown in Fig. 11.2. Some of the nanoparticle-based drug deliveries have found applications in certain clinical conditions like cancer or cancer therapies. Since nanoparticle-based drug delivery systems can pass through the blood–brain barrier, they facilitate targeting brain tumours (Saidi et al. 2018).

In medicine, the nano drug delivery method is increasingly used as nanomaterials can be used not only as carriers but also to nanoscale the drug particles themselves. Another key reason is the rise in surface atoms and molecules, which in turn causes the area of the surface to increase. This allows for drug and protein binding, adsorption, and a robust delivery agent along with them (Mukherjee et al. 2014). These nanoscale device systems can be targeted to access the inaccessible areas such as tumour cells and inflamed tissues. Their nanoscale size and shape improve the permeability and substantivity of drugs. In the treatment of intracellular infections, they are stated to be useful, e.g. it can be used to target reticulo-endothelial cells, which in turn allows passive drug delivery to liver and spleen macrophages (Mukherjee et al. 2014). The nano drug delivery system is therefore used to transfer genes and proteins via the oral route of administration as well (Figs. 11.1 and 11.2).

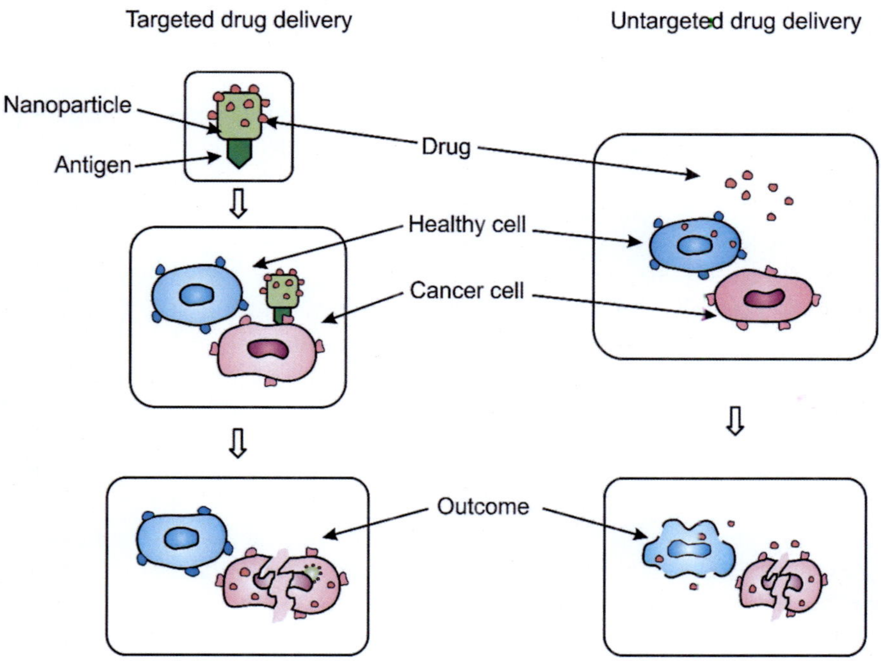


**Fig. 11.2** Targeted drug delivery using nanoparticles. (Picture adapted from Brakmane et al. 2012)

Nano drug delivery systems have distinct advantages like the ability to alter drug solubility, rate of release and diffusion, bioavailability and immunogenic profile (Patra et al. 2018). As a result, this can result in improving and developing convenient routes for administration, with lesser adverse effects and toxicity enhancing the drug life and bioavailability (Patra et al. 2018).

Nanoscale particles used for drug delivery should follow certain requirements such as they should be highly soluble, biocompatible and bioavailable (Patra et al. 2018). The toxicity associated with them should be low, so that particular diseased tissues can be targeted within a limited concentration. They should be non-invasive without occluding blood vessels (Patra et al. 2018). They should protect the drug from enzymatic and hydrolytic digestion (Patra et al. 2018). This, in turn, bypass the first-pass metabolism of the drug in the liver (Patra et al. 2018; Yu et al. 2016). Nano drugs coated with hydrophilic polymers persist in the circulation for longer times. They are ideal candidates for increasing the half-life of drugs with limited bioavailability and can also be used to monitor prolonged drug release. They are used for the transmission of DNA, too. These formulations have benefits such as improved drug solubility, increased therapeutic onset and dose reduction. The sudden disintegration of the drug through increased metabolism and clearance can be avoided because of the bioadhesive properties of the nanocarriers, thus increasing retention (Patra et al. 2018).

There are two mechanisms of drug delivery using nanostructures: passive and self-delivery. The passive delivery system comprises of the inclusion of the drugs into the inner surface of the nanomaterial by applying the hydrophobic effect (De Jong and Borm 2008). This hydrophobic environment results in a low concentration of the encapsulated drugs, and hence, when the nanomaterials are targeted to a particular site, instant delivery of the intended amount of drug is achieved (Patra et al. 2018; De Jong and Borm 2008; Deng et al. 2019). The attachment of the drug molecule to the nanoparticle is based on the properties of the drug carrier complex such as pH, temperature, molecular size and shape in passive targeting. This affinity of the drug carrier complex is instrumental in driving the complex to the target site (Patra et al. 2018). In contrast, in the active or the self-delivery system, drugs to be delivered are attached to the carrier nanostructures. The targets for these conjugated drug complexes are the receptors and lipid components on cell membranes and antigens present on the surfaces of the cell (Patra et al. 2018).

Nano-drug delivery systems include various systems such as 'biopolymeric nanoparticles, nano liposomes, dendrimers, fullerenes, nanopores, nanotubes, nanoshells, quantum dots, nanocapsule, nanosphere, nano vaccines, nanocrystals, etc.' as depicted in Fig. 11.3 (Bhavikatti et al. 2014). Additionally, the nanochips, nanorobotics and magnetic nanomaterials that are associated with a specific antibody, hollow viral capsids of nanoscale are recent advances in drug delivery system (Renugalakshmi et al. 2011). The materials used for designing and synthesis of these drug delivery system dictate the type of nanostructure, properties and release characteristics of the delivery system (Renugalakshmi et al. 2011).

Biopolymeric nanoparticles: Many biopolymeric materials have been described which include chitosan, alginate, xanthan gum, cellulose, liposomes, polymeric micelles and dendrimers whose properties are discussed below.

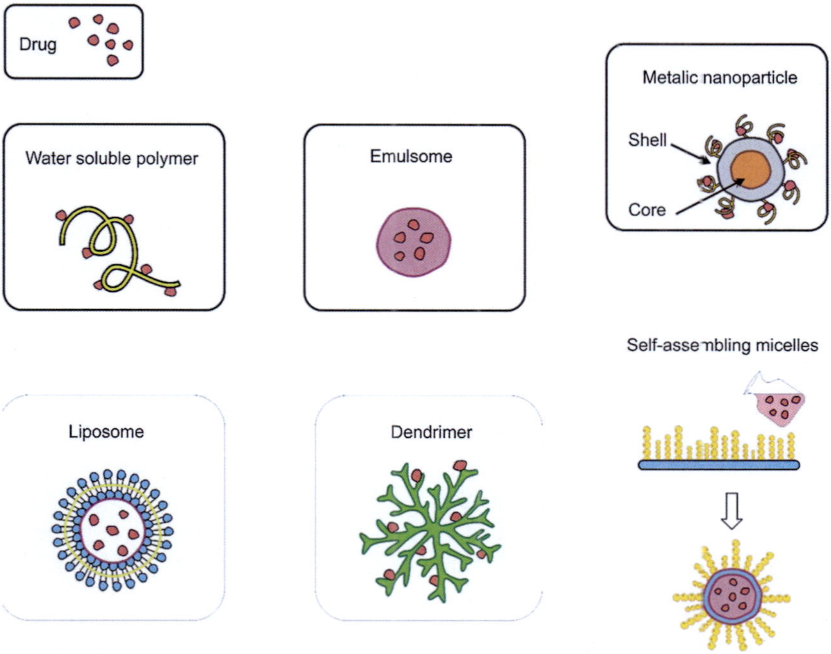


**Fig. 11.3** Different types of drug carriers. (Picture adapted from Brakmane et al. 2012)

## 11.2 Chitosan

Chitosan is commonly applied due to its mucoadhesive properties. Thus, chitosan-based nanomaterials find wide-ranging applications as sustained drug delivery systems for different types of epithelia of the buccal, intestinal, nasal, skin and pulmonary mucosa (Patra et al. 2018). A study conducted by Silva et al. to deliver an antibiotic to the eye conjugated with chitosan nanoparticles demonstrated better mucoadhesion, sustained release and increased life span of the antibiotic and hence proving its efficacy in the delivery of ocular drugs (Silva et al. 2015). Chitosan nanoparticles, as demonstrated by Pistone et al., were the most cytocompetitive when applied to the oral cavity when compared to alginate and pectin (Pistone et al. 2019). Liu et al. demonstrated the use of chitosan nanoparticles for the release of intranasal carbamazepine achieved higher concentrations of the drug in the brain than the plasma over a prolonged period (Liu et al. 2007; Patra et al. 2018).

## 11.3 Alginate

This biopolymeric material has been classified as an anionic mucoadhesive polymer due to its higher strength as opposed to cationic and neutral polymers (Patra et al. 2018). A study demonstrated that insulin-containing alginate nanoparticles with nicotinamide could transport insulin efficiently by the sublingual route and showed high bioavailability when used in diabetic rats (Ghavimishamekh et al. 2019; Heidarisasan et al. 2018). Haque et al. used alginate nanoparticles by the intranasal route for the management of depression (Haque et al. 2014). They demonstrated the direct release of the drug to the brain with higher blood–brain ratios when compared to the conventional intravenous or intranasal routes of delivery (Patra et al. 2018).

## 11.4 Xanthan Gum

It is a polyionic high molecular weight hetero-polysaccharide with excellent bioadhesive properties, nontoxic and nonirritating and hence used commonly as a pharmaceutical excipient (Patra et al. 2018). Thiolated xantham gums, as reported by Lafleur and Michalek et al., have proven their efficacy for decreasing the salivary flow in sialorrhea (Laffleur and Michalek 2017). A study reported by Huang et al. displayed the efficacy of modified xanthan injectable hydrogels in rebuilding the abdominal wall due to the presence of conjugated angiogenic factors (Huang et al. 2018; Patra et al. 2018). The bioadhesive nature of xanthan gums could be well appreciated in this study, as reported by Huang et al. (Huang et al. 2018; Patra et al. 2018).

## 11.5 Cellulose

Cellulose and its derivatives can modify the solubility and gelation of the drugs and can control their release profile (Patra et al. 2018). Elseoud et al. demonstrated that cellulose nanocrystals formed hydrogen bonds with the drug, which promoted the sustained release of the drug (Abo-Elseoud et al. 2018). The release rates of the drugs are also affected due to the presence of oxidized or native cellulose nanocrystals where lower release rates were produced by oxidized nanocrystals when compared to their native counterparts (Patra et al. 2018).

## 11.6 Nanomaterials for Drug Delivery

Quantum dots (QD) are fine nanocrystal particulate which is not greater than 10 nm in size. On ignition with the light, they fluoresce into various colours (Saidi et al. 2018). They have notable advantages like high photostability, narrow emission wavelength, strong luminescence intensity, broad excitation window and ease of preparation and surface modification. They are hence used extensively for drug delivery and bioimaging because of their exclusive optical and electronic characteristics (Matea et al. 2017). Their size-dependent emission enables ease of tuning

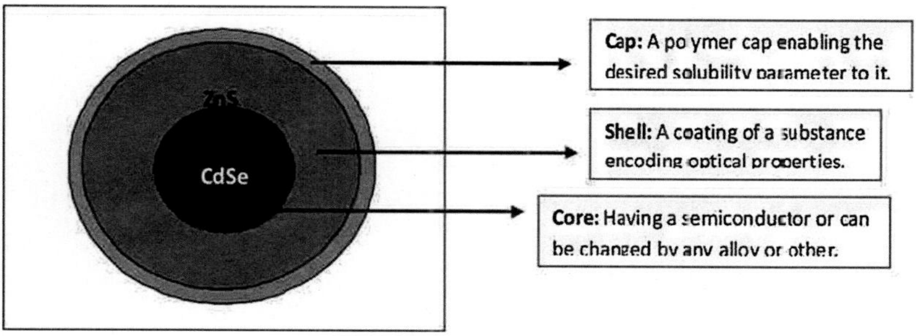


**Fig. 11.4** Structural composition of a semiconductor quantum dot. (Picture adapted from Yan and Chen 2014)

emission from visible to infrared wavelength by controlling their sizes. To make them ready for drug delivery, surface modifications have to be made (Wagner et al. 2019). The structure of a quantum dot is depicted in Fig. 11.4 (Yan and Chen 2014).

The quantum dot can be combined with other biomolecules to target other molecules. Attached with other biomolecules, they could behave as diagnostic biomarkers (Matea et al. 2017). QD is also proven to be used in diagnostics and imaging of sentinel lymph nodes. It is beneficial for clinical treatment planning and radiological staging of cancerous tissues (Saidi et al. 2018). QD is also used in a procedure called 'multiplexing', wherein they are designed into a plethora of luminescent colours that are tagged to either numerous proteins or gene sequence in the body. QD probes with different excitation and emission of varying wavelength are used for imaging. They are also used for real-time tracking of cancer markers which increases the specificity and sensitivity of cancer diagnosis.

Additionally, QD is used for viral diagnostics. The identification and analysis of the respiratory syncytial virus are critical for controlling viral diseases and the synthesis of new antiviral medicines. QD is linked to specific antibodies that enable the quantification of various tumour markers. It is an indispensable addition to the field of nanomedicine because of its special photonic properties. Due to its high sensitivity and precision, diagnostics and drug delivery systems add additional dimensions. They are designed in specific size and shape that can differ based on varying targeting principles (Mukherjee et al. 2014).

One of the greatest advantages of using nanomaterials for drug delivery is that these ultra-small materials can greatly enhance the intracellular delivery of a broad range of molecules for improved efficacy. As one example, QDs can be used for siRNA delivery. Interference RNA has great potential for biomedical applications because of its powerful ability to target specific mRNAs, resulting in mRNA destruction and gene silencing. However, the lack of stability and difficulty in cellular entry and endosome escape are the major problems that limit its application. Taking advantage of easy surface modification, QDs can be used for highly efficient and safe delivery of siRNA for mRNA interference (Mukherjee et al. 2014).

The adsorption, distribution, metabolism and excretion of QD, and also its cytotoxicity, depends on different factors, including its physicochemical characteristics and environmental factors. However, though very limited research has been conducted to analyse the cytotoxicity of QD. Few in vitro experiments have implicated the toxicity of QD to its surface characteristics (De Jong and Borm 2008).

Choi et al. in 2007 demonstrated that the non-modified QD stimulates lipid peroxidation in the cell while the QD whose surface was modified using *N* acetylcysteine was remarkably reduced (Wagner et al. 2019). A similar study reported by Cho et al. in 2005 revealed that the unmodified quantum dots caused cytotoxicity via the release of reactive oxygen species that results in intracellular damages within the plasma membrane, mitochondria and nucleus damage (Lovrić et al. 2005). Since the biological coating of quantum dots facilitates specific targeting of cells or cell organelles, it also needs to be used judiciously due to its ability to induce toxic effects. Release of free toxic $Cd^{+}$ ions was responsible for the toxic potential of cadmium containing quantum dots. Differing core compositions of the quantum dots also altered the cellular toxicity like the cellular toxicity was found in QD derived from cadmium/telluride however not in cadmium selenium/zinc sulphate QD (Mukherjee et al. 2014).

### *11.6.1 Magnetic Nanoparticles*

Magnetic nanoparticles have been used in magnetic resonance imaging (MRI), targeted drug/gene delivery, bioseparation, etc. Superparamagnetic is a unique property of magnetic nanoparticles. Superparamagnetic property refers to the ability of magnetic nanoparticles to be magnetized under an applied magnetic field, and such magnetization disappears once the magnetic field is removed. This property can prevent the agglomeration of magnetic nanoparticles after losing the magnetic field, and thus prolong their circulation half-time. With the rapid development of nanotechnology, magnetic nanoparticle-based drug delivery systems have been emerging for many cancer therapies (Yan and Chen 2014).

Magnetic nanoparticles conjugated with methotrexate (MTX, a chemotherapy anti-cancer drug) can increase the internalization by folate receptors overexpressed cancer cells and therefore improve the drug efficacy in comparison with free MTX drug. Engineered magnetic nanoparticles can also load cisplatin complexes (a therapeutic drug that binds and interrupts DNA), and studies showed that the magnetic conjugates were more toxic to cancer cells than free cisplatin. Porous hollow magnetic nanoparticles like $Fe_3O_6$ were also founded to be effective for targeted delivery. For instance, cisplatin was loaded into the hollow core of the magnetic nanoparticles, which were bonded with targeting molecules (Herceptin) (Yan and Chen 2014).

Moreover, magnetic nanoparticles coated with polyamidoamine (PAMAM) and polyethyleneimine (PEI) derivative are promising for gene delivery as well. Although effective, PEI molecules may have high toxicity due to their disruption to cell membranes. To address this limitation, magnetic nanoparticles can be further attached with PEG to reduce the toxicity and increase the circulation time (Yan and Chen 2014). The structure of a magnetic nanoparticle has been shown in Fig. 11.5.

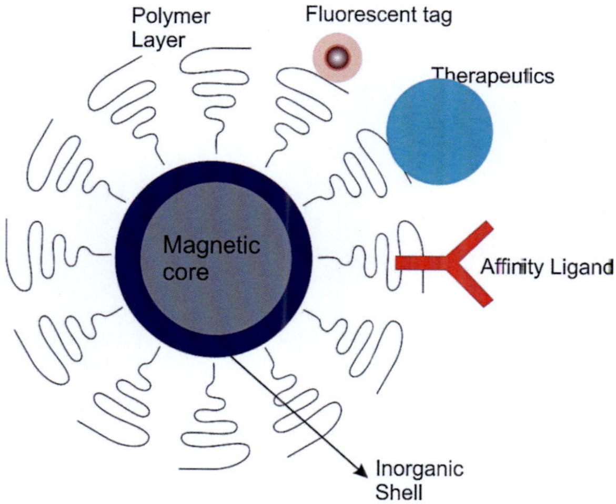


**Fig. 11.5** Structure of magnetic nanoparticle. (Picture adapted from Yan and Chen 2014)

## *11.6.2 Carbon Nanotubes*

Carbon nanotubes are single- or multiwall tubes that are derived from carbon with a benzene ring which can behave like biopersistent fibres. Nanotubes are composed of a network of carbon atoms branched to form a hexagon. Nanotubes aspect ratio is greater than 100, with length ranging in millimetres as depicted in Fig. 11.6. They are classified into two types, single-walled and multiwall nanotubes. Single-walled nanotubes range with a diameter of 0.7–1.5 nm while the multiwall carbon nanotubes rage with 2–50 nm (Sharma 2017).

Single-walled nanotubes are favoured over metallic nanoparticles due to their better drug-carrying ability and inherent stability (Bonifacio et al. 2014). Also, the ability to undergo easy surface adaptation improves its distribution in the blood. These properties make them ideal for the transportation of bioactive molecules such as nucleic acids, proteins and drug molecules. The nanotubes, when combined with antibodies and low molecular weight agents, can behave as targeting agents when internalized into cells. They have better loading capacity, and when internalized into cells, they allow the targeting agents to bio-integrate into the body to surpass the body defence mechanism (Sharma 2017).

Carbon nanotubes are also used for gene delivery. Functionalized carbon nanotubes formulated by the molecular dynamics technique, facilitated its application in gene therapy (Bonifacio et al. 2014).

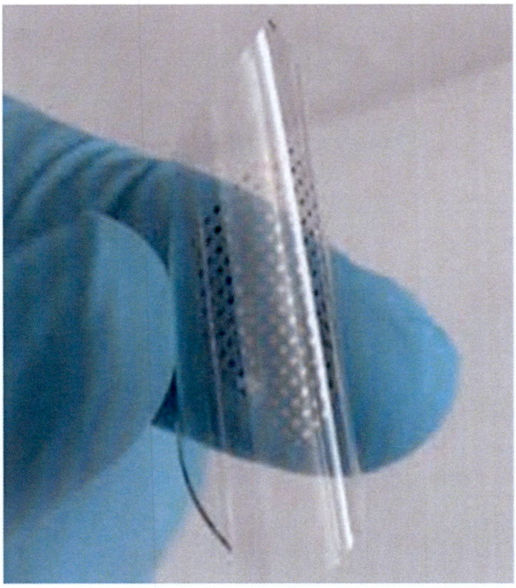

**Fig. 11.6** Carbon nanotubes having hexagonally shaped branched network of carbon atoms. (Picture adapted from https://www.nanowerk.com/nanotechnology/introduction/introduction_to_nanotechnology_22.php)

## 11.6.3 Gold Nanoparticles/Nanoshells

Metallic colloidal gold nanoparticles detectable at low concentrations are synthesized into various types such as rods, dots. They vary in size for commercial applications. They have found applications for drug and gene delivery, biomedical imaging. They are used in synthesizing novel anti-cancerous products. Some researchers have observed that gold nanoparticle are non-toxic to cells (De Jong and Borm 2008). However, the stabilizers such as CTAB in the gold nanorods could cause substantial toxicity in cells. Hence, PEG-modified gold nanorods excluded CTAB which did not show cell toxicity. On analysis of oral toxicity, it was observed that nanogold suspension with diameter approximately 50 nm did not show any signs of toxicity. The minimum lethal dose of $LD_{50}$ for an oral route is 5000 mg/kg body weight for nanogold suspension.

Additionally gold solutions are used as contrast agents for MRI (magnetic resonance imaging). They are generally composed of either gold and cooper or gold and silver. Gold and silica composition is used for photoablation of cancerous cells. Studies have shown that the non-targeted nanoshells did not show the toxicity of tumour cells (De Jong and Borm 2008).

## 11.6.4 Silica

In an in-vitro study, it was analysed that silica nanoparticles showed both toxic and nontoxic responses. There was an increase in toxicity not only due to higher dosages but also increased contact time with the silica nanoparticles (26, 68, and 72 h) (De Jong and Borm 2008). The toxic effects due to silica nanoparticle were reflected by a decrease in cell sustainability or proliferation, damage to cell membranes due to enzyme release, and release of reactive oxygen species. Cells that had a longer multiplication cycle had a higher probability of being affected by the toxic effects of the silica nanoparticles. Also, there was a difference in the manner in which various cell lines behaved to the exposure of silica nanoparticles. An experimental cell line study on alveolar macrophages showed more susceptibility of these cells to the toxic effects of silica nanoparticles when compared to cells in the pulmonary epithelium. It was suggested that the phagocytic properties in an alveolar macrophage were responsible for this response.

Further surface modification of these nanoparticles reduced the cytotoxic effects caused by them when compared to the particles without any surface modification (De Jong and Borm 2008). Silica nanoparticle exposure was found to be cytotoxic at a higher concentration, which was reflected in the reduction in cell viability and proliferation. Membrane disruption is often demonstrated by the secretion of lactate dehydrogenase from the cells. Higher levels of reactive oxygen and glutathione build a toxic environment with higher oxidative stress. Compared to cells with a lower doubling rate, cells with a higher doubling rate were sensitive and cytotoxic to silica nanoparticles. Different cell lines behaved differently to the exposure of silica nanoparticles, e.g., an alveolar macrophage cell line (MHS) was highly prone to nanoparticle-induced cytotoxicity which could be attributed to its phagocytic properties than a lung epithelial cell line (A569). Also, surface modification of cationic silica nanoparticles using amino-hexyl- amino-propyltrimethoxysilane caused no cytotoxicity when compared to the particles without any surface modification (De Jong and Borm 2008).

### 11.6.5 Fullerenes

Fullerenes have found applications as potent antimicrobial agents since they can produce reactive oxygen species following photoexcitation. However, these properties could cause ecotoxicity due to their deleterious effects on bacterial colonies when released as by-products into the ecosystem. Moreover, though no mortality has been reported, fullerenes have been reported to cause damage to the fat-rich brain tissue by the direct circulation or axonal transport (Mukherjee et al. 2014).

### 11.6.6 Dendrimers

Dendrimers are three-dimensional structures which are monodisperse and well defined. They are made up of a core, branch and surface to make up their structure. They are primarily monomers which undergo a process of polymerization to attain congregated or diversified step-growth structures (Sharma 2017; Patra et al. 2018). Their surface can be easily functionalized as they are highly bifurcated with a globular shape. This property makes excellent candidates for such structures for drug delivery, gene therapy, immune therapy and imaging agents (Sharma 2017; Patra et al. 2018).

Dendrimers can be synthesized in two ways, either it could begin from the core inside and then branch outwards or it could start from the exterior and then converge inwards. They are also classified based on their functionalization features. Among them, PAMAM is soluble in water and can pass through epithelial surfaces, making them feasible for oral drug delivery. Others have limited clinical applicability due to the cationic and positively charged toxic amine groups. To overcome these adverse effects, it is essential to modify the dendrimers to reduce this toxicity (Nikalje 2015; Sharma 2017; Patra et al. 2018).

Dendrimers could be employed for the transport of medications by different methods. They can be employed for the transport of medicines through they have been used as drug delivery agents for transdermal, ocular, oral and pulmonary routes. They can be either encapsulated or could be bound by electrostatic or covalent bonds. Drugs delivered could be assimilated by the disintegration of the binding bonds between the drug nanoparticle complex by cleaving enzymes covalent bonds between the drug and dendrimer by certain enzymes or the dissipation of the drug because of alterations in pH and temperature (Nikalje 2015; Sharma 2017).

Dendrimers have been used as a cancer prevention drug carrier model by Jain et al. A similar effect was also demonstrated by Kaur et al. who conjugated dendrimers with folic acid and transformed the nanocarriers for selective cell targeting, pH-mediated drug release in cancer therapy. Such mechanisms aid the tumour cells for selective uptake of the drug (Patra et al. 2018).

### *11.6.7 Micelles*

These nanostructures are formed by the polymerization process measuring approximately 100 nm in size. These polymeric structures are made up of amphiphilic blocks that fold upon themselves, resulting in the formation of these particles in an aqueous medium. Since they are amphiphilic, their hydrophobic core promotes the congregation of drugs like docetaxel and paclitaxel which are inherently water-repelling while the hydrophilic shell makes the drug nanocarrier conjugate soluble in water and stabilizes the core–shell structure at the same time. This makes them suitable for the transport of hydrophobic drugs, resulting in enhanced bioavailability and structural stability (Sharma 2017).

Since these micelles undergo a slow renal excretion, it permits a prolonged accumulation of the drug complex in tumour cells. Their polymeric nanostructure also prevents interactions with other biological tissues. They are produced by two techniques first where the polymer is dissolved in a solvent, and a dialysis process is commenced next and the other addition of a solvent resulting in the precipitation of the micelles (Sharma 2017).

Various factors like the size of the core molecules, their amphiphilic concentration, the agent dissolving the solutes and its temperature affect the formation of micelles. These micelles have found applications against cancer and ocular drug delivery systems (Sharma 2017) (Figs. 11.7 and 11.8).

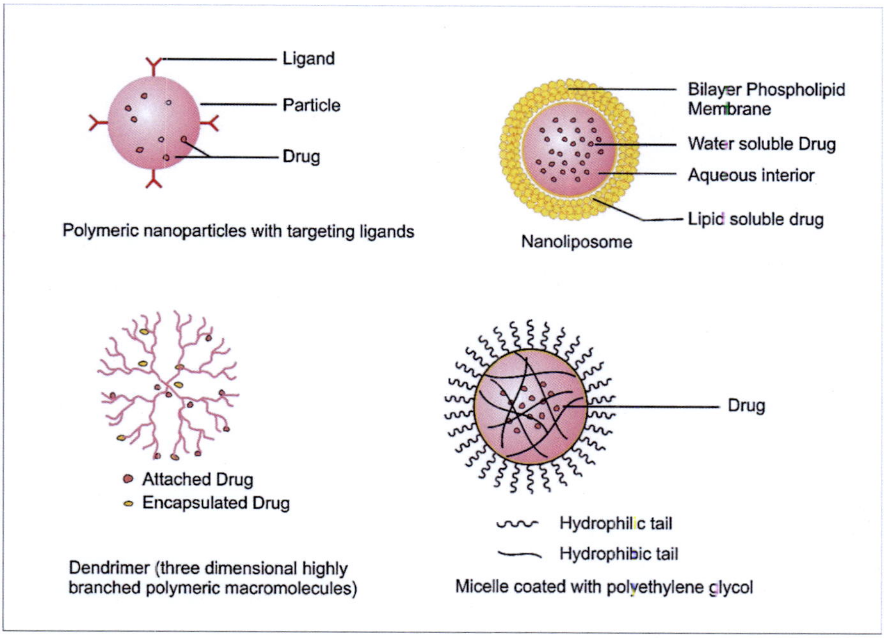


**Fig. 11.7** Different nanoparticles and their structural composition (Mukherjee et al. 2014)

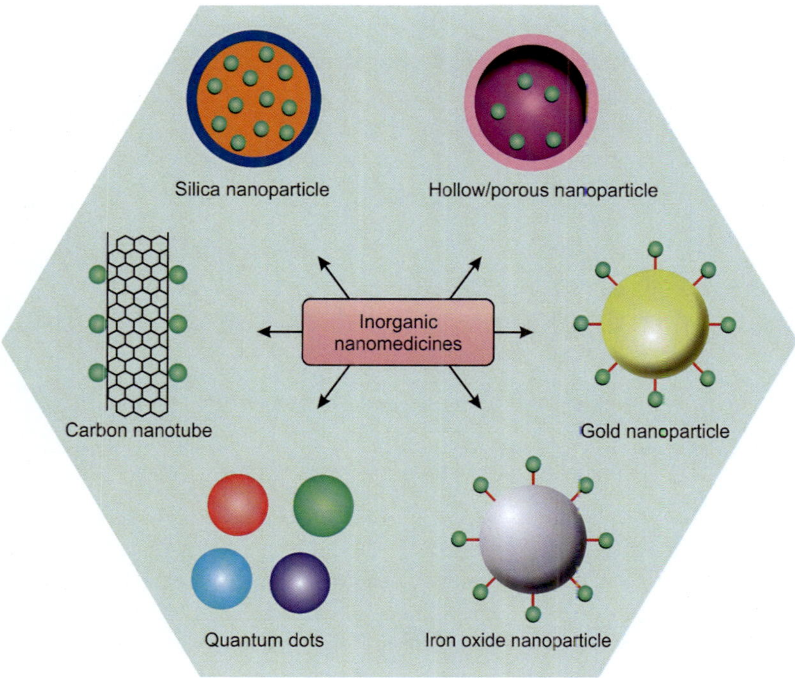


**Fig. 11.8** Different inorganic nanoparticles used for applications in nanomedicine (Tang and Cheng 2013)

Recently, the trend of research in the field of nanomaterials can be classified under two categories. The first approach, also known as the bottom-up approach involves the nanoscale structures to aggregate together in a regulated manner through different procedures to form higher intelligent macroscale structures. The second approach is also known as the top-down approach. Here the prevailing molecules in macrostructures are shrunk and miniaturized into nanoscale and further reorganized to gain desirable properties (Bhavikatti et al. 2014). Currently, various nanoparticles like nanorods, nanotubes, quantum dots and nanocapsules that are suitable for application in medicine and gene delivery are evaluated for its biocompatibility, safety and biological considerations (Bhavikatti et al. 2014).

Nanotubes are tubular structures measuring approximately 50–100 nm, which can carry small amounts of the drug. Since they are open at both the ends, these are attached with temperature or pH-sensitive caps which will disintegrate in suitable conditions and help in drug delivery. Similarly, fullerenes and polymeric nanoparticles are also used as means of transport for drug delivery. Their surfaces would be altered with antibodies which will assist in target-specific drug delivery that are otherwise difficult to reach through traditional drug delivery systems.

There have been various modifications by which nanoparticles can be made suitable for drug delivery. The iron oxide magnetic nanoparticles targeted for destroying malignant tissue works on the principle of increasing the local temperature in the tissue by applying a magnetic field. Physiological and biodegradable lipid nanospheres are being thought as innovative carriers for transdermal drug delivery. Nanospheres are also being considered as ideal candidates to enclose local anaesthetic agents within carrier vehicles to prolong the effect of the drug at the targeted site (Bhavikatti et al. 2014).

Lately, the anti-cancer properties of double-layered magnetic nanoparticles made of iron oxide have been examined. The nanoparticles of iron oxide are injected into cancerous tissues and stimulated by the magnetic field. The resulting exchange coupling causes temperature increase that destroys cancer tissue. Recently, a novel transdermal drug delivery system was developed that form nano-lipid carrier known as lipid nanospheres. They are biodegradable with no toxicity and enhanced penetrability. By improving their chemical stability, the shelf life of photosensitive compounds are increased (Bhavikatti et al. 2014).

### *11.6.8 Liposomes*

Liposomes are vesicular nanostructures made up of concentric layers of the bilayered phospholipid membrane. A central liquid core is enclosed by a membrane of a phospholipid bilayer (Panahi et al. 2017). The core facilitates the entrapment of hydrophilic drugs. A phospholipid bilayer attracts lipophilic drugs. The number of lipid bilayers and the size of the liposomes subclassify them into two types: multilamellar and unilamellar. The unilamellar variant is sub-classified into small and large unilamellar vesicles.

As the name suggests, multilamellar means composed of multiple layers of phospholipids interspersed with aqueous spaces. Unilamellar vesicles of the small category will have a single lipid layer surrounding the central aqueous media with diameters less than 100 nm. The large unilamellar vesicles differ from their small unilamellar counterparts in having single lipid bilayers surrounding the central aqueous media but diameters more than 100 nm (Patra et al. 2018).

Liposomes have been conjugated with the anti-cancerous drugs that are beneficial in the treatment of various malignancies. Liposomes are also tagged with antibodies and have been targeted for the management of Alzheimer's disease (Patra et al. 2018).

### 11.6.9 Nanoshells

Nanoshells are suggested as novel drug carriers for imaging and therapeutic purposes. They usually measure approximately 100–200 nm. They are made of a central core of silica surrounded by an outer metal covering. Their applications adopting various immune-based techniques in targeting tissues is well known. They can also be embedded in gel-based polymers before use. Nanoshells are helpful for the diagnosis of whole blood immunoassays. Their role in the management of diabetes and prevention of micrometastasis of tumours is currently under investigation (Mukherjee et al. 2014).

### 11.6.10 Nanopores

Nanopore wafers as their name suggests consist of highly porous structures which permit the passage of oxygen, glucose and chemicals like insulin to pass through them. Their size ranges ~20 nm in diameter. Nanowafers are thin discs which act as drug reservoirs and release drugs over a sustained period. Nanopore devices containing β-cells of the pancreas can be transplanted and have the advantage of protecting the transplanted tissues from the host immune system. These systems have also found applications in nucleic acid sequencing and differentiation of purines from pyrimidines (Mukherjee et al. 2014). They are used in drug delivery by implanting drugs which can measure ~20 nm in diameter consisting of high-density pores which permit entry of oxygen, glucose and other chemicals such as insulin to pass through. Nanopore devices can not only be used for transplantation but can protect transplanted tissues from the host immune system. Nanopore devices containing β-cells of the pancreas can be implanted in the recipient's body. Nanopores are also used for DNA sequencing and differentiation of purines from pyrimidines (Mukherjee et al. 2014).

### *11.6.11 Toxicity*

The cytotoxicity of nanomaterials is complex due to various factors like chemical composition, crystalline structure, size (since the variation in size can influence the basic physicochemical properties of materials) or aggregation (Renugalakshmi et al. 2011; Saidi et al. 2018). Various techniques have been employed to assess the cytotoxic properties of nanoparticles. The in vitro studies have been beneficial, considering the cost and time factors involved. Various assays have been used to assess cell sustainability and genotoxicity (De Stefano et al. 2012; Bahadar et al. 2016; da Silva Martins et al. 2017).

Several approaches have assessed the toxicity of engineered NPs. Among them, the in vitro studies have been beneficial considering the cost and time factors. Cell viability is assessed most commonly by tetrazolium reduction assays, cell membrane integrity with LDH assay, immunohistochemistry biomarkers for apoptosis, and comet assay for genotoxicity (De Stefano et al. 2012; Bahadar et al. 2016; da Silva Martins et al. 2017).

For assessing the intracellular localization of NPs, electron microscopy is employed. Compounds such as MTT, MTS, XTT and WST-1 are used to detect viable cells. MTT readily enters the viable eukaryotic cells as it is a positive compound while negative compounds such as MTS, XTT and WST-1 do not permeate cells rapidly. So among other assays, the MTT tetrazolium cell proliferation assay has been used widely in laboratories for the evaluation of cell toxicity (De Stefano et al. 2012; Bahadar et al. 2016).

Cell inflammatory response induced by NPs is analysed by measuring inflammatory biomarkers, such as IL-8, IL-6 and tumour necrosis factor, using ELISA. Lactate dehydrogenase (LDH) assay is used for cell membrane integrity and has been used to determine the cytotoxicity of many NPs produced from silica, iron oxide, titanium oxide and zinc oxide. Different types of cell cultures, including cancer cell lines, have been employed as in vitro toxicity models though they are associated with some risks and may not be able to mimic the in vivo conditions completely (De Stefano et al. 2012; Bahadar et al. 2016).

## 11.7 Selecting Nanoparticles for Toxicology Studies

The advent of nanoparticles has led to their incorporation into a wide variety of consumer products. The subsequent commercialization of these products has raised concerns over their safety and toxic effects. Among the nanotoxicology community, there is no single consensus regarding which of the nanomaterials are appropriate for defining toxicity (Lankveld et al. 2010; Monteiro-Riviere et al. 2010; Samberg et al. 2010).

Characterization of materials plays a very significant role to assess the toxicity of any chemical substance. Several techniques can be used to characterize nanomaterials and their interaction with biomolecules in powder, film or solution form. However, it is challenging in nanomaterials due to their variable shapes and sizes, charge, surface area, porosity, chemistry, crystallinity, agglomeration and solubility. Hence characterization of nanomaterials demands skilled human resources and highly sophisticated instrumentation (Gupta et al. 2019).

The complex nature of nanoparticles poses a major challenge for predicting their potential toxicity. Various characteristics of nanomaterials determine their toxicity. The primary characteristics of the particles (e.g. core chemistry, size, shape, crystallinity, surface and aggregation state), and secondary characteristics (e.g. protein corona, dissolution rate, biodistribution) determine the interaction of the nanoparticles with the target biological systems (Samberg et al. 2010) as shown in Fig. 11.9. Therefore, comprehensive characterization of nanomaterials will include size distribution, shape, surface area, surface chemistry, crystallinity, porosity, agglomeration state, surface charge, and solubility which could help correlate the physicochemical properties of the materials and their ensuing biological effects (Gupta et al. 2019).

Among these most of the commercially produced nanomaterials are nonhomogeneous, impure and do not carry any details regarding their process of manufacturing. Moreover, the interpretation of their toxicological profiles is challenging as characteristics of the material at the beginning, and their correlation with physicochemical properties of the end product is uncertain (Madkour 2020a; Meghea 2020). The biological response of modified properties of the material like size, shape or surface charge, core structure, etc. (e.g. core chemistry, size, shape or surface) could be understood when experiments were conducted with separate sets of nanomaterials which had only a single altered property. However, such sets of materials with exclusive modifiable properties are currently unavailable and are also difficult to harness for the researchers. They cause a lot of time delay and are tough to be characterized in laboratories during experiments (Lankveld et al. 2010; Monteiro-Riviere et al. 2010; Samberg et al. 2010).

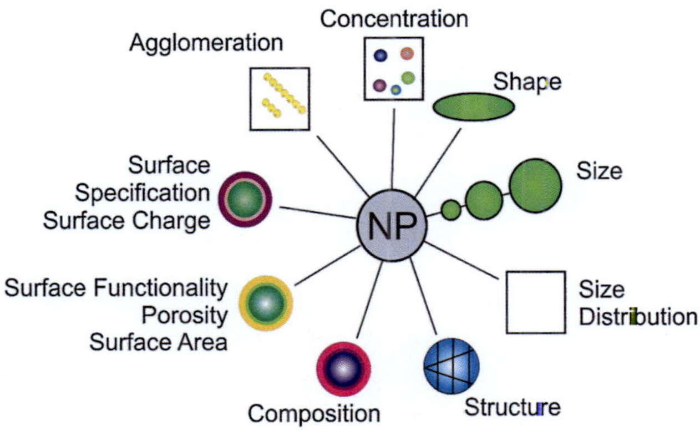


**Fig. 11.9** The primary and secondary characteristics of nanoparticles which influence their interaction and biological effects. (Picture adapted from De Stefano et al. 2012)

Nanomaterial toxicity can be assessed via in vivo, in vitro and in silico methods. In vitro assays like MTT assay, LDH leakage assay, trypan blue dye assay, apoptosis assay, 2,7-dichlorofluorescein diacetate (DCFH)-DA assay and comet assays are commonly used. In vivo methods are conducted on living animals where nanomaterials are administered to the testing animal and monitored through various methods. However, the scientists recommended the use of in silico assays for time constraints, ethical concerns and reliability of results. This novel approach is based on theoretical modelling and simulation of different properties of the nanomaterials. The in silico model provides a better and feasible model for testing the toxicity of nanomaterials (Gupta et al. 2019). Two types of nanoparticles are used for performing cytotoxicity tests: (a) the nanoparticles engineered in a detailed manner and (b) nanomaterials produced in bulk quantities. The engineered nanoparticles will have precisely altered physical and chemical properties unlike those manufactured for release in bulk quantities. A detailed description of both the systems is given below.

## *11.7.1 Nanoparticles Engineered in a Detailed Manner*

The ongoing research regarding toxic aspects of nanoparticles hints towards the interrelationship between the physical and chemical characteristics of the particles and their associated adverse effects. These precisely manufactured nanoparticles' individual chemical and physical properties must be probed to accomplish this (Lankveld et al. 2010; Samberg et al. 2010) (Fig. 11.10).

Bio-Pure silver and gold NPs are examples of precisely engineered nanoparticles. Each physicochemical property of these engineered particles is strictly controlled and monitored for in vitro and in vivo purposes. Besides the residual precursors of each formulation is washed. It could contribute to toxic effects that are unrelated to its individual physicochemical properties of the NPs. The purification and concentration of nanoparticles without segregation allows these particles to be distributed at higher concentrations than those usually available. Scientists performing experiments on acute exposure would be benefitted when colloidal silver conventionally available at 0.02 mg/mL will be provided at 1 mg/mL which is 50 times its original concentration (Lankveld et al. 2010; Monteiro-Riviere et al. 2010).

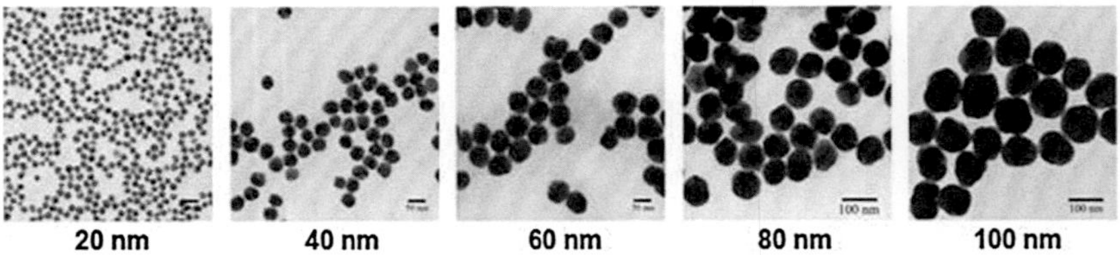


**Fig. 11.10** Precisely engineered nanoparticles and their different sizes. Picture adapted from Lankveld et al. 2010)

### 11.7.2 Understanding the Effects of Size

Size plays an important role in designing nanotoxicology experiments (Sosenkova and Egorova 2011). Size determination of nanoparticles can be done using different techniques (Shin et al. 2015). Further accurate determination of average sizes and size distribution varies with different methods and differing principle adopted in different techniques. Therefore, to assess appropriate size and distribution of the nanoparticle, a proficiency in the different techniques, the principle of working and other technical facts about the methods used are essential. Size affects how the cells are internalized and assimilated, which will further modify the outcome of the particle within the system. Experimental studies conducted to determine the effect of size on nanoparticle toxicity substantiates that degree of cytotoxicity is inversely proportional to its size (Hu et al. 2018; Gupta et al. 2019).

There are two types: primary size and hydrodynamic size. The size of an individual nanoparticle, when dissolved in a suspension, is known as the primary size. It plays an important role in determining the surface area, rate of dissolution and meting point of the particle. It serves as a benchmark to understand nano-dimensions regarding diffusion, uptake and biodistribution of nanomaterials. However, the greater the probability that particle cluster to form aggregates, the greater is the particle's reactive size. The nature of such agglomerated particles is different from the individual nanoparticles. The size of such agglomerated particles is usually referred to as the hydrodynamic size. The hydrodynamic size is more critical than the primary size for assessment of various parameters (Lankveld et al. 2010; Monteiro-Riviere et al. 2010; Samberg et al. 2010). To understand the effects of size, the suspensions of the nanoparticles, which are initially unagglomerated, should be used. The selection of particles with surfaces that covers varying sizes will yield better results. Understanding the aggregation rate and adjusting the protocols of testing to accommodate the rate of agglomeration is necessary (Lankveld et al. 2010; Monteiro-Riviere et al. 2010; Samberg et al. 2010).

### 11.7.3 Understanding the Effects of Shape

The shape of the material also influences toxic effects elicited by nanoparticles. For example, long fibres are known to have increased toxicity accompanied by the difficulty of clearance. So when the shape of the nanoparticles is altered, such engineered particles will have additional properties that allow absorption of various proteins in the solution. Numerous variants of shape for silver and gold nanoparticles in the form of nanowires and nanoplates are available. Altered properties of these materials could be attributed to the alteration in shape (Hao et al. 2016).

### 11.7.4 Understanding the Effects of Surface

Nanomaterials occur in different shapes like spheres, needles, cubes, tubes, rods, etc. Morphology of the nanoparticles affects the membrane interactions during internalization and cellular uptake mechanisms. Pore formation in cell membranes following interaction with nanoparticles resulting in an imbalance of ionic concentration inside and outside the cell has been reported by researchers (Gupta et al. 2019).

Surface changes can influence the structural characteristics of nanoparticles (Sukhanova et al. 2018). The charges on the surface of the particle can influence the positive or negative charge borne by the particle, the hydrophobicity or hydrophilicity, and the lipophobic or lipophilic properties of the particles. The surface charge of the molecules also influences their interaction with other molecules and their ability to initiate an immunogenic response. The surfaces of customized nanoparticles can be functionalized or altered by the adding various polymers like PEG, silica, phosphate and biomolecules like DNA or antibodies (Lankveld et al. 2010; Monteiro-Riviere et al. 2010).

Recognizing the surface state of a particle is vital in understanding its role in toxicity (Zhang et al. 2012). The exposure to the biological system can bring about significant changes on the surface which could be detrimental for the ultimate toxicity of the nanoparticles. Protein identification can be made by isolating the protein from the surface (Ajdary et al. 2018). To determine the surface changes that occur during experimentation, it is important to identify and quantify the condition of nanoparticle surfaces (Madkour 2020b). Different analytical tools such as 'matrix-assisted laser desorption ionization mass spectrometer (MALDI-MS), Fourier transform infrared spectroscopy (FTIR) and Raman spectroscopy' helps in determining the surface charges and isoelectric point of the particulate. Additionally, the proteins on the surface of nanoparticles can be identified using LC-MS/MS (Lankveld et al. 2010; Monteiro-Riviere et al. 2010).

Surface coating possesses distinct charges. It influences the intercellular contacts and inter-particle relations in the material that can cause cell toxicity (Patel and Shah 2017). Certain studies have reported that cationic nanomaterials were internalized with greater efficiency. In comparison, the neutral or anionic particles have fewer chances of toxicity. Based on their interactions, surface coatings are broadly divided into three types, namely covalent coatings having covalent bonding, the electrostatic surface coating having electrostatic interactions and atomic layer deposition where chemical bond formed between molecules and coating material. The surface coatings that hydrophobic are observed to be generally cytotoxic compared to hydrophilic surface coatings. For example CuNPs and $Fe_2O_3$ NPs coated with chitosan decreases the formation of reactive oxygen species, inflammatory reactions and hence inhibit the overall toxic reactions (Azandaryani et al. 2019).

### 11.7.5 Effects of Dissolution Rate and Their Relationship with Toxicity

The nanoparticles, when dissolved, release ions and ion complexes. These ion complexes are known to be responsible for the toxicity of some materials. For instance, the silver ions in silver nanocrystals are more toxic than the particles itself. The greater the surface area of material, the higher the release rate of the toxic metal ions from the surface. Toxicity studies have proven that smaller the particle greater the surface area leading to higher ion release rates and toxicity. This explains the increased toxicity at smaller primary sizes in particles like silver (Lankveld et al. 2010; Monteiro-Riviere et al. 2010; Samberg et al. 2010). Its shape and size influence the rate of discharge of these ions from the surface of the particles.

Additionally rate of discharge influences crystallinity, surface covering and ion concentration in solution. The environmental conditions that can affect storability like thermal conditions, exposure to oxygen/sulphur and light influence the ion concentration. These factors, in turn, affect the release rate. The concentration of ions in solution can be assessed by ultracentrifugation, chromatography or filtration/dialysis where the nanoparticles are separated from the suspending medium (Lankveld et al. 2010; Monteiro-Riviere et al. 2010; Samberg et al. 2010).

### 11.7.6 Residues Responsible for Toxicity

Depending on the method of manufacture, residual reactants may be seen in the solution of nanoparticles. During the wet manufacturing step of the nanomaterials, thorough cleansing and purification should be conducted to get rid of the by-products of these procedures. In addition to the cleaning nanoparticle, samples will be spun in a centrifuge which will separate the solids into a pellet, and the supernatant liquid will be left behind. This liquid could serve as a control for evaluating the effects of residual reactant effects. Sometimes, these supernatants can continue to cause adverse effects despite the removal of all residues from the system. This could be because of the discharge of ions from the surface of the nanoparticles (Lankveld et al. 2010; Monteiro-Riviere et al. 2010; Samberg et al. 2010).

### 11.7.7 Regulatory Aspect of Nano-drug Delivery Systems

Various governmental, academic and industry regulatory bodies have been established to assess the safety aspects of the nanomaterials. The USA established the NNI in the year 1996 to monitor other regulatory agencies across the world for the development of nanoscience and technology. Many other international organizations are being established for shouldering responsibilities in this field (Lankveld et al. 2010; Monteiro-Riviere et al. 2010; Samberg et al. 2010; D'Mello et al. 2017).

### *11.7.8 Future of Nanosize Drug Delivery Systems*

With the recent advances seen in the field of nano-drug delivery systems, the pharma industry will witness a paradigm shift. The amalgamation of science and technology has established a new platform for drug delivery where the disadvantages of conventional systems could be overcome, resulting in improved absorption with decreased toxicity. Use of nano-drug delivery modes requires a profound understanding of their properties and their systemic effects. The field of nanotechnology is still in its nascent stages with few drug delivery systems available commercially but many currently in clinical trials. Though nanocarriers have distinct advantages as vehicles for drug delivery systems, their clinical, toxicological and regulatory aspects still need to be addressed. When nanomaterials are used as nanomedicines, their biocompatibility is of utmost importance due to a range of effects on the human system.

It is therefore imperative that cost-effective, better and safer nanomaterials be introduced to provide effective and controlled delivery of drugs. With an increasing demand for novel applications in drug delivery systems, these nano-drug delivery systems hold promise for targeted drug therapy. The knowledge and understanding of toxicity and safety of nanomaterials and their biological responses will advance the applications of nanomaterials for drug delivery in the near future.

## References


Abo-Elseoud, W. S., Hassan, M. L., Sabaa, M. W., Basha, M., Hassan, E. A., & Fadel, S. M. (2018). Chitosan nanoparticles/cellulose nanocrystals nanocomposites as a carrier system for the controlled release of repaglinide. *International Journal of Biological Macromolecules, 111*, 604–613. https://doi.org/10.1016/j.ijbiomac.2018.01.044.

Ajdary, M., Moosavi, M. A., Rahmati, M., Falahati, M., Mahboubi, M., Mandegary, A., & Varma, R. S. (2018). Health concerns of various nanoparticles: A review of their in vitro and in vivo toxicity. *Nanomaterials, 8*(9), 634. https://doi.org/10.3390/nano8090634.

Azandaryani, A. H., Kashanian, S., & Jamshidnejad-Tosaramandani, T. (2019). Recent insights into effective nanomaterials and biomacromolecules conjugation in advanced drug targeting. *Current Pharmaceutical Biotechnology, 20*(7), 526–541. https://doi.org/10.2174/1389201020666190417125101.

Bahadar, H., Maqbool, F., Niaz, K., & Abdollahi, M. (2016). Toxicity of nanoparticles and an overview of current experimental models. *Iranian Biomedical Journal, 20*(1), 1. https://doi.org/10.7508/ibj.2016.01.001.

Bhavikatti, S. K., Bhardwaj, S., & Prabhuji, M. L. (2014). Current applications of nanotechnology in dentistry: A review. *General Dentistry, 62*(4), 72–77.

Bonifacio, B. V., da Silva, P. B., dos Santos Ramos, M. A., Negri, K. M. S., Bauab, T. M., & Chorilli, M. (2014). Nanotechnology-based drug delivery systems and herbal medicines: A review. *International Journal of Nanomedicine, 9*, 1.

Brakmane, G., Winslet, M., & Seifalian, A. M. (2012). Systematic review: The applications of nanotechnology in gastroenterology. *Alimentary Pharmacology & Therapeutics 36*(3), 213–221. https://doi.org/10.1111/j.1365-2036.2012.05179.x.

da Silva Martins, L. H., Rai, M., Neto, J. M., de Oliveira, J. A. R., da Silva Martins, J. H., Komesu, A., et al. (2017). Nanomaterials: Properties, toxicity, safety, and drug delivery. In *Nanotechnology applied to pharmaceutical technology* (pp. 363–381). Cham: Springer. https://doi.org/10.1007/978-3-319-70299-5_15.

De Jong, W. H., & Borm, P. J. (2008). Drug delivery and nanoparticles: Applications and hazards. *International Journal of Nanomedicine, 3*(2), 133. https://doi.org/10.2147/ijn.s596.

De Stefano, D., Carnuccio, R., & Maiuri, M. C. (2012). Nanomaterials toxicity and cell death modalities. *Journal of Drug Delivery, 2012*, 1–12. https://doi.org/10.1155/2012/167896.

Deng, Y., Zhang, X., Shen, H., He, Q., Wu, Z., Liao, W., & Yuan, M. (2019). Application of the nano-drug delivery system in treatment of cardiovascular diseases. *Frontiers in Bioengineering and Biotechnology, 7*, 489. https://doi.org/10.3389/fbioe.2019.00489.

D'Mello, S. R., Cruz, C. N., Chen, M. L., Kapoor, M., Lee, S. L., & Tyner, K. M. (2017). The evolving landscape of drug products containing nanomaterials in the United States. *Nature Nanotechnology, 12*(6), 523–529. https://doi.org/10.1038/nnano.2017.67.

Ghavimishamekh, A., Ziamajidi, N., Dehghan, A., Goodarzi, M. T., & Abbasalipourkabir, R. (2019). Study of insulin-loaded chitosan nanoparticle effects on tgf-β1 and fibronectin expression in kidney tissue of type 1 diabetic rats. *Indian Journal of Clinical Biochemistry, 34*(4), 418–426. https://doi.org/10.1007/s12291-018-0771-9.

Gupta, A., Kumar, S., & Kumar, V. (2019). Challenges for assessing toxicity of nanomaterials. In *Biochemical toxicology-heavy metals and nanomaterials*. London: IntechOpen. https://doi.org/10.5772/intechopen.89601.

Hao, N., Li, L., & Tang, F. (2016). Shape matters when engineering mesoporous silica-based nanomedicines. *Biomaterials Science, 4*(4), 575–591. https://doi.org/10.1039/c5bm00589b.

Haque, S., Md, S., Sahni, J. K., Ali, J., & Baboota, S. (2014). Development and evaluation of brain targeted intranasal alginate nanoparticles for treatment of depression. *Journal of Psychiatric Research, 48*(1), 1–12. https://doi.org/10.1016/j.jpsychires.2013.10.011.

Heidarisasan, S., Ziamajidi, N., Karimi, J., & Abbasalipourkabir, R. (2018). Effects of insulin-loaded chitosan-alginate nanoparticles on RAGE expression and oxidative stress status in the kidney tissue of rats with type 1 diabetes. *Iranian Journal of Basic Medical Sciences, 21*(10), 1035. https://doi.org/10.22038/ijbms.2018.28463.6899.

Hu, C. C., Wu, G. H., Lai, S. F., Shanmugam, M. M., Hwu, Y., Wagner, O. I., & Yen, T. J. (2018). Toxic effects of size-tunable gold nanoparticles on Caenorhabditis elegans development and gene regulation. *Scientific Reports, 8*(1), 1–10. https //doi.org/10.1038/s41598-018-33585-7.

Huang, J., Li, Z., Hu, Q., Chen, G., Ren, Y., Wu, X., & Ren, J. (2018). Bioinspired anti-digestive hydrogels selected by a simulated gut microfluidic chip for closing gastrointestinal fistula. *iScience, 8*, 40–48. https://doi.org/10.1016/j.isci.2018.09.011.

Kovvuru, S. K., Mahita, V. N., Manjunatha, B. S., & Babu, B. S. (2012). Nanotechnology: The emerging science in dentistry. *Journal of Orofacial Research, 2*(1), 33–36.

Laffleur, F., & Michalek, M. (2017). Modified xanthan gum for buccal delivery—A promising approach in treating sialorrhea. *International Journal of Biological Macromolecules, 102*, 1250–1256. https://doi.org/10.1016/j.ijbiomac.2017.04.123.

Lankveld, D. P., Oomen, A. G., Krystek, P., Neigh, A., Troost–de Jong, A., Noorlander, C. W., et al. (2010). The kinetics of the tissue distribution of silver nanoparticles of different sizes. *Biomaterials, 31*(32), 8350–8361. https://doi.org/10.1016/j.biomaterials.2010.07.045.

Liu, C., Tan, Y., Liu, C., Chen, X., & Yu, L. (2007). Preparations, characterizations and applications of chitosan-based nanoparticles. *Journal of Ocean University of China, 6*(3), 237–243. https://doi.org/10.1007/s11802-007-0237-9.

Lovrić, J., Cho, S. J., Winnik, F. M., & Maysinger, D. (2005). Unmodified cadmium telluride quantum dots induce reactive oxygen species formation leading to multiple organelle damage and cell death. *Chemistry & Biology, 12*(11), 1227–1234. https://doi.org/10.1016/j.chembiol.2005.09.008.

Madkour, L. H. (2020a). Improvement testing strategy for aligning nanomaterial safety assessments and oxidative stress responses. In *Nanoparticles induce oxidative and endoplasmic reticulum stresses* (pp. 521–556). Cham: Springer. https://doi.org/10.1016/b978-0-12-822481-6.00022-0.

Madkour, L. H. (2020b). Nanotoxicity, cytotoxicity, and genotoxicity mechanisms of nanomaterials. In *Nanoparticles induce oxidative and endoplasmic reticulum stresses* (pp. 47–98). Cham: Springer. https://doi.org/10.1016/b978-0-12-822481-6.00021-9.

Matea, C. T., Mocan, T., Tabaran, F., Pop, T., Mosteanu, O., Puia, C., et al. (2017). Quantum dots in imaging, drug delivery and sensor applications. *International Journal of Nanomedicine, 12*, 5421. https://doi.org/10.2147/IJN.S138624.

Meghea, A. (2020). Nanoparticles induce oxidative and endoplasmic reticulum stresses: Antioxidant therapeutic defenses, Loutfy H. Madkour Springer, 2020 752 pages, $142 (hardcover) ISBN 9783030372965. *MRS Bulletin, 45*(10), 868. https://doi.org/10.1557/mrs.2020.264.

Monteiro-Riviere, N. A., Oldenburg, S. J., & Inman, A. O. (2010). Interactions of aluminum nanoparticles with human epidermal keratinocytes. *Journal of Applied Toxicology: An International Journal, 30*(3), 276–285. https://doi.org/10.1002/jat.1494.

Moothedath, M., Moothedath, M., Jairaj, A., Harshitha, B., Baba, S. M., & Khateeb, S. U. (2019). Role of nanotechnology in dentistry: Systematic review. *Journal of International Society of Preventive & Community Dentistry, 9*(6), 535. https://doi.org/10.4103/jispcd.JISPCD_223_19.

Mukherjee, B., Dey, N. S., Maji, R., Bhowmik, P., Das, P. J., & Paul, P. (2014). Current status and future scope for nanomaterials in drug delivery. In *Application of nanotechnology in drug delivery*. London: IntechOpen. https://doi.org/10.5772/58450.

Nikalje, A. P. (2015). Nanotechnology and its applications in medicine. *Medicinal Chemistry, 5*(2), 81–89. https://doi.org/10.4172/2161-0444.1000247.

Panahi, Y., Farshbaf, M., Mohammadhosseini, M., Mirahadi, M., Khalilov, R., Saghfi, S., & Akbarzadeh, A. (2017). Recent advances on liposomal nanoparticles: Synthesis, characterization and biomedical applications. *Artificial Cells, Nanomedicine, and Biotechnology, 45*(4), 788–799. https://doi.org/10.1080/21691401.2017.1282496.

Patel, P., & Shah, J. (2017). Safety and toxicological considerations of nanomedicines: The future directions. *Current Clinical Pharmacology, 12*(2), 73–82. https://doi.org/10.2174/1574884712666170509161252.

Patra, J. K., Das, G., Fraceto, L. F., Campos, E. V. R., del Pilar Rodriguez-Torres, M., Acosta-Torres, L. S., et al. (2018). Nano based drug delivery systems: Recent developments and future prospects. *Journal of Nanobiotechnology, 16*(1), 71. https://doi.org/10.1186/s12951-018-0392-8.

Pistone, A., Iannazzo, D., Celesti, C., Piperopoulos, E., Ashok, D., Cembran, A., et al. (2019). Engineering of chitosan-hydroxyapatite-magnetite hierarchical scaffolds for guided bone growth. *Materials, 12*(14), 2321. https://doi.org/10.3390/ma12142321.

Renugalakshmi, A., Sekar Vinothkumar, T., & Kandaswamy, D. (2011). Nanodrug delivery systems in dentistry: A review on current status and future perspectives. *Current Drug Delivery, 8*(5), 586–594. https://doi.org/10.2174/156720111796642336.

Saidi, T., Fortuin, J., & Douglas, T. S. (2018). Nanomedicine for drug delivery in South Africa: A protocol for systematic review. *Systematic Reviews, 7*(1), 1–7. https://doi.org/10.1186/s13643-018-0823-5.

Samberg, M. E., Oldenburg, S. J., & Monteiro-Riviere, N. A. (2010). Evaluation of silver nanoparticle toxicity in skin in vivo and keratinocytes in vitro. *Environmental Health Perspectives, 118*(3), 407–413. https://doi.org/10.1289/ehp.0901398.

Sharma, K. (2017). Nanomaterials for drug delivery. In *Advances in personalized nanotherapeutics* (pp. 57–77). Cham: Springer. https://doi.org/10.1007/978-3-319-63633-7_5.

Shin, S. W., Song, I. H., & Um, S. H. (2015). Role of physicochemical properties in nanoparticle toxicity. *Nanomaterials, 5*(3), 1351–1365. https://doi.org/10.3390/nano5031351.

Silva, N. C., Silva, S., Sarmento, B., & Pintado, M. (2015). Chitosan nanoparticles for daptomycin delivery in ocular treatment of bacterial endophthalmitis. *Drug Delivery, 22*(7), 885–893. https://doi.org/10.3109/10717544.2013.858195.

Sosenkova, L. S., & Egorova, E. M. (2011, April). The effect of particle size on the toxic action of silver nanoparticles. In *Journal of Physics: Conference Series* (Vol. 291, No. 1, p. 012027). Bristol: IOP Publishing. https://doi.org/10.1088/1742-6596/291/1/012027.

Sukhanova, A., Bozrova, S., Sokolov, P., Berestovoy, M., Karaulov, A., & Nabiev, I. (2018). Dependence of nanoparticle toxicity on their physical and chemical properties. *Nanoscale Research Letters, 13*(1), 44. https://doi.org/10.1186/s11671-018-2457-x.

Tang, L., & Cheng, J. (2013). Nonporous silica nanoparticles for nanomedicine application. *Nano Today, 8*(3), 290–312. https://doi.org/10.1016/j.nantod.2013.04.007.

Tiwari, J. N., Tiwari, R. N., & Kim, K. S. (2012). Zero-dimensional, one-dimensional, two-dimensional and three-dimensional nanostructured materials for advanced electrochemical energy devices. *Progress in Materials Science, 57*(4), 724–803. https://doi.org/10.1016/j.pmatsci.2011.08.003.

Wagner, A. M., Knipe, J. M., Orive, G., & Peppas, N. A. (2019). Quantum dots in biomedical applications. *Acta Biomaterialia, 94*, 44–63. https://doi.org/10.1016/j.actbio.2019.05.022.

Yan, L., & Chen, X. (2014). Nanomaterials for drug delivery. In *Nanocrystalline materials* (pp. 221–268). Amsterdam: Elsevier. https://doi.org/10.1016/B978-0-12-407796-6.00007-5.

Yu, X., Trase, I., Ren, M., Duval, K., Guo, X., & Chen, Z. (2016). Design of nanoparticle-based carriers for targeted drug delivery. *Journal of Nanomaterials, 2016*, 1087250. https://doi.org/10.1155/2016/1087250.

Zhang, Y., Xu, D., Li, W., Yu, J., & Chen, Y. (2012). Effect of size, shape, and surface modification on cytotoxicity of gold nanoparticles to human HEp-2 and canine MDCK cells. *Journal of Nanomaterials, 2012*, 375496. https://doi.org/10.1155/2012/375496.

# Chapter 12
# Bio-inspired Materials in Nanobiotechnology Applications and Industrial Potential Scale

**Maria del Pilar Rodriguez-Torres and Kaushik Pal**

## 12.1 Introduction

Nanotechnology is defined as the technology developed at the nanoscale (1–100 nm) and additionally is intended to be applied in the real world. On December 29, 1959 Richard Feynman came up with the idea of the ability to manipulate and control atoms individually (molecular manufacturing), in his talk "There's plenty of room at the bottom" which took place at the California Institute of Technology (CALTECH) within an American Physical Society meeting. Such event would become the preamble for the birth of the nanotechnology concept. Fourteen years later (Sandhu 2006), Norio Taniguchi specifically coined the term nanotechnology in a talk related to his research on ultraprecision machining at a scientific conference (Drexler 2004) . It has to be mentioned, though, that Richard Zsigmondy, a chemist, was awarded the 1925 Nobel Prize for the proving the heterogeneous nature of colloid solutions and for approach he created and used for this purpose. In fact, he was the first person who used the term nanometer to refer to the $10^{-9}$ m scale (Mappes et al. 2012) . Nevertheless, it was not until 1981, when Heinrich Rohrer and his colleague Gerd Binnig at IBM (Zurich, Switzerland) built the first scanning tunneling microscope (STM). Such device allowed to see single atoms. This invention lead both of the Swiss scientists to share the 1896 Nobel Prize. Later on, in 1990, it was figured out how to use the STM to move single xenon atoms around on

M. del Pilar Rodriguez-Torres (✉)
Departamento de Ingenieria Molecular de Materiales, Centro de Fisica Aplicada y Tecnologia Avanzada, Universidad Nacional Autonoma de Mexico,
Queretaro, Mexico

K. Pal
Laboratório de Biopolímeros e Sensores, Instituto de Macromoléculas, Universidade Federal do Rio de Janeiro (LABIOS/IMA/UFRJ)–Cidade Universitária, Rio de Janeiro, Brazil

Wuhan University, Wuchang Dist., Hubei, China



K. Pal (ed.), *Bio-manufactured Nanomaterials*,
https://doi.org/10.1007/978-3-030-67223-2_12

a nickel surface in order to construct a sign that spelled out "IBM." This last event, in particular was the turning point that introduced the basis for the development of techniques such as atomic force microscopy and magnetic resonance imaging, for instance (Toumey et al. 2006). As for other materials at the nanoscale, besides the ones studied by Zsigmondy, chemists Richard E. Smalley, Robert F. Curl, and Harold W. Kroto discovered the buckminsterfullerene (also known as fullerenes) in 1985, and in 1996 all of them won the Nobel prize for such finding (Mason 2005). These structures are closed-shell carbon macromolecules. Subsequently, other fullerene structured such as nanotubes came into the scene. Some outstanding characteristics of these carbon-based nanomaterials are good heat and electricity conductivity, novel electrical properties, extreme tensile strength, and ability to penetrate membranes like cell walls.

Some nanomaterials properties can be listed briefly as follows (Jeevanandam et al. 2018; Khan et al. 2018):

- 1–100 nm dimension.
- Large surface to volume ratio.
- High percentage of either atoms or molecules on their surface.
- Chemical and physical properties can be tuned according to their size.
- Metal-based nanomaterials exhibit interesting scattering properties.
- Semiconductor-based nanomaterials display confined energy states.

Also, they can also be classified as shown in Fig. 12.1. Such representation is only taking into account dimensionality and chemical composition. Other classifications also exist (Buzea and Pacheco 2017; Saleh 2020; Darweesh 2018).

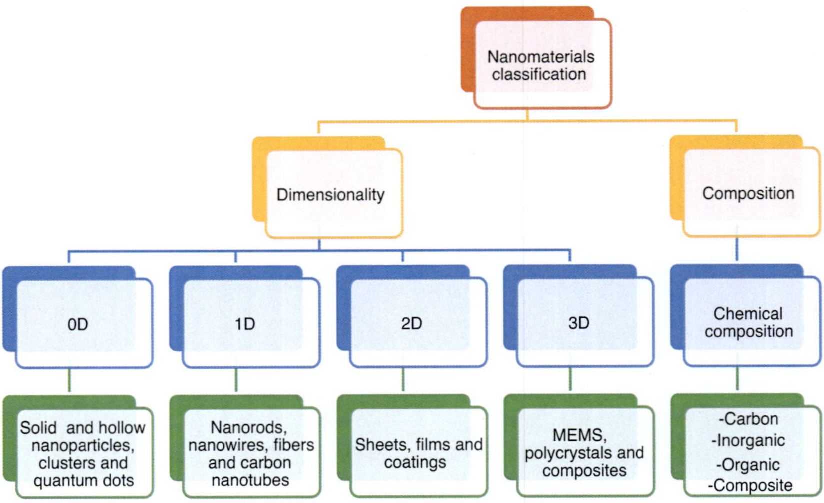


**Fig. 12.1** Nanomaterials classification scheme according to dimensionality and composition

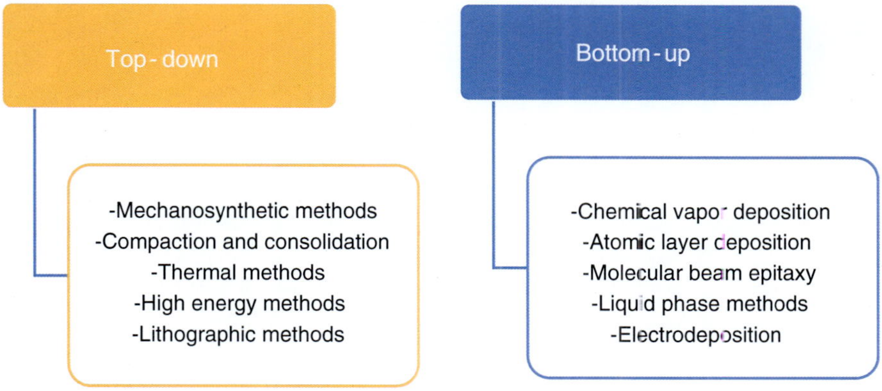


**Fig. 12.2** Approaches for nanomaterials synthesis

Nanomaterials are manufactured by means of two different approaches as in Fig. 12.2.

Top-down approaches refer to those methods in which size reduction is reached from bulk materials and they are usually utilized in the solid state procedures and also makes use of physical techniques, for example, milling and laser ablation (Kolahalam et al. 2019). On the other hand, the bottom-up route has to do with material synthesis from atoms and is usually used for the formulation of the majority of nanomaterials due to its capability to obtain uniform sizes, shapes and distributions. Bottom-up encloses chemical synthesis (Su and Chang 2018).

Different and diverse kinds of reactants have been used to synthesize nanomaterials, for example, surfactants, metallic salts, organic and inorganic solvents, cross-linking agents. Furthermore, different techniques (bottom-up) have also been developed for the same purpose. There are disadvantages linked to either reactants or synthesis methods: how harmful they can be to the environment or for certain applications, especially those involving biomedical and clinical ones. Therefore, work has been done for some time to overcome this problem by promoting green synthesis approaches. These methods aim for the usage of reactants and solvents that are environment-friendly and do not leave hazardous residues after they have been carried out. They might involve the usage of biomolecules, microorganisms, and plants (Virkutyte and Varma 2013; Gour and Jain 2019).

Nanobiotechnology is a term that is basically the joint of nanotechnology and biology, it is a hybrid field. It is particularly devoted to using biological materials as raw sources for nanomaterials assemblage, although other methodologies are not discarded completely; it also relies on biological design principles and is potentially beneficial for applications in biomedicine, agriculture, in the food sector, pharmacology to mention some (Fakruddin et al. 2012; Qamar et al. 2019).

This chapter is devoted firstly, to give the reader a general outlook on nanotechnology, nanomaterials, their synthesis and nanobiotechnology.

## 12.2 Nanomaterials Synthesis Methods

Throughout the years, nanomaterials synthesis methods have been developed, improved, and aimed at getting materials with different characteristics. In the previous section, the top-down and bottom-up routes for nanomaterials synthesis were mentioned, but this is rather quite a general description. Synthesis methods can be further subdivided in chemical, physical and biological as shown in Table 12.1. As for the biological routes they can be enclosed in the chemical approaches because the syntheses involve chemical processes such as reduction, crosslinking and functionalization, for instance (Thiruvengadathan et al. 2013; Rane et al. 2018).

It is pointed out that even though there are nanomaterials whose manufacturing involves other approaches other than the biological one, they are still considered part of the nanobiotechnology definition due to their applications.

## 12.3 Biogenic Synthesis of Nanomaterials and Their Applications

Biogenic approaches to synthesize nanoparticles involve the usage of biological agents as reactants, their role ranges from acting as reducing, stabilizing agents and templates. The biogenic synthesis can take place with only one method, for instance, by only using an extract or a solution from the biological agent or there can be the conjunction of several techniques to mediate the synthesis process (Patwardhan et al. 2018; Hasan et al. 2018).

These biological agents are biomolecules, microorganisms, plants, fungi, and organic waste. The importance these materials have gained in the last years is for the most part, due to them being green or environment-friendly unlike other traditional routes (physical and chemical) that imply hazards and pollution (Zan and Wu 2016).

**Table 12.1** Listing of nanomaterials synthesis classification methods

| Physical | Chemical |
|---|---|
| High energy ball-milling | Chemical reduction |
| Laser ablation | Micro-emulsion/colloidal |
| Melt mixing | Sonochemical |
| Nanolithography | Electrochemical |
| Sputtering | Solvothermal decomposition |
| Thermal decomposition | Sol-gel |
| Laser pyrolysis | Spinning |
| Electrospraying | Chemical vapor deposition |
| Inert gas condensation | Co-precipitation |
| Physical vapor deposition | Hydrothermal |
| Electron beam lithography | Microwave assisted |
| Pulsed wire discharge | Biological synthesis |

In addition to the aforementioned, several biological activities have been discovered and reported among which we can find antimicrobial, anti-inflammatory, anticancer, antidiabetic which are potentially beneficial in the area of nanobiotechnology. Besides biological activities, other traits such as biocompatibility, cell affinity and the like are also relevant (Salata 2004).

### *12.3.1 Plant-Based Synthesis Routes*

The nanomaterials generated by the plant-mediated approach are metal ones: special attention has been paid to silver and gold ones, although copper, platinum, and metal oxides are also studied.

These plant-based metal nanoparticles have been studied for long and it has been reported that the phytochemicals present in their structure, for example, polyphenols, alkaloids, phenolic acids, carbohydrates, and amino acids play an important role in the synthesis procedure as well as in the resulting biological activities of the synthesized nanoparticles. These phytochemicals are components in plants that act as a defense system to combat diseases, therefore exhibiting biological activities such as the antimicrobial, anti-cancerous, and anti-inflammatory ones, among others (Mohammadinejad et al. 2016; Iravani 2011).

The mechanism for nanoparticle formation that has been proposed indicates that phytoconstituents promote the reduction of metal ions. Also, it has been reported that the polyhydroxyl groups in the plants' secondary metabolites act as scavenging agents (Mittal et al. 2013; Kuppusamy et al. 2016).

It should be noted that in the plants types there are several species to consider, different research reports nanoparticles synthesized from the parts of trees (leaves, fruits), from herbs and from the other two types of plants, shrubs, algae (seaweed) and climbers (Rouhan and Gaudeul 2014). This division of plants is indicated in Fig. 12.3.

### *12.3.2 Microorganism-Based Nanoparticles Synthesis*

Syntheses using microorganisms is another branch of the bio-inspired nanomaterials area because these microorganisms are a source of biological activities that can be interesting for their future use in biomedicine or even in agriculture (Singh et al. 2016).

Additionally, there are assets like low-cost production, low-cost cultures, and easy controllable growth conditions such as oxygenation, temperature, pH, and incubation time that make these nanomaterials fascinating and potentially appropriate for further applications (Zhang et al. 2011).

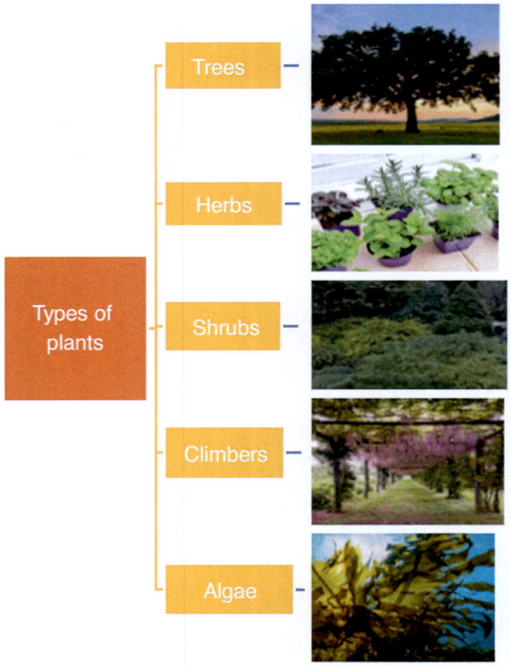


**Fig. 12.3** Scheme showing the types of plants and some examples

### *12.3.3 Fungi-Based Nanoparticle Synthesis*

Fungi belong to a kind of eukaryotic organisms that includes microorganisms such as yeasts and molds, as well as mushrooms, but it has to be considered, too that This section is devoted to the coverage of nanomaterials synthesis using yeasts, fungi, and mushrooms because they are the research works that have been the most worked on. Inorganic nanoparticles are the nanomaterials that have been synthesized the most with the aid of fungi. Unlike bacteria-based synthesis, the fungi-based approach offers advantages such as the easiness to be handled at the laboratory and the high nanoparticle yield that can be obtained (Halkai et al. 2017; Shende et al. 2017).

When using the fungal-based NP biosynthesis, a biomineralization process takes place via the reduction of metal ions by intracellular and extracellular enzymes and biomolecules. There is also the exchange of electrons from a donor molecule to ions that results in the ions precipitation that form the nanoparticles. The procedure for nanoparticle generation also displays some interesting processes, this is, it implies a defense mechanism against environmental contamination because fungi reduce ions toxicity through precipitation, immobilization, ion form modification, co-precipitation, and coupling to biological moieties. The resulting reduction of ions ends in the precipitation of metals in the form of nanomaterials in the intra- or extracellular spacings (Shende et al. 2017; Das et al. 2012).

Another important aspect to mention is the bioactive compounds present in fungi. Mushrooms chemical structure comprises lectins, polysaccharides (β-glucans), polysaccharide-peptides, polysaccharide-protein complexes, lanostanoids, other terpenoids, alkaloids, sterols, and phenolic whereas yeasts and fungi includes terpenoids, alkaloids, quinones, xanthones, peptides, steroids, flavonoids, phenols, and phenolic compounds. As it can be observed, they basically have compounds in common. These bioactive compounds become part of the final nanosized product and therefore, provide of diverse biological activities (Srivastava 2019; Rhanjani et al. 2020; Sen et al. 2013).

### *12.3.4 Nanomaterials Synthesis Using Bacteria*

In this type of synthesis, cellular extracts from bacterial or biomass are used. Metal ions are hazardous to bacteria, and the bio-reduction of metal ions is regarded as a defense mechanism in the same way it occurs when fungi are used, as mentioned in the previous section. Their bioactive compounds are quite similar to the fungi ones (Mukherjee and Nethi 2019; Anthony et al. 2014). Table 12.2 lists examples of nanoparticles derived from microorganisms, mushrooms and plants to give the reader a general outlook. The biological source will refer specifically to the organism used.

## 12.4 Biomolecule-Derived Nanomaterials

A biomolecule is defined as any organic in a living cell which includes carbohydrates, proteins, fats, aminoacids, and so on (Datta et al. 2020) . This section will deal with the processes for nanomaterials synthesis and some of the involved mechanism that have been reported as well as the potential applications of these materials.

Biomolecules can be incorporated into nanomaterials to generate functional nanosystems with novel and advanced properties, presenting great potential for applications in several areas (Willner and Willner 2010; Phogat et al. 2018). Biomolecules are divided into four types: carbohydrates, lipids, proteins, and nucleic acids (Rubach et al. 2019).

### *12.4.1 Carbohydrates*

They are plentiful biomolecules, with an acute tendency to form supramolecular networks. They have been studied for long in the generation of nanomaterials, especially for biomedical applications as reviewed in the literature (Stylianopoulos 2013).

**Table 12.2** List of works related to plants, mushrooms and microorganisms

| Biological source | Nanoparticle type | Size (nm) and shape | Biological activity or application | Reference |
|---|---|---|---|---|
| *Penicillium oxalicum* (fungi) | Ag | 60–80 nm, spherical | Antibacterial | Feroze et al. (2020) |
| *Periconium* sp. (fungi) | ZnO | 40 nm, quasispherical | Antibacterial, antifungal and antioxidant | Ganesan et al. (2020) |
| *Musa acuminata colla* (fruit) | Ag and Au | 12.6–15.7 nm (Ag) 10.1–15.6 nm (Au) | Antibacterial, anticancer and antioxidant | Valsalam et al. (2019) |
| *Agaricus bisporus* and *Ganoderma lucidum* (mushrooms) | Ag | 10–80 nm | Antioxidant, antibacterial and anti-inflammatory | Sriramulu and Sumathi (2017) |
| *Vitex negundo* (shrub) | Au | 40–70 nm, spherical | Antioxidant and antibacterial | Veena et al. (2019) |
| *Acorus calamus* (herb) | Au | <100 nm | Antibacterial and UV-blocking | Ganesan and Gurumallesh Prabu (2019) |
| *Bacillus cereus* (bacteria) | $Fe_2O_3$ | 29.3 nm, spherical | Imaging agents | Fatemi et al. (2018) |
| *Botryococcus braunii* (algae) | Ag and Cu | 40–100 nm (Ag) and 10–70 nm (Cu) | Antifungal and antibacterial | Arya et al. (2018) |
| *Rhodotorula glutinis* and *Rhodotorula mucilaginosa* (yeasts) | Ag | 15.45 ± 7.94 nm (*R. glutinis*) and 13.70 ± 8.21 nm (*R. mucilaginosa*), pseudospherical | Antifungal | Cunha et al. (2018) |

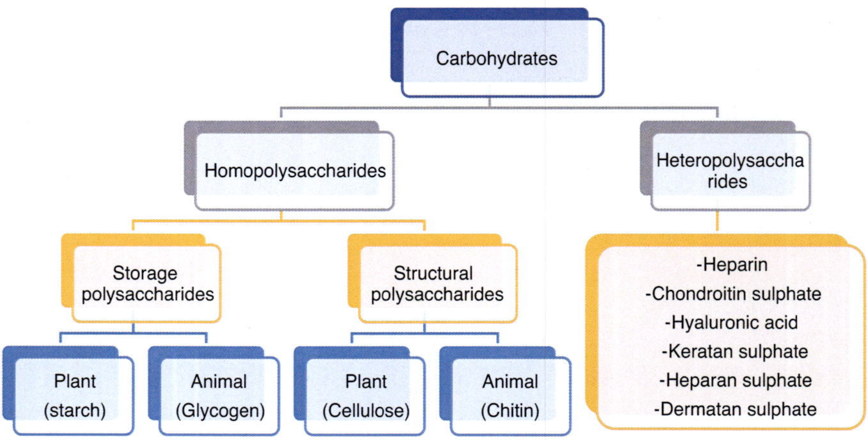


**Fig. 12.4** Classification of carbohydrates

Figure 12.4 shows a classification of carbohydrates aimed to give a hint on the wide range of materials that can be used in nanomaterials synthesis. All carbohydrates are abundant, biodegradable, and biocompatible. They are materials that can be easily modified by adding moieties to them (Gim et al. 2019).

Starch is a polymeric carbohydrate composed by numerous glucose units joined by glycosidic bonds; it also possesses a high amount of hydroxyl groups. It is produced by most green plants as energy storage and it is part of the diet of living beings (Stoddard 2016). There are reports on starch-based nanosystems applied to the manufacturing of coatings (it is commonly used already as a coating material because of its low-cost and its easiness of extraction from sources such as corn, potato, or rice) due to starch's ability to form films (Razavi and Amini 2016; Le Corre et al. 2010). A starch-based flexible coating enriched with ZnO nanoparticles was developed for food packaging papers with improved hydrophobicity and antimicrobial properties, by means of a casting method (Ni et al. 2018). Jung et al. synthesized a coating solution (starch and silver nitrate) using ultrasonication for its potential application in antibacterial packaging (Jung et al. 2018). These two works are interesting because they show potential for the quality enhancement of packagings in the food industry, there are also reports on the making of edible coatings to delay the ripening of fruits (Thakur et al. 2019) . Starch-based nanoparticles have been worked on as drug-delivery agents, the gelation methods were used to produce hollow starch nanoparticles first, then doxorubicin hydrochloride was used a test drug (30–300 nm, with a shell thickness of 5–10 nm). The findings in this research propose this system as being potentially promising for anticancer therapy, because they turned out to be highly cytotoxic to liver hepatocellular cells (Yang et al. 2017). *Triphala Churna*, a herb rich in polyphenols and vitamin C was encapsulated in nanoparticles, besides the demonstration of their good encapsulating efficiencies, these nanoproducts showed neuroprotective and antibacterial properties (Nallasamy et al. 2020). The Dispat group designed a composite system made up by Zinc oxide/TEOS modified starch (MS) via a sol-gel route; such nanocomposite displayed enhanced antibacterial activity, biodegradability, and reusability. Therefore, this material is a proposed to be beneficial as a soil conditioner aided against transient drought and as a growth promoter in the agriculture sector (Dispat et al. 2020).

Glycogen is a multibranched polysaccharide of glucose that serves as a form of energy storage in animals, fungi, and bacteria. Glycogen is found in the form of granules in the cytosol/cytoplasm in many cell types, and plays an important role in the glucose cycle and its structure is very similar to the amylopectine one (Roach and Zeeman 2016) . In the nanomaterials domain, it has been used in the synthesis of nanocomposites oriented to cancer theranostics (Gálisová et al. 2020). There are very few of works on them, most of them are reviews focused on their potential, synthesis protocols (Besford et al. 2020; Božanić et al. 2013; Shafiq et al. 2020; Tomanin et al. 2020).

Celullose is another carbohydrate that is used in nanomaterials synthesis. Its structure is a linear chain of several hundred to many thousands of β (1→4) linked D-glucose units. It is an important structural component of the primary cell wall of green plants, many forms of algae and of oomycetes. Cellulose is the most abundant organic polymer on Earth and is widely used in the textile industry (Hon 2001). Its application in the fabrication of nanomaterials possesses the following advantages: good mechanical properties due to their biodegradability, renewability, availability, sustainability, lower cost, lower weight, higher mechanical strength, biocompatibility, high hydrophilicity, and high surface area (Moon 2016; Mohammed 2018). As for reports on its usage in nanobiotechnological applications: functionalized water-soluble cellulose ester with bioactive steroids were used to synthesize cellulose conjugates through self-assembly to study their agrochemical activity or their plant growth stimulation activities which were found to be quite good (Pérez Quiñones et al. 2017) . Cellulose has also been coupled to nanoparticles for multifunctional purposes like in the case of Meng et al. whose $Cr_2O_3$/cellulose nanocomposite was synthesized by the ultrasonic method which exhibited a high responsive photocatalytic response that promotes the degradation of dyes in water, such outcome is encouraging because it is beneficial for water decontamination treatment (Lu et al. 2020). Some other works report similar findings (Li et al. 2017). Some more complex structures have been manufactured, for instance, an electrospun poly (lactic acid) fibrous scaffold was surface-modified with cellulose nanofibrils and Ag nanoparticles, afterwards conjunctival epithelial cell (CjECs) and corneal epithelial cells (CECs) were deposited on them. It was found that cell growth was promoted as well as antimicrobial activity. These composites could work as ocular bandages and as artificial ocular tissue equivalents for ocular wound healing (Yan et al. 2020) .Other fields of study in which cellulose-based nanomaterials are promising are tissue engineering (Hosseini et al. 2020; Zulkifli et al. 2017; Arias et al. 2018; Ju et al. 2020), drug delivery (Chin et al. 2018; Ching et al. 2019; Dai and Si 2017), pharmacology (Jatoi et al. 2019; Tsai et al. 2017; Smiechowicz et al. 2018) and in biosensing (Teodoro et al. 2019; Dong et al. 2012; Ahmadian et al. 2020).

Heparin, chondroitin sulfate, hyaluronic acid, keratan sulfate, heparan sulfate, and dermatan sulfate belong to the glycosaminoglycans (GAGs) family (Jones et al. 2010), as shown in Fig. 12.5. It is worth pointing out that not all glycosaminoglycans are used equally in nanomaterials production, but only some of them. They are interesting carbohydrates given the fact that they are produced within the body of

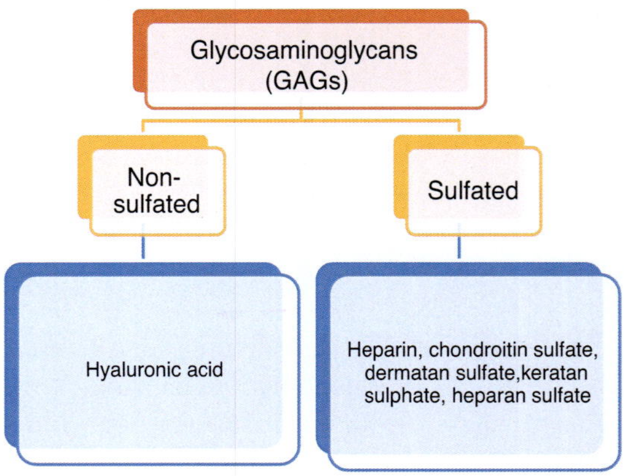


**Fig. 12.5** Glycosaminoglycan classification

many animals and the body of human beings and that because of their biocompatibility and negative charge that allows them to bind easily growth factors, proteases, chemokines, and anti-thrombin III (Gandhi and Mancera 2008), which makes them ideal candidates for applications related mainly to biomedicine (Scott and Panitch 2013) . Also, they add biomimetic traits to nanomaterials and improve their biofunctionality (Gandhi and Mancera 2008). As it can be seen in Table 12.3 the variety of materials listed in here range from nanoparticles (organic and inorganic), nanocomposites to scaffolds (Vassie et al. 2015). Literature on their usage in nanomaterials of diverse kinds is extensive, hence, a selection works was made.

### *12.4.2 Proteins, Lipids, and Nucleic Acids in Nanomaterials Production*

Proteins and lipids are examples of nutrients, molecules essential for the growth and development of life. They are organic substances, that have lots of carbon-hydrogen bonds. The difference is that lipids contain fatty acids and glycerol (Day 2016), while proteins contain amino acids, which have nitrogen (Tomii 2019) . Proteins structure is a 3D arrangement of atoms in an amino acid-chain molecule. Proteins are polymers known as polypeptides whose origin are sequences of amino acids, the monomers of the polymer (Sforza et al. 2016). On the other hand, lipids are defined as macrobiomolecules that are insoluble in water and soluble in nonpolar solvents. Unlike proteins, lipids are not polymers because they lack a repeating monomeric unit (Dupont 2005).

Nucleic acids are the biomolecules, essential to all living beings. The term nucleic acid is the overall name for DNA and RNA. Their structure comprises nucleotides, monomers made of three components: a 5-carbon sugar, a phosphate group and a nitrogenous base. If the sugar is a ribose, the polymer is RNA (ribonucleic acid); if the sugar comes from ribose as deoxyribose, the polymer referred to as DNA (Paleček and Bartošík 2019; Tinoco 2002).

Figure 12.6 shows in general terms the way each one of these biomolecules forms nanomaterials. Time, effort and budget have been invested in the exhaustive research of these three biomolecules with results that are promising, especially in the biomedical domain in the form of drug carriers, agents with varied biological activities and as biosensors. Figure shows the characteristics for assembly of each of one of them. Researchers take advantage of proteins and nucleic acids polymeric nature and their multiple functionalities in the design of nanomaterials (Xu et al. 2019; Diaz et al. 2018; García-Pinel et al. 2019).

Recombinant proteins with active domains and specific target peptide ligands have used for tumor targeting in cancer treatments research (Lv et al. 2020) .In other studies, such protein-based nanomaterials have been improved to use them in theranostics (Xie et al. 2019; Yang et al. 2016; Yu et al. 2018; Bruckman et al. 2016), in biosensing (Unser et al. 2017; Kairdolf et al. 2017; Lin et al. 2019; Chavan et al. 2019; Barbosa et al. 2018), in tissue engineering (Li et al. 2018; Azizian et al. 2018;

**Table 12.3** GAG-based nanomaterials examples

| Glycosaminoglycan | Nanomaterial type | Application | Reference |
|---|---|---|---|
| Hyaluronic acid | Honokiol by zein/hyaluronic acid core-shell *nanoparticles* | Breast cancer treatment due to the targeted delivery of honokiol that promotes tumor metastasis | Zhang et al. (2020) |
| Hyaluronic acid | PVA/hyaluronic acid/L-arginine *nanofibers* | Wound healing | Hussein et al. (2020) |
| Hyaluronic acid and chondroitin sulfate | Polymer *networks* of collagen, hyaluronic acid, and chondroitin sulfate | Brain tissue engineering (neurogenesis enhancement) | Li et al. (2020a, b) |
| Hyaluronic acid | *Scaffolds* of hyaluronic acid and collagen loaded with prednisolone | Osteoarthritis treatment | Tanideh et al. (2019) |
| Hyaluronic acid | Hyaluronic acid-based nanoparticles | *In vivo* imaging | Zhang et al. (2016) |
| Chondroitin sulfate | 3D structure of poly(D,L-lactic acid) loaded with chitosan/chondroitin sulfate *nanoparticles* | Drug delivery for tissue engineering | Santo et al. (2010) |
| Chondroitin sulfate | Gelatin-chondroitin sulfate/polycaprolactone nanofibrous *scaffolds* | Cartilage tissue engineering | Sharifi et al. (2020) |
| Chondroitin sulfate | Chondroitin sulfate-functionalized gold magnetic nanoparticles | Biosensing of farm animals pneumonia | Zhao et al. (2020) |
| Chondroitin sulfate | $TiO_2$–chondroitin-4-sulfate nanocomposites | Bone regeneration (enhanced osteocalcin production) | Kandiah et al. (2015) |
| Dermatan sulfate | Dermatan sulfate/chitosan nanoparticles | Therapy of vascular disease by the delivery of anti-inflammatory peptides with the nanoparticles | Blachman et al. (2020) |
| Dermatan sulfate | Dermatan sulfate-functionalized chitosan nanoparticles | Melanoma treatment (inhibition and lower myelosuppression rates) | Li et al. (2020a, b) |
| Heparin | Heparin-chitosan nanoparticles | Genetic drugs into epithelial cells (gene therapy) | Pilipenko et al. (2019) |
| Heparin | Heparin-based gold nanoparticles | Colorimetric sensing of antioxidant capacity measurement | Bener et al. (2018) |
| Heparin | Heparin/polypeptide nanoparticles | Anti-thrombus therapy | Chen et al. (2016) |
| Heparin | Heparin-based gold and silver nanoparticles | Antifungal therapy and uses | Rodriguez-Torres et al. (2020) |

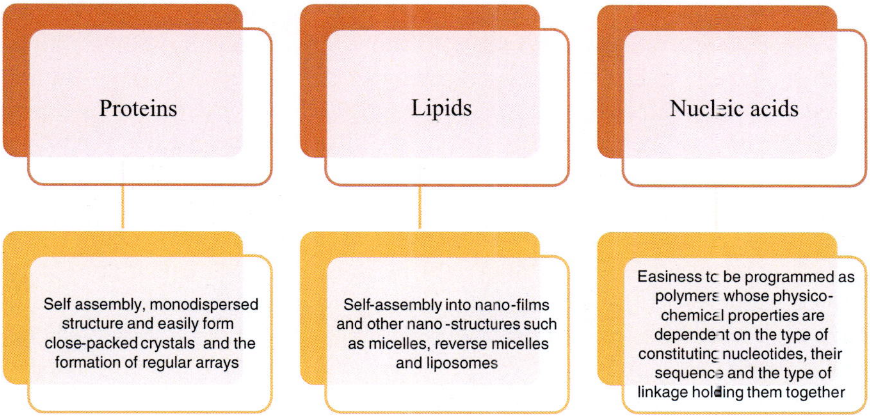


**Fig. 12.6** Scheme showing assemblage types for proteins, lipids, and nucleic acids in nanomaterial synthesis

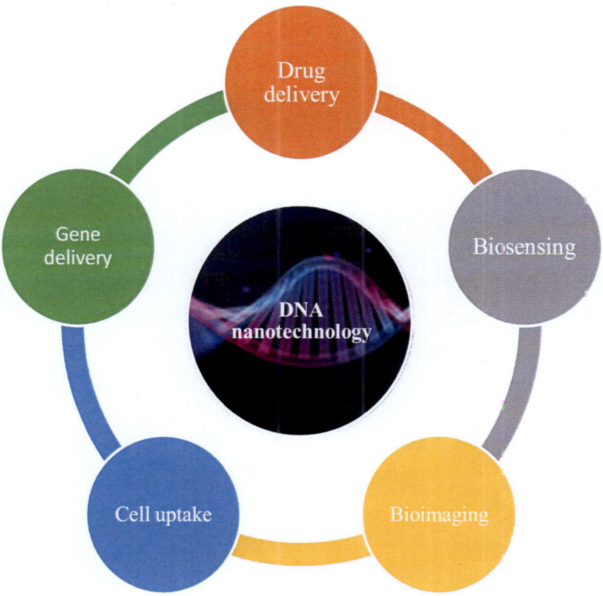


**Fig. 12.7** DNA nanotechnology applications

Brito-Pereira et al. 2018; Wang et al. 2017). In other biomedicine areas not related to cancer (Islam et al. 2019) and in potential food applications (Bourbon et al. 2019).

Nucleic acid-based nanomaterials are a special kind because of the highly specific Watson-Crick hydrogen bonding that allows properties and characteristics to be tuned as needed (Xu et al. 2020) The wide variety of nanostructures used in nanobiotechnology is huge, a whole book could be dedicated to this field of study. In consequence, only relevant and recent selected works related to nanobiotechnology will be covered for the sake of practicality.

Figure 12.7 shows diverse applications relevant to DNA nanotechnology. DNA-based nanomaterials have been manipulated with the purpose of attaining multiple functionalities. In the case of drug-delivery applications, there have been studies focused on ophthalmology using step-by-step synthesized systems, for example, beginning with lipid-modifies DNA strands functionalized nanoparticles, next by means of self-assembly such NPs are loaded with different drugs using aptamers (Willem de Vries et al. 2018), other examples of similar works include cancer (Yang et al. 2018; Li et al. 2017) and glaucoma treatment (Schnichels et al. 2020). Other type of DNA structure used for drug delivery is origami, which is an assembly approach that involves folding single-stranded DNA template molecules into target structures. These allow for the intercalating of drugs and the quick internalization by mammalian cells (Ge et al. 2018). These origami structures have proved to be excellent carriers that also overcome immediate drug resistance because of their enhanced binding (Halley et al. 2016).

When it comes to biosensing, DNA origami is also used, for example, in conjunction with organic fluorophores and the aid of fluorescence microscopy (Selnihhin et al. 2018). Other DNA platforms used for biosensing are hydrogels, the general mechanism is based on the trapping of a cargo using a synthetic polymer as the scaffold and DNA plays the role of the trigger agent; once a certain molecule triggers the dissociation of the hydrogel, the cargo is released generating a sensitive detection signal. Sun et al. prepared an aptamer-functionalized hydrogel for the detection of the T-2 toxin (Khajouei et al. 2020; Sun et al. 2020). Pandey developed a strong multifunctional platform that involved a luminescent hydrogel by the incorporation of DNA-derived carbon dots (Pandey et al. 2020), whose purpose was to work as a bioimaging agent and as a biosensor for dopamine.

Tetrahedral DNA (TDN) nanostructures have been reported as drug delivery and bioimaging systems for cancer therapy. These tetrahedral shapes are composed of only four oligonucleotides, that self-assemble into a well-defined tetrahedral shape (Duangrat et al. 2020). He et al. created a nanocomposite nanohydroxyapatite MRI Probe modified with tetrahedral DNA Nanostructures for tumor cell imaging in vitro. Adding the tetrahedral DNA structures enhanced the hydrophilicity and monodispersion of the nanohydroxyapatite which also resulted in the tumor targeting accessibility (He et al. 2020). Metal NPs have been coupled to DNA and PTEN (Phosphatase and tensin homolog) for targeted bioimaging as well as for cancer theranostics through up-regulation of tumor suppressor genes and downregulation of oncogenes (Wang et al. 2020). In situ monitoring of biological events in cells is also an opportunity area, branched catalytic hairpin assembly (bCHA)-reactions based nanomachine used for nucleic acid analysis were manufactured, too (Yue et al. 2019).

Applications in cell uptake are useful because in many situations it is necessary that certain molecules are delivered to cells. DNA does not only "take advantage of itself" as a programmable structure but also as a functionalizing agent: silica nanoparticles with a fluorescent dye within and functionalized with DNA oligonucleotides serve as high-affinity for MCF7 and A431 cancer cells, this outcome is interesting because no extra ligands were incorporated to the structures, also, these nanoparticles do not exhibit cytotoxicity which could be beneficial in cell signaling applications (Leidner et al. 2018). Other DNA origami studies on cell uptake can be consulted by the reader (Wang et al. 2018).

As for gene delivery, DNA-based nanomaterials have been developed to treat cancer and deliver therapeutic genes to multidrug resistant tumors (Liu et al. 2018), to be used as DNA vaccines (Ali et al. 2017), as a part of a multiblock system (Amani et al. 2019) and as DNA liposomes systems for transfection (Rasoulianboroujeni et al. 2017).

### *12.4.3 Biowaste as a Source of Nanomaterials*

Biowaste or biodegradable waste is mainly made up of organic materials including green, food, paper waste, and even biodegradable plastics. Biowaste comes from households and industry and consists of organic materials that generate carbon dioxide, water, methane or other organic molecules. As biowaste materials are low-cost resources and often considered as suitable for nanomaterials synthesis (Ashok kumar et al. 2013; Samaddar et al. 2018).

These biowaste derived nanomaterials have also found applications in the nano-biotechnology field. Nanocarbon structures have been obtained from the *Groundnut* shell with catalyst-free pyrolysis approach at different temperatures. The obtained products exhibited excellent antibacterial activity against Gram-positive and Gram-negative bacteria (Yallappa et al. 2017). Expended sugarcane bagasse has been used to synthesize graphene quantum dots interlaced with $SnO_2$ nanoparticles, with potential application as disinfectants given their long-term stability and great anti-bacterial activity against *Pseudomonas aeruginosa* (Mohan and Manoj 2019). Other biowaste-based quantum carbon dots have been produced for bioimaging (Shi et al. 2019; de Yro et al. 2019). Nanohydroxyapatite has also been acquired from mussel shell or eggshells through microwave irradiation (Kumar et al. 2017; Muthu et al. 2020), from tilapia fish scale using ultrasound (Sricharoen et al. 2020) for biomedical applications such as bone regeneration and implantology. Antibacterial metal nanoparticles have also been obtained by treating chemically fruit peels (R, J, A.S, K.P.B, & B, 2019). The aforementioned applications are limited, most of the reports on biowaste derived nanomaterials focus on renewable energies applications (Goswami et al. 2019; Sriram et al. 2019), wastewater treatment (Nguyen et al. 2020) and on synthesis protocols (Saeb et al. 2018).

## 12.5 Outlook and Future Aspects

Even though research on bio-inspired nanomaterials is actually at a really early stage, there a niche of opportunity for it to be scaled at industrial levels. Some of the nanomaterials covered in this chapter have some difficulties to be scalable to industrial levels, due to the difficulty to reproduce them like in the case of materials derived from plants and mushrooms because there are factors like environmental stress, season of the year and significant differences among the same batch of sources which alters the bioactive compounds action. In this case, a possible approach can be the synthetic production of compounds with similar characteristics

**Table 12.4** List of some nanobiotechnology-oriented companies in the world

| Company | Country | Research area/development in nanobiotechnology | Link or contact |
|---|---|---|---|
| Nanoretina | Israel | Devices intended to restore functional vision to the blind | https://www.nano-retina.com |
| Nanobridging molecules | Switzerland | Surface treatment for bone anchored implants | http://www.nbmolecules.com |
| Nanopharma | Czech Republic | Medical devices | https://www.nanopharma.cz |
| Nanoprobes, Inc. | USA | Nanoparticle immunogold labeling and immunoassay tests | https://www.nanoprobes.com |
| Nanobacterie | France | Cancer treatment platforms | http://www.nanobacterie.fr |

in the way it is done with the production of aspirin (Montinari, Minelli and De Caterina 2019).

Even though nanomaterials are still studied and not yet permitted for general use, good news is that there are formulations already approved by the FDA (Anselmo and Mitragotri 2019), there are not thousands of them, but still there is much hope for the effort and work invested in nanotechnology, they are mostly in the category of liposomes, micelles, nanocrystals and inorganic, polymeric and inorganic nanoparticles for the treatment of cancer, and fungal infections, as psychostimulants, antiemetics, bone substitutes, muscle relaxants, and more (Ventola 2017).

In a like manner, nanobiotechnology companies have been founded, particularly in developed countries and most of them are startups. Table 12.4 shows a brief list of such companies for reference.

The nanomaterials presented in this chapter are intended only for nanobiotechnology potential applications. Furthermore, information related to the field is presented for giving the reader a wider outlook on the potential horizons and possibilities that can be attained.

**Acknowledgments** *Maria del Pilar Rodriguez-Torres* wishes special Thanks to National Council for Science and Technology (CONACYT) for the postdoctoral scholarship granted.

**Conflicts of Interest** The authors declare no conflicts of interest for this chapter publication.

## References

Ahmadian-Fard-Fini, S., Ghanbari, D., Amiri, O., & Salavati-Niasari, M. (2020). Electrospinning of cellulose acetate nanofibers/Fe/carbon dot as photoluminescence sensor for mercury (II) and lead (II) ions. *Carbohydrate Polymers, 229*, 115428. https://doi.org/10.1016/j.carbpol.2019.115428.

Ali, A. A., McCrudden, C. M., McCaffrey, J., McBride, J. W., Cole, G., Dunne, N. J., et al. (2017). DNA vaccination for cervical cancer; a novel technology platform of RALA mediated gene delivery via polymeric microneedles. *Nanomedicine: Nanotechnology, Biology and Medicine, 13*(3), 921–932. https://doi.org/10.1016/j.nano.2016.11.019.

Amani, A., Kabiri, T., Shafiee, S., & Hamidi, A. (2019). Preparation and characterization of PLA-PEG-PLA/PEI/DNA nanoparticles for improvement of transfection efficiency and controlled release of DNA in gene delivery systems. *Iranian Journal of Pharmaceutical Research : IJPR, 18*(1), 125–141. Retrieved from https://pubmed.ncbi.nlm.nih.gov/31089350.

Amini,A.M., Razavi, S.M.A. (2016). A fast and efficient approach to prepare starch nanocrystals from normal corn starch, *Food Hydrocolloids 57*, 132–138. https://doi.org/10.1016/j.foodhyd.2016.01.022.

Anselmo, A. C., & Mitragotri, S. (2019). Nanoparticles in the clinic: An update. *Bioengineering & Translational Medicine, 4*(3), e10143. https://doi.org/10.1002/btm2.10143.

Anthony, K., Murugan, M., Jeyaraj, M., Rathinam, N., & Sangiliyandi, G. (2014). Synthesis of silver nanoparticles using pine mushroom extract: A potential antimicrobial agent against *E. coli* and *B. subtilis*. *Journal of Industrial and Engineering Chemistry, 20*, 2325–2331. https://doi.org/10.1016/J.JIEC.2013.10.008.

Arias, S. L., Shetty, A., Devorkin, J., & Allain, J. P. (2018). Magnetic targeting of smooth muscle cells in vitro using a magnetic bacterial cellulose to improve cell retention in tissue-engineering vascular grafts. *Acta Biomaterialia, 77*, 172–181. https://doi.org/10.1016/j.actbio.2018.07.013.

Arya, A., Gupta, K., Chundawat, T. S., & Vaya, D. (2018). Biogenic synthesis of copper and silver nanoparticles using green alga Botryococcus braunii and its antimicrobial activity. *Bioinorganic Chemistry and Applications, 2018*, 7879403. https://doi.org/10.1155/2018/7879403.

Azizian, S., Hadjizadeh, A., & Niknejad, H. (2018). Chitosan-gelatin porous scaffold incorporated with Chitosan nanoparticles for growth factor delivery in tissue engineering. *Carbohydrate Polymers, 202*, 315–322. https://doi.org/10.1016/j.carbpol.2018.07.023.

Balaganesh, A. S., Pavithra, N., RanjitKumar, R., Dinesh, K. P. B., & Chandar, S. B. (2019). Bio-Assisted Synthesis Of Potassium Doped Ferric Sulphide Nanoparticles For Agricultural Applications. *Kongunadu Research Journal, 6*(2). https://doi.org/10.26524/krj307.

Barbosa, A. J. M., Oliveira, A. R., & Roque, A. C. A. (2018). Protein- and Peptide-Based Biosensors in Artificial Olfaction. *Trends in Biotechnology, 36*(12), 1244–1258. https://doi.org/10.1016/j.tibtech.2018.07.004.

Bener, M., Şen, F. B., & Apak, R. (2018). Heparin-stabilized gold nanoparticles-based CUPRAC colorimetric sensor for antioxidant capacity measurement. *Talanta, 187*, 148–155. https://doi.org/10.1016/j.talanta.2018.05.021.

Besford, Q. A., Cavalieri, F., & Caruso, F. (2020). Glycogen as a building block for advanced biological materials. *Advanced Materials, 32*, 1904625. https://doi.org/10.1002/adma.201904625.

Blachman, A., Funez, F., Birocco, A. M., Saavedra, S. L., Lázaro-Martinez, J. M., Camperi, S. A., et al. (2020). Targeted anti-inflammatory peptide delivery in injured endothelial cells using dermatan sulfate/chitosan nanomaterials. *Carbohydrate Polymers, 230*, 115610. https://doi.org/10.1016/j.carbpol.2019.115610.

Bourbon, A. I., Martins, J. T., Pinheiro, A. C., Madalena, D. A., Marques, A., Nunes, R., & Vicente, A. A. (2019). 6—Nanoparticles of lactoferrin for encapsulation of food ingredients. In S. M. Jafari (Ed.), *Biopolymer nanostructures for food encapsulation purposes* (pp. 147–168). Cambridge: Academic. https://doi.org/10.1016/B978-0-12-815663-6.00006-9.

Božanić, D. K., Luyt, A. S., Trandafilovića, I. V., & Djoković, V. (2013). Glycogen and gold nanoparticlebioconjugates: controlled plasmon resonance viaglycogen-induced nanoparticle aggregation. *RSC Advances, 3*, 8705–8713. https://doi.org/10.1039/C3RA40189H.

Brito-Pereira, R., Correia, D. M., Ribeiro, C., Francesko, A., Etxebarria, I., Pérez-Álvarez, L., Vilas, J. L., Martins, P., & Lanceros-Mendez, S. (2018). Silk fibroin-magnetic hybrid composite electrospun fibers for tissue engineering applications. *Composites Part B: Engineering, 141*, 70–75. https://doi.org/10.1016/j.compositesb.2017.12.046.

Bruckman, M. A., Czapar, A. E., VanMeter, A., Randolph, L. N., & Steinmetz, N. F. (2016). Tobacco mosaic virus-based protein nanoparticles and nanorods for chemotherapy delivery targeting breast cancer. *Journal of Controlled Release, 231*, 103–113. https://doi.org/10.1016/j.jconrel.2016.02.045.

Buzea C., Pacheco I. (2017) Nanomaterials and their Classification. In: Shukla A. (eds) EMR/ESR/EPR Spectroscopy for Characterization of Nanomaterials. *Advanced Structured Materials, vol 62*. Springer, New Delhi. https://doi.org/10.1007/978-81-322-3655-9_1.

Chauhan, D., Kumar, R., Panda, A. K., & Solanki, P. R. (2019). An efficient electrochemical biosensor for vitamin-D3 detection based on aspartic acid functionalized gadolinium oxide nanorods. *Journal of Materials Research and Technology, 8*(6), 5490–5503. https://doi.org/10.1016/j.jmrt.2019.09.017.

Chen, C., Li, S., Liu, K., Ma, G., & Yan, X. (2016). Co-assembly of heparin and polypeptide hybrid nanoparticles for biomimetic delivery and anti-thrombus therapy. *Small, 34*, 4719–4725. https://doi.org/10.1002/smll.201600328.

Chin, S. F., Jimmy, F. B., & Pang, S. C. (2018). Size controlled fabrication of cellulose nanoparticles for drug delivery applications. *Journal of Drug Delivery Science and Technology, 43*, 262–266. https://doi.org/10.1016/j.jddst.2017.10.021.

Ching, Y. C., Gunathilake, T. M. S. U., Chuah, C. H., Ching, K. Y., Singh, R., & Liou, N. S. (2019). Curcumin/Tween 20-incorporated cellulose nanoparticles with enhanced curcumin solubility for nano-drug delivery: Characterization and in vitro evaluation. *Cellulose, 26*, 5467–5481. https://doi.org/10.1007/s10570-019-02445-6.

Cunha, F. A., Cunha, M. D. C. S. O., da Frota, S. M., Mallmann, E. J. J., Freire, T. M., Costa, L. S., et al. (2018). Biogenic synthesis of multifunctional silver nanoparticles from Rhodotorula glutinis and Rhodotorula mucilaginosa: Antifungal, catalytic and cytotoxicity activities. *World Journal of Microbiology and Biotechnology, 34*(9), 127. https://doi.org/10.1007/s11274-018-2514-8.

Darweesh, H. H. M. (2018). Nanomaterials: Classification and properties-part I. *Nanoscience, 1*(1), 1–11. https://doi.org/10.31058/j.nano.2018.11001.

Dai, L., Liu, R., Hu, L.-Q., & Si, C.-L. (2017). Simple and green fabrication of AgCl/Ag-cellulose paper with antibacterial and photocatalytic activity. *Carbohydrate Polymers, 174*, 450–455. https://doi.org/10.1016/j.carbpol.2017.06.107.

Das, S. K., Khan, M., Motiar, R., Guha, A. K., Das, A. R., & Mandal, A. B. (2012). Silver-nano biohybride material: Synthesis, characterization and application in water purification. *Bioresource Technology, 124*, 495–499. https://doi.org/10.1016/j.biortech.2012.08.071.

Datta, L. P., Manchineella, S., & Govindaraju, T. (2020). Biomolecules-derived biomaterials. *Biomaterials, 230*, 119633. https://doi.org/10.1016/j.biomaterials.2019.119633.

Day, L. (2016). Lipid chemistry. In C. Wrigley, H. Corke, K. Seetharaman, & J. Faubion (Eds.), *Encyclopedia of food grains* (2nd ed., pp. 248–256). Oxford: Academic Press. https://doi.org/10.1016/B978-0-12-394437-5.00099-1.

de Yro, P. A. N., Quaichon, G. M. O., Cruz, R. A. T., Emolaga, C. S., Que, M. C. O., Magdaluyo, E. R., Basilia, B. A., & Agarwal, R. K. (2019). Hydrothermal synthesis of carbon quantum dots from biowaste for bio-imaging. *AIP Conference Proceedings, 2083*(1), 20007. https://doi.org/10.1063/1.5094310.

Diaz, D., Care, A., & Sunna, A. (2018). Bioengineering strategies for Protein-Based Nanoparticles. *Genes, 9*(7), 370. https://doi.org/10.3390/genes9070370.

Dispat, N., Poompradub, S., & Kiatkamjornwong, S. (2020). Synthesis of ZnO/SiO2-modified starch-graft-polyacrylate superabsorbent polymer for agricultural application. *Carbohydrate Polymers, 249*, 116862. https://doi.org/10.1016/j.carbpol.2020.116862.

Dong, S., Hirani, A. A., Colacino, K. R., Lee, Y. W. O. O., & Roman, M. (2012). Cytotoxicity and cellular uptake of cellulose nanocrystals. *Nano Life, 02*(03), 1241006. https://doi.org/10.1142/S1793984412410061.

Drexler, K. E. (2004). Nanotechnology: From feynman to funding. *Bulletin of Science, Technology & Society, 24*(1), 21–27. https://doi.org/10.1177/0270467604263113.

Duangrat, R., Udomprasert, A., & Kangsamaksin, T. (2020). Tetrahedral DNA nanostructures as drug delivery and bioimaging platforms in cancer therapy. *Cancer Science, 111*(9), 3164–3173. https://doi.org/10.1111/cas.14548.

Dupont, J. L. (2005). LIPIDS | chemistry and classification. In B. Caballero (Ed.), *Encyclopedia of human nutrition* (2nd ed., pp. 126–132). Oxford: Elsevier. https://doi.org/10.1016/B0-12-226694-3/00192-7.

Fakruddin, M., Hossain, Z., & Afroz, H. (2012). Prospects and applications of nanobiotechnology: A medical perspective. *Journal of Nanobiotechnology, 10*(31). https://doi.org/10.1186/1477-3155-10-31.

Fatemi, M., Mollania, N., Momeni-Moghaddam, M., & Sadeghifar, F. (2018). Extracellular biosynthesis of magnetic iron oxide nanoparticles by Bacillus cereus strain HMH1: Characterization and in vitro cytotoxicity analysis on MCF-7 and 3T3 cell lines. *Journal of Biotechnology, 270*, 1–11. https://doi.org/10.1016/j.jbiotec.2018.01.021.

Feroze, N., Arshad, B., Younas, M., Afridi, M. I., Saqib, S., & Ayaz, A. (2020). Fungal mediated synthesis of silver nanoparticles and evaluation of antibacterial activity. *Microscopy Research and Technique, 83*(1), 72–80. https://doi.org/10.1002/jemt.23390.

Gálisová, A., Jirátová, M., Rabyk, M., Sticová, E., Hájek, M., Hrubý, M., & Jirák, D. (2020). Glycogen as an advantageous polymer carrier in cancer theranostics: Straightforward in vivo evidence. *Scientific Reports, 10*(1), 10411. https://doi.org/10.1038/s41598-020-67277-y.

Gandhi, N. S., & Mancera, R. L. (2008). The structure of glycosaminoglycans and their interactions with proteins. *Chemical Biology & Drug Design, 72*(6), 455–482. https://doi.org/10.1111/j.1747-0285.2008.00741.x.

Ganesan, R. M., & Gurumallesh Prabu, H. (2019). Synthesis of gold nanoparticles using herbal Acorus calamus rhizome extract and coating on cotton fabric for antibacterial and UV blocking applications. *Arabian Journal of Chemistry, 12*(8), 2166–2174. https://doi.org/10.1016/j.arabjc.2014.12.017.

Ganesan, V., Hariram, M., Vivekanandhan, S., & Muthuramkumar, S. (2020). Periconium sp. (endophytic fungi) extract mediated sol-gel synthesis of ZnO nanoparticles for antimicrobial and antioxidant applications. *Materials Science in Semiconductor Processing, 105*, 104739. https://doi.org/10.1016/j.mssp.2019.104739.

García-Pinel, B., Porras-Alcalá, C., Ortega-Rodríguez, A., Sarabia, F., Prados, J., Melguizo, C., & López-Romero, J. M. (2019). Lipid-based Nanoparticles: Application and Recent Advances in Cancer Treatment. *Nanomaterials, 9*(4), 638. https://doi.org/10.3390/nano9040638.

Ge, Z., Gu, H., Li, Q., & Fan, C. (2018). Concept and development of framework nucleic acids. *Journal of the American Chemical Society, 140*(51), 17808–17819. https://doi.org/10.1021/jacs.8b10529.

Gim, S., Zhu, Y., Seeberger, P. H., & Delbianco, M. (2019). Carbohydrate-based nanomaterials for biomedical applications. *WIREs Nanomedicine and Nanobiotechnology, 11*(5), e1558. https://doi.org/10.1002/wnan.1558.

Goswami, S., Dillip, G. R., Nandy, S., Banerjee, A. N., Pimentel, A., Joo, S. W., et al. (2019). Biowaste-derived carbon black applied to polyaniline-based high-performance supercapacitor microelectrodes: Sustainable materials for renewable energy applications. *Electrochimica Acta, 316*, 202–218. https://doi.org/10.1016/j.electacta.2019.05.133.

Gour, A., Jain, N. K. (2019). Advances in green synthesis of nanoparticles, *Artificial Cells, Nanomedicine, and Biotechnology, 47*(1), 844–851, https://doi.org/10.1080/21691401.2019.1577878.

Halkai, K. R., Mudda, J. A., Shivanna, V., Rathod, V., & Halkai, R. S. (2017). Biosynthesis, characterization and antibacterial efficacy of silver nanoparticles derived from endophytic fungi against P. gingivalis. *Journal of Clinical and Diagnostic Research : JCDR, 11*(9), ZC92–ZC96. https://doi.org/10.7860/JCDR/2017/29434.10681.

Halley, P. D., Lucas, C. R., McWilliams, E. M., Webber, M. J., Patton, R. A., Kural, C., et al. (2016). Daunorubicin-loaded DNA origami nanostructures circumvent drug-resistance mechanisms in a leukemia model. *Small, 12*(3), 308–320. https://doi.org/10.1002/smll.201502118.

Hasan, A., Morshed, M., Memic, A., Hassan, S., Webster, T. J., & Marei, H. E. (2018). Nanoparticles in tissue engineering: Applications, challenges and prospects. *International Journal of Nanomedicine, 13*, 5637–5655. https://doi.org/10.2147/IJN.S153758.

He, Y., Lv, C., Hou, X., & Wu, L. (2020). Mono-dispersed nano-hydroxyapatite based MRI probe with tetrahedral DNA nanostructures modification for in vitro tumor cell imaging. *Analytica Chimica Acta, 1138*, 141–149. https://doi.org/10.1016/j.aca.2020.09.006.

Hon, D. N. S., & Yan, H. (2001). Cellulose furoate. I. Synthesis in homogeneous and heterogeneous systems. *Journal of Applied Polymer Science, 81*, 2649–2655. https://doi.org/10.1002/app.1709.

Hosseini, H., Zirakjou, A., Goodarzi, V., Mousavi, S. M., Khonakdar, H. A., & Zamanlui, S. (2020). Lightweight aerogels based on bacterial cellulose/silver nanoparticles/polyaniline with tuning morphology of polyaniline and application in soft tissue engineering. *International Journal of Biological Macromolecules, 152*, 57–67. https://doi.org/10.1016/j.ijbiomac.2020.02.095.

Hussein, Y., El-Fakharany, E. M., Kamoun, E. A., Loutfy, S. A., Amin, R., Taha, T. H., et al. (2020). Electrospun PVA/hyaluronic acid/L-arginine nanofibers for wound healing applications: Nanofibers optimization and in vitro bioevaluation. *International Journal of Biological Macromolecules, 164*, 667–676. https://doi.org/10.1016/j.ijbiomac.2020.07.126.

Iravani, S. (2011). Green synthesis of metal nanoparticles using plants. *Green Chemistry, 13*, 2638–2650. https://doi.org/10.1039/c1gc15386b.

Islam, M. S., Reineke, J., Kaushik, R., Woyengo, T., Baride, A., Alqahtani, M. S., & Perumal, O. (2019). Bioadhesive food protein nanoparticles as pediatric oral drug delivery system. *ACS Applied Materials & Interfaces, 11*(20), 18062–18073. https://doi.org/10.1021/acsami.9b00152.

Jatoi, A. W., Kim, I. S., & Ni, Q. Q. (2019). Cellulose acetate nanofibers embedded with AgNPs anchored TiO2 nanoparticles for long term excellent antibacterial applications. *Carbohydrate Polymers, 207*, 640–649. https://doi.org/10.1016/j.carbpol.2018.12.029.

Jeevanandam, J., Barhoum, A., Chan, Y. S., Dufresne, A., & Danquah, M. K. (2018). Review on nanoparticles and nanostructured materials: History, sources, toxicity and regulations. *Beilstein Journal of Nanotechnology, 9*, 1050–1074. https://doi.org/10.3762/bjnano.9.98.

Jones, C. L., Liu, J., & Xu, D. (2010). 6.14—Structure, biosynthesis, and function of glycosaminoglycans. In H.-W. (Ben) Liu & L. Mander (Eds.), *Comprehensive natural products II* (pp. 407–427). Oxford: Elsevier. https://doi.org/10.1016/B978-008045382-8.00132-5.

Jung, J., Raghavendra, G. M., Kim, D., & Seo, J. (2018). One-step synthesis of starch-silver nanoparticle solution and its application to antibacterial paper coating. *International Journal of Biological Macromolecules, 107*, 2285–2290. https://doi.org/10.1016/j.ijbiomac.2017.10.108.

Kairdolf, B. A., Qian, X., & Nie, S. (2017). Bioconjugated nanoparticles for biosensing, in vivo imaging, and medical diagnostics. *Analytical Chemistry, 89*(2), 1015–1031. https://doi.org/10.1021/acs.analchem.6b04873.

Kandiah, K., Venkatachalam, R., Wang, C., Valiyaveettil, S., & Ganesan, K. (2015). In vitro and preliminary in vivo toxicity screening of high-surface-area TiO2–chondroitin-4-sulfate nanocomposites for bone regeneration application. *Colloids and Surfaces B: Biointerfaces, 128*, 347–356. https://doi.org/10.1016/j.colsurfb.2015.02.027.

Khajouei, S., Ravan, H., & Ebrahimi, A. (2020). DNA hydrogel-empowered biosensing. *Advances in Colloid and Interface Science, 275*, 102060. https://doi.org/10.1016/j.cis.2019.102060.

Khan, I., Saeed, K., & Khan, I. (2018). Nanoparticles: Properties, applications and toxicities. *Arabian Journal of Chemistry, 12*(7), 908–931. https://doi.org/10.1016/j.arabjc.2017.05.011.

Kolahalam, L. A., Kasi Viswanath, I. V., Diwakar, B. S., Govindh, B., Reddy, V., & Murthy, Y. L. N. (2019). Review on nanomaterials: Synthesis and applications. *Materials Today: Proceedings, 18*, 2182–2190. https://doi.org/10.1016/j.matpr.2019.07.371.

Kumar, B. Y. S., Isloor, A. M., Kumar, G. C. M., & Inamuddin, & Asiri, A. M. (2019). Nanohydroxyapatite reinforced chitosan composite hydrogel with tunable mechanical and biological properties for cartilage *Regeneration. Scientific Reports, 9*(1), 15957. https://doi.org/10.1038/s41598-019-52042-7.

Kuppusamy, P., Yusoff, M. M., Maniam, G. P., & Govindan, N. (2016). Biosynthesis of metallic nanoparticles using plant derivatives and their new avenues in pharmacological applications - An updated report. *Saudi pharmaceutical journal : SPJ : the official publication of the Saudi Pharmaceutical Society, 24*(4), 473–484. https://doi.org/10.1016/j.jsps.2014.11.013.

Le Corre, D., Bras, J., & Dufresne, A. (2010). Starch nanoparticles: A review. *Biomacromolecules, 11*(5), 1139–1153. https://doi.org/10.1021/bm901428y.

Leidner, A., Weigel, S., Bauer, J., Reiber, J., Angelin, A., Grösche, M., et al. (2018). Biopebbles: DNA-functionalized core–shell silica nanospheres for cellular uptake and cell guidance studies. *Advanced Functional Materials, 28*(18), 1707572. https://doi.org/10.1002/adfm.201707572.

Li, C., Armstrong, J. P. K., Pence, I. J., Kit-Anan, W., Puetzer, J. L., Correia Carreira, S., Moore, A. C., & Stevens, M. M. (2018). Glycosylated superparamagnetic nanoparticle gradients for osteochondral tissue engineering. *Biomaterials, 176*, 24–33 https://doi.org/10.1016/j.biomaterials.2018.05.029.

Li, G., Li, Y., Wang, Z., & Liu, H. (2017). Green synthesis of palladium nanoparticles with carboxymethyl cellulose for degradation of azo-dyes. *Materials Chemistry and Physics, 187*, 133–140. https://doi.org/10.1016/j.matchemphys.2016.11.057.

Li, F., Ducker, M., Sun, B., Szele, F. G., & Czernuszka, J. T. (2020a). Interpenetrating polymer networks of collagen, hyaluronic acid, and chondroitin sulfate as scaffolds for brain tissue engineering. *Acta Biomaterialia, 112*, 122–135. https://doi.org/10.1016/j.actbio.2020.05.042.

Li, S., Zhang, F., Yu, Y., & Zhang, Q. (2020b). A dermatan sulfate-functionalized biomimetic nanocarrier for melanoma targeted chemotherapy. *Carbohydrate Polymers, 235*, 115983. https://doi.org/10.1016/j.carbpol.2020.115983.

Lin, D., Pillai, R. G., Lee, W. E., & Jemere, A. B. (2019). An impedimetric biosensor for E. coli O157:H7 based on the use of self-assembled gold nanoparticles and protein G. *Microchimica Acta, 186*, 169. https://doi.org/10.1007/s00604-019-3282-3.

Liu, J., Song, L., Liu, S., Jiang, Q., Liu, Q., Li, N., et al. (2018). A DNA-based nanocarrier for efficient gene delivery and combined cancer therapy. *Nano Letters, 18*(6), 3328–3334. https://doi.org/10.1021/acs.nanolett.7b04812.

Lu, M., Cui, Y., Zhao, S., & Fakhri, A. (2020). Cr2O3/cellulose hybrid nanocomposites with unique properties: Facile synthesis, photocatalytic, bactericidal and antioxidant application. *Journal of Photochemistry and Photobiology B: Biology, 205*, 111842. https://doi.org/10.1016/j.jphotobiol.2020.111842.

Lv, X., Zhang, C., Shuaizhen, Q., Yu, R., & Zheng, Y. (2020). Design of integrin αvβ3 targeting self-assembled protein nanoparticles with RGD peptide. *Biomedicine & Pharmacotherapy, 128*, 110236. https://doi.org/10.1016/j.biopha.2020.110236.

Mappes, T., Jahr, N., Csaki, A., Vogler, N., Popp, J., & Fritzsche, W. (2012). The invention of immersion ultramicroscopy in 1912—The birth of nanotechnology? *Angewandte Chemie International Edition, 51*(45), 11208–11212. https://doi.org/10.1002/anie.201204688.

Mason, D. S. (2005). The world according to nanotechnology. *Journal of Chemical Education, 82*(5), 665. https://doi.org/10.1021/ed082p665.

Mittal, A. K., Chisti, Y., & Banerjee, U. C. (2013). Synthesis of metallic nanoparticles using plant extracts. *Biotechnology Advances, 31*(2), 346–356. https://doi.org/10.1016/j.biotechadv.2013.01.003.

Mohammadi, F., Tanideh, N., Mohammadi Samani, S., & Ahmadi, F. (2019). Efficacy of a hybrid system of hyaluronic acid and collagen loaded with prednisolone and TGF-β3 for cartilage regeneration in rats. *Journal of Drug Delivery Science and Technology, 51*, 55–62. https://doi.org/10.1016/j.jddst.2019.02.016.

Mohammadi, F., Tanideh, N., Mohammadi Samani, S., & Ahmadi, F. (2019). Efficacy of a hybrid system of hyaluronic acid and collagen loaded with prednisolone and TGF-β3 for cartilage regeneration in rats. *Journal of Drug Delivery Science and Technology, 51*, 55–62. https://doi.org/10.1016/j.jddst.2019.02.016.

Mohammadinejad, R., Karimi, S., Iravani, S., & Varma, R. S. (2016). Plant-derived nanostructures: Types and applications. *Green Chemistry, 18*, 20–52. https://doi.org/10.1039/C5GC01403D.

Mohammed, N., Grishkewicha, N., & Tam, K. C. (2018). Cellulose nanomaterials: Promising sustainable nanomaterials for application in water/wastewater treatment processes. *Environmental Science: Nano, 5*, 623–658. https://doi.org/10.1039/C7EN01029J.

Mohan, A. N., & Manoj, B. (2019). Biowaste derived graphene quantum dots interlaced with SnO2 nanoparticles—A dynamic disinfection agent against Pseudomonas aeruginosa. *New Journal of Chemistry, 43*(34), 13681–13689. https://doi.org/10.1039/C9NJ00379G.

Montinari, M. R., Minelli, S., & De Caterina, R. (2019). The first 3500 years of aspirin history from its roots—A concise summary. *Vascular Pharmacology, 113*, 1–8. https://doi.org/10.1016/j.vph.2018.10.008.
Moon, R. J., Schueneman, G. T., & Simonsen, J. (2016). Overview of cellulose nanomaterials, their capabilities and applications. *JOM, 68*, 2383–2394. https://doi.org/10.1007/s11837-016-2018-7.
Mukherjee, S., Nethi, S. K., & Patra, C. R. (2017). *Particulate technology for delivery of therapeutics* (pp. 359–393). Singapore. Green synthesized gold nanoparticles for future biomedical applications: Springer.
Muthu, D., Kumar, G. S., Kattimani, V. S., Viswabaskaran, V., & Girija, E. K. (2020). Optimization of a lab scale and pilot scale conversion of eggshell biowaste into hydroxyapatite using microwave reactor. *Ceramics International, 46*(16). https://doi.org/10.1016/j.ceramint.2020.06.288.
Nallasamy, P., Ramalingam, T., Nooruddin, T., Shanmuganathan, R., Arivalagan, P., & Natarajan, S. (2020). Polyherbal drug loaded starch nanoparticles as promising drug delivery system: Antimicrobial, antibiofilm and neuroprotective studies. *Process Biochemistry, 92*, 355–364. https://doi.org/10.1016/j.procbio.2020.01.026.
Nguyen, V. P., Nguyen, K. T. T., Ton, L. T., Nguyen, D. T., Nguyen, K. Q., Vu, M. T., & Tran, H. N. (2020). Dual-electronic nanomaterial (synthetic clay) for effective removal of toxic cationic and oxyanionic metal ions from water. *Journal of Nanomaterials, 2020*, 1783749. https://doi.org/10.1155/2020/1783749.
Ni, S., Zhang, H., Dai, H., & Xiao, H. (2018). Starch-Based Flexible Coating for Food Packaging Paper with Exceptional Hydrophobicity and Antimicrobial Activity. *Polymers, 10*(11), 1260. https://doi.org/10.3390/polym10111260.
Paleček, E., Bartošík, M. (2019). Electrochemistry of Nucleic Acids. *Chemical Reviews 112*(6), 3427–3481. https://doi.org/10.1021/cr200303p.
Pandey, P. K., Preeti, Rawat, K., Prasad, T., & Bohidar, H. B. (2020). Multifunctional{,} fluorescent DNA-derived carbon dots for biomedical applications: Bioimaging{,} luminescent DNA hydrogels{,} and dopamine detection. *Journal of Materials Chemistry B, 8*(6), 1277–1289. https://doi.org/10.1039/C9TB01863H.
Patwardhan, S. V., Manning, J. R. H., & Chiacchia, M. (2018). Bioinspired synthesis as a potential green method for the preparation of nanomaterials: Opportunities and challenges. *Current Opinion in Green and Sustainable Chemistry, 12*(2018), 110–116. https://doi.org/10.1016/j.cogsc.2018.08.004.
Pérez Quiñones, J., Cela Mardare, C., Walter Hassel, A., & Brüggemann, O. (2017). Self-assembled cellulose particles for agrochemical applications. *European Polymer Journal, 93*, 706–716. https://doi.org/10.1016/j.eurpolymj.2017.02.023.
Phogat, N., Kohl, M., Uddin, J., & A. (2018). Chapter 11 - interaction of nanoparticles with biomolecules, protein, enzymes, and its applications. In H.-P. Deigner & M. Kohl (Eds.), *Precision Medicine* (pp. 253–276, ISBN 9780128053645). Academic Press. https://doi.org/10.1016/B978-0-12-805364-5.00011-1.
Pilipenko, I., Korzhikov-Vlakh, V., Sharoyko, V., Zhang, N., Schäfer-Korting, M., Rühl, E., et al. (2019). pH-sensitive chitosan-heparin nanoparticles for effective delivery of genetic drugs into epithelial cells. *Pharmaceutics, 11*(7), 317. https://doi.org/10.3390/pharmaceutics11070317.
Qamar, S. A., Asgher, M., Khalid, N., & Sadaf, M. (2019). Nanobiotechnology in health sciences: Current applications and future perspectives. *Biocatalysis and Agricultural Biotechnology, 22*, 101388. https://doi.org/10.1016/j.bcab.2019.101388.
Rane, A. V., Kanny, K., Abitha, V. K., & Thomas, S. (2018). Chapter 5—Methods for synthesis of nanoparticles and fabrication of nanocomposites synthesis of inorganic nanomaterials. In S. M. Bhagyaraj, O. S. Oluwafemi, N. Kalarikkal, & S. Thomas (Eds.), (pp. 121–139). Woodhead Publishing and Elsevier.
Ranjani, S., Shariq Ahmed, M., Mohd, A., Senthil Kumar, N., Ruckmani, K., & Hemalatha, S. (2021). Synthesis, characterization and applications of endophytic fungal nanoparticles. *Inorganic and Nano-Metal Chemistry, 51*(2), 280–287. https://doi.org/10.1080/24701556.2020.1784231.

Ranjith Kumar, R., Manikantan, J., Balaganesh, A. S., Dinesh, K. P. B., & Chandar Shekar, B. (2019). Fruit biowaste mediated green route approach silver nanopartilces—As antibacterial material. *Kongunadu Research Journal, 6*(2), 81–86. https://doi.org/10.26524/krj307.

Rasoulianboroujeni, M., Kupgan, G., Moghadam, F., Tahriri, M., Boughdachi, A., Khoshkenar, P., et al. (2017). Development of a DNA-liposome complex for gene delivery applications. *Materials Science and Engineering: C, 75*, 191–197. https://doi.org/10.1016/j.msec.2017.02.012.

Roach, P. J., & Zeeman, S. C. (2016). Glycogen and starch. In R. A. Bradshaw & P. D. Stahl (Eds.), *Encyclopedia of cell biology* (pp. 263–270). Waltham: Academic Press. https://doi.org/10.1016/B978-0-12-394447-4.10031-8.

Rodriguez-Torres, M. D. P., Díaz-Torres, L. A., Millán-Chiu, B. E., García-Contreras, R., Hernández-Padrón, G., & Acosta-Torres, L. S. (2020). Antifungal and cytotoxic evaluation of photochemically synthesized heparin-coated gold and silver nanoparticles. *Molecules (Basel, Switzerland), 25*(12), 2849. https://doi.org/10.3390/molecules25122849.

Rouhan, G., & Gaudeul, M. (2014). Plant taxonomy: A historical perspective, current challenges, and perspectives. In P. Besse (Ed.), *Molecular plant taxonomy: Methods and protocols* (pp. 1–37). Totowa: Humana Press. https://doi.org/10.1007/978-1-62703-767-9_1.

Rubach, P., Zajac, S., Jastrzebski, B., Sulkowska, J. I., & Sułkowski, P. (2019). Genus for biomolecules. *Nucleic Acids Research, 48*(D1), D1129–D1135. https://doi.org/10.1093/nar/gkz845.

Saeb, M. R., Rastin, H., Nonahal, M., Paran, S. M. R., Khonakdar, H. A., & Puglia, D. (2018). Cure kinetics of epoxy/chicken eggshell biowaste composites: Isothermal calorimetric and chemorheological analyses. *Progress in Organic Coatings, 114*, 208–215. https://doi.org/10.1016/j.porgcoat.2017.10.018.

Salata, O. V. (2004). Applications of nanoparticles in biology and medicine. *Journal of Nanobiotechnology, 2*(1), 3. https://doi.org/10.1186/1477-3155-2-3.

Saleh, T. A. (2020). Nanomaterials: Classification, properties, and environmental toxicities. *Environmental Technology & Innovation, 20*, 101067. https://doi.org/10.1016/j.eti.2020.101067.

Samaddar, P., Ok, Y. S., Kim, K.-H., Kwon, E. E., & Tsang, D. C. W. (2018). Synthesis of nanomaterials from various wastes and their new age applications. *Journal of Cleaner Production, 197*, 1190–1209. https://doi.org/10.1016/j.jclepro.2018.06.262.

Sandhu, A. (2006). Who invented nano? *Nature Nanotechnology, 1*(2), 87. https://doi.org/10.1038/nnano.2006.115.

Santo, V. E., Duarte, A. R. C., Gomes, M. E., Mano, J. F., & Reis, R. L. (2010). Hybrid 3D structure of poly(d,l-lactic acid) loaded with chitosan/chondroitin sulfate nanoparticles to be used as carriers for biomacromolecules in tissue engineering. *The Journal of Supercritical Fluids, 54*(3), 320–327. https://doi.org/10.1016/j.supflu.2010.05.021.

Schnichels, S., Hurst, J., de Vries, J. W., Ullah, S., Gruszka, A., Kwak, M., et al. (2020). Self-assembled DNA nanoparticles loaded with travoprost for glaucoma-treatment. Nanomedicine: Nanotechnology, Biology and Medicine, 29, 102260. https://doi.org/10.1016/j.nano.2020.102260

Scott, R. A., & Panitch, A. (2013). Glycosaminoglycans in biomedicine. *WIREs Nanomedicine and Nanobiotechnology, 5*(4), 388–398. https://doi.org/10.1002/wnan.1223.

Selnihhin, D., Sparvath, S. M., Preus, S., Birkedal, V., & Andersen, E. S. (2018). Multifluorophore DNA origami beacon as a biosensing platform. *ACS Nano, 12*(6), 5699–5708. https://doi.org/10.1021/acsnano.8b01510.

Sen, I. K., Mandal, A. K., Chakraborti, S., Dey, B., Chakraborty, R., & Islam, S. S. (2013). Green synthesis of silver nanoparticles using glucan from mushroom and study of antibacterial activity. *International Journal of Biological Macromolecules, 62*, 439–449. https://doi.org/10.1016/j.ijbiomac.2013.09.019.

Sforza, S., Tedeschi, T., & Wierenga, P. A. (2016). Proteins: Chemistry, characterization, and quality. In B. Caballero, P. M. Finglas, & F. Toldrá (Eds.), *Encyclopedia of food and health* (pp. 548–553). Oxford: Academic Press. https://doi.org/10.1016/B978-0-12-384947-2.00575-4.

Shafiq, M., Anjum, S., Hano, C., Anjum, I., & Abbasi, B. H. (2020). An overview of the applications of nanomaterials and nanodevices in the food industry. *Food, 9*(2), 148. https://doi.org/10.3390/foods9020148.

Sharifi, F., Irani, S., Azadegan, G., Pezeshki-Modaress, M., Zandi, M., & Saeed, M. (2020). Co-electrospun gelatin-chondroitin sulfate/polycaprolactone nanofibrous scaffolds for cartilage tissue engineering. *Bioactive Carbohydrates and Dietary Fibre, 22*, 100215. https://doi.org/10.1016/j.bcdf.2020.100215.

Shende, S., Gade, A., & Rai, M. (2017). Large-scale synthesis and antibacterial activity of fungal-derived silver nanoparticles. *Environmental Chemistry Letters, 15*, 427–434. https://doi.org/10.1007/s10311-016-0599-6.

Shi, L., Yang, J. H., Zeng, H. B., Chen, Y. M., Yang, S. C., Wu, C., Zeng, H., Yoshihito, O., & Zhang, Q. (2016). Carbon dots with high fluorescence quantum yield: The fluorescence originates from organic fluorophores. *Nanoscale, 8*(30), 14374–14378. https://doi.org/10.1039/C6NR00451B.

Singh, P., Kim, Y.-J., Zhang, D., & Yang, D.-C. (2016). Biological synthesis of nanoparticles from plants and microorganisms. *Trends in Biotechnology, 34*(7), 588–599. https://doi.org/10.1016/j.tibtech.2016.02.006.

Smiechowicz, E., Niekraszewicz, B., Kulpinski, P., & Dzitko, K. (2018). Antibacterial composite cellulose fibers modified with silver nanoparticles and nanosilica. *Cellulose, 25*, 3499–3517. https://doi.org/10.1007/s10570-018-1796-1.

Sricharoen, P., Limchoowong, N., Nuengmatcha, P., & Chanthai, S. (2020). Ultrasonic-assisted recycling of Nile tilapia fish scale biowaste into low-cost nano-hydroxyapatite: Ultrasonic-assisted adsorption for Hg2+ removal from aqueous solution followed by "turn-off" fluorescent sensor based on Hg2+-graphene quantum dots. *Ultrasonics Sonochemistry, 63*, 104966. https://doi.org/10.1016/j.ultsonch.2020.104966.

Sriram, G., Supriya, S., Kurkuri, M., & Hegde, G. (2019). Efficient CO2 adsorption using mesoporous carbons from biowastes. *Materials Research Express, 7*(1), 15605. https://doi.org/10.1088/2053-1591/ab5f2c.

Sriramulu, M., & Sumathi, S. (2017). Photocatalytic, antioxidant, antibacterial and anti-inflammatory activity of silver nanoparticles synthesised using forest and edible mushroom. *Advances in Natural Sciences: Nanoscience and Nanotechnology, 8*(4), 45012. https://doi.org/10.1088/2043-6254/aa92b5.

Srivastava, N., & Mukhopadhyay, M. (2015). Green synthesis and structural characterization of selenium nanoparticles and assessment of their antimicrobial property. *Bioprocess and Biosystems Engineering, 38*(9), 1723–1730. https://doi.org/10.1007/s00449-015-1413-8.

Stoddard, F. L. (2016). Starch chemistry. In *Reference module in food science*. Amsterdam: Elsevier. https://doi.org/10.1016/B978-0-08-100596-5.00093-7.

Stylianopoulos, C. (2013). Carbohydrates: Chemistry and classification. In B. Caballero (Ed.), *Encyclopedia of human nutrition* (3rd ed., pp. 265–271). Waltham: Academic Press. https://doi.org/10.1016/B978-0-12-375083-9.00041-6.

Su, S. S., & Chang, I. (2018). Review of production routes of nanomaterials. In D. Brabazon, E. Pellicer, F. Zivic, J. Sort, M. Dolors Baró, N. Grujovic, & K.-L. Choy (Eds.), *Commercialization of nanotechnologies—A case study approach* (pp. 15–29). Cham: Springer International Publishing. https://doi.org/10.1007/978-3-319-56979-6_2.

Sun, Y., Li, S., Chen, R., Wu, P., & Liang, J. (2020). Ultrasensitive and rapid detection of T-2 toxin using a target-responsive DNA hydrogel. *Sensors and Actuators B: Chemical, 311*, 127912. https://doi.org/10.1016/j.snb.2020.127912.

Teodoro, K. B. R., Migliorini, F. L., Christinelli, W. A., & Correa, D. S. (2019). Detection of hydrogen peroxide (H2O2) using a colorimetric sensor based on cellulose nanowhiskers and silver nanoparticles. *Carbohydrate Polymers, 212*, 235–241. https://doi.org/10.1016/j.carbpol.2019.02.053.

Thakur, R., Pristijono, P., Bowyer, M., Singh, S. P., Scarlett, C. J., Stathopoulos, C. E., & Vuong, Q. V. (2019). A starch edible surface coating delays banana fruit ripening. *LWT, 100*, 341–347. https://doi.org/10.1016/j.lwt.2018.10.055.

Thiruvengadathan, R., Korampally, V., Ghosh, A., Chanda, N., Gangopadhyay, K., & Gangopadhyay, S. (2013). Nanomaterial processing using self-assembly-bottom-up chemical and biological approaches. *Reports on progress in physics. Physical Society (Great Britain), 76*(6), 066501. https://doi.org/10.1088/0034-4885/76/6/066501.

Tinoco, I., Jr. (2002). Physical chemistry of nucleic acids. *Annual Review of Physical Chemistry, 53*, 1–15. https://doi.org/10.1146/annurev.physchem.53.082001.144341.

Tomanin, P. P., Zhou, J., Amodio, A., Cimino, R., Glab, A., Cavalieri, F., & Caruso, F. (2020). Nanoengineering multifunctional hybrid interfaces using adhesive glycogen nanoparticles. *Journal of Material Chemistry B, 8*(22), 4851–4858. https://doi.org/10.1039/D0TB00299B.

Tomii, K. (2019). Protein properties. In S. Ranganathan, M. Gribskov, K. Nakai, & C. Schönbach (Eds.), *Encyclopedia of bioinformatics and computational biology* (pp. 28–33). Oxford: Academic Press. https://doi.org/10.1016/B978-0-12-809633-8.20266-5.

Toumey, C. P. (2006). Narratives for nanotech: Anticipating public reactions to nanotechnology. *Nanotechnology Challenges*, 383–411. https://doi.org/10.1142/9789812773975_0018.

Tsai, T. T., Huang, T. H., Chang, C. J., Ho, N. Y. J., Tseng, Y. T., & Chen, C. F. (2017). Antibacterial cellulose paper made with silver-coated gold nanoparticles. *Scientific Reports, 7*, 3155. https://doi.org/10.1038/s41598-017-03357-w.

Unser, S., Holcomb, S., Cary, R., & Sagle, L. (2017). Collagen-gold nanoparticle conjugates for versatile biosensing. *Sensors, 17*(2), 378. https://doi.org/10.3390/s17020378.

Valsalam, S., Agastian, P., Esmail, G. A., Ghilan, A.-K. M., Al-Dhabi, N. A., & Arasu, M. V. (2019). Biosynthesis of silver and gold nanoparticles using Musa acuminata colla flower and its pharmaceutical activity against bacteria and anticancer efficacy. *Journal of Photochemistry and Photobiology B: Biology, 201*, 111670. https://doi.org/10.1016/j.jphotobiol.2019.111670.

Vassie, J. A., Whitelock, J. M., & Lord, M. S. (2015). Glycosaminoglycan functionalized nanoparticles exploit glycosaminoglycan functions. In K. Balagurunathan, H. Nakato, & U. R. Desai (Eds.), *Glycosaminoglycans: Chemistry and biology* (pp. 557–565). New York: Springer. https://doi.org/10.1007/978-1-4939-1714-3_44.

Veena, S., Devasena, T., Sathak, S. S. M., Yasasve, M., & Vishal, L. A. (2019). Green synthesis of gold nanoparticles from vitex negundo leaf extract: Characterization and in vitro evaluation of antioxidant–antibacterial activity. *Journal of Cluster Science, 30*(6), 1591–1597. https://doi.org/10.1007/s10876-019-01601-z.

Ventola, C. L. (2017). Progress in nanomedicine: Approved and investigational nanodrugs. *P & T: A Peer-Reviewed Journal for Formulary Management, 42*(12), 742–755. Retrieved from https://pubmed.ncbi.nlm.nih.gov/29234213.

Virkutyte, J., & Varma, R. K. (2013). Green synthesis of nanomaterials: Environmental aspects. Chapter 2. In N. Shamim & V. K. Sharma (Eds.), *Sustainable nanotechnology and the environment: Advances and achievements, ACS symposium series*, ISBN13:9780841227842, eIBSN:9780841227859. American Chemical Society, Washington, DC, 1124 (pp. 11–39).

Wang, M., Chen, Y., Cai, W., Feng, H., Du, T., Liu, W., et al. (2020). In situ self-assembling Au-DNA complexes for targeted cancer bioimaging and inhibition. *Proceedings of the National Academy of Sciences, 117*(1), 308–316. https://doi.org/10.1073/pnas.1915512116.

Wang, P., Rahman, M. A., Zhao, Z., Weiss, K., Zhang, C., Chen, Z., Hurwitz, S. J., Chen, Z. G., Shin, D. M., & Ke, Y. (2018). Visualization of the cellular uptake and trafficking of DNA origami nanostructures in cancer cells. *Journal of the American Chemical Society, 140*(7), 2478–2484. https://doi.org/10.1021/jacs.7b09024.

Wang, Z., Wang, Z., Lu, W. W., Zhen, W., Yang, D., & Peng, S. (2017). Novel biomaterial strategies for controlled growth factor delivery for biomedical applications. *NPG Asia Materials, 9*(10), e435–e435. https://doi.org/10.1038/am.2017.171.

Willner, I., & Willner, B. (2010). Biomolecule-based nanomaterials and nanostructures. *Nano Letters, 10*(10), 3805–3815. https://doi.org/10.1021/nl102083j.

Willem de Vries, J., Schnichels, S., Hurst, J., Strudel, L., Gruszka, A., Kwak, M., et al. (2018). DNA nanoparticles for ophthalmic drug delivery. *Biomaterials, 157*, 98–106. https://doi.org/10.1016/j.biomaterials.2017.11.046.

Xie, W., Deng, W.-W., Zan, M., Rao, L., Yu, G.-T., Zhu, D.-M., Wu, W.-T., Chen, B., Ji, L.-W., Chen, L., Liu, K., Guo, S.-S., Huang, H.-M., Zhang, W.-F., Zhao, X., Yuan, Y., Dong, W., Sun,

Z.-J., & Liu, W. (2019). Cancer cell membrane camouflaged nanoparticles to realize starvation therapy together with checkpoint blockades for enhancing cancer therapy. *ACS Nano, 13*(3), 2849–2857. https://doi.org/10.1021/acsnano.8b03788.

Xu, W., He, W., Du, Z., Zhu, L., Huang, K., Lu, Y., & Luo, Y. (2019). Functional nucleic acid nanomaterials: Development, properties, and applications. *Angewandte Chemie International Edition*. https://doi.org/10.1002/anie.201909927.

Xu, W., He, W., Du, Z., Zhu, L., Huang, K., Lu, Y., & Luo, Y. (2019). Functional nucleic acid nanomaterials: Development, properties, and applications. *Angewandte Chemie International Edition*. https://doi.org/10.1002/anie.201909927.

Yallappa, S., Deepthi, D. R., Yashaswini, S., Hamsanandini, R., Chandraprasad, M., Ashok Kumar, S., & Hegde, G. (2017). Natural biowaste of groundnut shell derived nano carbons: Synthesis, characterization and itsin vitro antibacterial activity. *Nano-Structures & Nano-Objects, 12*, 84–90. https://doi.org/10.1016/j.nanoso.2017.09.009.

Yan, D., Yao, Q., Yu, F., Chen, L., Zhang, S., Sun, H., et al. (2020). Surface modified electrospun poly(lactic acid) fibrous scaffold with cellulose nanofibrils and Ag nanoparticles for ocular cell proliferation and antimicrobial application. *Materials Science and Engineering: C, 111*, 110767. https://doi.org/10.1016/j.msec.2020.110767.

Yang, C., Bromma, K., Di Ciano-Oliveira, C., Zafarana, G., van Prooijen, M., & Chithrani, D. B. (2018). Gold nanoparticle mediated combined cancer therapy. *Cancer Nanotechnology, 9*(1), 4. https://doi.org/10.1186/s12645-018-0039-3.

Yang, J., Li, F., Li, M., Zhang, S., Liu, J., Liang, C., et al. (2017). Fabrication and characterization of hollow starch nanoparticles by gelation process for drug delivery application. *Carbohydrate Polymers, 173*, 223–232. https://doi.org/10.1016/j.carbpol.2017.06.006.

Yang, Y., Wan, J., Niu, Y., Gu, Z., Zhang, J., Yu, M., & Yu, C. (2016). Structure-dependent and glutathione-responsive biodegradable dendritic mesoporous organosilica nanoparticles for safe protein delivery. *Chemistry of Materials, 28*, 9008–9016. https://doi.org/10.1021/acs.chemmater.6b03896.

Yu, X., Gou, X., Wu, P., Han, L., Tian, D., Du, F., Chen, Z., Liu, F., Deng, G., Chen, A. T., Ma, C., Liu, J., Hashmi, S. M., Guo, X., Wang, X., Zhao, H., Liu, X., Zhu, X., Sheth, K., Chen, Q., Fan, L., & Zhou, J. (2018). *Advanced Materials, 30*, 1705383. https://doi.org/10.1002/adma.201705383.

Yue, S., Song, X., Song, W., & Bi, S. (2019). An enzyme-free molecular catalytic device: Dynamically self-assembled DNA dendrimers for in situ imaging of microRNAs in live cells. *Chemical Science, 10*(6), 1651–1658. https://doi.org/10.1039/C8SC04756A.

Zan, G., & Wu, Q. (2016). Biomimetic and bioinspired synthesis of nanomaterials/nanostructures. *Advanced Materials, 28*(11), 2099–2147. https://doi.org/10.1002/adma.201503215.

Zhang, X., Yan, S., Tyagi, R. D., & Surampalli, R. Y. (2011). Synthesis of nanoparticles by microorganisms and their application in enhancing microbiological reaction rates. *Chemosphere, 82*(4), 489–494. https://doi.org/10.1016/j.chemosphere.2010.10.023.

Zhang, X., Yao, M., Chen, M., Li, L., Dong, C., Hou, Y., et al. (2016). Hyaluronic acid-coated silver nanoparticles as a nanoplatform for in vivo imaging applications. *ACS Applied Materials & Interfaces, 8*(39), 25650–25653. https://doi.org/10.1021/acsami.6b08166.

Zhang, Q., Wang, J., Liu, D., Zhu, W., Guan, S., Fan, L., & Cai, D. (2020). Targeted delivery of honokiol by zein/hyaluronic acid core-shell nanoparticles to suppress breast cancer growth and metastasis. *Carbohydrate Polymers, 240*, 116325. https://doi.org/10.1016/j.carbpol.2020.116325.

Zhao, S., Zhou, Y., Wei, L., & Chen, L. (2020). Low fouling strategy of electrochemical biosensor based on chondroitin sulfate functionalized gold magnetic particle for voltammetric determination of mycoplasma ovipneumonia in whole serum. *Analytica Chimica Acta, 1126*, 91–99. https://doi.org/10.1016/j.aca.2020.06.015.

# Chapter 13
# Phyto-fabricated Metal Oxide Nanoparticles as Promising Antibacterial Agents

**Mallikharjuna Reddy Bogala**

## 13.1 Background

Nanoparticles (NPs) or nanomaterials are small size (1–100 nm) materials of different shapes and they exhibit many interesting physicochemical characteristics because of high surface area per unit volume and quantum confinement effect (Chen et al. 2018). Usually, NPs are made from pure elements such as carbon, metals (Au, Ag, Cu, Fe, Pd, Pt, and Zn), and from metal oxides like $Ag_2O$, CuO, $CeO_2$, FeO, NiO, $TiO_2$, and ZnO. Many biological functions of the metal NPs and metal oxide NPs are directly related to their diverse chemistry and some of them include: antibacterial, anticancer, antifungal, anti-inflammatory antioxidant, antiparasitic, and antiviral activities (Teow et al. 2018; Cagno et al. 2018). As shown in Fig. 13.1, NPs can be synthesized using several physical, chemical and biological methods (Teow et al. 2018).

Examples include (a) physical methods (cathodic arc or physical vapor deposition, ion etching or suttering, thin film deposition, melt blending), (b) chemical methods (co-precipitation, photoinduced reduction, UV-initiated photoreduction, sol-gel, microemulsions, sonochemical) (c) biological methods (fabrication from plants, microbes, eukaryotes, yeasts, molds, enzymes, Metazoas, biological waste), and (d) hybrid methods (Teow et al. 2018; Ali et al. 2020; Saratale et al. 2018).

All these fabrication methods have both advantages and disadvantages; therefore, the selection of synthesis method is mainly dependent on the intended biomedical applications. The broad range of physical and chemical properties of NPs allows them to exhibit variety of biological tasks. The NPs are very adaptable and

M. R. Bogala (✉)
Department of Materials Science, Central University of Tamil Nadu, Thiruvarur, TN, India

Department of Metallurgical and Materials Engineering, The University of Alabama, Tuscaloosa, AL, USA
e-mail: mrbogala@crimson.ua.edu


K. Pal (ed.), *Bio-manufactured Nanomaterials*,
https://doi.org/10.1007/978-3-030-67223-2_13

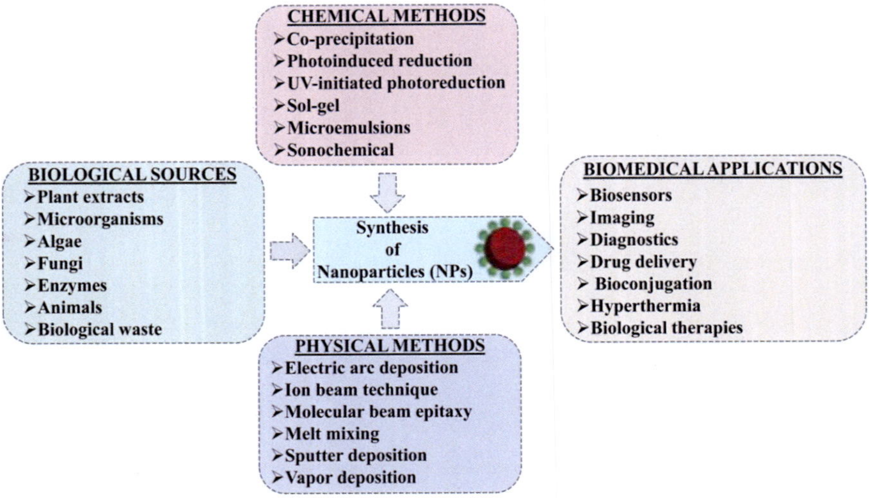


**Fig. 13.1** A schematic of synthesis methods and biomedical applications of nanoparticles

can be used in several biomedical applications. NPs play a very critical role in biomedical studies, as new biomaterials with improved performance and low toxicity, their necessity is steadily increasing in recent years. Because of low dimensions and better cell-permeating ability, nanoparticles are used in several biomedical engineering like bio-sensing, medical-imaging, disease recognition, delivery of pharmaceuticals, bio-coupling, heat stroke, and physiological therapeutics (Teow et al. 2018; Han et al. 2019). Recently, NPs have been extensively used in sensing and imaging applications, primarily because of their extraordinary localization ability in therapies (Han et al. 2019). Other benefits of nanoparticles are versatility in surface alteration, facile size regulation, and generation of considerably diminishable nanoparticles for in vivo therapeutics (Teow et al. 2018; Zhu et al. 2018). Furthermore, nanoparticles are used in bio-coupling and by conjugating with pharmaceuticals, they ease the targeted drug delivery process.

Detailed chemistry underlying bio-coupling of several pharmaceuticals and nanoparticles have been reported by Wadhawan et al. (2019). Also, the authors explained the application of NPs in drug delivery system for targeted therapies. The possible utilization of nanoparticles in heat stroke treatment is mostly to kill the tumor cells (Liu et al. 2020). With magnetic NPs, hyperthermia can be cured by targeting and killing the specific cancerous cells, while minimizing the damage to the healthy tissues (Jose et al. 2020; Shakil et al. 2019). Hence, NPs have a very crucial role in medical and biomedical research. Several investigations have been made in assessing the antibacterial properties of NPs. The potential antibacterial outcomes revealed by NPs emphasize the necessity to develop them into next generation antibacterial agents. In addition to cancer treatment, NPs are effective against diseases caused by *Escherichia coli* and *Pseudomonas aeruginosa* infections (Happy et al. 2019; Liao et al. 2019).

The exclusive properties of metal NPs (or MNPs) and metal oxide NPs (or MONPs) such as huge area, excellent elastic modulus, stability, high melting temperature makes them suitable for clinical applications like therapy, drug delivery, biosensors, anti-biofilms, biomedical imaging and infections (Teow et al. 2018; Khatami et al. 2018). Hydrophilic and hydrophobic properties of NPs make them appropriate for drug delivery applications (Fratoddi 2018).

MONPs can be synthesized by various physicochemical processes (Fig. 13.1). But, requirement of intensive power, complicated equipment, huge cost, less productivity, etc., are limitations of the physical methods (Jamkhande et al. 2019). The chemical methods for synthesis of MONPs are inexpensive, provides good product yield, and do not need complex experimental instrumentation (Akbari et al. 2018). Nevertheless, chemical methods make use of volatile and toxic chemical reagents (like aromatic amines, thiols, sodium borohydrate and hydrazine) that are very harmful, causes air pollution and also are non-ecofriendly in nature (Teow et al. 2018; Jamkhande et al. 2019; Wu and Tang 2018). Researchers are developing novel fabrication methods of MONPs that are sustainable and eco-friendly in nature (Singh et al. 2018). Therefore, the green nanotechnology protocols for fabrication of MONPs are gaining importance. The present study gives recent updates on antibacterial effects of MONPs fabricated from plants using green nanotechnology. Underlying challenges in developing MONPs as antibacterial agents in clinics, applications, and future prospective are also discussed.

## 13.2 Phyto-fabrication of MONPs

### *13.2.1 Green Nanotechnology*

There are different methods to synthesize MONPs and every procedure has important inference on the biological role of MONPs (Fig. 13.1). Among all the available nanoparticle synthesis methods, MONPs produced from green nanotechnology methods are less toxic, inexpensive, and efficient. These procedures are specifically appropriate for biomedical research. In general, fabrication of MONPs using biological methods has additional advantages, while compared to their synthesis using physicochemical processes. The biological methods for the synthesis of NPs are facile, efficient, simple to step-up (manufacture), and they require less power consumption (Teow et al. 2018).

The "greener" nanotechnology methods are sustainable and they employ non-hazardous reagents and produces environmentally harmless substances (Singh et al. 2018; El Shafey 2020). The developed green nanotechnology methods are appropriate and can be useful for the production of a wider range of biomedical products like pharmaceuticals, food, and cosmetics (Teow et al. 2018; El Shafey 2020; Soni et al. 2018). Relatively, chemical and physical methods are expensive and frequently employ hazardous and toxic chemicals that are more harmful to human cells. Moreover, the green nanotechnology methods do not rigorously need more energy, high pressure, and temperature conditions for the preparation of MONPs (Teow et al. 2018). Nevertheless, parameters like chemical concentration, pH, reaction time, and temperature are very crucial for constant production of biologically important MONPs (El Shafey 2020; Hasan et al. 2018). Moreover, fabrication of MONPs from the plants does not require other maintenance costs like for growth of microbial culture and thus, reduces the unnecessary costs for isolation of microorganisms and preparation of microbial culture media (El Shafey 2020). As shown in Table 13.1, majority of MONPs obtained from plants have sizes <100 nm. Some larger size particles (100–500 nm) are also observed (Teow et al. 2018; Thatoi et al. 2016; Tippayawat et al. 2016). The phyto-fabricated nanoparticles are nontoxic to many cells like HEK293 (human embryonic kidney), NIH3T3 (mouse embryonic fibroblast), VSMCs (rat aortic vascular smooth muscle cells), and PBMCs (peripheral blood mononuclear cells).

### 13.2.2 MONPs Synthesis from Plants

MONPs fabrication from plants is effective in achieving high productivity, while compared to their production from other biological sources like fungi and microbes. Plants have many biochemicals (like polyphenols) and metabolites, which can function as reducing and stabilizing agents during the production of biogenic MONPs. As shown in Fig. 13.2, phyto-fabrication of MONPs is simple, economical, and eco-friendly, thus prohibiting the use of harmful chemicals (Singh et al. 2020). MONPs obtained from the plant extracts were found to be relatively stable, while compared to those obtained from fungal and microbial extracts. Phyto-fabrication of MONPs is broadly categorized into (a) within cell synthesis, (b) extracellular generation, and (c) fabrication using plant chemicals. Intracellular synthesis of MONPs occurs within the cells of plant tissues via specific enzymes present in the cells.

After the synthesis, the MONPs are recovered upon rupture of plant cell membranes. Other methods that use plant-based extracts as raw materials are preferred over the intracellular synthesis methods for production of MONPs. Fabrication of MONPs from the plant extracts is moderately cheaper process and it often gives higher yields because of the availability of significant quantities of plant chemicals in the extracts, which reduces or stabilizes the oxidation state of metal in MNPs (Mohammadinejad et al. 2019). Plant chemicals dependent fabrication of metal-based nanoparticles is difficult because of involvement of specific phytochemicals required for producing stable nanoparticles (Dauthal and Mukhopadhyay 2016). Spherical silver nanoparticles were fabricated using the leaf extracts from *Azadirachta indica* by Ahmed et al. (2016a, b). These investigations revealed that the extracts containing phytochemicals and bioflavonoids function like stabilizing and reducing agents during the production of silver nanoparticles. They showed high potential to be used as antimicrobial agents. Gold nanoparticles were obtained from the extracts of *Morinda citrifolia* roots that also exhibited antipathogenic activity (Suman et al. 2014).

**Table 13.1** Phyto-fabricated (MONPs) and their antibacterial effects (Teow et al. 2018)

| MONPs | Particle length (nm) | Plant part | Scientific nomenclature | Common nomenclature | Bacteria type | Reference |
|---|---|---|---|---|---|---|
| $Ag_2O$ | ~43 | Root | *F. benghalensis* | Banyan | *S. mutans* | Manikandan et al. (2017) |
| $Ag/Ag_2O$ | 8–21 | Leaf | *E. odoratum* | Christmas bush | *E. coli, S. typhi, S. aureus, B. subtilis* | Elemike et al. (2017) |
| $Ag\text{-}MnO_2$ | 5–40 | Leaf | *C. pepo* | Summer squash | *E. coli, S. aureus, L. monocytogenes, B. cereus, S. typhi, S. enterica* | Ahmed et al. (2016a, b) |
| $CeO_2$ | 24 | Leaf | *O. europaea* | Olive | *E. coli, S. aureus, K. pneumonia, P. aeruginosa* | Maqbool et al. (2016) |
| | 45 | Peel | *M. oleifera* | Horseradish | *E. coli, S. aureus* | Surendra and Roopan (2016) |
| $Ce_2O_3$ | 8–11 | Crude | *E. amygdaloides* | Wood spurge | *P. acidilactici* | Nadaroglu et al. (2017) |
| CuO | 30–223 | Leaf | *S. rosmarinus* | Keliab | *E. coli, S. aureus* | Rezaie et al. (2017) |
| FeO | – | Peel | *P. granatum* | Pomegranate | *P. aeruginosa* | Irshad et al. (2017) |
| NiO | 10–20 | Leaf | *E. globulus* | Blue glum | *E. coli, S. aureus, MRSA, P. aeruginosa* | Saleem et al. (2017) |
| | ~10 | Crude | *M. oleifera* | Drumstick | *S. pneumonia, E. hermannii, E. coli, S. aureus* | Ezhilarasi et al. (2016) |
| ZnO | 27–85 | Pulp seed and fruit | *C. colocynthis L.* | Schrad | *MRSA, P. aeruginosa, E. coli, B. subtilis* | Azizi et al. (2017) |
| | – | Leaf | *L. leschenaultiana* | Lobelia | *P. aeruginosa, S. sonnei, P. vulgaris, V. parahaemolyticus* | Banumathi et al. (2017) |
| | 50 | Fruit | *R. canina* | Dog rose | *E. coli, L. monocytogenes, P. aeruginosa* | Jafarirad et al. (2016) |
| | 400–500 | Leaf | *S. apetala* | Sonneratia mangrove | *S. flexneri* | Thatoi et al. (2016) |
| | ~47 | Leaf | *L. nobilis* | Bay tree | *S. aureus, P. aeruginosa* | Vijayakumar et al. (2016) |
| | ~20 | Leaf | *P. × yedoensis* | Yoshino cherry | *B. liners, S. epidermidis* | Velmurugan et al. (2016) |

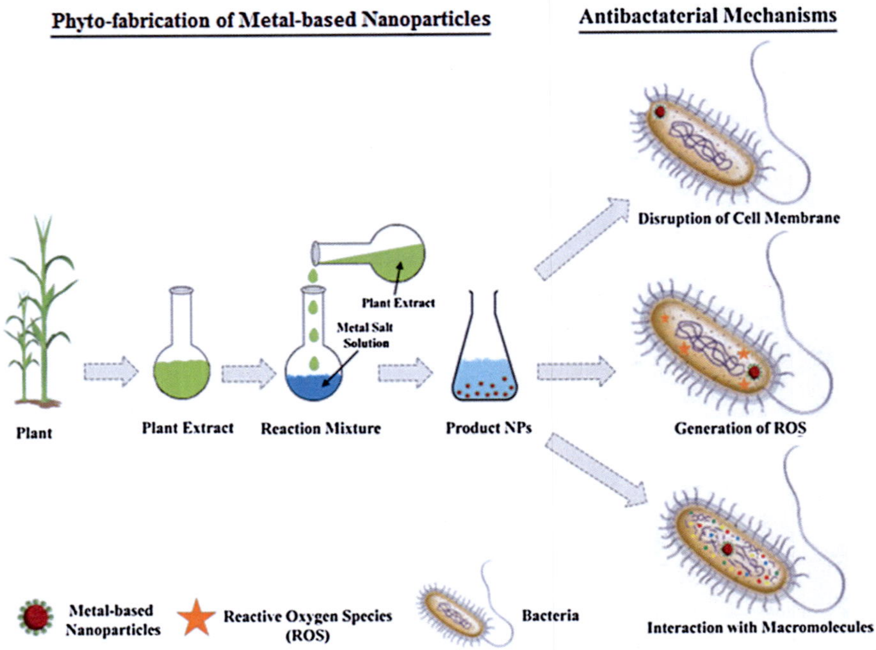


**Fig. 13.2** A depiction of phyto-fabrication process and antibacterial mechanisms of nanoparticles

### *13.2.3 Purification of MONPs*

After phyto-fabrication, appropriate purification of biocompatible MONPs is required prior to their utilizing in biomedical engineering research. For many years, centrifugation techniques have been used to isolate MONPs because of easy and fast operation steps. Hence, multiple washings and high-speed centrifugations are employed to separate and purify biologically synthesized metal-based NPs and to remove any unreacted bioactive molecules from the products (Dauthal and Mukhopadhyay 2016). But, the centrifugation method have some limitations like the centrifugation step can often result in agglomeration of MONPs and destabilization of MONPs because of detachment of inert agents from the MONPs surface, thus changing the MONPs inherent properties. Alternatively, dialysis technique uses suitable cutoff membrane for separation of MONPs. Dialysis membrane separates mixture of molecules present in the plant extracts by selectively retaining larger biomolecules inside the membrane and expelling the smaller unwanted biomolecules out of the membrane. The larger biomolecules that act as inert agents on the surface of MONPs are coupled tightly with them. Later, MONPs are collected from the dialysis bag during the washing step. Dialysis purification process usually requires a day (time consuming) to isolate the MONPs from the plant extracts. But, the diafiltration technique cannot be employed for MONPs because of their aqueous insolubility. MONPs such as $Fe_2O_3$ and $Fe_3O_4$ can be detached simply by using the field of magnet (physical method). But, eliminating chemically bound molecules from the MONPs surface is another key task that needs significant knowledge of chemistry.

### *13.2.4 MONPs Processing Parameters*

Considerable number of research works is being carried out globally to investigate the effect of temperature on phyto-fabrication of MONPs. Temperature is the single most controlling factor that influences the size, morphology and shape of the MONPs and the rate at which they are synthesized. The different shapes (octahedral platelets, triangular, rod-like, and spherical structures), change in the size and reaction time for MONPs preparation are related to the temperature. With the increase in temperature, a simultaneous rise in rate of the reaction is observed and formation of nucleation centers of NPs is observed (Sneha et al. 2010). Acidity or basicity of the reactant mixture is critical in phyto-fabrication of MONPs. It influences the creation of seeding centers for growth of MONPs. With the increase in pH, an instantaneous growth in centers for nucleation process was noticed, thus improving the metal-based NPs formation. The pH takes a vital role in regulating the morphology and size of the MONPs. In addition to temperature and pH, reaction time is another major factor that affects the morphology of MONPs. The role of the time on dimensions of pure zinc oxide and cerium-doped zinc oxide particles was investigated by Flor et al. (2004). The authors reported that the particle size rises linearly with the reaction time. Also, the authors observed that the size of cerium-doped ZnO particles is relatively larger than the size of ZnO particles for a given constant reaction time.

## 13.3 Antibacterial Behavior of MONPs

### *13.3.1 MONPs Antibacterial Activities*

MONPs fabricated from plants illustrated bactericidal properties and have greater prospective for becoming next generation antibacterial agents, primarily because of their low toxicity (Gupta et al. 2017). Earlier reports indicates that NPs inhibit gram-positive bacteria like *Bacillus* spp., *Streptococcus* spp. and *Staphylococcus* spp. and gram-negative bacteria like *Escherichia* spp., *Proteus* spp., *Salmonella* spp., *Vibrio* spp., *Shigella* spp., and *Pseudomonas* spp. (Teow et al. 2018; Singh et al. 2020; Kanjikar et al. 2018; Manikandan et al. 2017; Vijayakumar et al. 2016; Banumathi et al. 2017). In addition, NPs inhibit antibiotic-resistant bacteria like anti-drug *MRSA* and *E. coli* (Teow et al. 2018; Jadhav et al. 2016; Azizi et al. 2017). Table 13.1 summarizes the bactericidal nature of MONPs, data obtained from literature

(PubMed journal directory) published during 2016–2017 (Teow et al. 2018). About 107 articles was found using the online literature search engine (NCBI database) with the input words like green synthesis, nanoparticles, plants, and antibacterial (NCBI 2018; Teow et al. 2018). Among the 107 searched articles, 17 papers were eliminated because of their irrelevant data for the current book chapter, which comprised of reports, retracted articles and non-phyto related nanoparticles information. About 90 articles are considered for the present study, with the relevant data of classified MONPs is listed in Table 13.1 (Teow et al. 2018). Because of many fabrication techniques for producing MONPs (Fig. 13.1), the phytosynthesis source of the MONPs (Table 13.1) differs with the type of metal oxide, although they are obtained from same common precursor i.e., plant extracts. In general, extensively studied nanoparticles include silver, gold, transition metals based MNPs or MONPs and their composites. Transition metals are anticipated to be the ideal candidates for the synthesis of metal-based nanoparticles because they possess partially filled d-orbitals that makes them redox-active (cations are reduced to zero valence states), a quality that assists aggregation of MONPs (Sánchez-López et al. 2020).

From Table 13.1, MONPs have the size of less than 100 nm except for a couple of them (Teow et al. 2018; Thatoi et al. 2016; Rezaie et al. 2017). Irrelevant of the MONPs type, majority of them were exclusively obtained from the leaf extracts, except in few cases from other plant sources like fruits, seeds, crude, etc. Table 13.2 presents bacteria (gram +ve and gram −ve) that are targeted using different MONPs along with the number of bacteria-related research articles (Teow et al. 2018). Gram −ve bacteria considered for current study include *Escherichia coli* and *Pseudomonas aeruginosa*, while gram +ve bacteria investigated include *B. subtilis, B. cereus* and *S. aureus*. Relatively, gram −ve bacteria are affected by metal-based NPs in comparison to gram +ve bacteria. Metal-based nanoparticles act as antibacterial agents because they can efficiently penetrate gram −ve bacteria cell membrane (well-known for impermeable multilayered cell surface) and kill them (Zgurskaya et al. 2015). Recently, the potential of AgNPs in killing *K. pneumonia* was demonstrated by Acharya et al. (2018).

Scanning electron microscopy (SEM) images of *K. pneumonia* after treating with Ag NPs illustrate the disruption caused by the Ag NPs to the cell surfaces and resulted in cell death. Figure 13.3a shows the FESEM image of untreated and intact *K. pneumonia* cells. After treatment with Ag NPs, the FESEM image of Fig. 13.3b clearly shows the disruption (black arrows) caused by Ag NPs to the *K. pneumonia* cell surfaces. Also, in another study, the SEM images showed disruption/lysis of cell surface by AgNPs and outflow of the cellular fluids of *Escherichia coli* and their random distribution into different regions (Das et al. 2017). Previous reports reveal that metal-based NPs can be improved as antibacterial agents against skin infection–linked bacteria (*S. aureus*) and gut-associated bacteria (*P. aeruginosa*, *E. coli*, *K. pneumoniae*, *S. typhi, S. flexneri,* and *B. cereus*). Silver nanoparticles are most effective against the skin bacterial infection (*S. epidermidis* and *S. aureus*) and the gut-associated bacteria. Au NPs are efficient in killing *S. aureus* and *E. coli*. Some metal-based NPs are efficient against specific type of bacteria, like ZnO NPs are efficient against *S. aureus* and *P. aeruginosa*.

**Table 13.2** Effect of phyto-fabricated MONPs on bacteria type (Teow et al. 2018)

| Bacteria type | Metal oxide ($M_xO_y$) nanoparticles | | | | | | |
|---|---|---|---|---|---|---|---|
| | $Ag_2O$ | $CeO_2$ | $Ce_2O_3$ | CuO | FeO | NiO | ZnO |
| Gram-negative species | | | | | | | |
| *E. hermannii* | ... | ... | ... | ... | ... | 1 | ... |
| *E. coli* | ... | 2 | ... | 1 | ... | 2 | 2 |
| *P. aeruginosa* | ... | 1 | ... | ... | 1 | 1 | 4 |
| *P. vulgaris* | ... | ... | ... | ... | ... | ... | 1 |
| *K. pneumoniae* | ... | 1 | ... | ... | ... | ... | ... |
| *S. flexneri* | ... | ... | ... | ... | ... | ... | 1 |
| *Lactobacilli* sp. | 1 | ... | ... | ... | ... | ... | ... |
| *V. parahaemolyticus* | ... | ... | ... | ... | ... | ... | 1 |
| *S. sonnei* | ... | ... | ... | ... | ... | ... | 1 |
| Gram-positive species | | | | | | | |
| *S. mutans* | 1 | ... | ... | ... | ... | ... | ... |
| *S. aureus* | ... | 2 | ... | 1 | ... | 2 | 1 |
| *S. epidermidis* | ... | ... | ... | ... | ... | ... | 1 |
| *Methicillin-resistant S. aureus* | ... | ... | ... | ... | ... | 1 | 1 |
| *B. linens* | ... | ... | ... | ... | ... | ... | 1 |
| *S. pneumonia* | ... | ... | ... | ... | ... | 1 | ... |
| *P. acidilactici* | ... | ... | 1 | ... | ... | ... | ... |
| *B. subtilis* | ... | ... | ... | ... | ... | ... | 1 |
| *L. monocytogenes* | ... | ... | ... | ... | ... | ... | 1 |

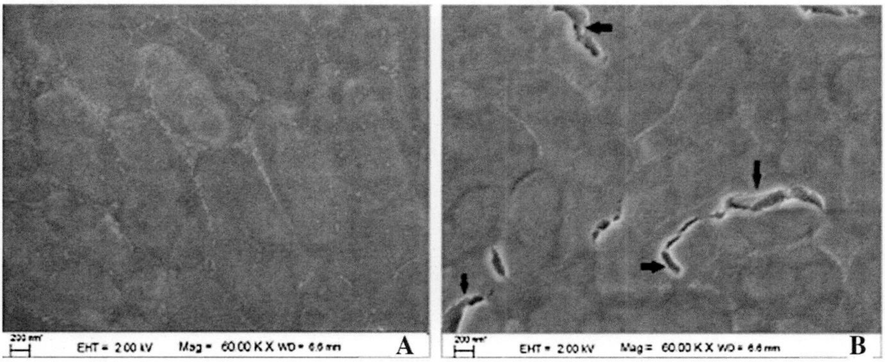


**Fig. 13.3** FESEM images of (**a**) untreated and (**b**) Ag NPs treated *K. pneumoniae* cells. (Adapted from Acharya et al. 2018)

## 13.3.2 MONPs-Based Antibiotics

In addition, phyto-fabricated metal-based NPs can deliver drugs or antibacterial compounds in cells as nanoemulsions and by via bioconjugation, thus symbiotically improving bactericidal results (Pagar and Darekar 2019; Wadhawan et al. 2019). Ag

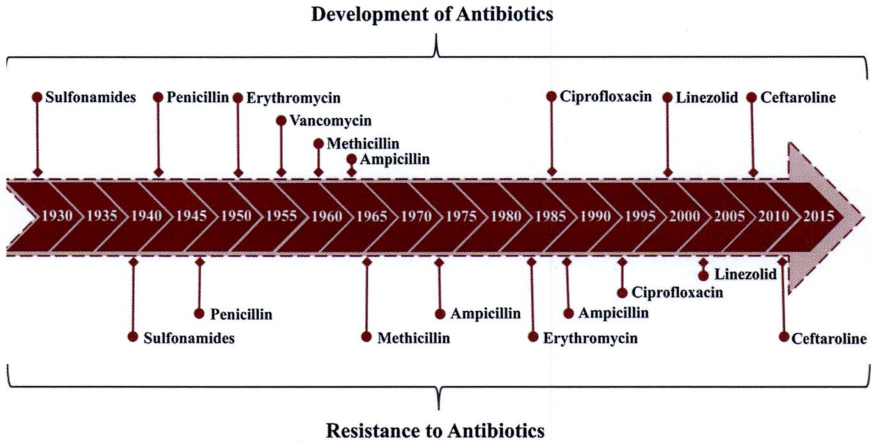


**Fig. 13.4** A timeline showing development of antibiotics and bacterial resistance to antibiotics

NPs fabricated using plant extracts resulted in killings of food borne pathogenic bacteria, while they were used in association with other antibiotics (Thatoi et al. 2016). Au NPs improve the Amoxicillin killings of gram +ve bacteria (*Bacillus* and *Staphylococcus*) and gram +ve bacteria (*Escherichia coli*) (Kalita et al. 2016). The metal-based NPs reversed the progress of resistance induced from antibiotics via death of *MRSA* bacteria using mice *MRSA* disease prototypes (Kalita et al. 2016). Figure 13.4 shows a schematic of the timeline for the development of antibiotics in treatment of different bacterial infections and subsequent timeline for the development of resistance by the bacteria to the antibiotics (Sánchez-López et al. 2020).

### *13.3.3 MONPs Antibacterial Mechanism*

The exact controlling mechanisms of metal-based NPs synergism with antibiotics still remain unknown. Some possibilities include: (a) production of bactericidal hydroxyl radicals by the NPs, (b) generation of more bactericidal metal ions by the NPs, and (c) efficient blocking of the flow for killings of drug-resistant bacteria (Gupta et al. 2017; Deng et al. 2016; Hwang et al. 2012). Other possible antibacterial effects of metal-based NPs include: (a) facilitation of columbic attraction (+ve nanoparticles and −ve bacteria), (b) many nanoparticles binding to the surface of bacteria that stops respiration and permeability-related processes, (c) damage or breakage of bacterial genes after effective penetration of the NPs into the cells, and (d) degradation and/or inactivation of vital proteins present in the bacteria (Manikandan et al. 2017; Reidy et al. 2013; Guzman et al. 2012). Ag NPs enter bacterial cells and results in considerable damage to DNA via interacting with phosphorus- and sulfur-containing molecules (Swamy et al. 2015; Ramesh et al. 2015). Also, Ag NPs release $Ag^{+}$ ions and radicals in bacteria cells. They are very reactive enough to cause antibacterial effects (Ovington 2004). The $Ag^+$ ions in association with sulfur-rich proteins of bacteria surface result many damages to cells (Reidy et al. 2013). Ag nanoparticles deactivate enzymes present in bacteria and produce harmful substance ($H_2O_2$) that causes killings of the bacteria (Raffi et al. 2008).

### *13.3.4 MONPs-ROS Induced Cell Damage*

The precise mechanism of metal-based nanoparticles against bacteria is still not fully understood and several investigations are being carried out across the globe. Scientists have revealed that metal-based nanoparticles can induce harmful effects by disrupting the outside surface of bacteria cells (Fig. 13.3). Silver nanoparticles create dents and cracks on the outer cell layers of bacteria and eject charged species that promote interaction of sulfhydryl/disulfide moieties of proteins and charged species, thus resulting in disturbance of metabolic pathway and ultimately leading to the bacterial cell death (Kailasa et al. 2019). As shown in Fig. 13.2, metal oxide nanoparticles generate reactive oxygen species (ROS), which stimulates oxidative stress inside bacteria cells and causes the cell death in bacteria (Dizaj et al. 2014). Ag NPs synthesized using green nanotechnology kills drug-resistant bacteria through ROS mediated damage of bacterial cell membrane (Das et al. 2017). ZnO NPs enhances the ROS production on the surface of bacterial cell membrane, thus resulting in the malfunction of the membrane and finally leading to the death of bacterial cells (Santhoshkumar et al. 2017). Similarly, examination of the mechanistic aspects of stress induced from reactive oxygen species production due to treatment of bacteria with titania nanoparticles was done by Allahverdiyev et al. (2011). The ROS production induced by $TiO_2$ NPs can result in lipid peroxidation, which further influences the integrity and fluidity of bacterial cell membrane. The toxic and inhibitory effects of MONPs against different bacteria depend on the concentration of ejected charged species inside the cells by MONPs. The concentration of charged species increases with the amount of ROS that is generated within the cells.

The precise cell signaling mechanism mediated via MONPs for their bactericidal role is unknown. But, metal-based NPs stimulate two different pathways in exhibiting the bactericidal influence; producing charged species (by MNPs) and ROS (by MONPs), as shown in Fig. 13.2. The freed ROS/charged species via oxidation cause harm to cell surface, lipids, or nucleic acids (Sintubin et al. 2011). During MONPs fabrication, the reactive compounds present in the microbial and plant extracts can also reduce them (Gautam et al. 2019). These molecules give high stability to the phyto-fabricated MONPs and also attach diverse functional moieties on to the MONPs surface. The moieties facilitate chemical bonding and provide specific regions that enhance MNPs-cell physical interaction, which is required for bactericidal activity (Baker et al. 2017). Growth of resistance to antibiotics by bacteria is the main problem of research for the scientists (Sánchez-López et al. 2020). In another work, zinc oxide nanoparticles based antibiotics was found to efficiently kill *MRSA* therapeutical segregates (Ali et al. 2016).

## 13.4 Applications of Antibacterial MONPs

To date, EMA and FDA have approved many metal-based NPs for different clinical studies that include cure of anemia (Vifor-based Fe-substitute treatment), imaging (Resovist), delivery of anticancer drugs (Mepact and Onivyde), fungal infection (AmBisome), vaccines for viral diseases like hepatitis A (Epaval), influenza (Inflexal V) and many others are in pipeline (Carnevale and Ko 2016; Bovier 2008; Herzog et al. 2009; Rivnay et al. 2019). But, none of these approved metal-based NPs are utilized for the treatment of bacterial infections. At present, only one liposomal NP formulation (CAL02) is going through clinical trials for the treatment of bacterial pneumonia (Anselmo and Mitragotri 2019).

### 13.4.1 Challenges for Antibacterial MONPs

Several challenges exist during the development of MONPs into antibacterial agents that are clinically tested and approved by EMA and/or FDA. Table 13.3 summarizes several limitations and challenges of using phyto-fabricated MONPs as antibacterial agents. Certainly, the nanoscale dimensions of metal-based NPs assist them in penetrating through the cell membrane and even in crossing of the blood–brain barrier (BBB) (Nowak et al. 2020; Tosi et al. 2020). This improves the biological activity and target specificity of the MONPs. But, the small size of the MONPs can often be a challenge during the clinical studies.

Metal-based NPs have shorter life time, low stability, and poor bioavailability under biological conditions of host cells (Schubert and Chanana 2019). Several enzymes and proteins present in the human blood attacks the MNPs and degrade them before they reach to the intended target cells (Bondarenko et al. 2020; Jain et al. 2018). To counter this drawback, various strategies to modify the MNPs have been implemented like alteration of MNPs surface chemistry, production of stable nanoemulsions of MNPs, conjugation of the MNPs with protein stabilizer (serum albumin), functionalization and improvements of the combined nanoparticles (Pagar and Darekar 2019; Guo et al. 2020; Ling et al. 2019). Moreover, the morphology and shape of the MNPs play a crucial role in their toxicity and biological actions against the bacteria (Teow et al. 2018; Ling et al. 2019). Certain nanoparticles combine together and form aggregates due to their surface chemistry. As a result, unnecessary harmfulness to the healthy cells and also considerable reduction in entrance of metal-based nanoparticles into the intended cells can be observed (Enea et al. 2019). Lethality accounts for many characteristics of metal-based nanoparticles (chemistry, stability, retention rate, specificity, and biodistribution), target cells and method of their administration (Enea et al. 2019).

### 13.4.2 Challenges in MONPs Production

As listed in Table 13.3, numerous setbacks exist during the production of MONPs for biomedical application (Teow et al. 2018). Because the biological roles of MONPs vastly rely on their size, shape, cell-permeability, physical and chemical properties, fabrication of MONPs at industrial scale must firmly stick to the tightly-regulated, reliable, and repeatable SOPs and GMPs. An additional challenge to MONPs research is the heterogeneous nature of several bacterial infections in humans. MONPs therapeutic studies on diseased patients are complex, since diverse patients have different immunological response although they have been infected from the same common bacterial source. Considering the huge list of potential

**Table 13.3** Phyto-fabricated metal oxide nanoparticles and their antibacterial effects

| Source type and factors | Challenges in developing MONPs as antibacterial agents |
|---|---|
| *Structural* | |
| Size | Small size of MONPs improves their penetration into bacterial cells, however they may possess low bioavailability or stability |
| Shape | Specific shape of MONPs may enhance their biological role because the total surface area exposed is high and surface area is a shape dependent property |
| Aggregation | MONPs that participate in aggregate formation increases the total particle size, thus restricting their cell penetration and enhance the cell harmfulness |
| *Physiological* | |
| Distribution | Low diffusion of MONPs because of limited cell permeation (skin barrier) |
| Availability | Low bioavailability of MONPs results in quick loss of antibacterial activity |
| Specificity | Good MONPs specificity permit low off-intended actions and they are efficient |
| Approval | Better withhold of MONPs guarantees good efficacy and easy clinical approval |
| Toxicity | MONPs agglomeration cause harm to healthy cells |
| *Technological* | |
| Diversity | Heterogeneous nature of disease may cause difficulties in its cure |
| Step-up | Reforming of MONPs fabrication process and scale-up of manufacturing uniform sized and non-aggregate MONPs in regulated and reproducible mode |
| Productivity | Synthesis of MONPs is multistep, difficult and do not allows optimization of high-throughput process |
| Forecast | Prediction of MONPs effectiveness in clinical trials using computer models is a very challenging task |
| *Industrial* | |
| Quantity | Bulk production may result nonuniform size and physical and chemical properties of MONPs |
| Processing | Consistency and reproducibility of industrialized processes requires latest instrumentation and nanotechnology |
| Quality | Constant production of highly uniform, good quality, and functional MONPs |

MONPs, the demand for clinical laboratory testing facility that involves immense performance operation for physiological sorting out of the candidate MONPs also raises proportionately. This is directly related to the automation of the processes that permits cost-efficient production of MONPs, modeling and computing technologies that would assess the toxicities or efficacies of MONPs on the intended cells. Developments of immense performance computation as well as screening strategies certainly further metal oxide nanoparticles usage in biomedical research.

## 13.5 Conclusions and Future Outlook

The current book chapter presents the new developments of phyto-fabricated nanoparticles that act as potential antibacterial agents. The MONPs obtained using green nanotechnology methods are promising as next generation antibacterial agents and they are cable of synergistically improving the efficiency of the antibiotics. Although the precise controlling mechanism(s) remain still unidentified, a considerable research is in progress to introduce potential antibacterial MONPs in therapeutic trials. Knowledge on antibacterial mechanisms of metal-based NPs is very crucial in controlling and overcoming the growing concern of bacterial resistance to the MONPs. As outlined, several drawbacks and impending challenges have to be surmounted to maximize the usage of MONPs in clinical research. Biological properties such as homogeneity, specificity, stability, and toxicity of MONPs are the primary controlling factors that decide the MONPs scope in therapeutic trials. Considering the manufacturing aspects, growth of green nanotechnology techniques and equipments are critical for manufacturing MONPs at industrial scale amounts for the clinical purposes. New advancements in nanotechnology and biotechnology are expected to further the usage of phyto-fabricated metal oxide-based nanoparticles as antibacterial agents in clinical and biomedical industries. Very few reports exist on green nanotechnology that uses plants as a chief source for the production of MONPs. In this study, brief background on green nanotechnology methods for the phyto-fabrication of MONPs, the controlling antibacterial mechanisms of MONPs, challenges in the development and future prospective of MONPs as antibacterial agents were discussed.

**Acknowledgments** The author would like to thank Department of Materials Science at Central University of Tamil Nadu for providing the necessary facilities to complete the work. No external financial support was received for carrying out the current project.

Ethical Issues The work is original investigation by the author and there are no ethical issues.

**Conflict of Interest** The author declares that there is no conflict of interest.

## References

Acharya, D., Singha, K. M., Pandey, P., Mohanta, B., Rajkumari, J., & Singha, L. P. (2018). Shape dependent physical mutilation and lethal effects of silver nanoparticles on bacteria. *Scientific Reports, 8*, 201.

Ahmed, S., Ahmad, M., Swami, B. L., & Ikram, S. (2016a). A review on plants extract mediated synthesis of silver nanoparticles for antimicrobial applications: A green expertise. *Journal of Advanced Research, 7*, 17–28.

Ahmed, S., Saifullah, Ahmad, M., Swami, B. L., & Ikram, S. (2016b). Green synthesis of silver nanoparticles using *Azadirachta indica* aqueous leaf extract. *Journal of Radiation Research and Applied Science, 9*, 1–7.

Akbari, A., Amini, M., Tarassoli, A., Eftekhari-Sis, B., Ghasemian, N., & Jabbari, E. (2018). Transition metal oxide nanoparticles as efficient catalysts in oxidation reactions. *Nano-Struct Nano-Object, 14*, 19–48.

Ali, K., Dwivedi, S., Azam, A., Saquib, Q., Al-Said, M. S., Alkhedhairy, A. A., & Musarrat, J. (2016). Aloe vera extract functionalized zinc oxide nanoparticles as nanoantibiotics against multi-drug resistant clinical bacterial isolate. *Journal of Colloid and Interface Science, 472*, 145–156.

Ali, M., Ahmed, T., Wu, W., Hossain, A., Hafeez, R., Islam Masum, M., Wang, Y., An, Q., Sun, G., & Li, B. (2020). Advancements in plant and microbe-based synthesis of metallic nanoparticles and their antimicrobial activity against plant pathogens. *Nanomaterials, 10*(6), 1146.

Allahverdiyev, A. M., Abamor, E. S., Bagirova, M., & Rafailovich, M. (2011). Antimicrobial effects of $TiO_2$ and $Ag_2O$ nanoparticles against drug-resistant bacteria and leishmania parasites. *Future Microbiology, 6*, 933–940.

Anselmo, A. C., & Mitragotri, S. (2019). Nanoparticles in the clinic: An update. *Bioengineering & Translational Medicine, 4*(3), e10143.

Azizi, S., Mohamad, R., & Mahdavi Shahri, M. (2017). Green microwave-assisted combustion synthesis of zinc oxide nanoparticles with *Citrullus colocynthis* L. schrad: Characterization and biomedical applications. *Molecules, 22*, E301.

Baker, S., Pasha, A., & Satish, S. (2017). Biogenic nanoparticles bearing antibacterial activity and their synergistic effect with broad spectrum antibiotics: Emerging strategy to combat drug resistant pathogens. *Saudi Pharmaceutical Journal, 25*, 44–51.

Banumathi, B., Vaseeharan, B., Ishwarya, R., Govindarajan, M., Alharbi, N. S., Kadaikunnan, S., Khaled, J. M., & Benelli, G. (2017). Toxicity of herbal extracts used in ethno-veterinary medicine and green-encapsulated ZnO nanoparticles against *Aedes aegypti* and microbial pathogens. *Parasitology Research, 116*, 1637–1651.

Bondarenko, L. S., Kovel, E. S., Kydralieva, K. A., Dzhardimalieva, G. I., Illés, E., Tombácz, E., Kicheeva, A. G., & Kudryasheva, N. S. (2020). Effects of modified magnetite nanoparticles on bacterial cells and enzyme reactions. *Nanomaterials, 10*(8), 1499.

Bovier, P. A. (2008). Epaxal®: A virosomal vaccine to prevent hepatitis A infection. *Expert Review of Vaccines, 7*, 1141–1150.

Cagno, V., Andreozzi, P., D'Alicarnasso, M., Silva, P. J., Mueller, M., Galloux, M., Le Goffic, R., Jones, S. T., Vallino, M., Hodek, J., & Weber, J. (2018). Broad-spectrum non-toxic antiviral nanoparticles with a virucidal inhibition mechanism. *Nature Materials, 17*(2), 95–203.

Carnevale, J., & Ko, A. H. (2016). MM-398 (nanoliposomal irinotecan): Emergence of a novel therapy for the treatment of advanced pancreatic cancer. *Future Oncology, 12*, 453–464.

Chen, Q., Tong, R., Chen, X., Xue, Y., Xie, Z., Kuang, Q., & Zheng, L. (2018). Ultrafine ZnO quantum dot-modified $TiO_2$ composite photocatalysts: The role of the quantum size effect in heterojunction-enhanced photocatalytic hydrogen evolution. *Catalysis Science & Technology, 8*(5), 1296–1303.

Das, B., Dash, S. K., Mandal, D., Ghosh, T., Chattopadhyay, S., Tripathy, S., Das, S., Dey, S. K., Das, D., & Roy, S. (2017). Green synthesized silver nanoparticles destroy multidrug resistant bacteria via reactive oxygen species mediated membrane damage. *Arabian Journal of Chemistry, 10*, 862–876.

Dauthal, P., & Mukhopadhyay, M. (2016). Noble metal nanoparticles: Plant-mediated synthesis, mechanistic aspects of synthesis, and applications. *Industrial and Engineering Chemistry Research, 55*, 9557–9577.

Deng, H., McShan, D., Zhang, Y., Sinha, S. S., Arslan, Z., Ray, P. C., & Yu, H. (2016). Mechanistic study of the synergistic antibacterial activity of combined silver nanoparticles and common antibiotics. *Environmental Science & Technology, 50*, 8840–8848.

Dizaj, S. M., Lotfipour, F., Barzegar-Jalali, M., Zarrintan, M. H., & Adibkia, K. (2014). Antimicrobial activity of the metals and metal oxide nanoparticles. *Materials Science and Engineering: C, 44*, 278–284.

El Shafey, A. M. (2020). Green synthesis of metal and metal oxide nanoparticles from plant leaf extracts and their applications: A review. *Green Processing and Synthesis, 9*(1), 304–339.

Elemike, E. E., Onwudiwe, D. C., Ekennia, A. C., Sonde, C. U., & Ehiri, R. C. (2017). Green synthesis of Ag/$Ag_2O$ nanoparticles using aqueous leaf extract of *Eupatorium odoratum* and its antimicrobial and mosquito larvicidal activities. *Molecules, 22*, E674.

Enea, M., Peixoto de Almeida, M., Eaton, P., Dias da Silva, D., Pereira, E., Soares, M. E., Bastos, M. D., & Carmo, H. (2019). A multiparametric study of gold nanoparticles cytotoxicity, internalization and permeability using an in vitro model of blood–brain barrier. Influence of size, shape and capping agent. *Nanotoxicology, 13*(7), 990–1004.

Ezhilarasi, A. A., Vijaya, J. J., Kaviyarasu, K., Maaza, M., Ayeshamariam, A., & Kennedy, L. J. (2016). Green synthesis of NiO nanoparticles using *Moringa oleifera* extract and their biomedical applications: Cytotoxicity effect of nanoparticles against HT-29 cancer cells. *Journal of Photochemistry and Photobiology B: Biology, 164*, 352–360.

Flor, J., de Lima, S. M., & Davolos, M. R. (2004). Effect of reaction time on the particle size of ZnO and ZnO:Ce obtained by a Sol–gel method. In *Surface and colloid science* (pp. 239–243). Berlin: Springer.

Fratoddi, I. (2018). Hydrophobic and hydrophilic Au and Ag nanoparticles. Breakthroughs and perspectives. *Nanomaterials, 8*(1), 11.

Gautam, P. K., Singh, A., Misra, K., Sahoo, A. K., & Samanta, S. K. (2019). Synthesis and applications of biogenic nanomaterials in drinking and wastewater treatment. *Journal of Environmental Management, 231*, 734–748.

Guo, X., Wei, X., Chen, Z., Zhang, X., Yang, G., & Zhou, S. (2020). Multifunctional nanoplatforms for subcellular delivery of drugs in cancer therapy. *Progress in Materials Science, 107*, 100599.

Gupta, A., Saleh, N. M., Das, R., Landis, R. F., Bigdeli, A., Motamedchaboki, K., Campos, A. R., Pomeroy, K., Mahmoudi, M., & Rotello, V. M. (2017). Synergistic antimicrobial therapy using nanoparticles and antibiotics for the treatment of multidrug-resistant bacterial infection. *Nano Futures, 1*, 015004.

Guzman, M., Dille, J., & Godet, S. (2012). Synthesis and antibacterial activity of silver nanoparticles against gram-positive and gram-negative bacteria. *Nanomedicine: Nanotechnology, Biology and Medicine, 8*, 37–45.

Han, X., Xu, K., Taratula, O., & Farsad, K. (2019). Applications of nanoparticles in biomedical imaging. *Nanoscale, 11*(3), 799–819.

Happy, A., Soumya, M., Kumar, S. V., Rajeshkumar, S., Sheba, R. D., Lakshmi, T., & Nallaswamy, V. D. (2019). Phyto-assisted synthesis of zinc oxide nanoparticles using Cassia alata and its antibacterial activity against *Escherichia coli. Biochemistry and Biophysics Reports, 17*, 208–211.

Hasan, M., Ullah, I., Zulfiqar, H., Naeem, K., Iqbal, A., Gul, H., Ashfaq, M., & Mahmood, N. (2018). Biological entities as chemical reactors for synthesis of nanomaterials: Progress, challenges and future perspective. *Mater Today Chemistry, 8*, 13–28.

Herzog, C., Hartmann, K., Künzi, V., Kürsteiner, O., Mischler, R., Lazar, H., & Glück, R. (2009). Eleven years of inflexal® V—A virosomal adjuvanted influenza vaccine. *Vaccine, 27*, 4381–4387.

Hwang, I. S., Hwang, J. H., Choi, H., Kim, K. J., & Lee, D. G. (2012). Synergistic effects between silver nanoparticles and antibiotics and the mechanisms involved. *Journal of Medical Microbiology, 61*, 1719–1726.

Irshad, R., Tahir, K., Li, B., Ahmad, A., Siddiqui, A. R., & Nazir, S. (2017). Antibacterial activity of biochemically capped iron oxide nanoparticles: A view towards green chemistry. *Journal of Photochemistry and Photobiology B: Biology, 170*, 241–246.

Jadhav, K., Dhamecha, D., Bhattacharya, D., & Patil, M. (2016). Green and ecofriendly synthesis of silver nanoparticles: Characterization, biocompatibility studies and gel formulation for treatment of infections in burns. *Journal of Photochemistry and Photobiology B: Biology, 155*, 109–115.

Jafarirad, S., Mehrabi, M., Divband, B., & Kosari-Nasab, M. (2016). Biofabrication of zinc oxide nanoparticles using fruit extract of Rosa canina and their toxic potential against bacteria: A mechanistic approach. *Materials Science & Engineering. C, Materials for Biological Applications, 59*, 296–302.

Jain, A., Singh, S. K., Arya, S. K., Kundu, S. C., & Kapoor, S. (2018). Protein nanoparticles: Promising platforms for drug delivery applications. *ACS Biomaterials Science & Engineering, 4*(12), 3939–3961.

Jamkhande, P. G., Ghule, N. W., Bamer, A. H., & Kalaskar, M. G. (2019). Metal nanoparticles synthesis: An overview on methods of preparation, advantages and disadvantages, and applications. *Journal of Drug Delivery Science and Technology, 53*, 101174.

Jose, J., Kumar, R., Harilal, S., Mathew, G. E., Prabhu, A., Uddin, M. S., Aleya, L., Kim, H., & Mathew, B. (2020). Magnetic nanoparticles for hyperthermia in cancer treatment: An emerging tool. *Environmental Science and Pollution Research, 27*(16), 19214–19225.

Kailasa, S. K., Park, T. J., Rohit, J. V., & Koduru, J. R. (2019). Antimicrobial activity of silver nanoparticles. *Nanoparticles in Pharmacotherapy, 1*, 461–484.

Kalita, S., Kandimalla, R., Sharma, K. K., Kataki, A. C., Deka, M., & Kotoky, J. (2016). Amoxicilin functionalized gold nanoparticles reverts MRSA resistance. *Materials Science & Engineering. C, Materials for Biological Applications, 61*, 720–727.

Kanjikar, A. P., Hugar, A. L., & Londonkar, R. L. (2018). Characterization of phyto-nanoparticles from *Ficus krishnae* for their antibacterial and anticancer activities. *Drug Development and Industrial Pharmacy, 44*, 377–384.

Khatami, M., Alijani, H. Q., & Sharifi, I. (2018). Biosynthesis of bimetallic and core–shell nanoparticles: Their biomedical applications—A review. *IET Nanobiotechnology, 12*(7), 879–887.

Liao, S., Zhang, Y., Pan, X., Zhu, F., Jiang, C., Liu, Q., Cheng, Z., Dai, G., Wu, G., Wang, L., & Chen, L. (2019). Antibacterial activity and mechanism of silver nanoparticles against multidrug-resistant *Pseudomonas aeruginosa*. *International Journal of Nanomedicine, 14*, 1469–1487.

Ling, W., Wang, M., Xiong, C., Xie, D., Chen, Q., Chu, X., Qiu, X., Li, Y., & Xiong, X. (2019). Synthesis, surface modification, and applications of magnetic iron oxide nanoparticles. *Journal of Materials Research and Technology, 34*(11), 1828–1844.

Liu, D., Hong, Y., Li, Y., Hu, C., Yip, T. C., Yu, W. K., Zhu, Y., Fong, C. C., Wang, W., Au, S. K., & Wang, S. (2020). Targeted destruction of cancer stem cells using multifunctional magnetic nanoparticles that enable combined hyperthermia and chemotherapy. *Theranostics, 10*(3), 1181.

Manikandan, V., Velmurugan, P., Park, J. H., Chang, W. S., Park, Y. J., Jayanthi, P., Cho, M., & Oh, B. T. (2017). Green synthesis of silver oxide nanoparticles and its antibacterial activity against dental pathogens. *3 Biotech, 7*(1), 72.

Maqbool, Q., Nazar, M., Naz, S., Hussain, T., Jabeen, N., Kausar, R., Anwaar, S., Abbas, F., & Jan, T. (2016). Antimicrobial potential of green synthesized $CeO_2$ nanoparticles from *Olea europaea* leaf extract. *International Journal of Nanomedicine, 11*, 5015–5025.

Mohammadinejad, R., Shavandi, A., Raie, D. S., Sangeetha, J., Soleimani, M., Hajibehzad, S. S., Thangadurai, D., Hospet, R., Popoola, J. O., Arzani, A., & Gómez-Lim, M. A. (2019). Plant molecular farming: Production of metallic nanoparticles and therapeutic proteins using green factories. *Green Chemistry, 21*, 1845–1865.

Nadaroglu, H., Onem, H., & Gungor, A. A. (2017). Green synthesis of $Ce_2O_3$ NPs and determination of its antioxidant activity. *IET Nanobiotechnology, 11*, 411–419.

NCBI (National Center for Biotechnology Information). (2018). Bethesda (MD): National Library of Medicine (US), National Center for Biotechnology Information. Retrieved from https://www.ncbi.nlm.nih.gov/.

Nowak, M., Brown, T. D., Graham, A., Helgeson, M. E., & Mitragotri, S. (2020). Size, shape, and flexibility influence nanoparticle transport across brain endothelium under flow. *Bioengineering & Translational Medicine, 5*(2), e10153.

Ovington, L. G. (2004). The truth about silver. *Ostomy/Wound Management, 50*, 1S–10S.

Pagar, K. R., & Darekar, A. B. (2019). Nanoemulsion: A new concept of delivery system. *Asian Journal of Research in Pharmaceutical Sciences, 9*(1), 39–46.

Raffi, M., Hussain, F., Bhatti, T. M., Akhter, J. I., Hameed, A., & Hasan, M. M. (2008). Antibacterial characterization of silver nanoparticles against *E. coli* ATCC-15224. *Journal of Materials Science and Technology, 24*, 192–196.

Ramesh, P. S., Kokila, T., & Geetha, D. (2015). Plant mediated green synthesis and antibacterial activity of silver nanoparticles using *Emblica officinalis* fruit extract. *Spectrochimica Acta. Part A, Molecular and Biomolecular Spectroscopy, 142*, 339–343.

Reidy, B., Haase, A., Luch, A., Dawson, K. A., & Lynch, I. (2013). Mechanisms of silver nanoparticle release, transformation and toxicity: A critical review of current knowledge and recommendations for future studies and applications. *Materials, 6*, 2295–2350.

Rezaie, A. B., Montazer, M., & Rad, M. M. (2017). Photo and biocatalytic activities along with UV protection properties on polyester fabric through green in-situ synthesis of cauliflower-like CuO nanoparticles. *Journal of Photochemistry and Photobiology B: Biology, 176*, 100–111.

Rivnay, B., Wakim, J., Avery, K., Petrochenko, P., Myung, J. H., Kozak, D., Yoon, S., Landrau, N., & Nivorozhkin, A. (2019). Critical process parameters in manufacturing of liposomal formulations of amphotericin B. *International Journal of Pharmaceutics, 565*, 447–457.

Saleem, S., Ahmed, B., Khan, M. S., Al-Shaeri, M., & Musarrat, J. (2017). Inhibition of growth and biofilm formation of clinical bacterial isolates by NiO nanoparticles synthesized from *Eucalyptus globulus* plants. *Microbial Pathogenesis, 111*, 375–387.

Sánchez-López, E., Gomes, D., Esteruelas, G., Bonilla, L., Lopez-Machado, A. L., Galindo, R., Cano, A., Espina, M., Ettcheto, M., Camins, A., & Silva, A. M. (2020). Metal-based nanoparticles as antimicrobial agents: An overview. *Nanomaterials, 10*, 292.

Santhoshkumar, J., Kumar, S. V., & Rajeshkumar, S. (2017). Synthesis of zinc oxide nanoparticles using plant leaf extract against urinary tract infection pathogen. *Resource-Efficient Technologies, 3*, 459–465.

Saratale, R. G., Saratale, G. D., Shin, H. S., Jacob, J. M., Pugazhendhi, A., Bhaisare, M., & Kumar, G. (2018). New insights on the green synthesis of metallic nanoparticles using plant and waste biomaterials: Current knowledge, their agricultural and environmental applications. *Environmental Science and Pollution Research, 25*(11), 10164–10183.

Schubert, J., & Chanana, M. (2019). Coating matters: Review on colloidal stability of nanoparticles with biocompatible coatings in biological media, living cells and organisms. *Current Medicinal Chemistry, 25*(35), 4556.

Shakil, M. S., Hasan, M., & Sarker, S. R. (2019). Iron oxide nanoparticles for breast cancer theranostics. *Current Drug Metabolism, 20*(6), 446–456.

Singh, J., Dutta, T., Kim, K. H., Rawat, M., Samddar, P., & Kumar, P. (2018). 'Green' synthesis of metals and their oxide nanoparticles: Applications for environmental remediation. *Journal of Nanbiotechnology, 16*(1), 84.

Singh, A., Gautam, P. K., Verma, A., Singh, V., Shivapriya, P. M., Shivalkar, S., Sahoo, A. K., & Samanta, S. K. (2020). Green synthesis of metallic nanoparticles as effective alternatives to treat antibiotics resistant bacterial infections: A review. *Biotechnology Reports, 25*, e00427.

Sintubin, L., De Gusseme, B., Van der Meeren, P., Pycke, B. F., Verstraete, W., & Boon, N. (2011). The antibacterial activity of biogenic silver and its mode of action. *Applied Microbiology and Biotechnology, 91*, 153–162.

Sneha, K., Sathishkumar, M., Kim, S., & Yun, Y. S. (2010). Counter ions and temperature incorporated tailoring of biogenic gold nanoparticles. *Process Biochemistry, 45*, 1450–1458.

Soni, M., Mehta, P., Soni, A., & Goswami, G. K. (2018). Green nanoparticles: Synthesis and applications. *IOSR Journal of Biotechnology and Biochemistry, 4*, 78–83.

Suman, T. Y., Rajasree, S. R., Ramkumar, R., Rajthilak, C., & Perumal, P. (2014). The green synthesis of gold nanoparticles using an aqueous root extract of *Morinda citrifolia L. Spectrochimica Acta Part A: Molecular and Biomolecular Spectroscopy, 118*, 11–16.

Surendra, T. V., & Roopan, S. M. (2016). Photocatalytic and antibacterial properties of phytosynthesized $CeO_2$ NPs using *Moringa oleifera* peel extract. *Journal of Photochemistry and Photobiology B: Biology, 161*, 122–128.

Swamy, M. K., Sudipta, K. M., Jayanta, K., & Balasubramanya, S. (2015). The green synthesis, characterization, and evaluation of the biological activities of silver nanoparticles synthesized from *Leptadenia reticulate* leaf extract. *Applied Nanoscience, 5*, 73–81.

Teow, S. Y., Wong, M. M., Yap, H. Y., Peh, S. C., & Shameli, K. (2018). Bactericidal properties of plants-derived metal and metal oxide nanoparticles (NPs). *Molecules, 23*(6), 1366.

Thatoi, P., Kerry, R. G., Gouda, S., Das, G., Pramanik, K., Thatoi, H., & Patra, J. K. (2016). Photo-mediated green synthesis of silver and zinc oxide nanoparticles using aqueous extracts of two mangrove plant species, Heritiera fomes and Sonneratia apetala and investigation of their biomedical applications. *Journal of Photochemistry and Photobiology B: Biology, 163*, 311–318.

Tippayawat, P., Phromviyo, N., Boueroy, P., & Chompoosor, A. (2016). Green synthesis of silver nanoparticles in aloe vera plant extract prepared by a hydrothermal method and their synergistic antibacterial activity. *PeerJ, 4*, e2589.

Tosi, G., Duskey, J. T., & Kreuter, J. (2020). Nanoparticles as carriers for drug delivery of macromolecules across the blood-brain barrier. *Expert Opinion on Drug Discovery, 17*(1), 23–32.

Velmurugan, P., Park, J. H., Lee, S. M., Yi, Y. J., Cho, M., Jang, J. S., Myung, H., Bang, K. S., & Oh, B. T. (2016). Eco-friendly approach towards green synthesis of zinc oxide nanocrystals and its potential applications. *Artificial Cells, Nanomedicine, and Biotechnology, 44*, 1537–1543.

Vijayakumar, S., Vaseeharan, B., Malaikozhundan, B., & Shobiya, M. (2016). *Laurus nobilis* leaf extract mediated green synthesis of ZnO nanoparticles: Characterization and biomedical applications. *Biomedicine & Pharmacotherapy, 84*, 1213–1222.

Wadhawan, A., Chatterjee, M., & Singh, G. (2019). Present scenario of bioconjugates in cancer therapy: A review. *International Journal of Molecular Sciences, 20*(21), 5243.

Wu, T., & Tang, M. (2018). Review of the effects of manufactured nanoparticles on mammalian target organs. *Journal of Applied Toxicology, 38*(1), 25–40.

Zgurskaya, H. I., Lopez, C. A., & Gnanakaran, S. (2015). Permeability barrier of gram-negative cell envelopes and approaches to bypass it. *ACS Infectious Diseases, 1*, 512–522.

Zhu, N., Ji, H., Yu, P., Niu, J., Farooq, M. U., Akram, M. W., Udego, I. O., Li, H., & Niu, X. (2018). Surface modification of magnetic iron oxide nanoparticles. *Nanomaterials, 8*(10), 810.

# Chapter 14
# A Unique Perspective in Precision of Nano-biotechnology for Sustainable Agricultural Fields

**Praseetha P. Nair**

## 14.1 Background of Nano-biotechnology and Agricultural Thematic Research

Nano-biotechnology is the perfect fusion of nanotechnology with biological system which provides the research sphere with flood gate of opportunities and commercial world with potential applications. Nanotechnology is an interdisciplinary field which blends with many areas and provides exotic and amazing results. The legendary statement, "There is a plenty of room at the bottom", given in American Physical Society by "Richard P. Feynman"—The Father of nanotechnology states that abundant and variety of properties can be exploited from materials when the size is reduced. This implies that when we observe from big to small, there are plenty of opportunities. Nanotechnology is not a new knowhow, it has been there from time immemorial. We can see a lot of examples of nanotechnology in nature itself. The hydrophobicity of lotus leaves, the beautiful colours of butterfly wings, etc. are due to the crystal nanostructures incorporated on their surfaces (Pradeep, 2007). Nanotechnology is the production, manipulation and control of materials at nano dimensions that will change most of the properties of the materials like weight, strength, value and make it more defined (Poonam, Prem, Ajay, Sudheer, & Gyanendra, 2018; Qureshi, Singh, & Dwivedi, 2018). Nanomaterials may show exactly opposite property from its parent element, for example, a nanomaterial may be conductor while its parent element is an insulator. These unusual changes in properties of nanomaterials make them the material of this century. The change in properties exhibits by nanomaterials are due to large surface area and hence surface energy, quantum confinement and reduced imperfections. Many techniques are

P. P. Nair (✉)
Department of Chemical Engineering, Government Engineering College, Thrissur, Kerala, India
e-mail: praseetha@gectcr.ac.in



K. Pal (ed.), *Bio-manufactured Nanomaterials*,
https://doi.org/10.1007/978-3-030-67223-2_14

adopted by research community for the synthesis of nanoparticles. However, environmental friendly processes of nanoparticle synthesis have ever growing demand. Remarkable contribution is imparted by natural extracts and microbes for nanoparticle synthesis by biological methods (Luis et al., 2020). As the rate of reaction with biological systems is very low, uniformity in size and growth of nanoparticles can be attained through biosynthesis (Mukherjee et al., 2008). Unlike chemical and physical methods, biosynthesis offers higher yield of nanoparticles at less expense. Stability of synthesized nanoparticles is a major issue faced which can be better solved by biosynthesis because nanoparticles in aqueous suspension are stable due to the presence of biomolecules (Parikh et al., 2008). Moreover, biosynthesis requires mild experimental conditions and has attracted attention mainly due to its clean, reliable, nontoxic nature. They exhibit high surface reactivity and biocompatibility due to its high specific surface area which are the required properties for in vivo medicinal applications. Medicinal applications of nanotechnology are in the areas of molecular imaging, early stage cancer detection, targeted drug therapy inside the body, in vivo sensors and X-ray absorbers.

Nanotechnology is emerging out as a powerful tool which embraces agriculture and food industry with its promising applications. Direct application of nanotechnology can be observed in cultivation, production, processing and package of food (Safdar et al., 2018). Thus nanotechnology made its mark in every walk of life. In most parts of the world, people depend on agriculture for their livelihood (Poonam et al., 2018). Agriculture industry contributes a major part in the economy of almost all countries. Of course it is the backbone of developing countries and can reduce poverty, improve the living standards and food security of people (Manjunatha, Biradar, & Aladakatti, 2016; Poonam et al., 2018). Due to population density, the future generation may suffer from scarcity of land and resources which will create deficiencies of production in agriculture industry. So it is vital to improve the present scenario in agriculture sector so as to enhance crop production. The enormous use of chemical fertilizers and pesticides for massive production have resulted in environmental pollution, reduce soil fertility, depletion of ground water and top soil quality (Gahukar & Das, 2020). Nanotechnology can contribute extensively to alleviate these problems to a remarkable extend. It is high time to exploit the possibilities of nanotechnology tools in precision agriculture and farming so as to protect our ecosystem. All the processes in biology happens in nanometre range. The intersection of biotechnology, nanotechnology and bioinformatics leads to the introduction of a very special branch, nano-biotechnology. This has reformed the advancement and growth in biology which directs towards many prospective applications directly applied to mankind. This will help in numerous ways in industrial biotechnology with solutions to many of the existing challenges.

This chapter focus on how the application of nano-biotechnology can be beneficially used in various stages of crop production and management. Remarkable contribution of nano-biotechnology starts from seed germination to post-harvest yield storage techniques (Poonam et al., 2018). Nano-engineered smart tools for high-tech agricultural production systems like intelligent and smart delivery systems for nutrients and agrochemicals, nano-formulations for agrochemicals, advanced usage of nano-biosensors, etc. are described. Sustainable intensification in agriculture can be realized with nano-biotechnology. This system intends to improve production without adversely affecting the environment.

## 14.2 Significant Properties of Nanomaterials

Nanomaterials can be defined based on their size, shape and structural morphologies. These are not the only parameters which determine whether the particle is nano but also its characteristics. The properties of the nanomaterials are different from their parent element from which it is formed. Size and size distribution, shape, aspect ratio, hydrophobicity, porosity, solubility, surface morphology, structure, surface charge, aggregation/agglomeration, etc. are some of the vital properties of nanomaterials (Manjunatha et al., 2016).

The prominent features of nanomaterials are the following;

1. Due to their small size, the accumulation of charge per unit area, volume or length for nanomaterials, are higher compared to their parent element. This increases the reactivity of nanoparticles.
2. Due to their high specific surface area, atoms and molecules in the interior of the surface will become available and active due to the exposed surface. This also increases the activity of the particle compared to their parent element especially when applied for their catalytic property.
3. Higher surface to volume ratio is the primary motivation for the development of nanomaterials. It is an indication of the quantity of the interfacial region compared to the bulk material in the nano-composite. As by Feynman's exploration, maximizing the interfacial region, maximizes the potential of defining new material properties. Consider spherical particles with radius $r$. The ratio of surface area to volume is given by $\frac{3}{r}$. Hence as the radius decreases, the availability of the surface increases. The total interfacial area to volume is given by $\frac{3\phi}{r}$ where $\phi$ is the sphericity of the particle. If $\phi$ is constant which depends on the shape of the particle, decreasing $r$ increases the availability of surface area for interfacial interaction. For a cylindrical particle of radius $r$ and length $L$, the surface area to volume ratio of the particle is given by $\frac{2}{r}+\frac{2}{L}$. This shows that as size decreases the availability of surface for interaction increases. Nanoparticles act as reinforcement during the synthesis of composite materials. Due to the availability of surface for interaction which is the predominant requirement of reinforcement, nanomaterials can act as a very effective filler for the preparation of nano-composites.
4. The large surface to volume ratio enhances the strength and resistivity of nanomaterials. It also decreases the melting point of nanomaterials.
5. The availability of exposed surface improves the adsorption rate of nanoparticles on any adsorbed species. The larger specific surface makes the nanoparticles highly active which improves the catalytic activity where the size as well as shape affects the activity. Tetrahedral, cubicle and spherical structures show high reactivity at their sharp edges and corners (Adhikari, Biswas, & Kundu, 2010).

### 14.2.1 Antimicrobial Property of Nanomaterials

The high surface to volume ratio of nanomaterials imparts them the amazing antimicrobial property. Due to its high surface area, it can easily penetrate into the cell wall of microbes and alter their properties. The ability of nanomaterials to destroy microbes is highly specific both with respect to nanomaterials and with respect to microbes and it is not easy to generalize. Silver and gold nanoparticles are having very good antibacterial properties (Luis et al., 2020). Silver in nano form is antibacterial agent to *Escherichia coli* and *Staphylococcus aureus* than copper nanoparticles while zinc oxide and nickel oxide nanoparticles are more active against *Staphylococcus aureus* and *Bacillus subtilis* (Baek & An, 2011). Generally nanoparticles exhibit their antibacterial effect more towards fast growing bacteria than slow growing ones (Lu, Brauer, & Botstein, 2009). Basically most bacteria have the ability to form biofilms (Mahmoudi & Serpooshan, 2012). Biofilms are multifarious contagious community that form by bonding to a solid surface and by exudation of DNA, proteins, etc., which give protection to microbes against antibiotics. This leads to the advancement of long-lasting contagions (Landini, Antoniani, Burgess, & Nijland, 2010). The electrostatic and magnetic properties of nanoparticles help them to interact with the biofilm, penetrate into it, alter their characteristics, thereby restrict or prevent the formation of community of microbes and produce toxicity to the pathogens (Mahmoudi & Serpooshan, 2012). The ultrafine size of nanomaterials makes them efficient candidates for pharmaceutical applications. Researchers showed that the toxicity of silver nanoparticles depends on the size while the toxic effect of gold nanoparticles to microbes or pathogens is independent of size of the nanoparticle (Chen et al., 2006). For certain nanomaterials like cyanoacrylates the toxicity depends on chemical properties (Lherm, Muller, Puisieux, & Couvreur, 1992). Thus not only the size but the chemistry also contributes to the inbuilt toxic effect of nanomaterials.

The research studies conducted by Shrivastava and coworkers (2007) about the antibacterial activity of silver nanoparticles used four bacterial strains viz., *Escherichia coli*, *Staphylococcus aureus*, ampicillin resistant *Escherichia coli*, *Salmonella typhi* for the analysis and found that the activity was concentration dependent and was more prominent against gram-negative bacteria when compared to other microorganisms. Antibacterial activity of silver nanoparticles synthesized by *Fusarium oxysporum* strain (Macarto, De Souza, Alves, Esposito, & Duran, 2005) was analysed in hospitals where the application of sterile cloths and materials are essential. Often the cloths we use in hospitals will be contaminated with microorganisms. When silver nanoparticles are incorporated into the cloths, it shows excellent antibacterial activity. The same experiments were conducted in silk cloth, but antibacterial activity was not observed because of the futility of silk cloth in the incorporation of the silver nanoparticles into it due to its small pore size (Macarto et al., 2005).

Nanomaterials are available in various shapes like spherical, oval, cylindrical, tubes, wires, cubicle, polygons, helical, etc. The shape and aspect ratio of the nanomaterials also play a major role in imparting exotic properties to them. The shape and morphology of certain nanoparticles like carbon nanotubes, nano forms of gold, nickel, titanium, etc. reported to affect the properties of nanomaterials. Aspect ratio of nanomaterials is directly proportional to the toxicity of nanomaterials (Fubini, Fenoglio, Tomatis, & Turci, 2011).

### *14.2.2 Other Diverse Properties of Nanomaterials*

Mechanical properties like tensile strength, impact strength, flexural strength, flexural modulus and elastic modulus of nanomaterials show substantial up gradation from their bulk materials. Nano-composites synthesized by the reinforcement with nano-fillers in polymers or polymer composites exhibit remarkable and commendable change in mechanical properties which make them suitable for high strength applications without compromising the light weight quality of polymers.

The biomedical properties of nanoparticles are influenced by the size, shape, composition, charge density, morphology and aggregation of nanomaterials (Navya & Hemant, 2016). The biocompatibility of carbon materials like fullerenes, graphene, diamond, quantum dots and nanotubes are suitable for their biomedical potentials like nano-biosensors, drug targeting and delivery (Chen et al., 2019). Quantum dots entrapped in polymers are also applicable in drug targeting and gene delivery.

The electrical, optical and magnetic properties of nanomaterials exhibit considerable deviation from their parent element. The studies conducted with polymer metal nano-composites observed that there is improvement in electrical and magnetic properties of the nano-composites and the properties depend on the size and loading of nanoparticles.

## 14.3 Biosynthesis of Nanomaterials

Nanoparticles can be synthesized by chemical routes like sol gel method, chemical and physical vapour deposition, precipitation, sputtering, laser ablation, etc. However, there is a demand for biological synthesis of nanoparticles due to its green and environmental friendly approach. Basically there are two approaches for the synthesis of nanomaterials. They are top down approach and bottom up approach. The schematic diagram for the synthesis of nanoparticles is shown in Fig. 14.1. Size and homogeneity of the nanoparticles are very much influenced by the method of synthesis. The method of synthesis also plays a crucial role in the control of size of nanoparticles. Top down approach involves reduction in size of the materials to nanoscale through mechanical size reduction techniques. Metallic and ceramic nanoparticles and nano-composites can be prepared by top down approach. Building up of numerous small subunits like atoms, molecules, etc. by self-assemblies to a well-ordered arrangement in nanoscale range forms bottom-up synthesis. We get

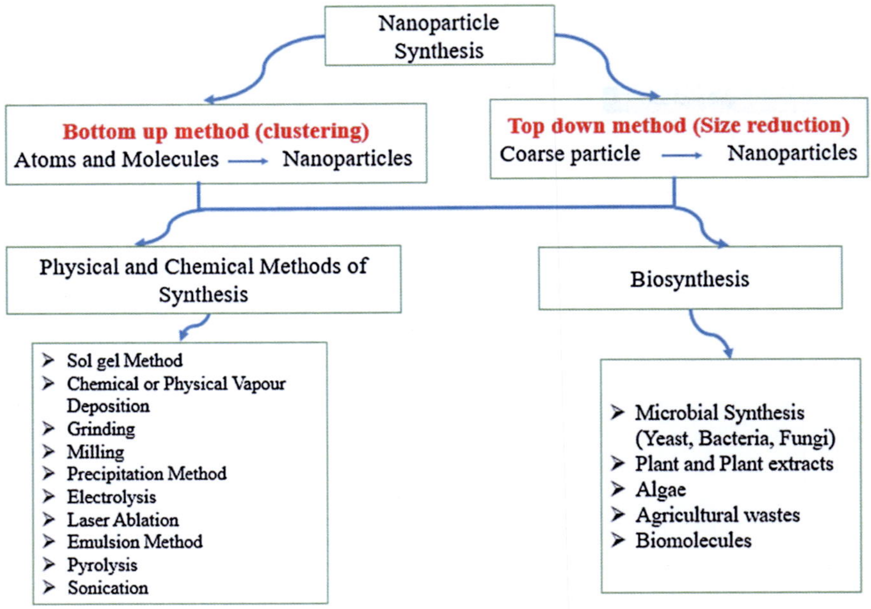


**Fig. 14.1** The schematic illustration of the synthesis of nanoparticles

highly monodispersed or uniform size particles through bottom up approach because of the freedom of control over the dimension during synthesis. Homogeneity determines the required property and hence application of the nanomaterial.

### *14.3.1 Microbial Synthesis of Nanoparticles*

Microorganisms exhibit a unique characteristic that it can detoxify metals through the reduction of metal ions. This property can be exploited for the synthesis of nanoparticles. The mechanism of reduction is through the release of reductase or enzymes which can convert metal substrate to nanoparticles under maintained pH and temperature (Ahmad et al., 2003). Silver has been known for its excellent wound healing properties that it had been used in the form of ointments during world wars. This is due to it's highly toxic nature towards a wide range of bacteria and thus silver based compounds have been used extensively in bactericidal applications (Rai, Yadav, & Gade, 2009).

Nanoparticles are sufficiently small enough to pass through outer cell membranes and enter inner cell membranes. Microbial synthesis of nanoparticles can be divided mainly into two major categories depending on the location of formation of nanoparticles as either extracellular formation (nanoparticles forms outside the cell) or intracellular formation (nanoparticles forms inside the cell). Extracellular formation of nanoparticles is more advantageous of the fact that it makes the downstream processing much easier although intracellular process offers good control over the size and shape of the nanoparticles (Bhainsa & D'Souza, 2006). The applications of the nanoparticles will be better understood if produced extracellular (Ahmad et al., 2003). The various microorganisms responsible for the production of nanoparticles are yeast, bacteria and fungi. Kowishik and coworkers (2003) worked on yeast strain MKY3 for the production of nanoparticles.

The bactericidal production of nanoparticles namely *Pseudomonas stutzeri* were carried out by Klaus, Joerger, Olsson, and Granquist (1999), *Aeromonas* sp. SH 10 by Mouxing et al. (2006), *Plectonema boryanum* by Lengke, Fleet, and Southam (2007). Fungal stains like *Fusarium oxysporum*, *Aspergillus fumigatus* and *Cladosporium cladosporioide* were used for the synthesis of nanoparticles by Ahmad et al. (2003), Bhainsa and D'Souza (2006) and Balaji, Basavaraja, Bedre, Prabhakar, and Ventaraman (2009) respectively. Microbial synthesis of silver nanoparticles was reviewed by Subin, Tapobrata, Praseetha, and Thomas (2014) as silver is the most mentioned nanomaterial ever used with variety of applications (Subin et al., 2014). Microbial stain is treated with a precursor silver compound, for example $AgNO_3$. The enzymes secreted by the microbes reduce the substrate to silver nanoparticles through extracellular or intracellular synthesis (Praseetha & Panda, 2012). Pictorial representation of microbial synthesis of nanomaterials is given in Fig. 14.2. In addition to microbes, plant extracts like neem, tulsi, etc. can be used for the synthesis of nanoparticles.

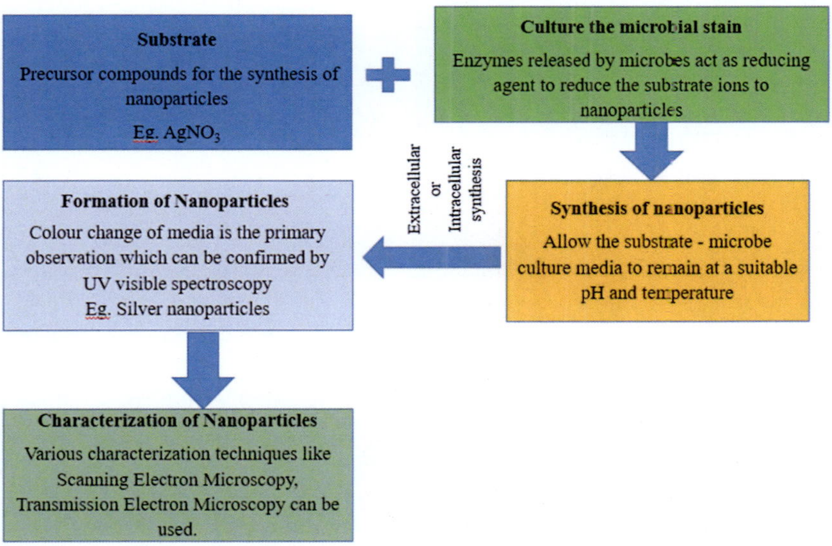


**Fig. 14.2** Diagrammatic representation of microbial synthesis of nanoparticles

## 14.4 Applications of Nano-biotechnology in Agriculture

There are many developments and modernization in agricultural industry. The combined influence of nanotechnology with agriculture delivers furtive solutions to the numerous challenges faced by agriculture sector. One of the major problems that is encountering by farmers is the pathogens that affect the crops which necessitates the over dosage of pesticides. Fertilizer usage is also very high for supplementing soil richness and for increasing crop production (Saxena, Jain, Upadhyay, & Gauba, 2018). With the help of sensor nanotechnology, smart delivery systems can be employed to diagnose the diseases at early stage, reduce the dosage of pesticides, slow and controlled release of fertilizers.

The way of adopting modern technology for controlling the utilization of resources and for efficient production is called Controlled Environment Agriculture (CEA). This technology, commonly used in many of the European countries, is an excellent platform for introducing nanotechnology in agricultural sector. CEA can address various challenges and can optimize the usage of resources, crop harvesting time, etc. CEA is someway related to the concept of precision farming. Conceptually precision farming is maximizing the yield and harvest from agriculture and farming through optimized and measured usage of fertilizers, pesticides, etc. by monitoring environmental variations and actions. Precision agriculture adopts the facilities of satellite systems and remote sensing devices for the localized monitor and control of agriculture and farming (Tanveer, Bisma, Tariq, & Reiaz, 2019). CEA and Precision agriculture are the dream concepts in agriculture which can be fulfilled with the advent of nanotechnology and with the culmination of nanotechnology with biotechnology. Various application of nano-biotechnology is concisely given in Fig. 14.3.

### *14.4.1 Fertilizers and Smart Delivery System*

#### 14.4.1.1 Nano-fertilizers

The role of fertilizers are very crucial and inevitable in agricultural production (Manikant, Suresh, Amit, Poonam, & Shailendra, 2018; Panpatte, Jhala, Shelat, & Vyas, 2016). About 60–70% of the conventional chemical fertilizers and agrochemicals supplied to vegetation are lost due to decomposition, leaching and evaporation which leads to an extensive application of large quantity of fertilizers. This will unfavourably affect the soil nutrition and health as well as increase the expenses of cultivation of crops. The use of nano-fertilizers is the best substitute for conventional fertilizers which will reduce the soil and ground water pollution. Gradual and targeted release of nano-fertilizers prevent the loss of fertilizers during its application. The slow release of nano-fertilizers can be maintained during the entire growth period of the crop which reduces the excessive usage of fertilizers thereby reducing

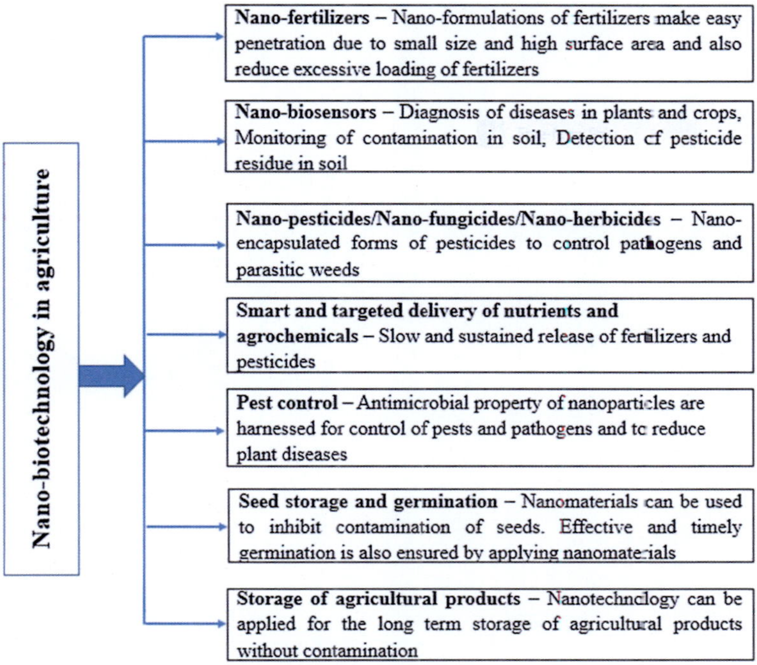


**Fig. 14.3** Application of nano-biotechnology in agriculture

its toxic effect in the ecosystem. Controlled and sustained delivery of nutrients to vegetation and soil can be regarded as a means to attain sustainable agriculture and environment (Banotra et al., 2017). The studies conducted by Subramanian and Sharmil (2009) show that nano-fertilizers are proficient in supplying nutrients slowly and steadily for more than 50 days while the nutrient supply by a conventional fertilizer last up to 12 days (Subramanian & Sharmil, 2009). Thus fertilizers can be used effectively, the cost can be minimized and profit can be maximized for the same yield. Usage of nanomaterials coated fertilizers reduces the erosion of nutrients and it contributes to the disease resistant capacity of plants. It prevents the tendency of bending of roots while promotes the deep rooting growth of vegetation (Singh et al., 2017). The commonly used carrier materials for nano-enabled fertilizers and their characteristic features are represented in Fig. 14.4 (Guo, White, Wang, & Xing, 2018). Carrier materials like nano-clays, hydroxyapatite nanoparticles, mesoporous silica, carbon-based nanomaterials and polymer nanoparticles are used for nano-formulation. Nano-fertilizers are highly soluble, durable, ensures targeted and easy delivery with less loading, less toxic to the ecosystem when compared to conventional chemical fertilizers. Nano-agrochemicals provide very effective interactions with plant and soil and prevent the loss of components. The scalability, environmental considerations, socioeconomic impacts are needed to be addressed (Kah, Kookana, Gogos, & Bucheli, 2018).

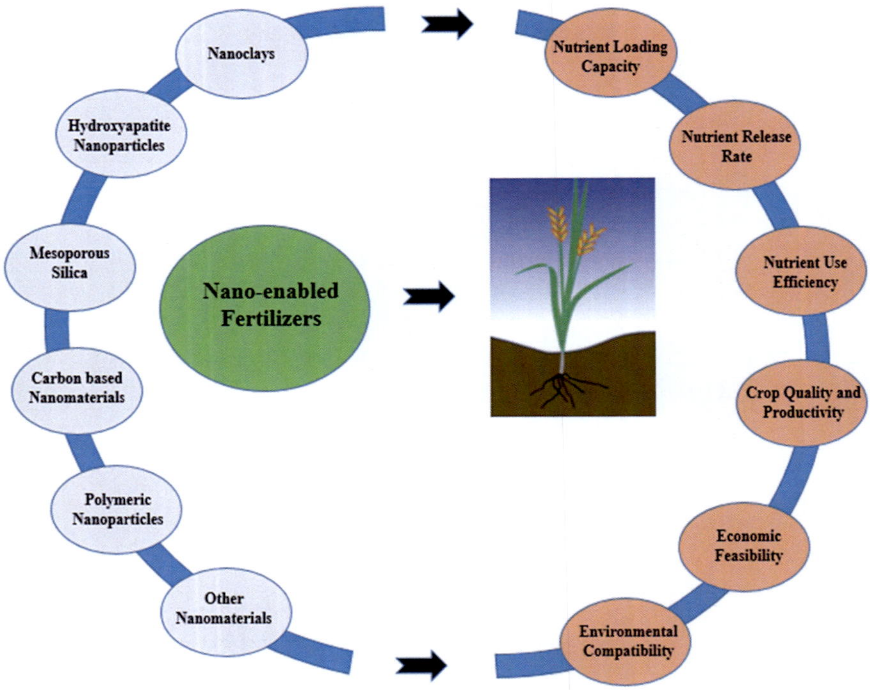


**Fig. 14.4** Nano-enabled fertilizers for sustained release. (Copyright@ Elsevier, 2018)

The surface areas of nanoparticles are so high that it can be coated on fertilizers. Nano-fertilizers are required in very low quantity and it releases nutrients in a slow and controlled manner. Nanomaterial coatings on the fertilizers can firmly hold the materials on plant due to high surface tension (Yang et al., 2012) and it improves the nutrient consumption efficiency of plants (Liu & Lal, 2015).

#### 14.4.1.2 Nano-encapsulation of Agrochemicals and Nutrients

Advanced techniques like micro or nano-encapsulation approaches are employed for ensuring the bioavailability of agrochemicals and nutrients to vegetation. Nano-encapsulation proposes many potentials to the encapsulation processes. The production of nano-fertilizers is by entangling or entrapping fertilizers within nanoparticle. Different methods are used for encapsulation of fertilizers with nanoparticles. They are entrapping of nutrients inside nanomaterials, coating with thin polymeric films, releasing as nanoparticles or nano-emulsions (Rai, Acharya, & Dey, 2012). The process of entrapping active components or core materials inside the carrier vehicles is called encapsulation. The selection of carrier agents and techniques for encapsulation are very crucial for the preparation of a particular encapsulated system with appropriate application and it determines whether the encapsulation is micro or nano. Biocompatible compounds like chitosan, starch, cellulose, biopolymers, gelatines, etc. can be used as carriers. Popularly used techniques are freeze drying, spray drying, emulsification, electro-spinning, dispersion, precipitation, deposition, etc. (Shishira, Xiea, Sunb, Zhenga, & Chena, 2018). Nano-encapsulation techniques produce nano-fertilizers and nano-pesticides for agricultural application. It ensures controlled and targeted release, reduces wastages through loss of chemicals by evaporation and improves stability and bioactivity. Coating or binding of fertilizers within nano-membranes guarantees slow and stable release of nutrients to the soil (Subramanian & Tarafdar, 2011). Naturally occurring nano-kaolinite clays and zeolites can be used with fertilizers for slow release of NPK nutrients (Panpatte et al., 2016). The micro or nano-porous structure of zeolites makes them very good adsorbents. The surface of zeolites can be modified with surfactants which will increase their surface area, enhancing its adsorption capacity and this will make it a good fertilizer carrier for gradual, controlled and targeted release of nutrients especially nitrate and phosphate group (Liu & Lal, 2015). Many studies have proved that nano-composites made by the intercalation of polymeric materials in nano-kaolinite clay layers act as reinforcing materials for slow and regulated delivery of fertilizers. It shows tremendous improvement in growth of crop and increase in productivity (Panpatte et al., 2016). The effectiveness of photosynthesis by plants can be improved by carbon nanotubes and $TiO_2$ nanoparticles which will improve the growth of the plants as well as germination of seeds (Iavicolia, Lesoa, Beezholdb, & Shvedova, 2017). There is a scope of developing nanomaterials for nutrient delivery through roots of plants (Panpatte et al., 2016). This system of application of nano-fertilizers ensures the slow and controlled delivery of deficient nutrients in soil. If a soil is lacking in a particular nutrient, fertilizers can be coated with the nano form of that specific element and allowed for delivery. Since it is coated with nutrients in nanoscale the rate of mass transfer will be low and it will allow a slow and sustained release of fertilizer as nutrients to the soil. Studies show that nanomaterials like sulphur, biodegradable polymeric nanomaterials, etc. can contribute to the slow and steady delivery of fertilizers.

#### 14.4.1.3 Bio-fertilizers

Bio-fertilizers are produced by the microbial fermentation of organic and biological waste into manures that promises eco-friendly and green production of fertilizers. Blue green algae, azotobacter, azospirillum, rhizobium, etc. are used nowadays. It provides nutrition to plants and agriculture, enhances soil richness and ultimately increases crop production. The use of bio-fertilizers should be encouraged because it promotes green protocol of environmental friendly application of fertilizers. But there are certain drawbacks for the conventionally used ones. The most important disadvantage is that it has very short shelf life.

Some technologies should be applied to improve their life, to reduce the rancidity, and to reduce the storage and transportation issues (Jha & Prasad, 2006). Potential application of nanotechnology can alleviate many issues regarding the

**Table 14.1** Comparison between nano-formulated and conventional fertilizers

| Characteristics | Nano-formulated fertilizers | Conventional fertilizers |
| --- | --- | --- |
| Fertilizer loading | Limited quantity of nano-fertilizers or biological nano-fertilizers provide effective nutrient for the same product development and yield | Quantity of fertilizer required is more for the same yield and plant growth |
| Mode of release of fertilizer | Controlled and sustained release of nutrients in nano form is possible. Sustained release extends the period of delivery of nutrients | At the time of supply, crops and vegetation utilizes it. Massive release causes wastage of fertilizers. This damages the soil health |
| Undesirable loss of nutrients | Due to the large surface area, fast rate of mass transfer and hence easy attachment is ensured. Loss rate is less | Loss rate is high due to drain off of nutrients |
| Effectiveness in the intake of nutrients | Nutrient intake efficiency is high | Less intake efficiency |
| Solubility of nutrients | Improved solubility of nutrients and availability is more due to small size | Solubility of nutrients is less |

application of conventional bio-fertilizers. They can be coated with nanomaterials like polymeric particles and can be applied. They may be applied in the form of nano sprays. It should be suitably developed for enhancing the intake capacity of nutrients by plants. Encapsulation can be provided with nanomaterials and it ensures the slow and sustained release of nutrients to crop and soil (Peteu, Oancea, Sicuia, Constantinescu, & Dinu, 2010). The advantages of nano-fertilizers over conventional fertilizers are given in Table 14.1.

### *14.4.2 Pesticides*

Pesticides like fungicides, insecticides and herbicides are used to kill pests and pathogens. It is quite disastrous that pesticide usage has become a necessary evil for maintaining production and preventing economic losses in agriculture industry. The situation is really alarming for the indiscriminate application of chemical pesticides have posed many problems worldwide. It disturbs the biodiversity of nature, makes the ecosystem unsustainable, pollutes the environment especially ground water and soil and are not easily degradable. Even if they decompose to some extent, their products are highly toxic and the residues will remain for years. The extensive usage of chemical pesticides are contaminating the soil and reduces its fertility (Gahukar & Das, 2020). They are actually giving birth to too many pathogens which can withstand the high strength chemical pesticides thereby enhance their immunity towards pesticides. This necessitates the development of more powerful pesticides.

There is a remarkable elevation of pesticide accumulation in the food chain. A few years back, the pathetic situation of impairment of health of people is witnessed due to the over usage of an acute toxic insecticide, endosulfan. Many people died, children were born with neuro-functional disorders due to the persistent exposure of the inhabitants to endosulfan. Even though the activity of bio-pesticides is inferior compared to chemical fertilizers, application of bio-pesticides is to be promoted as part of green revolution and to minimize the adverse effect of chemical fertilizers. The dosage requirement of bio-pesticides are higher compared to chemical fertilizers, but it will not cause harm to the ecosystem as chemical fertilizers. The best alternative is to apply nano-formulations in pesticides. Any active ingredient in a pesticide is of nano-size or any modification in its structure with the properties of pesticides formulates nano-pesticides (Pandey, Giri, Kumar, Mishra, & Rishi, 2018). Pesticide formulations based on nanoparticles is represented in Fig. 14.5 (Huang et al., 2018). Polymer materials can be used to formulate nano or micro-encapsulation for the intelligent delivery of pesticides. Nano-vehicle systems for loading pesticides are prepared by adsorption, coupling, encapsulation and embedding. The reduced size during nano-modification ensures their enhanced dispersion, spread and adhesion of droplets on plant surfaces. The bioavailability and efficiency of pesticides are increased by nano-based pesticide formulation. Moreover, it protects the active ingredients in pesticides, improves their stability and decreases volatility. Nano-pesticides are very stable and highly compatible with the environment, have large specific surface area, possess high solubility compared to bio-pesticides. They are less toxic to the environment and require very low dosage while the area of coverage will be the maximum (Gahukar & Das, 2020; Kumar et al., 2019; Manjunatha et al., 2016).

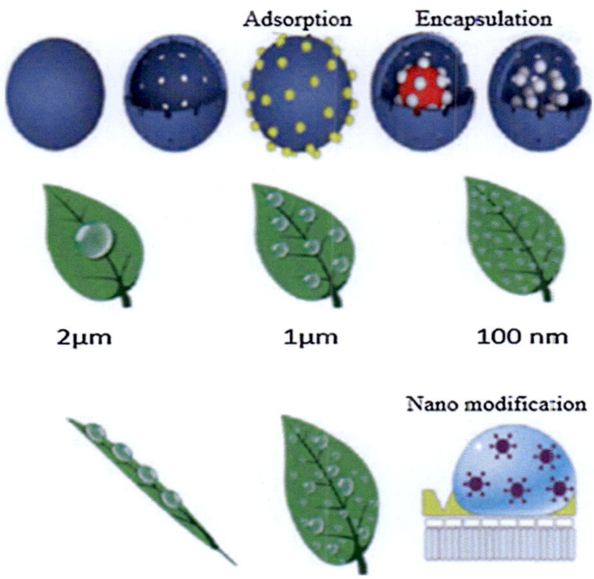


**Fig. 14.5** Nano-based modification of pesticide formulation. (Copyright@ MDPI, 2018)

#### 14.4.2.1 Nano-herbicides

Excessive growth of weeds is a major problem that causes a reduction in crop production. Conventional methods of destruction of weeds using commercially available herbicides are time consuming. Moreover, they are harmful to the crops as well as to the plants and soil. Prolonged application of same herbicide results in resistance by weeds to the herbicide. An environmental friendly way of removal of weeds can be attained by nano-herbicides. Polymer encapsulated nano-herbicides can be used (Kumar, Bhanjana, Sharma, Sarita, & Dilbaghi, 2015). Targeted application of herbicides entrapped in nanoparticles to the roots of weeds, hinder food reserve, eventually kills the weeds by starving the food resource (Joginder et al., 2017; Manjunatha et al., 2016). Nano forms of silver, copper, zinc, iron, manganese and their oxides are the commonly used nano-herbicides for effective weed control (Joginder et al., 2017). Enormous application of herbicides over a period of time may leave its remaining in the soil which will retain in the soil without decomposition. Silver modified with nanoparticles of magnetite and carboxy methyl cellulose is effective in the detoxification of herbicide remaining in soil (Manjunatha et al., 2016).

#### 14.4.2.2 Nano-fungicides

Commercially available fungicides are causing harm not only to the fungus which attacks the vegetation but they adversely affect the plants also. Nano-fungicides can take up this challenge and can act as antifungal mediators. Many nanomaterials like silver nanoparticles, zinc oxide nanoparticles and titanium dioxide nanoparticles are the proven examples of nanoparticles for their antifungal activity. Unlike Titanium dioxide and Zinc oxide nanoparticles, silver nanoparticles are efficient in their antifungal activity at low loading (Shyla, Natarajan, & Nakkeeran, 2014). Studies showed that nano-silica is effective in fighting against *Fusarium oxysporum* and *Aspergillus niger* in maize (Joginder et al., 2017). Nano-silver is the most cited nanomaterial for its antifungal activity. The action of nano-silver against fungus depends on its size and shape.

#### 14.4.2.3 Delivery of Pesticides

The main problem with conventional pesticides is that they are sparingly soluble in water. As we have discussed in nano-fertilizers, nanotechnology has potential application in the smart delivery of pesticides also (Panpatte & Jhala, 2019). Existing technologies available for distribution of pesticides is not very effective because they cannot provide a stable supply of pesticides to vegetation without undergoing degradation. Action of pesticides can be intensified if the size is very small which will increase their surface area and hence the solubility. Solubility of nano-pesticides can also be achieved by structural modifications like nano-dispersions, nano-emulsions, etc. (Suresh et al., 2013).

Nano-encapsulation and hence slow and sustained release of pesticides can be achieved through polymeric nanoparticles, nano-gels and nano-fibres (Kah & Hofmann, 2014). Research works are carried out to evaluate the possibilities of highly complex nano-formulations with the combination of organic and inorganic nanoparticles like silicon, $TiO_2$, alumina and copper nanoparticles (Kang et al., 2012). Another area where nanomaterials can contribute is the targeted pesticide delivery so that pesticide dosage can be minimized. Slow and sustained release of nutrients and pesticides is described in Fig. 14.6 (Joginder et al., 2017). The nano-material coating on the nutrients and pesticides protects large particles and enables slow release due to the high surface tension of nanoparticles which holds the nutrients and pesticides strongly.

### 14.4.3 Modulation of Sensors Designing and Applications

Effective technologies are indispensable to track, locate and regulate pathogens. After detection of pathogens it will be easy to use pesticides and fertilizers. There are variety of pathogens which are not easy to detect with conventional mechanisms. Fast and speedy detection tools are very much required for the early stage and smart diagnosis of fungal, viral and bacterial pathogens. Sensors based on nanoparticles can easily detect microbial damage with a wide tolerance limit in plants and crops. In-situ tracking of farming and agriculture can be done with wireless nano-sensors along with Global Positioning System (GPS) in cultivated lands. These sensors can take care of the growth of crops and the condition of soil throughout the growth period of crop. Nano-silica is highly efficient in identifying bacterial spot disease in plants. The effect of gold nanoparticles based sensors was studied by

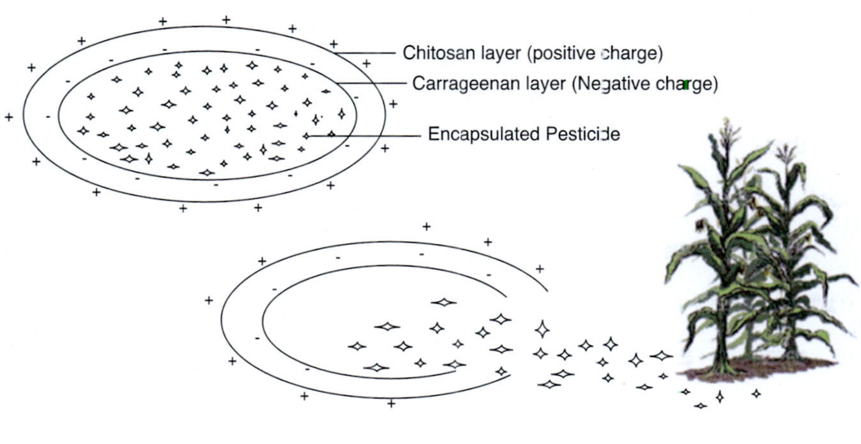


**Fig. 14.6** Slow and sustained release of nutrients/pesticides. (Copyright@ Elsevier, 2017)

Singh, Singh, Agrawal, and Kumar (2010) for detecting diseases in wheat. Plant hormones like jasmonic acid, methyl jasmonate and salicylic acid controls the resistance capability of the plants to various stress environments. Microbial content in a crop can be determined with the help of sensors that could observe and identify the levels of these plant hormones. One such study has been conducted by Wang et al. (2010). They developed an electrochemical sensor productively, with copper nanoparticles and observed the content of salicylic acid in oil seeds to detect a fungus. The concept of precision farming can be best achieved by applying nanosensors. As we have discussed in earlier sections, precision farming and Controlled Environment Agriculture are the ideal and dream concepts which can be well realized with the integration of nanotechnology with agriculture. It is easy to detect pathogens, deficiency of nutrients, level of moisture and fertilizers in soil, etc. with nano-sensors. With the application of nano-sensors farmers can get knowledge of harvesting time of crops, infection of crops and soil, strength of crops and soil. This will help to the optimum application of fertilizers and pesticides at the time of need thereby prevent the unwanted loss of fertilizers and pesticide and reduce the expenses. Also it will help to improve the production and income.

#### 14.4.3.1 Nano-biosensor

Biosensors are devices that measure a biological molecule like DNA sequence or protein using any electronic, optical or magnetic technology through a compact analysis (Scrinis & Lyons, 2007). Nano-biosensors are developed through the integration of developments in the field of nanotechnology as well as electronic fabrication industry for the diagnosis of contaminations (Rai et al., 2012). As in human beings, if early stage detection of unhealthy conditions is possible in plants also, there will be a remarkable reduction of failure in the cultivation. Biomarkers are biological molecules which can be used for health diagnosis by giving indication of pathological processes. Analysis system of these biomolecules requires quick and timely diagnosis and continuous monitoring during the entire growth period which can be attained by real time monitoring with a biosensor. A hybrid analysis system of biosensor with nanomaterials can be employed which give rise to nano-biosensors. The action and performance of the biosensors can be improved with the incorporation of nanomaterials. Antibodies, pathogens, metabolites, etc. can be detected by nano-biosensor. The working principle of biosensor is the intonation of a physiochemical signal accompanying with the binding of bio-analytes onto bio-receptors. They selectively bind the target or bio-analyte. The signal is apprehended and transformed to electrical signal by a transducer. Any variation in the signal is constantly monitored. The variation can be in the form of voltage, current, conductance, impedance, intensity, and phase of electromagnetic radiation, mass, temperature viscosity, etc. Existence or absenteeism of bioagents can be quantified by analysing these parameters.

Nano-biosensors are made with the combination of nanomaterials with transducers (Adamson & Gast, 1997). The performance of biosensors can be upgraded by using nanoparticles because of the ability to modify their size and structure thereby the design of sensor. Nanomaterial such as nanowires, nano-membranes, carbon

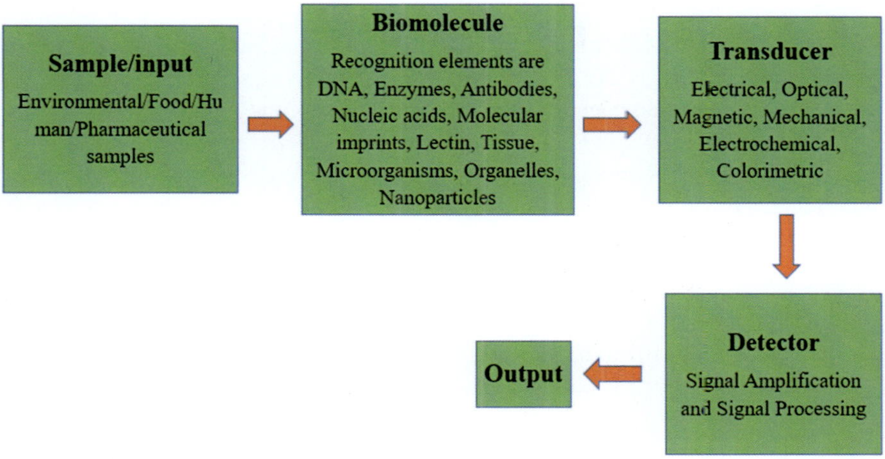


**Fig. 14.7** Components of a nano-biosensor

nanotubes, etc. can be employed. The cost of diagnosis will be reduced and fastness of measurements will be enhanced using biochip technology in nanoscale range.

The main components of a nano-biosensor are the following and are represented in Fig. 14.7.

1. Probe—A biomolecule or biomimetic component—The function of a probe is to accept signals from the sample, modulate it and communicate to the transducer. The biomolecule can be enzymes, antibodies, nucleic acids, molecular imprints, lectins, tissue, microorganisms, or- ganelles, etc. (Shalini, 2014).
2. Transducer—Plants react to different stresses in different ways. The transducer measures the signal from the reaction at the biological element and converts it into measurable electrical signal. Different transducer systems are there depending on the applications.
3. Detector—Detector takes the electrical signal from transducer, amplify it and send for data processing in a microprocessor. The data are analysed and stored (Vineeta, Sefali, & Nrisingha, 2012). Nano-biosensors filled with various microorganisms can be dipped in soil sample—buffer suspension and can measure the oxygen consumption rate by microorganisms. Based on the analysis, the microorganism favourable for the soil can be found out. These sensors can also be used to find out the health condition of the soil and thus soil quality and infection can be assessed before cultivation (Vineeta et al., 2012).

Slow and sustained release of nano-fertilizers can also be attained by the application of nano-sensors. Nano-fertilizer can be attached with nano-biosensors which helps for the measured discharge of nutrients to the soil. Zeolites have the capability to remain in soil for a long time and hold the nutrients to the plant root. It helps in the measured release of fertilizers if it is linked to fertilizers during application. If zeolites are linked to biosensors and applied, it will be very effective to measure the soil deficiencies and also for a controlled release of nutrients to the soil. The same thing can be applied to pesticides also (Vineeta et al., 2012).

Carbon nanotubes have plenty of applications in bio-sensing chips. Biosensors based on liposome are able to observe the presence of pesticides like dichlorvos and paraoxon (Vineeta et al., 2012). Biosensors based on Ti nanoparticles are used to detect bacteria. Researchers developed carbon electrode-based sensors using Bismuth nano-film to detect *E. coli*. DNA damages are detected by nano-biosensors with carbon nanotubes and chitosan bio-nano-composites. Nano-biosensors based on protein nanomaterials can specifically detect protein molecules, plant pathogens, nutrient deficiency, etc. (Vineeta et al., 2012).

### *14.4.4 Seed Germination and Plant Growth*

The primary step in farming is the use of efficient and perfect seedlings which decide the destiny of plants and crop yield. Even though agriculture department states that 80–90% of the seed supplied to the farmers are effective in germination, actually in large scale farming this does not happen due to many technical reasons (Lav, Sindhuja, Joe, Reza, & Edmund, 2012). In most of the places irrigation may not be proper due to scarcity of water which cause poor germination of seeds. Innovative technologies have to be developed and adopted with the help of nanotechnology to overcome these challenges (Subramanian & Tarafdar, 2011). Carbon nanotubes can infiltrate into tomato seeds and are good in penetration of moisture, thus helping for effective and timely germination of seeds (Fernandez & Eichert, 2009). They are efficient in penetrating into hard seed coatings and route the moisture to interior of seeds, thereby enhance seed germination. Carbon nanotubes can act as carrier of preferred molecules into seeds at the time of germination which helps the seed to develop and grow without any disease. However, if the concentration of nanomaterials provided is more, there will be a negative effect in the germination and growth of seedling (Qureshi et al., 2018). Nanotechnology based practices can be used for post-harvest seed storage applications.

## 14.5 Conclusion, Outlook, Industrial Aspects and Future Development

The usage of extensive land areas for agriculture will be declining in future due to the vast population density and scarcity of lands. Only option available is to make use of the existing resources to the fullest extent for sustainable crop production. Sustainable growth in agriculture sector is possible only with the introduction of modern innovative technologies. The application of nano-biotechnology in agriculture industry is opening up many research areas. What we are experiencing is only a tip of the iceberg, yet more have to be figured out. Nanotechnology combined with the tools of green biotechnology develops many smart and efficient systems which are eco-friendly, harmless and environmentally benevolent. The purpose of nano-biotechnology in agriculture is to develop attractive technologies for the intelligent, targeted and sustained delivery of agrochemicals, the use of nano-biosensors in early stage identification of diseases in vegetation, new methods for weed control, pesticide residue detection and degradation, seed germination and plant growth. Remaining of pesticides in soil can be detected by nano-sensors. Smart delivery system of input ingredients to vegetation is a remarkable contribution of nanotechnology which is a solution to many challenges. Government policies and schemes aiming for the maximum production in agriculture industries should promote the implementation of these modern technologies. Initiatives should be taken to create awareness of these technologies among farmers which help in the acceptance and usage of powerful tools of nanotechnology in agriculture.

While we are discussing about the potential uses of nanotechnology, it is to be noticed that safety aspects of nanomaterials are not properly and adequately addressed. Future researches should be emphasized on the areas of the consequences of the interaction of nanomaterials with biological components which is known as nano-bio interface, a most complicated and least understood system. The answers to these questions can be better understood by knowing the mechanism of nano-bio interface which may lead to the safer synthesis of nanoparticles. Nevertheless, it is quite exciting that most of the strategies, that we have discussed in this chapter, for harnessing the potential of nano-biotechnology in agriculture industry, are less harmful to the environment and health. The concept of nano-encapsulation in agriculture is the application of biocompatible coatings on nanoparticles that prevents various undesirable reactions and helps in the smart delivery of agrochemicals. Mixing of bio-fertilizers and agrochemicals with nanomaterials for effective application are less toxic to the ecosystem as well as health. All these aspects are discussed in detail in this chapter. However, furthermore researches are mandatory in elucidating the mechanism of the exciting properties of nanomaterials as well as their interaction with biological system. These studies will pave the way for an unexpected exploration of the possibilities of nanotechnology and its usage in agricultural sector to its fullest extent.

**Acknowledgments** The corresponding author would like to sincerely acknowledge the support from colleagues of Government Engineering College, Thrissur for the successful completion of the work. Copyright permission for reusing some figures published by MDPI and Elsevier are greatly acknowledged.

Ethical issues This article does not contain any studies with human participants or animals performed by the author.

**Conflict of Interest** The corresponding author states that there is no conflict of interest.

## References

Adamson, A., & Gast, A. (1997). *Physical chemistry of surfaces* (6th ed.). New York: John Wiley & Sons.

Adhikari, T., Biswas, A. K., & Kundu, S. (2010). Nanofertilizer—A new dimension in agriculture. *Indian Journal of Fertilisers, 6*(8), 22–24.

Ahmad, A., Mukherjee, P., Senapati, S., Mandal, D., Khan, M. I., Kumar, R., & Sastry, M. (2003). Extracellular biosynthesis of silver nanoparticles using the fungus Fusarium oxysporum. *Colloids and Surface B: Biointerfaces, 28*, 313–318.

Baek, Y. W., & An, Y. J. (2011). Microbial toxicity of metal oxide nanoparticles (CuO, NiO, ZnO, and Sb2O3) to Escherichia coli, Bacillus subtilis, and Streptococcus aureus. *Science of Total Environment, 409*, 1603–1608.

Balaji, D. S., Basavaraja, S., Bedre, M. D., Prabhakar, B. K., & Ventaraman, A. (2009). Extracellular biosynthesis of functionalized silver nanoparticles by strains of Cladosporium cladosporioides fungus. *Colloids and Surfaces B: Biointerfaces, 68*, 88–92.

Banotra, M., Kumar, A., Sharma, B. C., Nandan, B., Verma, A., Kumar, R., Gupta, V., & Bhagat, S. (2017). Prospectus of use of nanotechnology in agriculture—A review. *International Journal of Current Microbiology and Applied Sciences, 6*(12), 1541.

Bhainsa, K. C., & D'Souza, S. F. (2006). Extracellular biosynthesis of silver nanoparticles using the fungus Aspergillus fumigates. *Colloids and Surfaces B: Biointerfaces, 47*, 160–164.

Chen, M., Suna, Y., Lianga, J., Zenga, G., Lia, Z., Tanga, L., Zhub, Y., Jianga, D., & Song, B. (2019). Understanding the influence of carbon nanomaterials on microbial. *Communities Environment International, 126*, 690–698.

Chen, Z., Meng, H., Xing, G., Chen, C., Zhao, Y., Jia, G., Wang, T., Yuan, H., Ye, C., Zhao, F., Chai, Z., Zhu, C., Fang, X., Ma, B., & Wan, L. (2006). Acute toxicological effects of copper nanoparticles in vivo. *Toxicology Letters, 163*(2), 109–120.

Fernandez, V., & Eichert, T. (2009). Uptake of hydrophilic solutes through plant leaves: Current state of knowledge and perspectives of foliar fertilization. *Critical Reviews in Plant Sciences, 28*, 36–68.

Fubini, B., Fenoglio, I., Tomatis, M., & Turci, F. (2011). Effect of chemical composition and state of the surface on the toxic response to high aspect ratio nanomaterials. *Nanomedicine, 6*(5), 899–920.

Gahukar, R. T., & Das, R. K. (2020). Plant-derived nano-pesticides for agricultural pest control: Challenges and prospects. *Nanotechnology for Environmental Engineering, 5*, 1–10.

Guo, H., White, J. C., Wang, Z., & Xing, B. (2018). Nano-enabled fertilizers to control the release and use efficiency of nutrients. *Current Opinion in Environmental Science & Health, 6*, 77–83.

Huang, B., Chen, F., Shen, Y., Qian, K., Wang, Y., Sun, C., & Cui, H. (2018). Advances in targeted pesticides with environmentally responsive controlled release by nanotechnology. *Nanomaterials, 8*(2), 102.

Iavicolia, I., Lesoa, V., Beezholdb, D. H., & Shvedova, A. A. (2017). Nanotechnology in agriculture: Opportunities, toxicological implications, and occupational risks. *Toxicology and Applied Pharmacology, 15*(329), 96–111.

Jha, M. N., & Prasad, A. N. (2006). Efficacy of new inexpensive cyanobacterial biofertilizer including its shelf-life. *World Journal of Microbiology and Biotechnology, 22*, 73–79.

Joginder, S. D., Ravinder, K., Naresh, K., Pawan, K., Kiran, N., & Surekha, D. (2017). Nanotechnology: The new perspective in precision agriculture. *Biotechnology Reports, 15*, 11–23.

Kah, M., & Hofmann, T. (2014). Nanopesticide research: Current trends and future priorities. *Environment International, 63*, 224–235.

Kah, M., Kookana, R. S., Gogos, A., & Bucheli, T. D. (2018). A critical evaluation of nanopesticides and nano-fertilizers against their conventional analogues. *Nature Nanotechnology, 13*, 677–684.

Kang, M. A., Seo, M. J., Hwang, I. C., Jang, C., Park, H. J., Yu, Y. M., & Youn, Y. N. (2012). Insecticidal activity and feeding behavior of the green peach aphid, Myzus persicae, after treatment with nano types of pyrifluquinazon. *Journal of Asia-Pacific Entomology, 15*, 533–541.

Klaus, T., Joerger, R., Olsson, E., & Granquist, C. G. (1999). Silver-based crystalline nanoparticles microbially fabricated. *Proceedings of the National Academy of Sciences USA, 96*, 13611–13614.

Kowishik, M., Ashtaputre, S., Kharrazi, S., Vogel, W., Urban, J., Kulkarni, S. K., & Patnikar, K. M. (2003). Extracellular synthesis of silver nanoparticles by a silver-tolerant yeast strain MKY3. *Nanotechnology, 14*, 95–100.

Kumar, S., Bhanjana, G., Sharma, A., Sarita, Sidhu, M. C., et al. (2015). Herbicide loaded carboxymethyl cellulose nanocapsules as potential carrier in agrinanotechnology. *Science of Advanced Materials*, *7*, 1143–1148.

Kumar, S., Nehra, M., Dilbaghi, N., Marrazza, G., Hassan, A. A., & Kim, K. H. (2019). Nano-based smart pesticide formulations: Emerging opportunities for agriculture. *Journal of Controlled Release, 294*, 131–153.

Landini, P., Antoniani, D., Burgess, J. G., & Nijland, R. (2010). Molecular mechanisms of compounds affecting bacterial biofilm formation and dispersal. *Applied Microbiology and Biotechnology, 86*, 813–823.

Lav, R. K., Sindhuja, S., Joe, M. M., Reza, E., & Edmund, W. S. (2012). Applications of nanomaterials in agricultural production and crop protection: A review. *Crop Protection, 35*, 64–70.

Lengke, M. F., Fleet, M. E., & Southam, G. (2007). Biosynthesis of silver (I) nitrate complex. *Langmuir, 23*, 2694–2699.

Lherm, C., Muller, R. H., Puisieux, F., & Couvreur, P. (1992). Alkylcyanoacrylate drug carriers: II. Cytotoxicity of cyanoacrylate nanoparticles with different alkyl chain length. *International Journal of Pharmaceutics, 84*(1), 13–22.

Liu, R., & Lal, R. (2015). Potentials of engineered nanoparticles as fertilizers for increasing agronomic productions. *Science of Total Environment, 514*, 131–139.

Lu, C., Brauer, M. J., & Botstein, D. (2009). Slow growth induces heat-shock resistance in normal and respiratory-deficient yeast. *Molecular Biology of Cell, 20*, 891–903.

Luis, C. H., Karla, A. A., Jeisson, U. A., Laura, V. F., Gabriela, M. O. V., & Jose, R. V. B. (2020). Green synthesis of gold and silver nanoparticles from plant extracts and their possible applications as microbial agents in agriculture area. *Nanomaterials, 10*, 1763–1785.

Macarto, P. D., De Souza, G. I. H., Alves, O. L., Esposito, E., & Duran, N. (2005). *Antibacterial activity of silver nanoparticles synthesized by Fusarium oxysporum strain*. 2nd Mercosur Congress on Chemical Engineering.

Mahmoudi, M., & Serpooshan, V. (2012). Silver-coated engineered magnetic nanoparticles are promising for the success in the fight against antibacterial resistance threat. *ACS Nano, 6*, 2656–2664.

Manikant, T., Suresh, K., Amit, K., Poonam, T., & Shailendra, K. (2018). Agro-nanotechnology: A future technology for sustainable agriculture. *International Journal of Current Microbiology and Applied Sciences, 7*, 196–200.

Manjunatha, S. B., Biradar, D. P., & Aladakatti, Y. R. (2016). Nanotechnology and its applications in agriculture: A review. *Journal of Farm Sciences, 29*(1), 1–13.

Mouxing, F., Qingbiao, L., Daohua, S., Yinghua, L., Ning, H., Xu, D., Huixuan, W., & Jiale, H. (2006). Rapid preparation process of silver nanoparticles by bioreduction and their characterizations. *Chinese Journal of Chemical Engineering., 14*, 114–117.

Mukherjee, P., Roy, M., Mandal, B. P., Dey, G. K., Mukherjee, P. K., Ghatak, J., Tyagi, A. K., & Kale, S. P. (2008). Green synthesis of highly stabilized nanocrystalline silver particles by a non-pathogenic and agriculturally important fungus *T. asperellum*. *Nanotechnology, 19*, 075103–075110.

Navya, P. N., & Hemant, K. D. (2016). Rational engineering of physicochemical properties of nanomaterials for biomedical applications with nanotoxicological perspectives. *Nano Convergence, 3*, 1–16.

Pandey, S., Giri, K., Kumar, R., Mishra, G., & Rishi, R. R. (2018). Nano-pesticides: Opportunities in crop protection and associated environmental risks. *Proceedings of National Academy of Sciences, India Section B: Biological Sciences, 88*, 1287–1308.

Panpatte, D. G., & Jhala, Y. K. (Eds.). (2019). *Nanotechnology for agriculture advances for sustainable agriculture*. Singapore: Springer.

Panpatte, D. G., Jhala, Y. K., Shelat, H. N., & Vyas, R. V. (2016). Nanoparticles: The next generation technology for sustainable agriculture. In D. P. Singh et al. (Eds.), *Microbial inoculants in sustainable agricultural productivity* (pp. 289–300). New Delhi: Springer.

Parikh, R. Y., Singh, S., Prasad, B. L. V., Pathole, M. S., Sastry, M., & Shouche, Y. S. (2008). Extracellular synthesis of crystalline silver nanoparticles and molecular evidence of silver resistance from Morganella sp.: Towards understanding biochemical synthesis mechanism. *Chembiochem, 9*, 1415–1422.

Pradeep, T. (2007). *Nano: The essentials—Understanding nanoscience and nanotechnology*. Noida: Tata McGraw-Hill.

Praseetha, P. N., & Panda, T. (2012). Kinetics of biosynthesis of silver nanoparticles using Fusarium oxysporum. *Current Trends in Technology and Sciences, 1*(1), 47–52.

Peteu, S. F., Oancea, F., Sicuia, O. A., Constantinescu, F., & Dinu, S. (2010). Responsive polymers for crop protection. *Polymers, 2*, 229–251.

Poonam, J., Prem, L. K., Ajay, K. B., Sudheer, K., & Gyanendra, P. S. (2018). Scope and applications of nanotechnology for wheat production: A review of recent advances. *Wheat and Barley Research, 10*(1), 1–14.

Qureshi, A., Singh, D. K., & Dwivedi, S. (2018). Nano-fertilizers: A novel way for enhancing nutrient use efficiency and crop productivity. *International Journal of Current Microbiology and Applied Sciences, 7*(2), 3325–3335.

Rai, M., Yadav, A., & Gade, A. (2009). Silver nanoparticles as a new generation of antimicrobials. *Biotechnology Advances., 27*, 76–83.

Rai, V., Acharya, S., & Dey, N. (2012). Implications of nano-biosensors in agriculture. *Journal of Biomaterials and Nano-biotechnology, 3*, 315–324.

Safdar, A., Obaidullah, S., Tariq, M., Muhammad, A. H., Ijaz, A., & Bashir, A. K. (2018). A review about perspectives of nanotechnology in agriculture. *Pakistan Journal of Agricultural Research, 30*(2), 116–121.

Saxena, A., Jain, A., Upadhyay, P., & Gauba, P. G. (2018). Applications of nanotechnology in agriculture. *Journal of Nanoscience Nanoengineering and Applications, 8*(1), 20–27.

Scrinis, G., & Lyons, K. (2007). *International Journal of Sociology of Food and Agriculture, 15*(2), 22–44.

Shalini, P. (2014). Nano-biosensors: The future for diagnosis of disease? *Dovepress Journal, 2014*, 1–10.

Shrivastava, S., Bera, T., Roy, A., Singh, G., Ramachandrarao, P., & Dash, D. (2007). Characterization of enhanced antibacterial effects of novel silver nanoparticles. *Nanotechnology, 18*, 103–111.

Singh, M. D., Chirag, G., Prakash, P. O., Mohan, M. H., Prakasha, G., & Vishwajith (2017). Nano-fertilizers is a new way to increase nutrient efficiency in crop production. *International Journal of Agricultural Sciences*, *9*(7), 3831–3833.

Singh, S., Singh, M., Agrawal, V. V., & Kumar, A. (2010). An attempt to develop surface plasmon resonance based immune sensor for Karnal bunt (Tilletia indica) diagnosis based on the experience of nano-gold based lateral flow immune dipstick test. *Thin Solid Films, 519*, 1156–1159.

Shishira, M. R. I., Xiea, L., Sunb, C., Zhenga, X., & Chena, W. (2018). Advances in micro and nano-encapsulation of bioactive compounds using biopolymer and lipid-based transporters. *Trends in Food Science & Technology, 78*, 34–60.

Shyla, K. K., Natarajan, N., & Nakkeeran, S. (2014). Antifungal activity of zinc oxide, silver and titanium dioxide nanoparticles against Macrophomina phaseolina. *Journal of Mycology and Plant Pathology, 44*, 268–273.

Subramanian, K. S., & Sharmil, R. C. (2009). *Nano-fertilizer formulations for balanced fertilization of crops*. Paper presented at the Platinum Jubilee Celebrations of ISSS, New Delhi. 21–25.

Subramanian, K. S., & Tarafdar, J. C. (2011). Prospects of nanotechnology in Indian farming. *Indian Journal of Agricultural Sciences, 81*(10), 887–893.

Subin, P., Tapobrata, P., Praseetha, P. N., & Thomas, T. (2014). Biosynthesis of silver nanoparticles. *Journal of Nanoscience and Nanotechnology, 14*, 2038–2049.

Suresh, K. R. S., Shiny, P. J., Anjali, C. H., Jerobin, J., Goshen, K. M., Magdassi, S., Mukherjee, A., & Chandrasekaran, N. (2013). Distinctive effects of nano-sized permethrin in the environment. *Environmental Science and Pollution Research, 20*, 2593–2602.

Tanveer, B. P., Bisma, M., Tariq, M., & Reiaz, U. R. (2019). *Development of nano-bioformulations of nutrients for sustainable agriculture* (pp. 381–396). Boca Raton: CRC Press, Taylor and Francis Group.

Vineeta, R., Sefali, A., & Nrisingha, D. (2012). Implications of nano-biosensors in agriculture. *Journal of Biomaterials and Nano-biotechnology, 3*, 315–324.

Wang, Z., Wei, F., Liu, S. Y., Xu, Q., Huang, J. Y., Dong, X. Y., Yua, J. H., Yang, Q., Zhao, Y. D., & Chen, H. (2010). Electrocatalytic oxidation of phytohormone salicylic acid at copper nanoparticles-modified gold electrode and its detection in oilseed rape infected with fungal pathogen Sclerotinia sclerotiorum. *Talanta, 80*, 1277–1281.

Yang, Y., Hong-tao, Z., Jian, W., Meng, X., Yang, L., & Yu-long, Z. (2012). Preparation and properties of modified polyvinyl alcohol film for encapsulation of fertilizer. *Acta Metallurgica Sinica, 18*, 1295–1301.

# Chapter 15
# Key Challenges and Scopes of Biomaterials Commercialization: Therapeutic Delivery

**Tean Zaheer, Muhammad Imran, Amjad Islam Aqib, Kaushik Pal, Alisha Tahir, Iqra Zaheer, and Rao Zahid Abbas**

## 15.1 Overview of Bio-nanomaterials: Global Significance in Therapeutic Mechanism

Nano-biomaterials or bio-nanomaterials or green nanomaterials or biogenic materials are usually more or less synonymous terms applied to the composites that completely or partially arise from biological entities (plants, algae, viral/human/animal cell lines, bacteria, fungi, or any other biomolecule) having at least one dimension less than 100 nm in size (Honek, 2013; Rahman et al., 2019; Vasquez et al., 2016). Biological materials of varying physiochemical properties have been tested and validated as effective therapeutics and vehicles/cargo for therapeutics. The concept of utilizing biomaterials for therapy is not nascent; rather it comes from prehistoric times, when extensive list of plants was employed to treat human/animal maladies (Spałek et al., 2019). The unique biochemical constituents of green nanomaterials

T. Zaheer (✉) · M. Imran · R. Z. Abbas
Department of Parasitology, University of Agriculture, Faisalabad, Pakistan
e-mail: 2013ag2605@uaf.edu.pk

A. I. Aqib
Department of Medicine, Cholistan University of Veterinary and Agriculture Sciences, Bahawalpur, Pakistan

K. Pal (✉)
Laboratório de Biopolímeros e Sensores, Instituto de Macromoléculas, Universidade Federal do Rio de Janeiro (LABIOS/IMA/UFRJ)–Cidade Universitária, Rio de Janeiro, Brazil

Wuhan University, Wuchang Dist., Hubei, China

A. Tahir
Department of Physical Chemistry, Quaid e Azam University, Islamabad, Pakistan

I. Zaheer
Department of Pathology, University of Agriculture, Faisalabad, Pakistan


K. Pal (ed.), *Bio-manufactured Nanomaterials*,
https://doi.org/10.1007/978-3-030-67223-2_15

are more biologically compatible, less hazardous, least toxigenic, eco-friendly, renewable and sustainable natural resources possess potential scope against disease pathogens (Sarli, Kalani, & Moradi, 2020). Chemical ingredients that make biomaterials standout the other comparable treatments have also been extensively studied. It is, therefore, well-established that nanomaterials derived from biological resources may be more efficacious, least toxic, safer, and green alternative to synthetic/artificial compounds in tackling pathogens of concern.

Bio-nanomaterials could be synthesized through different routes, each having its own merits and limitations. Usually, the synthesis involves the addition of precursor chemical, biological extract/molecules/composites and the provision of required set of temperature and atmospheric conditions (Ovais et al., 2018). The addition of surfactant may not be required due to intrinsic surfactants present in the biological raw materials used as mediators for nanoparticle synthesis (Bhattarai, Zaker, & Bigioni, 2018). Scientists have shifted their research focus from purely chemical to green or bio-nanomaterials, keeping in view the safety profile and ease of synthesis. Additionally, the end-consumers are more comfortable with the products labeled as of originating from biological/natural/organic origins. However, this bias toward green nanomaterials is to be weighed for its applicability at commercial scales.

Nano-biotechnology holds great promise in treatment of drug-resistant and even multidrug-resistant pathogens. It is possible to use them as alternative to otherwise expensive and/or in-effective treatment options. Green nanomaterials at laboratory scales present some challenges associated with scalability of synthesis, quantitation, shelf-stability, poor size control, and degree of effectiveness against target pathogens (Do & Friščić, 2017). These factors may also affect the reproducibility of laboratory-scale experiments. Moreover, the fate of core-ions within the host biological systems is also questionable. Even if excreted out of the target host, if unchanged, they still might pose threats to associated ecological assemblages. This chapter deals with the concerns and possible ways to appropriately address these issues.

## 15.2 Challenges in Commercialization of Biomaterials

### *15.2.1 Brief Recap of Tested Bio-nanomaterials*

The applications of green nanomaterials are multidisciplinary. It has secured a promising stance in electronics, medical sciences, water management, robotics, food packaging and preservation, mechatronics, agriculture, genetic engineering, textiles, nutraceuticals, aeronautical sciences, machine learning, cosmetics, energy enrichment, artificial intelligence, construction industry and most importantly, waste valorization and environment remediation (Huang, Mei, Chen, & Wang, 2018; Vasquez et al., 2016). A comprehensive listing of how nano-biomaterials standout the modern therapeutics has been given in Fig. 15.1.

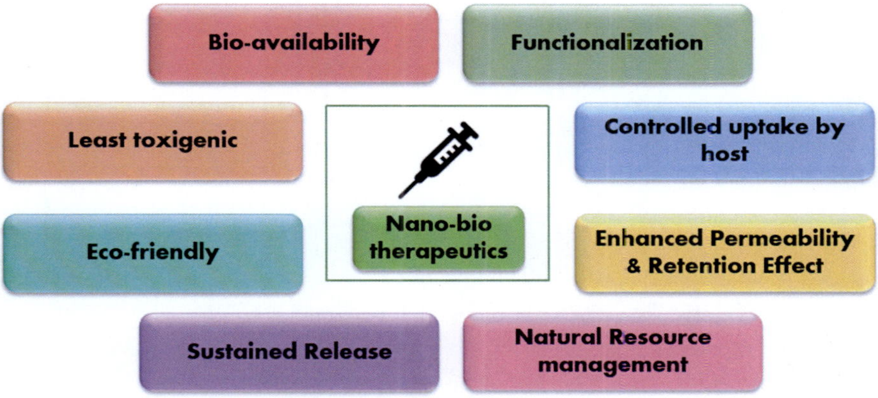


**Fig. 15.1** Unique properties of nano-biomaterials for therapeutic applications

**Table 15.1** Summary of most recent biogenic nanomaterials against pathogens

| Biogenic nanomaterial | Model of study | Applications | References |
|---|---|---|---|
| Ginsenoside Rh2 nanoniosomal | In vitro | Anticancer | Zare-Zardini et al. (2020) |
| Plant (Malva Sylvestris) mediated $Fe_3O_4$ NP | In vitro | Antibacterial, anticancer | Mousavi et al. (2019) |
| Plant (*Olax nana*) mediated Ag and Au NP | In vitro and in vivo | Antibacterial, anticancer, antiparasitic, enzyme inhibition, anti-inflammatory, antinociceptive | Ovais et al. (2018) |
| Plant (*Arisaema flavum*) mediated Ag NP | | Antibacterial against multidrug-resistant bacteria | Rahman et al. (2019) |
| Algae (*Spatoglossum asperum*) mediated Ag NP | In vitro | Antibacterial | Ravichandran et al. (2018) |
| Nanoniosomes | In vivo | Anti-biofilm, antifungal | Haque, Alfatah, Ganesan, and Bhattacharyya (2016) |

The applications in biomedicine include those in sensing, theranostics, therapeutics, precision medicine, vaccinology, organ/tissue transplants, disease management, and prevention of many diseases (Ovais et al., 2018). Major disease pathogens of bacterial, viral, cancerous, parasitic, auto-immune, of metabolic and non-infectious/functional origin have been shown to be adversely affected by green nanomaterials with or without combination therapeutics (Chopra et al., 2013; Khoobchandani et al., 2020; Khoobchandani, Katti, Maxwell, Fay, & Katti, 2016; Ovais et al., 2017; Takeda et al., 2016). Moreover, the effective and highly selective absorption of green nano-composite medications within heterogeneous solid tumor masses have been reported (Al-Yasiri et al., 2017). Table 15.1, presents a comprehensive inventory of variety of biogenic materials against several pathogens of concern.

## 15.2.2 Lessons from Laboratory-Scale Testing

### 15.2.2.1 Analyses and Testing

Modern approaches to facile synthesis of green nanomaterials have been tested widely. Idea of using green materials for synthesizing nanoparticles revolves around the concept of minimum toxicity and hazards associated. One-pot and single-step approaches have been optimized at laboratory scales.

An approach focusing on pullulan-silver nanoparticles (P-Ag NPs) featuring the use of UV radiation instead of any toxic chemical, was described (Rahman et al., 2019). Topical nano-therapeutics may be tested for permeation ability, which could be helpful in further rationalizing the desired viscosity of preparation (Fetih, 2016).

### 15.2.2.2 Ecotoxicological Profiles During In Vivo Testing

There is a list of viability, overall integrity, and organelle-specific functionality testing for the evaluation of nontarget toxico-pathological effects of nanoparticles. However, due to intrinsic variation among nanomaterials, in their ability to adhere to different organelles/systems of the cell, it is difficult to rationalize and declare a single or a set of assays quantifying cytotoxic effects. Nanomaterials have shown a large promise in effective prophylaxis of fish diseases, when administered within aquatic systems. This fact indicates the higher probability of least or minimally toxigenic potential of green nanomaterials which arise from natural resources. The tea-mediated silver nanoparticles showed nontoxic effects, as revealed by mitochondrial function, cell viability and integrity in human keratinocytes (Moulton et al., 2010). Therefore, it is important to develop very sensitive and specific toxicological assays targeting the quantitation of nanomaterials.

## 15.2.3 Obstacles in Scaling up of Biomaterials

Procuring biomaterials for the biogenic synthesis of nanomaterials could be laborious. The identification of plants up to desired genera also demands high expertise. Moreover, materializing the extracts from roots/barks/seeds may be rendered difficult as they require variant protocols. Quantity and purity of the extracts from plants, fungi, and algae may also be variant. Refinement in terms of quality enhancement requires more research attention.

Distribution of biogenic flora and fauna throughout the world is variant. Moreover, the phytochemical profiles, owing to age of flora, area of cultivation, seasonal fluctuation, may differ to that of other comparable groups. Similarly, the genetic variations among bacterial, viral and yeasts strains may dictate the quality of biogenic nanoparticles procured. Therefore, it is important to analyze and quantify the chemical recipes of biological materials being utilized for green nanoparticle synthesis. The heterogeneity of infectious samples is difficult to achieve at laboratory scales. True representative clinical samples may be procured to ensure the efficacy of laboratory-scale materials in challenge clinical infections (Kelley, 2017). Appropriate tools may be optimized and used in the studies to compare the effect of chemical make-up on the overall yield and physicochemical properties of nanomaterials.

## 15.3 Significance in Therapeutic Biomaterial Commercialization

### *15.3.1 Need for Alternative and Personalized Therapeutics*

Protection of bacteria from being harmed depends upon osmotic and mechanical pressure of cell wall. Composition and structural moiety of cell wall in case of gram negative being abundantly encrypted with lipopolysaccharides settled Criss cross way makes it hard for antibiotics to work compared to the cell walls of gram-positive bacteria. Resistance against antibiotics is not limited to these phenomena, however. Bacteria use quorum sensing phenomena to modify their genetics in response to antibacterial drugs. Being complicated phenomena, the antimicrobial resistance has appeared top trending in scientific community in that it spreads across 22 countries having approximately 500,000 people, are affected round the globe (Izquierdo-Barba et al., 2011). Antimicrobial resistance is thought to cause higher deaths than cancer by 2050 if solutions are not sought (Colilla et al., 2014). Biofilm appears to be significant limiting factor in the process of treatment due to uncontrolled bacterial population. This can be covered if chemical/physical properties of surfaces are changed (anti-fouling surface) with to make surfaces unfavorable for bacterial attachments (Rodriguez-Palomo, Monopoli, Afonso, Izquierdo-Barba, & Vallet-Regí, 2016). Nanoparticles on the other hands may serve directly as antibacterial while indirectly in the form of nanocarriers (Castillo et al., 2019). In case of surgical implants, the biofilm may form reversibly by nonspecific interaction while irreversible by protein mediated specific and nonspecific interactions. Once the biofilm is structured, suppression of bacteria becomes hard due to inability of antibiotics to reach bacteria. Hence single treatment does not work but alternative approaches are required to resist bacterial colonization, adherences, and biofilm formation. Advent of nanoparticles have appeared as promising alternative in the form of their direct antibacterial application or indirectly impregnate antibiotics. Examples of salient nanoparticles of these attributes are silver, copper, titanium dioxide films, and NO-releasing materials.

#### 15.3.1.1 Zwitterionic Surfaces in Nanomaterials

Zwitter ionization uses covalent grafting of various parts resulting in uniform number of positive and negative charges on surface to maintain electrical interraciality. In such conditions, bacteria are unable to adhere due to highly hydrophobicity of surface made of tightly bound water layers. Lysin has been used for this purpose and trials have shown *S. aureus* unable to bound with surface have lysin coating. Nano biomaterials have been used to get similar findings in a way that positively and negatively charges were grafted directly and simultaneously. Silica based SBA-15 zwitterion (-NH3 + /COO- groups) mesoporous bioceramics have proved effective against *E. coli* (Magiorakos et al., 2012). Using mesoporous structures dual antibacterial capability-based SBA-15 bio ceramics have been created. Additionally, the pores are filled with broad spectrum antibiotics to eliminate biofilm. Ions like -NH3 +/− SiO- and = NH2 +/− SiO- are used as zwitterionic ion pairs that following incubation (90 minutes) causes inhibition of growth of *S. aureus* by 99.9%. Use of cephalexin in mesopores for 15 days resulted in slow release of antibiotics thus removal of planktonic bacteria from the environment (Colilla et al., 2014). Some of alloys such as Ti6Al4V have also gone through zwitter ionization (Rodriguez-Palomo et al., 2016) but clinical trials are still needed to find efficacy studies.

#### 15.3.1.2 Textural Modifications

On the basis manufacture of various nano patterns, such as nanotubes, nanoparticles, and nanopillars, various methods have been studied to obtain artificial antibacterial surfaces. In this respect, the $TiO_2$ nanotube coating has been coated on the titanium surface by the anodic oxidation circuit. Their antibacterial property is closely related to contact angle, nanotube size and crystallinity (rutile or anatase phase) (Mei et al., 2014). Another instance is the use of magnetron sputtering (MS-GLAD) on titanium alloy surface that can produce nanostructured coatings with various morphologies on a large area. Recently, it has been reported to form nano-patterned coatings on Ti6Al4V substrates. The coating is made by nanopillars arranged nearly vertical, the length of the nanopillars is 250–350 nm, the diameter is 40–60 nm, and the center-to-center distance is 100–200 nm (Izquierdo-Barba et al., 2015). After 90 min of cultivation, the bacterial adhesion of *Staphylococcus aureus* was reduced by 70%. The engineering of surface properties of materials could lead to many useful applications at the cellular levels of pathogens.

#### 15.3.1.3 Inherent Antibacterial Properties

Metals and metal oxides are widely known for their inherent antibacterial properties, so they have been utilized in fungal and bacterial infections treatment before the revelation of antibiotics. In addition to silver, other metallic nanoparticles (e.g., gold) and nanoparticles of metal oxides such as zinc oxide (ZnO), copper oxide (CuO), iron oxide (FeO), and titanium dioxide ($TiO_2$), etc.) are also profoundly studied for antibacterial therapy (Gold, Slay, Knackstedt, & Gaharwar, 2018). The synthesis and preparation of nanoparticles of silver is a key area of concern today due to its bactericidal properties. In addition to traditional techniques based on chemical reduction processes (reducing agents using Ag + ions in the existence of stabilizers in appropriate solvents), updated methods rooted on green synthetic chemistry are also becoming more popular. The green synthesis of nanoparticles is environmentally friendly; it uses plants, biological materials, and microbial entities as reducing and capping agents. Silver nanoparticles gained through green biosynthesis provide an up-to-date potential substitute to chemically developed nanoparticles (Roy, Bulut, Some, Mandal, & Yilmaz, 2019). In ruminants, Ag-NPs have been evaluated in vivo. For postoperative treatment of silver nanoparticles, they have been successfully applied to the postoperative treatment of case intestinal lymphadenitis in small ruminants and have achieved positive results. The use of Ag-NPs-based ointments for postoperative treatment of caseous lymphadenitis can lead to faster healing, reduced wound contamination, and no obvious toxicity (Santos et al., 2019). Polydopamine and chitosan showed dual chelation in hybrid coated silver nanoparticles and significantly reduced ion release. The coating shows more than 90% antimicrobial efficacy against *Staphylococcus epidermidis*, *Staphylococcus aureus*, and *Escherichia coli*. In an in vivo study, it was also found that the overlay can promote the differentiation of osteogenic MC3T3-E1 cells and showed promoted bone implant sustainment and osseointegration. Thus, composite coatings are excellent antibacterial agents and accelerators for osseointegration of bone implants, providing broader insights for the use of nanoparticles (Xie et al., 2019). Recently, when used in combination with NIR photothermal therapy, non-antibiotic-based nano-formulations containing gold nanorods have shown significant antibacterial effects in the treatment of drug-resistant pneumonia. The gold nanorods of 50–100 nm length are enriched with glucose-oxidized polymers for blocking the lectins of bacterial origin particularly; a vital for the formation of bacterial biofilms. The new formula manifests the most potent elimination potential for bacterial habitat and microbial infections such as *Pseudomonas aeruginosa* through blocking the bacterial lectins and NIR light-induced photothermal effect of gold nanorods (Zhao et al., 2019). Lately, amine-functionalized zinc oxide nanocrystals have also been created as a highly compatible and osteo-inductive nano-based antibiotic delivery for bony tissue grafting and engineering. A contemporary, rapid and reproducible microwave-assisted synthesis method is used to prepare ZnO nanocrystals with a diameter of 20 nm. After chemical functionalization by immobilizing the aminopropyl groups on the surface of ZnO, the biocompatibility of ZnO-NH2 nanocrystals was tested. The crystals can promote osteoblasts and Langerhans negative bacteria (including gram-positive dermal Bacteria) cell proliferation and differentiation. Such as *Escherichia coli* and *Staphylococcus aureus*. In vitro results indicate ZnO-NH2 nanocrystals as potent candidates for handling the infectious diseases in bony implants and promoting bone tissue proliferation (Garino et al., 2019).

### *15.3.2 Nanomaterials as Nanocarriers*

A substitutive policy to fight infection using nanoparticles is to manipulate them as vehicles to distribute antimicrobial agents such as antibiotics or other bactericide nanoparticles. Such nanocarriers can be of great use in cancer therapy because of their multifaceted function, ability to create smart nanomaterials with prompt response components (Castillo et al., 2019) have targeting capability against cancer cells (Montalvo-Quirós et al., 2019) and great penetration power against the deepest areas of solid tumors.

Nanocarriers of antimicrobial agents must be able to guard the active ingredient from degeneration and to intensify the potency of the active compound or supplement its bioavailability for treatment purpose. The nanocarrier may facilitate the controlled and sustained release of the loaded antimicrobial drugs; that will be of great use in maintaining an optimum drug concentration level in the bloodstream for a certain period of time. They can also offer this delivery to several other antibiotics at the same time or can be used in combined therapy if other stimuli responsive nanoparticles are used in a single similar nano-system. Furthermore, these nanocarriers can also act as a platform for surface modification that can allow target specification at the site of infection, if previously infection has occurred. Table 15.2 summarizes the mechanism of action of nanomaterials in various applications:

**Table 15.2** Mechanism of activity and application of salient nanoparticles

| Nanoparticles | Mode of action/range of activity | Applications |
|---|---|---|
| Silver nanoparticles (Ag-NPs) | • The interlinkage of NP with intracellular proteins (especially membrane proteins consisting of sulfur and microbial DNA) can hinder with cell division and cause cell death. Because of Ag ions liberation from Ag-NPs, bacterial replication is impaired (Chen & Schluesener, 2008).<br>• Antibacterial, antiviral, antifungal. | • Medical device coating.<br>• Medical equipment sterilization.<br>• Household gadgets and water purifiers.<br>• Coating of refrigerators and food containers (Jain & Pradeep, 2005).<br>• Nano sensors, biotags, and optical data storage (El-Sayed, 2001). |
| Zinc nanoparticles (ZnO-NPs) | • NPs liberate metal ions, produce reactive oxygen species (ROS) and amass in the bacterial cell membrane, resulting in damage to the cell wall, increased membrane penetration, and internalization of NPs due to lack of proton power and incorporation of toxic dissolved zinc ions (Sirelkhatim et al., 2015). | • Food industry in color preservation and spoilage prevention (Seil & Webster, 2012).<br>• Self-cleaning glass and ceramics made up of ZnO-NPs. |
| Titanium nanoparticles ($TiO_2$-NPs) | • $TiO_2$-NP has ability to destroy microorganisms under light due to its photocatalytic properties.<br>• $TiO_2$-NPs oxidizes cellular components by ROS that results in damage.<br>• In the absence of light, direct contact with cell wall and adsorption causes cellular integrity (Seil & Webster, 2012).<br>• Titanium nanoparticles are antibacterial, anti-biofilm activity, antiviral, and antiparasitic. | • Pharmaceutical industries, Food items, and cosmetic products.<br>• Air filtration, water filtration, antibiotic covering on medical devices (Seil & Webster, 2012). |

(continued)

**Table 15.2** (continued)

| Nanoparticles | Mode of action/range of activity | Applications |
|---|---|---|
| Copper nanoparticles (Cu-NPs) | • The synthesis of copper nanotubes is challenging because it is extremely sensitive to air, leading to the oxide layer formation, that can result in a significant decrease in antibacterial property.<br>• Chelate is a highly stable product that can maintain metal ions surrounded by organic molecules (chelating agents).<br>• The antibacterial activity of copper amino acid chelate is about ten times that of copper nanoparticles.<br>• Cu's bactericidal activity based on the degree of agglomeration, which is an equivalent problem with Cu-NPs. Lowering agglomeration will result in less NPs, thereby yielding high surface area for dissolving copper ions and interacting with bacterial membranes, leading to greater toxicity. The metallic and ionic forms of copper produce hydroxyl radicals, which harm the key proteins and DNA (Chatterjee & Chakra, 2014). | • Coating of medical device.<br>• Antimicrobial preparation. |
| Gold nanoparticles (AuNPs) | • Photothermal therapy (PTT) gives an intriguing method for enhancing the bactericidal potential of AuNPs (Zheng et al., 2011).<br>• The gold nanoparticles irradiated by laser energy liberate heat because of excitation and oscillation of electrons, forming them anticancerous and antibacterial agents (Riley & Day, 2017). | • Potential candidate in gene delivery systems in case of cancer therapy. |
| Miscellaneous metallic oxide (silicon, calcium, magnesium oxide, aluminum oxide) | • Silicon dioxide nanowires can have a biocidal effect due to hinderance in cell functions like; cell differentiation, adhesion, and bacterial spread. Furthermore, Si nanoparticles may hinder the attachment of bacteria with oral biofilms. Silica nanoparticles have antibacterial properties and help in biomedical applications by working as carriers of antibacterial agents.<br>• CaO and MgO have strong antibacterial properties, that are mediated by the generation of superoxide on the particle surface and the increase in pH because of nanoparticle hydration with water.<br>• $Al_2O_3$NP can firmly bind and destroy the bacterial cell wall, thereby increasing the permeability of the cell wall. | |

(continued)

**Table 15.2** (continued)

| Nanoparticles | Mode of action/range of activity | Applications |
|---|---|---|
| Carbon-based nanomaterials (fullerene, graphene and diamond-like carbon, carbon nanotubes (CNTs and DLC)) | • Fullerenes induce cell membrane damage, DNA damage and affect energy metabolism pathways, thereby inactivating microorganisms. It also has photochemical property, that when exposed to light will cause the production of ROS, which contributes to antibacterial activity (Al-Jumaili et al., 2017).<br>• Carbon nanostructures may cause physical scratches and impair bacterial cell walls and cell membranes.<br>• During the antibacterial process, CNTs longer than 50μm may be wrapped around bacterial cells.<br>• Graphene sheets can discrete bacteria and microenvironment, which is nearly harmless to survival.<br>• The chemical interaction between nanomaterials and the bacterial cells surface can help in generation of reactive oxygen species.<br>• It is also recommended to remove electrons from the surface of microorganisms into carbon nanomaterials, which is the source of oxidative stress independent of ROS. | |

## 15.3.3 Multidisciplinary Collaboration

Field of biomaterials has been modified in which nanomaterials are divided into three different generations based on biomedical applications (Navarro, Michiardi, Castano, & Planell, 2008). First generation contains bio inert materials, while there are biodegradable and bioactive materials in second generation and third generation contains those materials used to produce definite response at the molecular level. Third generation is bioactive as well as bioresorbable and it can activate cellular function. Biomaterials used for orthopedic implants should possess the following properties including biocompatibility, bioactivity, osteoconduction, stress protecting, and osteo induction (Hench & Jones, 2005). There are 3D spongy structures which can stimulate wound healing through functional planes containing extracellular matrix representing peptide structures to activate cellular functions. Biomaterials are responsible to deliver growth factors or drugs to control cell function via mechano-transduction. Tissue can be repaired by two different routes including tissue engineering and in-situ tissue regeneration. In tissue engineering, progenitor cells are planted together with adjusted frameworks and permitted to grow outside the body to simulate natural tissues followed by implantation into the patient body to exchange injured tissues. Second, in-situ tissue regeneration involves limited healing of tissue by using biomaterials in the form of solutions, powders, or fixed units. Biomaterials are combined with bone morphogenetic protein (BMP) which release ions in solution form to activate attached cells. These cells produce their natural growth factors to activate growth of adjacent cells.

Many implant surgeries become complex due to infections mostly caused by *Staphylococcus aureus* and *Staphylococcus epidermidis* (Goodman, Yao, Keeney, & Yang, 2013). So, implants must remain non-infected to increase the existence and satisfactory working of implants including its osteointegration ability. Infection can hinder the bone healing process. The reformative property of growth factors can be improved by simultaneous use of fibronectin-based hydrogels and heparin-based glycosaminoglycans (Kose & Kose, 2015). Nano-formulations have been made in the shape of frameworks for continuous release of antibiotics during bone formation. PMMA (Polymethyl methacrylate) cement loaded with amino glycoside is frequently used nano bead based antibiotic therapy. It is useful for continuous release of antibiotics, but major drawbacks of its use include non-biodegradability, toxic byproducts, inflammation, aggravation of immune response damage host tissues and delay in bone formation. So, to overcome this biodegradable nanomaterial was used which can stimulate bone redevelopment without requiring any other surgery (Uskoković & Desai, 2014). HA (Hydroxyapatite) nanoparticles are made by wet synthesis and processed at low temperature can be used for enhanced distribution of drugs in comparison to other materials used in orthopedics. The combination of HA and chitosan helps in relieving crystallization problem, prompt release of antibiotics, and helps in treating infection by enhancing bone biocompatibility. Nowadays, new techniques like electro spinning can be used to produce nano-fibers which act as backbone to release antibiotics e.g., the release of amoxicillin was improved by using nanofibers of PLGA (polylactic-co-glycolic acid) with HA (Zheng et al., 2013).

### *15.3.4 Direction of Future Research Focus*

Nanomedicine is a relatively young field, which is increasingly and exponentially growing, characterized by emerging ethical issues and suggestions (Bragazzi, 2019). Due to certain limitation in conventional pharmaceutical mediators, and older constructions and delivery systems, nanotechnology plays a significant role in the nanomedicine field and drug delivery. Conventional drug delivery system has main drawback that it is difficult to remove the residual parts of such systems, thus departure nonbiodegradable substantial within the patient's body and can cause toxicity. Due to the crumbliness, instability and reactivity of nanoscale materials new regulatory consultant strategies must be established quickly to ensure safe and reliable transfer of new advances in nanomedicine from laboratory to marketplace. Additionally, nanomaterials are more complex to synthesize in comparison with bulk materials, so their preparation and storage could be more complicated and expensive (Balakumar, Vijaya, Tamil, Hari, & Abdu, 2013; Farjadian et al., 2018). Another important challenge of nanomedicine is the instantaneous detection of several molecules, investigation of all sub-cellular components at the molecular level, and replacement of antibodies as detection reagents (Karimi et al., 2017).

Novel group of biological materials having intrinsic ability to act as surfactants and anti-pathogenic may be experimented as nanomaterials. Sophorolipids, for instance, naturally secreted by some actinomycete genera have shown to act against bacteria, fungi and their biofilms, as well as carcinomas (Haque et al., 2016).

### *15.3.5 Sensitizing the End-Consumer Along with the Industry*

Nanotechnology is exciting and favorable and its development is related to many others field of nanoscience. Although nanotechnology slows to move forward, there are numerous technical and non-technical trials in the critical path to understand nano-related consumer products. The nanotechnology ability in producing new manufacturing procedures and products in vastly diverse areas from agriculture to drugs could create changes in society as a whole. Nanotechnology has ability to satisfying consumer's request and security and environmental protection (Production & Nanomaterial, 2018). Research studies show that consumer had insufficient knowledge or awareness of nanotechnology. Consumer need direct communication, which undergo discussing, experiencing and making right decision. Through communication between consumers, producers, and experts, it is important to educate the public and thereby decrease consumer nervousness toward nano-related materials (Altenstetter, 2011; Epa & Osa, 2007). A summary of how nano-biomaterials could be effectively commercialized has been presented in Fig. 15.2.

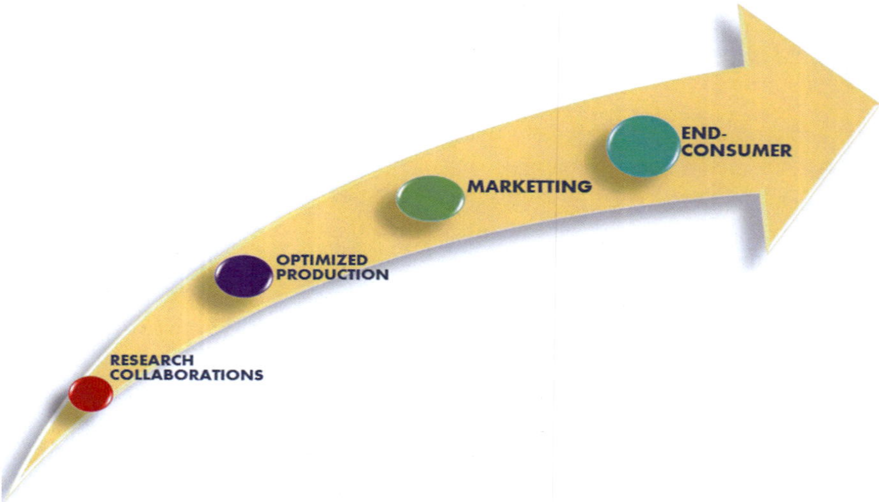


**Fig. 15.2** Pathway leading to effective commercialization of nano-biomaterials

Due to poor reproducibility in humans and lack of proper analytical tools nanopharmacology are slow on developing. Inadequate attention has been paid to differences between animal and human models of disease and their response to potential therapies. Critical and unintended methods of measurements can lead to uncertainty in drug measurement which could be one of the many reasons that mostly drugs fail in clinical trials. Nanoparticle drug delivery development and conversion to a product in the market is very difficult and extremely long and expensive (Bhardwaj, Kaushik, Khatib, Nair, & Mcgoron, 2019). Improved drug delivery systems can decrease the associated side effects and enhance the anticancer efficacy and quality of patient's life. Information of the activity and poisonousness of the free medicine, the role of different delivery system and drug bioavailable enable scheme sides to select an appropriate range of nanomedicines to examine (Altenstetter, 2011; Liu et al., 2016).

Ethical attentions will likely play important role in the development and custom of nanotechnological products in medical care. Primarily, there are some significant ethical concerns which are continued to focus on risk assessment and environmental organization. Later on, novel ethical problems and unexpected dilemmas will rise as the field develop further and interrupts other sectors of biomedical field, such as genomics, personalized medicine, bioinformatics, and neurobiology (Bawa & Johnson, 2007).

Consumers with low monthly domestic earnings had high nervousness and expectations for nanotechnology and low awareness of nanotechnology. In addition, these consumers supposed they needed to obtain relevant information about nanotechnology through education and promotion, as they did not have sufficient knowledge of nanotechnology and nanomaterials. A communication and promotion system pointing consumers should be recognized to overcome the differences among consumers and experts in their nanotechnology awareness and expectation. The establishment of knowledge for nanotechnology is required instantaneously. Extraordinary attention should be paid to new nanoproducts that use nanotechnology (Hare et al., 2017). New analytical technologies should be established that predict the complete process, from tracking of the NPs in human body system, to ejection drug at the disease site, and finally the clearance of the NPs without any harmful effect. This pathway could help convince the end consumers for routinely utilizing the useful bio-nanomaterial products.

## 15.4 Conclusions, Outlook, and Future Perspectives

Biogenic synthesis and medical applications of nanomaterials is a multidisciplinary science. For technical and applied transformations in this field, it is imperative to develop stronger research nexuses. For instance, fruit/vegetable peels or sometimes seeds may be utilized to synthesize biogenic nanoparticles, their procurement and shipment may be well-coordinated with farms/culinary set ups actively utilizing them as such or processing other parts of the fruit. Nano-biomaterials possess tremendous potential in various biomedical systems, with inherent ability to co-op up with biological systems more efficiently than non-biogenic materials. It is thus important to scoop-up the ways to uniformly develop and utilize clinically applicable bio-nanomaterials. Further, an intelligible safety evaluation system must be set up, the results of which should be treated by some reliable expert groups for dispersion to the public or consumers. This could further help make informed decisions in selecting nanoproducts in the market.

## References

Al-Jumaili, A., Alancherry, S., Bazaka, K., et al. (2017). Review on the antimicrobial properties of carbon nanostructures. *Materials, 10*(9), 1066. https://doi.org/10.3390/ma10091066.

Altenstetter, C. (2011). Medical device regulation and nanotechnologies determining the role of patient safety concerns in policymaking. *Law & Policy, 33*(2), 227–255.

Al-Yasiri, A. Y., Khoobchandani, M., Cutler, C. S., et al. (2017). Mangiferin functionalized radioactive gold nanoparticles (MGF-(198)AuNPs) in prostate tumor therapy: Green nanotechnology for production, in vivo tumor retention and evaluation of therapeutic efficacy. *Dalton Transactions, 46*(42), 14561–11457.

Balakumar, K., Vijaya, C., Tamil, N., Hari, R., & Abdu, S. (2013). Self nanoemulsifying drug delivery system (SNEDDS) of Rosuvastatin calcium: Design, formulation, bioavailability and pharmacokinetic evaluation. *Colloids Surfaces B Biointerfaces, 112*, 337–343.

Bawa, R., & Johnson, S. (2007). The ethical dimensions of nanomedicine. *Medical Clinics of North America, 91*(5), 881–887.

Bhardwaj, V., Kaushik, A., Khatib, Z. M., Nair, M., & Mcgoron, A. J. (2019). Recalcitrant issues and new frontiers in nano-pharmacology. *Frontiers in Pharmacology, 10*, 1369.

Bhattarai, B., Zaker, Y., & Bigioni, T. P. (2018). Green synthesis of gold and silver nanoparticles: Challenges and opportunities. *Current Opinion in Green and Sustainable Chemistry, 12*, 91–100.

Bragazzi, N. L. (2019). Nanomedicine: Insights from a bibliometrics-based analysis of emerging publishing and research trends. *Medicina, 55*(12), 785.

Castillo, R. R., Lozano, D., González, B., Manzano, M., Izquierdo-Barba, I., & Vallet-Regí, M. (2019). Advances in mesoporous silica nanoparticles for targeted stimuli-responsive drug delivery: An update. *Expert Opinion on Drug Delivery, 16*, 415–439.

Chatterjee, A., & Chakra, R. (2014). Mechanism of antibacterial activity of copper nanoparticles. *Nanotechnology, 25*(13), 1–23. https://doi.org/10.1088/0957-4484/25/13/1351014.

Chen, X., & Schluesener, H. (2008). Nanosilver: A nanoproduct in medical application. *Toxicology Letters, 176*(1), 1–12. https://doi.org/10.1016/j.toxlet.2007.10.004.

Chopra, A., Saluja, M., Tillu, G., et al. (2013). Ayurvedic medicine offers a good alternative to glucosamine and celecoxib in the treatment of symptomatic knee osteoarthritis: A randomized, double-blind, controlled equivalence drug trial. *Rheumatology, 52*(8), 1408–1417.

Colilla, M., Martinez-Carmona, M., Sánchez-Salcedo, S., Ruiz-González, M. L., González-Calbet, J. M., & Vallet-Regí, M. (2014). A novel zwitterionic bioceramic with dual antibacterial capability. *Journal of Materials Chemistry B, 2*, 5639–5651.

Do, J.-L., & Friščić, T. (2017). Mechanochemistry: A force of synthesis. *ACS Central Science, 3*, 13–19.

El-Sayed, M. A. (2001). Some interesting properties of metals confined in time and nanometer space of different shapes. *Accounts of Chemical Research, 34*(4), 257–264. https://doi.org/10.1021/ar960016n.

Epa, U. S, & Osa, A. (2007). *Nanotechnology white paper*. Washington DC: EPA.

Farjadian, F., Ghasemi, A., Gohari, O., Roointan, A., Karimi, M., & Hamblin, M. R. (2018). Nanopharmaceuticals and nanomedicines currently on the market: Challenges and opportunities. *Nanomedicine, 14*(1), 93–126.

Fetih, G. (2016). Fluconazole-loaded niosomal gels as a topical ocular drug delivery system for corneal fungal infections. *Journal of Drug Delivery Science and Technology, 35*, 8–15.

Garino, N., Sanvitale, P., Dumontel, B., Laurenti, M., Colilla, M., Izquierdo-Barba, I., Cauda, V., & Vallet-Regì, M. (2019). Zinc oxide nanocrystals as a nanoantibiotic and osteoinductive agent. *RSC Advances, 9*, 11312–11321.

Gold, K., Slay, B., Knackstedt, M., & Gaharwar, A. K. (2018). Antimicrobial activity of metal and metal-oxide based nanoparticles. *Advances in Therapy, 1*, 1700033.

Goodman, S. B., Yao, Z., Keeney, M., & Yang, F. (2013). The future of biologic coatings for orthopaedic implants. *Biomaterials, 34*(13), 3174–3183.

Haque, F., Alfatah, M., Ganesan, K., & Bhattacharyya, M. S. (2016). Inhibitory effect of sophorolipid on Candida albicans biofilm formation and hyphal growth. *Scientific Reports, 6*, 23575.

Hare, J. I., et al. (2017). Challenges and strategies in anti-cancer nanomedicine development: An industry perspective. *Advanced Drug Delivery Reviews, 108*, 25–38.

Hench, L., & Jones, J. (Eds.). (2005). *Biomaterials, artificial organs and tissue engineering*. Cambridge: Elsevier.

Honek, J. F. (2013). Bio-nanotechnology and bionanomaterials: John Honek explains the good things that can come in very small packages. *BMC Biochemistry, 14*, 29.

Huang, Y., Mei, L., Chen, X., & Wang, Q. (2018). Recent developments in food packaging based on nanomaterials. *Nanomaterials, 8*, 830.

Izquierdo-Barba, I., Sánchez-Salcedo, S., Colilla, M., Feito, M. J., Ramírez-Santillán, C., Portolés, M. T., & Vallet-Regí, M. (2011). Inhibition of bacterial adhesion on biocompatible zwitterionic SBA-15 mesoporous materials. *Acta Biomaterialia, 7*, 2977–2985.

Izquierdo-Barba, I., García-Martín, J. M., Alvarez, R., Palmero, A., Esteban, J., Pérez-Jorge, C., Arcos, D., & Vallet-Regí, M. (2015). Nanocolumnar coatings with selective behavior towards osteoblast and Staphylococcus aureus proliferation. *Acta Biomaterialia, 15*, 20–28.

Jain, P., & Pradeep, T. (2005). Potential of silver nanoparticle-coated polyurethane foam as an antibacterial water filter. *Biotechnology and Bioengineering, 90*(1), 59–63. https://doi.org/10.1002/bit.20368.

Karimi, M., Sahandi Zangabad, P., Baghaee-Ravari, S., Ghazadeh, M., Mirshekari, H., & Hamblin, M. R. (2017). Smart nanostructures for cargo delivery: Uncaging and activating by light. *Journal of the American Chemical Society, 139*(13), 4684–4610.

Kelley, S. O. (2017). What are clinically relevant levels of cellular and biomolecular analytes? *ACS Sensors, 2*(2), 193–197.

Khoobchandani, M., Katti, K., Maxwell, A., Fay, W. P., & Katti, K. V. (2016). Laminin receptor-avid nanotherapeutic EGCg-AuNPs as a potential alternative therapeutic approach to prevent restenosis. *International Journal of Molecular Sciences, 17*(3), 316.

Khoobchandani, M., Katti, K. K., Karikachery, A. R., Thipe, V. C., Srisrimal, D., Dhurvas Mohandoss, D. K., et al. (2020). New approaches in breast cancer therapy through green nanotechnology and nano-ayurvedic medicine—Pre-clinical and pilot human clinical investigations. *International Journal of Nanomedicine, 15*, 181–197.

Kose, N., & Kose, A. A. (2015). Application of nanomaterials in prevention of bone and joint infections. In *Nanotechnology in diagnosis, treatment and prophylaxis of infectious diseases* (pp. 107–117). New York: Academic Press.

Liu, L., Ye, Q., Lu, M., Lo, Y. C., Hsu, Y. H., Wei, M. C., & Ho, C. A. (2016). New approach to reduce toxicities and to improve bioavailabilities of platinum-containing anti-cancer nanodrugs. *Scientific Reports, 5*(1), 1–11.

Magiorakos, A.-P., Srinivasan, A., Carey, R., Carmeli, Y., Falagas, M. E., Giske, C., Harbarth, S., Hindler, J., Kahlmeter, G., Olsson-Liljequist, B., et al. (2012). Multidrug-resistant, extensively drug-resistant and pandrug-resistant bacteria: An international expert proposal for interim standard definitions for acquired resistance. *Clinical Microbiology and Infection, 18*, 268–281.

Mei, S., Wang, H., Wang, W., Tong, L., Pan, H., Ruan, C., Ma, Q., Liu, M., Yang, H., Zhang, L., et al. (2014). Antibacterial effects and biocompatibility of titanium surfaces with graded silver incorporation in titania nanotubes. *Biomaterials, 35*, 4255–4265.

Montalvo-Quirós, S., Aragoneses-Cazorla, G., Garcia-Alcalde, L., Vallet-Regí, M., González, B., & Luque-Garcia, J. L. (2019). Cancer cell targeting and therapeutic delivery of silver nanoparticles by mesoporous silica nanocarriers: Insights into the action mechanisms using quantitative proteomics. *Nanoscale, 11*, 4531–4545.

Moulton, M. C., Braydich-Stolle, L. K., Nadagouda, M. N., Kunzelman, S., Hussain, S. M., & Varma, R. S. (2010). Synthesis, characterization and biocompatibility of "green" synthesized silver nanoparticles using tea polyphenols. *Nanoscale, 2*(5), 763.

Mousavi, S. M., Hashemi, S. A., Zarei, M., Bahrani, S., Savardashtaki, A., Esmaeili, H., Lai, C. W., Mazraedoost, S., Abassi, M., & Ramavandi, B. (2019). Data on cytotoxic and antibacterial activity of synthesized Fe3O4 nanoparticles using *Malva sylvestris*. *Data in Brief, 28*, 104929.

Navarro, M., Michiardi, A., Castano, O., & Planell, J. A. (2008). Biomaterials in orthopaedics. *Journal of the Royal Society Interface, 5*(27), 1137–1158.

Ovais, M., Nadhman, A., Khalil, A. T., et al. (2017). Biosynthesized colloidal silver and gold nanoparticles as emerging leishmanicidal agents: An insight. *Nanomedicine, 12*(24), 2807–2819.

Ovais, M., Khalil, A. T., Raza, A., et al. (2018). Multifunctional theranostic applications of biocompatible green-synthesized colloidal nanoparticles. *Applied Microbiology and Biotechnology, 102*(10), 4393–4408.

Production, M., & Nanomaterial, O. F. (2018). Nanotechnology and society: From lab to consumer. https://doi.org/10.1016/B978-0-323-48057-4/00017-7.

Rahman, A. U., Khan, A. U., Yuan, Q., et al. (2019). Tuber extract of *Arisaema flavum* eco-benignly and effectively synthesize silver nanoparticles: Photocatalytic and antibacterial response against multidrug resistant engineered E. coli QH4. *Journal of Photochemistry and Photobiology. B, 193*, 31–38.

Ravichandran, A., Subramanian, P., Manoharan, V., et al. (2018). Phyto-mediated synthesis of silver nanoparticles using fucoidan isolated from *Spatoglossum asperum* and assessment of antibacterial activities. *Journal of Photochemistry and Photobiology. B, 185*, 117–125.

Riley, R., & Day, E. (2017). Gold nanoparticle-mediated photothermal therapy: Applications and opportunities for multimodal cancer treatment. *WIREs Nanomedicine and Nanobiotechnology, 9*(4), e1449. https://doi.org/10.1002/wnan.1449.

Rodriguez-Palomo, A., Monopoli, D., Afonso, H., Izquierdo-Barba, I., & Vallet-Regí, M. (2016). Surface zwitterionization of customized 3D Ti6Al4V scaffolds: A promising alternative to eradicate bone infection. *Journal of Materials Chemistry B, 4*, 4356–4365.

Roy, A., Bulut, O., Some, S., Mandal, A. K., & Yilmaz, M. D. (2019). Green synthesis of silver nanoparticles: Biomolecule-nanoparticle organizations targeting antimicrobial activity. *RSC Advances, 9*, 2673–2702.

Santos, L. M., Stanisic, D., Menezes, U. J., Mendonça, M. A., Barral, T. D., Seyffert, N., Azevedo, V., Durán, N., Meyer, R., Tasic, L., et al. (2019). Biogenic silver nanoparticles as a post-surgical treatment for corynebacterium pseudotuberculosis infection in small ruminants. *Frontiers in Microbiology, 10*, 824.

Sarli, S., Kalani, M. R., & Moradi, A. (2020). A potent and safer anticancer and antibacterial taxus-based green synthesized silver nanoparticle. *International Journal of Nanomedicine, 15*, 3791–3801.

Seil, J. T., & Webster, T. J. (2012). Antimicrobial applications of nanotechnology: Methods and literature. *International Journal of Nanomedicine, 7*, 2767.

Sirelkhatim, A., Mahmud, S., Seeni, A., et al. (2015). Review on zinc oxide nanoparticles: Antibacterial activity and toxicity mechanism. *Nano-Micro Letters, 7*(3), 219–242. https://doi.org/10.1007/s40820-015-0040-x.

Spałek, K., Spielvogel, I., Proćków, M., et al. (2019). Historical ethnopharmacology of the herbalists from Krummhübel in the Sudety mountains (seventeenth to nineteenth century), Silesia. *Journal of Ethnobiology and Ethnomedicine, 15*, 24.

Takeda, T., Tsubaki, M., Kino, T., et al. (2016). Mangiferin enhances the sensitivity of human multiple myeloma cells to anticancer drugs through suppression of the nuclear factor kappa B pathway. *International Journal of Oncology, 48*(6), 2704–2712.

Uskoković, V., & Desai, T. A. (2014). In vitro analysis of nanoparticulate hydroxyapatite/chitosan composites as potential drug delivery platforms for the sustained release of antibiotics in the treatment of osteomyelitis. *Journal of Pharmaceutical Sciences, 103*(2), 567–579.

Vasquez, R. D., Apostol, J. G., de-Leon, J. D., Mariano, J. D., Mirhan, C. M. C., Pangan, S., & Zamora, E. T. (2016). Polysaccharide-mediated green synthesis of silver nanoparticles from Sargassumsiliquosum J.G. Agardh. Assessment of toxicity and hepatoprotective activity. *OpenNano, 1*, 16–24.

Xie, K., Zhou, Z., Guo, Y., Wang, L., Li, G., Zhao, S., Liu, X., Li, J., Jiang, W., Wu, S., et al. (2019). Long-term prevention of bacterial infection and enhanced osteoinductivity of a hybrid coating with selective silver toxicity. *Advanced Healthcare Materials, 8*, e1801465.

Zare-Zardini, H., Alemi, A., Taheri-Kafrani, A., Hosseini, S. A., Soltaninejad, H., Hamidieh, A. A., Haghi Karamallah, M., Farrokhifar, M., & Farrokhifar, M. (2020). Assessment of a new ginsenoside Rh2 nanoniosomal formulation for enhanced antitumor efficacy on prostate cancer: An in vitro study. *Drug Design, Development and Therapy, 14*, 3315–3324.

Zhao, Y., Guo, Q., Dai, X., Wei, X., Yu, Y., Chen, X., Li, C., Cao, Z., & Zhang, X. (2019). A biomimetic non-antibiotic approach to eradicate drug-resistant infections. *Advanced Materials, 31*, 1806024.

Zheng, K., Setyawati, M., Leong, D. T., et al. (2011). Antimicrobial gold nanoclusters. *ACS Nano, 11*(7), 6904–6914. https://doi.org/10.1021/acsnano.7b02035.

Zheng, F., Wang, S., Wen, S., Shen, M., Zhu, M., & Shi, X. (2013). Characterization and antibacterial activity of amoxicillin-loaded electrospun nano-hydroxyapatite/poly (lactic-co-glycolic acid) composite nanofibers. *Biomaterials, 34*(4), 1402–1412.

# Chapter 16
# Synthesis and Application of Nanomaterials for Biomedical Anticancer Therapy

**Sugumari Vallinayagam and Karthikeyan Rajendran**

## 16.1 Introduction

The nanomaterials, can be grouped into nanoparticles, nanorods, nanosphere, nanoshells, and nanostars according to their structure and these have been usually employed in various medical purposes like biomedical imaging and treatment of cancer. These nanomaterials are considered to be good carriers for drug molecules. The technology using nanomaterials have developed in such a way that these are employed in the fields of functional imaging, tumor therapies, and synergistic combinational platforms. In this chapter, we will about numerous purposes of nanomaterials in the fields of biomedical imaging and tumor therapies. The clinical imaging modalities incorporate attractive reverberation imaging, imaging by using any kind of penetrating wave, antielectron discharge penetrating wave, single antielectron outflow automated penetrating wave and optoacoustic imaging. Different disease restorative techniques will likewise be incorporated, with photothermal treatment, PDT, cancer treatment, and biological therapy. This survey likewise covers both therapeutics and diagnostics, which utilize a similar specialist in analysis and treatment. This remembers ongoing advances for multimodality imaging, picture guided treatment, and mix treatment. We establish that the ceaseless advance of union and plan of new nanoparticles will improve the future improvement of clinical imaging and malignant growth treatment. Notwithstanding, more assets ought to be accessible to inspect results and cell harmfulness when utilizing nanomaterials in people (Siddique & Chow, 2020).

S. Vallinayagam (✉) • K. Rajendran
Department of Biotechnology, Mepco Schlenk Engineering College, Sivakasi, Tamil Nadu, India
e-mail: sugumari@mepcoeng.ac.in


K. Pal (ed.), *Bio-manufactured Nanomaterials*, https://doi.org/10.1007/978-3-030-67223-2_16

The utilization of nanoparticles has upgraded practically all significant imaging methods, especially MRI, PET, etc. A portion of the significant achievements are the utilization of Fe3O4 nanoparticles in T1 weighted as well as T2-weighted MRI, the plan of radioisotope fluorescent free antielectron discharge penetrating wave, and the improvement of fluorescent nanoparticles (Rosado-De-Castro, Morales, Pimentel-Coelho, Mendez-Otero, & Herranz, 2018). Then again, novel sorts of optical nanoprobes, for example, determined glow nanoparticles (PLNPs), are being created to exploit enduring close infrared (NIR) radiance capacity (Lecuyer et al., 2016). This permits optical imaging without steady excitation and auto fluorescence (Liu et al., 2019). The most recent examination and progression in nanotechnology direct to the advancement of different nanoparticles for symptomatic and helpful application. Despite the fact that clinically, the quantity of utilizations of nanomaterials is restricted by the mind boggling requests on their pharmaceutical features, nanodiagnostics boost the comprehension of significant physiological standards of different illnesses and medicines. Then again, NPs are broadly utilized in the center for remedial purposes. Restorative nanoparticles boost the aggregation and arrival of pharmacologically dynamic operators at the obsessive site, which generally speaking, expands remedial viability and diminishes the frequency and power of the results. NPs hold extraordinary guarantee for incorporating symptomatic and restorative operators into a solitary nanoparticle for various medical applications. A genuine model would screen biodistribution and target site collection, evaluating and envisioning discharge of the drug, and longitudinally surveying helpful adequacy. These nanoparticles are utilized for customized nanomedicine-based treatments (Baetke, Lammers, & Kiessling, 2015). Nanoparticles' natural extraordinary attractive or optical parameters help in making them perfect for using in different bioimaging purposes. Nanoparticles make incredible differentiation specialists because of their high affectability, little size, and piece. Nanoparticles are regularly formed with reasonable focusing on ligands on the outside of the particles. Nanoparticles with various functions can be created by joining different useful materials, and this empowers multimodal imaging and treatment at the same time, otherwise called theranostics (Kim, Lee, & Hyeon, 2017). Albeit every one of the imaging and treatment modalities has improved essentially in the course of there are admonitions in nanoparticle properties that are hindering its uses. For instance, no sole atomic imaging methodology can proffer all the necessary information completely portraying the properties of a regulated operator. Each imaging methodology has a significant weakness, for example, MRI has great-goal yet less affectability, optical strategies have restricted infiltration and procedures have generally helpless goal yet high affectability. Consolidating different imaging procedures can empower these applications to supplement each other, and a multimodal imaging operator turns into the way to upgrading those imaging frameworks (Burke, Cawthorne, & Archibald, 2017).

This section investigated the various functions of nanomaterials, for example, contrast specialist and portion enhancer, in biomedical imaging and disease treatment. Additionally, the section examined the fundamental instruments of nano particles including physical, compound, and natural frameworks. Some new employments of nano particles as theranostic pros are discovered.

## 16.2 Synthesis of Nanoparticles for Anticancer Therapy

### *16.2.1 Gold Nanoparticles*

#### 16.2.1.1 Synthesis Through Ultrasound Irradiation

This method is rapid and the synthesis is carried out under appropriate situation without the use of more reducing agents. The heating in ultrasound is produced from the acoustic cavitation. Usually, water is employed in the production using ultrasound. The acoustic cavitation develops through the sonolysis of water which will result in the production of numerous huge number of $H^-$ radicals, the combinations of which results in the production of hydrogen (Bang & Suslick, 2010; Duan, Wang, & Li, 2015).

#### 16.2.1.2 Synthesis Through Microwave Irradiation

The effectiveness of microwave irradiation is due to high dielectric solvents induced by localized superheating and pressure effect, which has capacity of higher heating rate when compared with the usual heating which will result in the decrease in the time period. One of the major drawbacks of microwave irradiation synthesis is that microwave irradiation pose difficulty in penetrating fluid in huge reactors and thus difficult to use for industrial purpose (Duan et al., 2015).

#### 16.2.1.3 Synthesis by Arc-Discharge Method

This is another method for nanoparticles production, where we employ two electrodes made of gold to produce gold nanomaterials. These gold electrodes are made to be in contact with an arc and separated as soon as possible to maintain the arc inside the salt solution. These two electrodes can be subjected to high temperature and can be melted and this can result in the plasma discharge and now it can be changed into gold nanomaterials (Ashkarran, 2012).

#### 16.2.1.4 Seed-mediated Synthesis

Seed-mediated synthesis of nanoparticles is yet another and most widely adopted method to synthesize gold nanoparticles for the particles other than spherical shape. The very general principle of the seeded growth method is, first these particles formed by the reduction of the gold salts. Then to the solution the seed particles are added which contains metal salts which is done usually in the presence of a weak reducing agent which prevents nucleation and will speed up the growth of the Au NPs (Shah et al., 2014).

#### 16.2.1.5 Synthesis Through Chemicals

Au NPs synthesis utilizing chemical, is a cruel way to obtain nanoparticles of various types. Lessening specialists, for example, sodium tetrahydridoborate, caustic soda, ascorbic corrosive, $Na_3C_6H_5O_7$ and $H_2O_2$ are utilized in the decrease of Au(III) particles to Au(0). The union happens within the sight of at least one water-dissolvable polymers, surfactants, or covering specialists with the inclusion of outside energy like warmth, photograph light, and ultrasound illumination. The advantage of this method is mono-dispersed materials and the figure and dimension of the nanomaterial can be inhibited through this method (Brown & Brown, 1962; Chandra et al., 2013; Schlesinger et al., 1953; Xiao, Shlyahovsky, Popov, Pavlov, & Willner, 2005).

#### 16.2.1.6 Biological Synthesis

This biological method includes the use of prokaryotic cells to eukaryotic fungus and complex plants. Biological synthesis methods have been suggested as safe, inexpensive, possible environment friendly ways and alternate options to chemical and physical kind of synthesis (Manikprabhu et al., 2016). Generally, plant adds its therapeutic properties to the nanoparticles, thereby enhancing its activities (Muthukumar et al., 2016).

These microbes result in the production of inorganic materials either inside their cells or outside their cells which synthesize the nanoparticles. In the intracellular methodology, it needs a unique ion transportation in the microbial cell. There are several proteins which are resident on the cell wall and these can decrease the ions. Once these particles are synthesized they can get diffused out. These microorganisms are well known for their ability to produce cofactors and enzymes.

### *16.2.2 Gold Nanoparticles for Cancer Therapy*

Gold nanoparticles found great advantage in cancer therapy based on: (1) It's non-toxic nature (2) the specific physical parameters and (3) their potential to interact only with cancer cells and tumor masses which are damaged. Gold nanoparticles for killing tumor in vivo have to overcome many obstacles (Kodiha, Wang, Hutter, Maysinger, & Stochaj, 2015). Gold nanoparticles destroy cancer cells, either by mechanical damage or due to hyperthermia (Kodiha et al., 2015).

Destroying the cancer cells via hyperthermia depends on the heating sensitivity of the tissues leading to cell death. First the nanoparticles are directed to the cancerous tissue. These particles can get accumulated in the tumor mass. Using an external source these particles can be heated up and the temperature or heat around the cancerous cells increases and thereby these cancerous cells will be selectively killed.

### 16.2.3 *Silver Nanoparticles*

#### 16.2.3.1 Production of Silver Nanomaterials Through Physical and Chemical Methodologies

In the physical methodology, nanoparticles are developed with the evaporation–condensation technique (Gurav, Kodas, Wang, Kauppinen, & Joutsensaari, 1994; Kruis, Fissan, & Rellinghaus, 2000). In olden days other kinds of physical methodologies were employed for the preparation of silver nanoparticles (Pluym et al., 1993; Tien et al., 2008). Physical methods are considered to be better because of these methods are rapid, and the radiation is done using reducing agents. Another advantage in physical method is that these methods do not involve any poisonous chemicals. Limitations of physical method is that the yield will be less and so much energy is consumed for the methodology, the solvent gets contaminated, and at times it is not uniformly distributed (Elsupikhe, Shameli, Ahmad, Ibrahim, & Zainudin, 2015; Shameli et al., 2010).

In the chemical methodology, water or organic solvents are employed for the synthesis of the AgNPs (Tao, Sinsermsuksakul, & Yang, 2006). This chemical method involves three major compounds, which are metal precursors, reducing agents, and stabilizing agents. Usually, these silver nanomaterials can be produced by two mechanisms, which are grouped as "top-down" and "bottom-up" (Deepak et al., 2011). In the first mechanism that is "top-down" method, it involves the mechanical grinding of the heavy metals simultaneously will stabilize with the help of colloidal protecting agents (Amulyavichus, Daugvila, Davidonis, & Sipavichus, 1998). In the second method that is the "bottom-up" method, it involves a chemical reduction, electrochemical methods, and sono-decomposition. Chemical methods have more preference because this method gives higher yield, in comparison with the physical methods, which have shown to give low yield.

Different researchers around the globe are keen on delivering nanomaterials as an elective device to make details that can target explicit dangerous masses as it were. Gopinath et al. in one of the examinations took care of the subatomic system of silver nanoparticles. That examination bunch saw that customized cell passing was fixation subordinate under different conditions. In these investigations, the analysts found that silver nanoparticles actuate apoptosis as well as sharpen malignancy cells (Gopinath, Gogoi, Chattopadhyay, & Ghosh, 2008). Silver nanoparticles actuated changes in cell structure, diminished cell suitability and biological movement, and expanded pressure related to oxidation prompting mitochondrial harm and expanded creation of ROS, finishing with issues in cells.

Silver nanoparticles are employed in both diagnosis and therapies of cancer. Various clinical testing have shown the application of silver nanoparticles as nanocarriers for target site delivery of drugs. These nanoparticles are also employed as agents which has therapeutic effects and also as enhancers for radiation and photodynamic therapy.

### 16.2.4 Titanium Oxide Nanoparticles

Titanium dioxide photocatalyst is a notable and well-informed photocatalyst because of its intriguing properties which incorporate strength, non-harmfulness, compatibility and other parameters. The ongoing advancement of nanotechnology has demonstrated that nanomaterials, can have high movement in the photodegradation of a wide scope of natural and inorganic pollutants present in water bodies. It is fit for corrupting poisons (Beydoun, Amal, & Low, 1999). For it to be extremely compelling, it ought to have certain properties, for example, reasonable molecule size, shape, crystallinity, and a decent proportion of anatase to rutile. This is the motivation behind why most analysts have been attempting to utilize various techniques to get particles with reasonable attributes for natural remediation or for different uses of interest. As of late, packages and varieties of titanium dioxide nanotubes (TNTs) with various characteristics have been blended by a wide range of strategies, for example, layout helped, sol-gel, aqueous, electro-anodization, compound fume affidavit and actual fume statement (Macák, 2008). There are different procedures for getting ready titanium dioxide ($TiO_2$) nanoparticles and these incorporate converse micelles, the sol-gel measure, the metal-natural compound fume affidavit (MOCVD) (Pratsinis, 2011), gas stage (airborne) blend (Chen & Mao, 2007), wet-substance union by precipitation of hydroxides from salts, microemulsion-interceded techniques (Chabra, Pillai, Mishra, Morrone, & Shah, 1995) and electrochemical amalgamation. These strategies can be isolated into five general gatherings to be specific sol-gel, affidavit techniques, sonochemical, and microwave-helped strategies, hydro/solvothermal strategies and oxidation strategies.

These nanomaterials were explored as an upgrade operator for processed tomography imaging (CT) and radiation treatment. In radiation treatment, ionizing radiation targets tumor and makes the harm DNA of tumor cells which brings about cell demise. Double mode picture differentiation and improvement treatment is another territory wherein nanoparticles are utilized. Harada et al. arranged in vivo examinations to assess conveyance of micelles through tissues and ultrasound effectiveness in hindering development of tumors (Harada, Ono, & Yuba, 2013). In the examination with CT26 cells, it was discovered that $TiO_2$ nanoparticles have created oxidative pressure without UV. $TiO_2$ NPs incite creation of responsive oxygen species (ROS), this is another anticancer component of these nanoparticles (Fujiwara, Luo, & Sasaki, 2015).

### 16.2.5 Iron Oxide Nanoparticles

Basically there are two methods by which these nanoparticles can be produced: The first method is known as the "top-down" method. In this particular method large pieces are broken down in order to produce these nanomaterials (Harris & Chess, 2003; Veiseh, Gunn, & Zhang, 2010). 15 base up combination systems for IONPs including coprecipitation, warm deterioration, microemulsion, aqueous, and solvothermal union. The IONPS incorporated with these techniques permit the control of their size and shape, expanding their strength, dissolvability, and biocompatibility (Tombácz, Turcu, Socoliuc, & Vékás, 2015; Wu, He, & Jiang, 2008).

The metabolism of iron in tumorous mass is altered. Tumor cells have an increased need for iron to facilitate their high proliferating rate (Buss, Torti, & Torti, 2003; Terman & Kurz, 2012). The iron uptake into cancer cells is increased by the elevation of iron regulatory proteins such as transferrin receptors (Bystrom, Guzman, & Rivella, 2014). The need of iron is important for various cellular processors such as cell division, synthesis of genetic material and mitochondrial electron transport. Iron is important for the function of ribonucleotide reductase, which is an enzyme that reduces ribonucleotides to deoxyribonucleotides for DNA synthesis and repair. This is the rate-controlling stage of genetic material biosynthesis. This enzyme is closely correlated with cellular replication in the G1-S phase of the cell cycle. Since deoxyribonucleoside triphosphates are in low concentrations in cells, it would be a good strategy to inhibit the enzyme, ribonucleotide reductase. This, in turn, causes embarrassment of cellular production in tumor tissues (Rao et al., 2009).

Manipulating the iron content in the cancer cells could be a good strategy to destroy cancer cells. This can be done by the use of iron chelators. Iron chelators will have a double impact: Firstly, they will be depleting iron through chelation and inhibiting ribonucleotide reductase, thus inhibiting cell proliferation. The second method is that iron chelators can be used to bind to intrasomal iron to form redox active chelators to produce cytotoxic reactive oxygen species through the Fenton reaction promoting programmed cell death (apoptosis) in tumor masses (Foy & Labhasetwar, 2011; Liu, Lin, & Sartorelli, 1992).

### *16.2.6 Silica Nanoparticles*

Silica NPs can likewise be delivered during maximum heat fire decay of metal-natural forerunners. This cycle is additionally alluded to as compound fume buildup (CVC) (Silva, 2004). In a regular Central venous pressure measure, silica NPs are created by responding tetrachlorosilane with $H_2$ and $O_2$ (Vansant, Voort, & Vrancken, 1995). Trouble in controlling the molecule dimension, shape, and stage structure is the fundamental impediment of the fire union (Klabunde, 2001). In any case, this is the conspicuous technique that has been utilized to economically create silica NPs in fine powder structure.

Enormous Si compounds with a normal size of 2.2 m for the most part have bigger polarization than NPs. Notwithstanding, an ongoing report indicated that a lot more modest silicon molecule. For magnetic resonance imaging application, a Si-based difference specialist can be delivered by consolidating progress metal particles into a molecule's body. This differentiation operator abbreviates the atomic turn grid unwinding time (T1) of the protons of close by tissues, and at last, intensifies the sign in T1 imaging. Direct recognition of the Si signal is preposterous because of its low affectability of Si cores, which prompts long obtaining occasions. Be that as it may, this impediment can be illuminated through hyperpolarization. Using this strategy, the imaging window length keeps going around 50–130 s, which permits fast responses and digestion without oxygen to be examined and additionally be utilized to describe the diagnosis of disease. Upside of utilizing a Si-based differentiation specialist is its adaptability of science; the connection of utilitarian natural atoms on the outside of the compounds doesn't altogether lessen any of the ideal atomic attractive reverberation properties (Kwiatkowski et al., 2017).

### *16.2.7 Polymeric Nanomaterials*

These nanomaterials can be set up by scattering performed polymers utilizing various techniques as dissolvable dissipation, nanoprecipitation, salting-out, dialysis and supercritical liquid innovation. These systems have comparative properties that they include a natural stage containing the nanoparticle segments and a water stage containing stabilizers. The other comparability is the helpless epitome of halfway water solvent and uninhibitedly water-dissolvable medications (involving proteins and peptides), which escape from the natural stage to the watery stage (Quintanar-Guerrero, 1998).

To defeat the constraints of chemotherapy like the off objective medication conveyance it was important to locate another option. Polymeric nanoparticles have been seen to be better medication carriers as they can convey the medications to the objective site and furthermore considers have announced their capacity to perform both finding and treatment of tumor (Parveen and Sahoo, 2008). These polymeric nanomaterials have different focal points like soundness, biocompatibility, biodegradability, can be changed, and are economical.

## 16.3 Influence of Physicochemical Properties on Nanocarriers

### *16.3.1 Dimension and Nature of the NPs*

For the determination of nanocarriers one significant boundary is the dimension of the NPs. The size of nanoparticles can impact the cell take-up, and because of the inclination to frame groups in the arrangement, the size of nanoparticles can likewise expand (Benne, van Duijn, Kuiper, Jiskoot, & Slutter, 2016; Silva, Soema, Slutter, Ossendorp, & Jiskoot, 2016). The perception and investigation of three sorts of attractive NPs by Ge et al. showed that NPs with various dimensions and surface properties can result in diverse reactions, particularly in total of MNPs, and the bigger particles would create higher cell takes-up (Ge, Zhang, Xia, et al., 2009; Kyrychenko, Pasko, & Kalugin, 2017; Limbach, Li, & Grass, 2005). A comparable report on the connection among the dimension of ceria NPs and cell take-up demonstrated a straight relationship inside a specific reach. The cell take-up of bigger compounds was fundamentally larger than that of more modest compounds in a similar focus. The dimension of the nanoparticles will likewise influence their leeway from the flow (Longmire, Choyke, & Kobayashi, 2008). Compounds with 200 nm or bigger are generally taken out by the macrophage system framework, interceded by cells in different body organs (Moghimi, Hunter, & Murray, 2001). At 100 nm, NPs include helpless dispersion inside the thick collagen framework of the interstitial space, subsequently bringing about helpless entrance into the tumor tissue and limited NP collection about tumor veins (Moghimi et al., 2001).

The state of the nanomaterial is likewise an unmistakable boundary that influences the action of the molecule (Culver et al., 2016; Rampersaud, Fang, & Wei, 2016). Reports have indicated to pathways by which compounds go through the tissues, duration for cycling, focusing on impact, capacity to beat natural boundary, and different properties rely generally upon molecule shape and size, on the grounds that these attributes are probably going to impact the compounds in the blood carrying, particularly in little nerves and cancer cell associated nerves, and how cells see and react (Moghimi et al., 2001).

### *16.3.2 Surface Charge of Nanoparticles*

Surface charges are firmly identified with different natural exhibitions of the nanoparticles, such as solvency, bio distribution, security, cell take-up, cytotoxicity (Peng, Lu, Wang, Li, & Chen, 2017; Mou, Xing, & Ren, 2017), for example, solvency, bio distribution, security, cell take-up, cytotoxicity, and such. The charge reaction among particles and cells is a significant reason for these natural exhibitions (Pittella, Zhang, & Lee, 2011). The exploratory after effects of Tang et al. demonstrated that the NPs are scattered in the way of life means, just decidedly stimulating compounds can be ingested by the tissues; if the compounds are associated with the protein, among the emphatically and adversely stimulating compounds can be totally wiped out (Tang et al., 2015). Despite the fact that protein-covered compounds may help distinguish receptors (Holgate, 2010; Mou et al., 2017) principally in light of the fact that the cell layer is adversely charged, now and again with a limited quantity of the fix, consequently, the decidedly charged compounds are added effectively than contrarily unbiased compounds by cell layer adsorption. Graf et al., in view of the investigation of Si NPs, found that maximum sure charge compounds can instigate viable cell disguise, while contrarily charged compounds and the PEG functionalized compounds express diminished cell take-up (Leonenko, Finot, & Amrein, 2007).

### *16.3.3 Cytotoxic Impacts*

The toxic level will depend on the reactions which are between the nanomaterials and the surrounding where the nanoparticles get into inside the body of humans. The structure of the nanomaterial has a great role in deciding the toxic property, especially the core material. The mechanism of toxicity is such that the toxic molecules get leaked from these nanomaterials when they decompose and these molecules gets deposited (Pelaz, Charron, & Pfeiffer, 2013). This decomposition process of the nanomaterials can be avoided with the help of an inorganic core or a shell. When it is coated with such coatings it can prevent the nanoparticle decomposition and these can stay on the surface or can be embedded in a cross-shaped polymer and thereby these coatings can stay stable.

These numerous properties of nanomaterials are prominent parameters that help in the determination of the nanomaterial's characteristics, which are proved by the strong outcomes in both theories and in practicals. These outcomes have proved their application in the area of medical fields. The various parameters of nanomaterials can be modified accordingly and can be used as the nanodrugs to treat tumor, which means, these kind of treatments can be used to cure the cancerous cells accordingly to the patients conditions.

## 16.4 Cancer Therapy

Cancer therapy is used to get rid of the tumor cells or to stop the growth of such cancerous cell masses. There are numerous methods which are available and each one of them is comparatively better than the other therapies. Application of nanoparticles have enhanced most of the cancer treatments and these therapies are discussed in the below section:

### *16.4.1 Photothermal Therapy*

This type of treatment is a hyperthermia-dependent treatment. Mechanism of photothermal treatment is to destruct the cancer masses meanwhile the healthy tissues are not heated and protected. The tissue cells do not have NIR-absorbing chromophores. When the lasers are used they have a "tissue optical window" which is around 700–1000 nm and thereby it reduces the tissue heating, whereas the silver nanoparticles highly absorb the heat in the NIR region. This is the main mechanism to achieve selective heating of the cancerous cells (Hirschberg & Madsen, 2019). Another advantage of the gold nanomaterials is that these particles can be altered with numerous functional groups. The colloidal gold showcase localized plasmon surface resonance. This helps the gold to take in light at various frequencies or wavelengths. This property helps them to be applied for the hyperthermic tumor therapy application. The plasmon surface resonance can be altered by changing the gold nanoparticle's structure shape and size. This modification can result in the change of other characteristics like its photothermal and photoacoustic parameters, which will then help these particles to utilize various range of wavelengths. Due to size in nanometer scale, these particles can get localized in the cancerous cells and can be excreted through the urine (Vines, Yoon, Ryu, Lim, & Park, 2019). One of limitations of this photothermal therapy is that heat which is generated by the nanoparticles are heterogeneous in the cancer cells, that is only a part of the cell mass which gets heated and will be treated whereas some portion of tumor gets untreated. Researchers came up with a new proposal which involved the application of silica gold nanoshells for photothermal therapy (Simón, Norregaard, Jørgensen, Oddershede, & Kjaer, 2019).

Usually nanoparticles which are gold based are used in the therapy due to the reasons that these gold nanoparticles are very compatible bioparticles, are very efficient in converting light to heat energy and also they can NIR light which penetrates tissue more deeply. Another highlight using nanoparticles is that due to their small diameter, they can easily go inside or penetrate to cancerous cells and due to bioconjugation any functional groups can be attached to these nanoparticles. Usually Nanoshells, nanocages, nanorods, and nanostars are employed as photothermal transducers. Most of the gold nanoparticles have been developed to take in maximum within the first NIR window, which has been reported to be about 2–3 cm of the tissue (Riley & Day, 2017).

It was recently reported that PET are reliable to be used with photothermal therapy (Jørgensen et al., 2016). In this method the temperature was increased to 42 °C and due to this high temperature the tumor cells were destroyed. To make it more efficient an agent capable of absorbing light or materials with photothermal affect can also be put inside the cancer tissues (Estelrich & Antònia Busquets, 2018).

### 16.4.2 Photodynamic Therapy

In this therapy, light therapy is used and it can produce molecular oxygen, visible light, and photosensitizers (PS) which will result in the destruction of the tumor cells and pathogenic bacteria. This therapy is noninvasive mode of therapy. The advantage of this therapy is that it can selectively kill the malignant cells. When light is used the molecules are produced and these can directly cause damage. The photosensitizer is present throughout the cancer tissue. When light at a specific wavelength is given, due to the oxygen present in the molecular form that is as reactive oxygen species (ROS) will be produced and this will lead to oxidative damage of the components inside the tissue or cell and thereby leads to death of the tissue or cells. Thereby the tumor cells get damaged. Commonly gold nanomaterials are employed in photodynamic therapy (Mokoena, George, & Abrahamse, 2019). Porphyrins are reported to be effective in treating tumor tissues in photodynamic therapy (Bretin et al., 2019). The mechanism of photodynamic therapy is by the deposition of the photosensitizers in the cancer cells. Structure like bacteriochlorins, porphyrins, and phthalocyanines which are modified with functional groups are employed in photodynamic therapies. Most of these compounds have passed various clinical trials. In photodynamic therapy, photosensitizers are designed in such a way that these are conjugated with various functional groups. One of the main purposes of nanotechnology is target-based delivery of drug (Abrahamse & Hamblin, 2016). When the supply of nutrients is inhibited it would lead to starvation therapy and the hydrogen peroxide can be employed to improve the photodynamic therapy (Yu, Zhou, Pan, Li, & Tang, 2018).

### *16.4.3 Chemotherapy*

One of the drugs DOX comes under the division of anthracycline. Studies have shown that they are widely used in various cancer therapies. It has been reported that these drugs are widely used in treating tumor in breast. One of the other common drugs applied in breast cancer is paclitaxel. Some of the other drugs include cisplatin, tamoxifen, trastuzumab, and docetaxel. It is fact that if the drug molecule reaches the exact target site then the drug can be very effective at the site of delivery. As nanotechnology progressed drug were enclosed in nanoparticle-based carriers and later reaches the target site. Few of the nanomaterials which are used in therapy of breast tumor are nanoparticles based on polymers, liposome nanoparticles, metal-based nanoparticles based on gold, nanoparticles made of carbon atoms, mesoporous silica nanoparticles, and nanoparticles with proteins (Liyanage et al., 2019).

New developments have occurred in the field of nanomaterial vehicles and clinicians are currently using these and also few of these vehicles are under clinical trials for treating tumor. These consist of dendrimers, liposomes, and polymeric micelles. Researchers are doing various trial levels for the development of new nano-based drug formulations, and most of these include activation when a particular stimulus comes in contact with the particles.

Various nanoparticle combinations are accepted by the FDA and EMA for the treating numerous cancers (Lee, Dees, & Wang, 2017). Numerous kinds of proteins molecules or short peptides are added on to the outer area of the nanomaterials to enhance the efficiency of the drugs. This helps in the action of drugs only at various specific parts. One such example is of serum glycoprotein, which is applied along with the nanoparticles for the delivery of drugs (Press, 2018). The drug chloroquine can help in reducing the immunological clearance of nanoparticles by macrophages which are present in the liver, which will result in the accumulation of the drug at that site (Pelt et al., 2018).

DOX-BLM-PEG gold nanomaterials and EpCAM-RP gold nanomaterials are two nanoparticles of gold which have been reported to have high efficiency in various therapies (Singh et al., 2018). Cisplatin is a genotoxic agent which can be employed along with radiation or can be applied alone in various chemotherapies. This has been reported to be used in various kinds of cancers. The limitation is that the resistance which are both acquired and intrinsic and dose to normal tissue. Chemotherpy can lead to toxicity because it can also affect the healthy normal cells along with the cancerous tumor cells or tissues. So to overcome this problem, these nanomaterials can be employed which helps in delivering the drugs such as cisplatin to the exact target site unlike the normal conventional drugs. Few organic nanomaterials can be employed to transport drugs like cisplatin. Few inorganic nanoparticles are gold nanomaterials, ferromagnetic nanoparticles, and mesoporous silica nanoparticles. Few of the hybrid nanoparticles are carbon nanotubes, nanoscale coordination polymers, and polysilsesquioxane nanoparticles (Duan, He, Kron, & Lin, 2016). For delivering drug molecules usually organic nanoparticles or organic nanomaterials are commonly and popularly used. It is because of the reason that they have higher time period of life and gets accumulated in the cancer cells in comparison with other nanoparticles. Therapy involves a combination of various drugs used in treating tumors (Meng, Han, & Yeo, 2017). Nanoparticles can be loaded with numerous drugs which belong to various classes. Literature tells that the preclinical models are in the final trials. One such trial is going on for DOX and mitomycin C drugs which are encapsulated inside the nanoparticles. It was observed that these kind of drug delivery have better efficiency and also it can act on the cells which have become resistant to these drugs. For the trials with nanoparticles encapsulated with various drugs, the main drugs used were: paclitaxel, 17-AAG (Triolimus), and rapamycin. The study was conducted on MDA-MB-231 cancer containing animal model (Fisusi & Akala, 2019). Hypoxia promotes the invasiveness of cancer masses and resistance to particular drugs. Recently a nanocarrier have been designed which has a combination of curcumin and baicalin. It was observed that these can overcome the resistance developed by the cancer cells against the usual drugs. Mannose will get bound to the receptors which are present on the outer portion of the macrophages which are associated with the cancer tissues. To boost the cellular uptake by the cancer cells, oligomeric hyaluronic acid can also be employed. These will act as targeting molecules. Nanodandelions are small molecules which can get inside the cancer cells or tissues easily. The outcomes showed better anticancer activity and it was also observed that side effects were less and these anticancer experiments were conducted and confirmed in A549 mouse which had cancer cells (Wang et al., 2019).

These nanoparticles helps in sustaining the amount of drug at the site of cancer and this can make the treatment more effective. These particles can also enhance the penetration of drug inside the cancer tissues or cells. This comparatively had better anticancer activity with respect to the conventional drugs (Mangal, Gao, Li, & Zhou, 2017).

### 16.4.4 Immunotherapy

As time went by, the field of nanotechnology improved and this had promised a favorable promise in the area of tumor immune therapy. Immunotherapy is a technique through which the immune system of the patient will be stimulated or activated. This activation will help in the destruction of the tumor mass (Saleh & Shojaosadati, 2016). The five divisions of this therapy includes: growth factors trigger the production of white blood cells, agonistic Ab which acted beside various proteins, cancer immunotherapy, immune cells, and vaccines for cancer. To target the T cells nanomaterials are employed. Nanoparticles are also employed in the transportation of RNA to the tumor masses, or for the transportation of various vaccines in this therapy (Riley, June, Langer, & Mitchell, 2019). Studies have shown that nanoparticle systems are efficient in delivering the antigens at particular site.

It can be seen that these nanoparticles have various roles in modern immunotherapy (Wen, Umeano, Kou, Xu, & Farooqi, 2019). It has been reported that this therapy is one of the most efficient method for treating tumors. It is because in this therapy, the tumor sites are treated along with the combination of immunology (Park, Heo, & Han, 2018). Very recently cancer immunotherapy was proposed employing plant virus-based nanoparticles. In this for the production of vaccine, the virus particles which infect plant were employed (Sau et al., 2018).

Metallic nanoparticles also have great roles in this therapy. Various metallic nanoparticles like gold nanoparticles are employed in such therapies and can be used in combination with the therapeutic agents like ovalbumin (OVA). Metallic nanoparticles have been reported to enhance anticancer cytotoxic T cell response (Evans, Bugga, Asthana, & Drezek, 2018).

### *16.4.5 Metalloid Boron as an Enzyme Inhibitor*

This is one of the most common proteins or peptides which are produce proteolysis effect in human prostate cells and it is a well-known biomarker of prostate tumor (Emanet Ciofani, Şen, & Culha, 2020). PSA is thought to cleave insulin-like growth factor-binding protein-3 (IGFBP 3) (Emanet Ciofani et al., 2020; Gallardo-Williams et al., 2004). In addition foods rich in boron can prevent several cancers due to their role in inhibition (Scorei & Popa, 2012). All these studies have actually developed a novel path to treat cancer cells or tissues by developing boron based nanoparticles. Most of the studies related to boron based nanoparticles and cancer cells suggest that these cells are being arrested by the boron compounds (Tian et al., 2019). In nature of hydrogen borate and BN are the common forms. These can be a very fine supply of boron compounds and can be used for such therapeutic applications (Li et al., 2017; Mateti et al., 2017). The researcher suggests that the biocompatibility solely based on their dimension, outline, formation, chemical nature and that apoptosis is affected by the unsaturated boron atoms. In addition replacing the rest of the atoms with boron atoms can be a good method to block the cancer growth as mentioned before.

### 16.4.6 Oxygen Capturing Approach

The cancer cells can develop to a size of about 1–2 $mm^3$ when it is given the proper nutrients and oxygen. These are supplied by diffusion. But later the tumor cells will need new vessels to grow beyond this size. This process of forming blood vessels is known as vascularization (Orme & Chaplain, 1997). When the cancer cells become angiogenic in nature, the mass of cells will attract the blood vessels (Hillen & Griffioen, 2007). This process of attracting the blood vessels is taken care by various stimulators and inhibitors. The stimulator do up regulation of the genes required for the angiogenesis and the inhibitors help in down regulation (Nishida, Yano, Nishida, Kamura, & Kojiro, 2006). Anti-angiogenic plans only cannot stop or inhibit the cancer cells growth because these cancer cells can find new mechanisms to form the blood vessels (Abunahla et al., 2019; Kerbel & Kamen, 2004). But reports signifies that most of these drugs which inhibits angiogenesis can create a toxic environment and there is a need for novel plans to deliver these drugs at the target site (Yoncheva & Momekov, 2011). Nanoparticles can also make the membranes of the cancer cells more leaky and therefore nanomedicine plays a very prominent role in cancer therapies (Desai, 2012; Ma et al., 2001). The production of various compatible nanoparticles has become very demanding in the field of therapies and medicine. Any method in order to produce nanoparticles is very worthy given that the starting materials are compatible or if it can safely release from the body (Zhang et al., 2017).

### 16.4.7 Self-therapeutic Organic Nanomaterials

These are like peptide-based nanomaterials and are employed in therapies for various medical uses. One of the greatest advantage using peptide-based therapeutic systems is that they can act only on the tumor cells and remain stable in the healthy normal tissues without leaving a toxic trail in the healthy normal tissues (Wu et al., 2014). The (KLAKLAK)2 (KLAK) amphiphilic peptide is a good anticancer amino acid commonly utilized in tumor treatment and that repress the development of severe cancer forms by break the cell organelles (Qi, Gao, Wang, & Wang, 2018). Wang et al. created a library of KLAK-based NPs for tumor therapy by in situ polymerization and optimized the NPs with different formulations and concentrations of compounds (Qiao et al., 2016). The beclin1 peptide is a different tumor inhibits peptide; (Wang et al., 2015). For restoring p53 cancer cell inhibition effect, the disruption of interactions between the MDM2 and p53 axis is a promising strategy. However, due to its low proteolytic constancy and poor cell-membrane penetration, the cancer-suppressing effect of PMI is severely controlled (Frappier, Duran, & Keating, 2018; Yan et al., 2017).

## 16.5 Nanomaterial as Drug Delivery Agents

### *16.5.1 Advantages of Nanomaterials as Drug Carrier*

During times when chemotherapeutic drugs were mostly used. All together it was observed that such drugs can reduce the therapeutic efficiency when compared to the nano drugs. The Nanocarrier-based platforms are systems in which nanoparticles can be employed for the target site specific drug delivery. The size of these nano carriers are around 500 nm. It has been seen that these nanostructured carriers have great efficiency in delivering a drug in the diseased area which makes these nanocarriers efficient drug carriers. The drugs enclosed can be even anticancer drugs.

The nanoparticles systems can be grouped into two namely *organic* and *inorganic* nanocarriers. These particles characteristics change according to their compositions such as organic, inorganic, or hybrid, various parameters which include size, shapes, and their outer layer parameters. These nanocarriers have been tested for various clinical trials and many of them are also in the clinical phase of testing (Tran, DeGiovanni, Piel, & Rai, 2017; Ventola, 2017). Some of the characteristics of nanocarriers are discussed below. With the help of nanotechnology drug delivery can get the following advantages:

1. It helps in delivery of drugs which are hydrophobic.
2. It is a target site drug delivery there by protecting the healthy cells.
3. It will help in retaining the drug or chemical molecules with in the body for longer time thereby increasing the efficiency of the therapy.
4. The drug must be protected from the surrounding biological stimuli.
5. It can help the drug molecules to reach various tissues by helping them cross the epithelial and endothelial barriers.
6. Nanotechnology helps in combining both diagnosis and the therapy for a particular disease.

### *16.5.2 Inorganic Nanoparticles*

The Inorganic nanomaterials involve silver, gold, iron oxide, and silica nanomaterials. Much of the reports based on these are not there but still few reports discuss about their potential uses. For the clinical use only few of these nanomaterials are only used and some of the other nanoparticles are still on different phases of clinical study. Metal nanoparticles, silver and gold, have specific characteristics such as surface plasmon resonance, that various other nanoparticles don't have. These nanoparticles show various benefits like their biocompatibility and the property of modification which helps in attaching various functional groups.

The reports on the drug delivery related activities are not clear but the mechanism of action has been proposed (Choi, Lee, Jeong, & Choy, 2013). The particular drugs for a particular disease can be conjugated to the surface of the gold nanomaterial. The basic force between the drug and nanomaterial is ionic or covalent bonding and physical absorption. This helps in proper delivery of the drug molecule to the desired target site. Then the drug release can be controlled when there is a proper biological stimuli or it can be activated in a particular wavelength of the light (Kong et al., 2017). Silver nanomaterials have been reported for their antimicrobial properties. But much reports the role as drug carriers are not been conducted, for example, Prusty and Swain (2018), produced a conjugated and soft PAM/D nano-hydrogels (Prusty & Swain, 2018). In one of the studies, the iron oxide nanoparticles were produced with the help of laser pyrolysis method (Marcu et al., 2013).

#### 16.5.2.1 Metal Nanoparticle and Metal Oxides

It has been few years, that researchers started focusing on the metallic nanoparticles and various applications in the field of medicine (McNamara & Tofail, 2015; McNamara & Tofail, 2017). The basic concept is that, these synthesized nanoparticles can be adapted or various functional groups can be linked to the nanoparticles. These groups can help these nanoparticles bind to the antibodies, drugs and other ligands. This property makes the nanoparticles very promising agent in the field of medicine (Kudr et al., 2017).

#### 16.5.2.2 Carbon-based Materials

These carbon-based nanoparticles are also involved in drug delivery and thereby release the drug at the target site. This is possible by attaching the functional groups to the surface or the outside of these nanomaterials. The drug which is entrapped inside can be released by various methods like thermal or enzymatic disengagement. Both these two mechanisms i.e., in thermal and enzymatic mechanism, there will be a damage for the drug molecules, so it is said that a better mechanism is by filling the hollow region of the carbon nanotube with the desired drug molecules. The nanotubes can be filled with metals (Huang et al., 2004; Kumar, Ramesh, Lin, & Fey, 2004). This technique can only be suitable for the metals but it cannot be used for other chemotherapeutic agents to the inside portion of the tubes. The most common and simple method used to fill the interior of the nanotubes is by capillary action (Monthioux, 2002). By oxidation reaction, nanotubes can be opened and it can help in forming holes along the wall of the tube and this can lead to the enhancement of the permeability for filling the drug inside the nanotubes (Sitharaman et al., 2005). Opened nanotubes without any functional groups commonly will have carboxyl groups to make the drugs soluble (Ajima et al., 2005; Matsumura, Ajima, Yudasaka, Iijima, & Shiba, 2007).

## 16.5.3 Organic Nanomaterials

### 16.5.3.1 Liposomes

In the year 1960, liposomes were discovered by Alec Bangham. These are employed in the pharma industries for the delivery of molecules and are learned the most. These were developed for improving the drug to be delivered properly. They are reported to be very efficient in drug delivery (Bozzuto & Molinari, 2015). Problems have been reported with the use of liposomes for the purpose of transporting drugs (Sercombe et al., 2015). Dimov, Kastner, Hussain, Perrie, and Szita (2017), have submitted works on the work plan of synthesis, functionalization and cleansing of these materials. It is prominent because the expense of production have a major role in the determination of commercialization. These are now accepted by the FDA (Sapsford et al., 2013; Zhang et al., 2008; Zylberberg & Matosevic, 2016).

### 16.5.3.2 Polymeric Nanoparticles

These particles are in the nano dimensions and these nanoparticles are constituted with copolymers which are amphiphilic in nature. These come together to create a shield like configuration when kept in a liquid environment. The interior portion will be hydrophobic in nature and the interior portion can be filled with drugs which are hydrophobic in nature for example camptothecin. Meanwhile the outside portion will be hydrophilic and stabilize the whole structure. These particles are under 100 nm in size (Miyata, Christie, & Kataoka, 2011; Xu et al., 2017).

The drugs are entrapped within these polymers of micelles usually by three methodologies. These strategies incorporate right off the bat direct disintegration measure, at that point dissolvable vanishing measure, lastly by the dialysis cycle. As of the immediate disintegration measure, the copolymer and the medications join with one another without anyone else in the water medium and structures a medication stacked with the aggregate of surfactant molecules. While on account of dissolvable dissipation procedure, the copolymer and the medication that must be consolidated are broken down utilizing an unstable natural dissolvable lastly, if there should be an occurrence of the dialysis cycle, both the medication in arrangement and the copolymer in the natural dissolvable are joined in the dialysis pack and afterward dialyzed with the development of the aggregate of surfactant molecules (Xu, Ling, & Zhang, 2013). The focusing of the medications utilizing distinctive polymeric micelles as set up by different component of activity including the helped vulnerability and the holding impact improvements; complexing of an unmistakable pointing ligand atom to the outside of the aggregate of surfactant molecules (Wakaskar, 2017). Polymeric micelles are accounted for to be pertinent for both medication conveyance against disease (Kulthe, Choudhari, Inamdar, & Mourya, 2012) and furthermore for visual medication conveyance (Mandal, Bisht, Rupenthal, & Mitra, 2017).

#### 16.5.3.3 Dendrimers

These are branched or bifurcated, colloid, definite and 3D forms. These dendrimers are shaped in the form of globules and the outer portion is attached with functional groups and these can make them very potent in delivering drugs (Kesharwani et al., 2015; Madaan, Kumar, Poonia, Lather, & Pandita, 2014; Zhu & Shi, 2013). These nanoparticles can be prepared by two methods. Firstly, dendrimer synthesis begins form the inner region and then it is extensive outwards. The other method is the combined one, which begins from the outer of the dendrimer (Cheng, Xu, Ma, & Xu, 2008). Depending on the functional groups attached the dendrimers are classified into: Poly(amidoamine), Proton-pump inhibitors, LCs, core–shell, chirality effect, amino acids, architectures of dendrimers and Poly(amidoamine-organosilicon). Most of study reports are on PAMAM, for the oral delivery of the drugs due to their solubility of the drugs and also these can pass through the epithelial cells (Noriega-Luna et al., 2014). Due to the amine functional groups present on these nanoparticles these dendrimers are not used much in clinical studies. The amine groups are charged positive and dude to this charge it makes them toxic, so these nanoparticles are altered to make it less toxic or to remove the particles from the body. The drug is encapsulated in these nanoparticles through various methods (Tripathy & Das, 2013). The drug which is encapsulated in these dendrimers are delivered by various mechanisms, initially by the in vivo corruption of medication cascade molecules' covalent holding based on accessibility of reasonable chemicals or positive climate that could separate the securities and furthermore by release of the medication because of changes in the actual climate like pH, temperature, and so on, and cascade molecules have been created for systemic distribution, oral, visual, aspiratory and in focused medication conveyance (Kesharwani, Jain, & Jain, 2014). Jain, Gupta, and Jain (2014) have reported one of the studies with folate conjugated with poly-L-lysine dendrimers and they have shown to prevent tumor using the drug encapsulated in them.

## 16.6 Nanomaterials for Targeted Delivery

### *16.6.1 Passive Targeting*

It has been reported that under several states of conditions such as in inflammation or hypoxia, which is usually seen in association with cancer, the endothelium of blood vessels will become leaky in comparison with the healthy vessels (Torchilin, 2011). During condition like hypoxia, the growing masses of cancer cells will start making new vessels or the cells will engulf existing blood vessels. These vessels which are leaky and new can help in the entry of various nanoparticles and molecules to the cancer sites.

It means that when the membranes become leaky the nanoparticles can enter easily into these cancer cells and also these particles will get retained inside these cancer tissues in comparison with the normal healthy tissues. This special property, in any case, isn't pertinent to small particle drugs which have practically short course time and quick waste of time from the cancer cells. Along these lines, the embodiment of small-molecule tranquilizes in small size drug transporters improves their pharmacokinetics, gives some tumor selectivity and diminishes results. This sort of cancer focusing on named "aloof" depends on transporter qualities and cancer science, however doesn't have a functional group for explicit tissue or organ authoritative (Fang, 2011; Maeda, 2001).

However, EPR effect provides rather modest tumor specificity with 20–30% in delivery increase compared to normal organs. The EPR effect is highly dependent on the intrinsic tumor biology (Kobayashi, 2013). All of these factors, together with the physical and chemical properties of nanocarriers, will determine its drug delivery effectiveness.

Be that as it may, it is conceivable to regulate enhanced permeability and retention impact synthetically or precisely to accomplish vascular standardization prompting higher amassing of nanocarriers. Among synthetic enhanced permeability and retention enhancers, one could discover $C_{50}H_{73}N_{15}O_{11}$, NO, $NO_3$, group of lipids, VPF, VEGF, and different growth factors (Fang, 2011; Maeda, 2001). These atoms prompt hypertension or vascular standardization, which could impermanent upgrade cancer perfusion. Different methodologies use ultrasound, radiation, high body temperature, or PI to weak cancer cells and increment nano systems penetration. In any case, all depicted techniques have restrictions and contra-indications and consequently require cautious thought (Arap, 1998; Huynh & Zheng, 2015).

### *16.6.2 Active Targeting*

In this kind of targeting, it has been seen that drug can be delivered to the exact target site when compared to the other modes of drug delivery.

Once the nanoparticle reaches the target site then the efficiency of the drug can be improved with the method called as active targeting. The cancer tissues or the cells will have particular biomarkers on the membranes which can bind to the functional groups present on the surface of the nanocarriers. When this binding happens the nanoparticles gets activated and also this interaction improves the strength of association and there by helps in better diffusion. The main proof of this wonder was proposed in 1980 with antibodies united in the outside of liposomes, (Leserman, 1980) trailed by different sorts of functional groups (Kamaly, 2012; Wang, 2014).

Numerous numbers of receptors have been identified as well as their antibodies have been identified. These can be produced and investigated both in vitro and in vivo. Initiating solid receptor, subsequently filling in as possible models to advance dynamic focusing on innovation. It has been discovered that arginine-glycine-aspartate peptide ties to $\alpha V\beta 3$ principal receptors. These receptors are exceptionally introduced on both the brain cancer cells and on the cellular arrangement of tumor microenvironment (Bello, 2001). F3 peptide was found to tie to nucleolar protein communicated on formation of new blood vessels endothelial cells in the tumor microenvironment (Christian, 2003). In like manner, enzymes has been distinguished as expected receptor in the tumor microenvironment (Pasqualini, 2000) and has been demonstrated to be focused by a peptide derived from three amino acids (Arap, 1998). Among the old style instances of functional groups, we can refer to the folic corrosive that explicitly ties to the B-vitamin receptor just as present in TME. All things considered, various methodologies have been accounted for, through blend of FA-drug forms and through FA-grafting onto nanocarriers advancing their actively transporting molecules in malignancy cells.

### *16.6.3 siRNA Nanomedicine for Cancer Therapy*

siRNA are reported to be very efficient in treating viral diseases as well as cancer by inhibiting the disease-causing genes (Farrow et al., 2003). The discovery was in nematode *Caenorhabditis elegans*. In the mammalian cells, scientists have reported that constructed siRNAs were also able to promote RNAi and thereby it can block the expression of the changed proteins (Fire et al., 1998; Schwarz, Hutvágner, Haley, & Zamore, 2002).

The RNAi for in vivo uses have been modified through numerous chemicals and it has been observed that this can increase the efficacy and potency (Behlke, 2008). In such manner, normal compound plans have permitted siRNA traveler strands to be bound to adjustment than siRNA manage strands. These engineered systems have empowered to supplant either non-crossing over oxygen on the phosphate linkage with a sulfur molecule, the 2′-hydroxyl bunch change of the sugar ring with a methyl gathering and ethyl gathering, among others (Chiu & Rana, 2003; Watts, Deleavey, & Damha, 2008). Moreover, different methodologies have been created to convey siRNAs securely in the cytoplasm. While most stripped siRNAs have been successful for a decent number of tumor cells in vitro, these siRNAs have sadly bombed when have infused in vivo by foundational organizations (Xu & Wang, 2015).

Immunotherapy have been focused too much these days due to their potency in activating the host immune system. The mechanism of activation is through by simply introducing cytokines, or using antigen-presenting cells (APC) (Ghafouri-Fard & Ghafouri-Fard, 2012). Dendritic cells (DCs) are considered to be the strongest antigen exhibiting cells (Banchereau & Steinman, 1998; Klippstein & Pozo, 2010).

## 16.7 Drug Release Strategy

### *16.7.1 Redox-Activated Drug Release*

The unique feature with tumor tissues is that they have a reducing environment inside the cells. This environment will act as a biological stimuli for the redox-responsive nanocarriers and thereby the cancerous tissues or cells will be degraded and release the drug with in these carriers. These type of nanocarriers have three basic advantages when compared to the other nanocarriers. Firstly, these redox-activated drugs are stable in normal tissues, which can protect the cells from the toxic effects of the drugs. Second, they show a brief reaction to high Glutathione focus in tumor cells to deliver cargoes (typically a couple of moments to hours). At last, contrasted with other possible locales of freight discharge, the delivery in cytoplasm is regularly expected to have better remedial impacts (Meng, Cheng, Deng, & Zhong, 2012; Meng, Hennink, & Zhong, 2009). In this survey, we summed up presently existing DDSs transporters into the accompanying classifications dependent on their disparities in structure. For example, DDSs conveyance frameworks with disulfide bonds and DDSs conveyance frameworks with diselane bonds.

### *16.7.2 pH-Mediated Drug Release*

When the polymeric micelles are used in delivering a drug, various parameters are taken into consideration. These include the rate of diffusion of the drugs, partition coefficient, micelle stability, and rate of biodegradation of the copolymers. For cancer chemotherapy, it is very important to take note on the amount of the drug which is released at the cancer tissues and also minimal release rate is advised during its movement in the blood or through the healthy normal tissues or cells (Felber, Dufresne, & Leroux, 2012).

For the purpose of developing new treatments for cancer therapy, the cancer microenvironment has to be well observed. In the case where the drugs will be released during in response to a stimuli, the drug which is can be encapsulated in the nanoparticles and can be given a particular stimuli like the temperature or pH difference. In comparison with all the stimuli, acidic pH is considered to be a perfect trigger for releasing the drug because of the reason that the cancer tissues will have a comparatively lower pH with respect to the normal healthy cells. This implies that pH based approach is better that the use of traditional drugs (Manchun, Dass, & Sriamornsak, 2012).

pH-sensitive polymeric micelles are employed and these are used for target site drug delivery to cancerous tissues. These micelles are stable at physiological pH, but once they encounter acidic conditions then these micelles will be deformed and this deformation is responsible for the release of the drug which was entrapped. The cancerous cells are under mild acidic conditions outside or inside, as a consequence the cancerous cells will be treated with minimum effects on the surrounding healthy cells or tissue. These polymeric micelles are pH-sensitive, that is these micelles will release the drugs when there is a change in the pH level. In comparison with the healthy tissues, the cancer cells will have comparatively lower pH, thereby these micelles releases the drug only in the tumor cells rather than releasing in the healthy tissues (Zhang, Lin, & Gillies, 2010).

## 16.8 Nanomaterials in Diagnosis of Cancer

### 16.8.1 *Magnetic Resonance Imaging*

It is one form a noninvasive technique which helps in creating or forming images (Yousaf, Dervenoulas, & Politis, 2018). It was in the 1980s, this technique developed and it altered the medical and clinical imaging technology (Hemond & Bakshi, 2018). The contrast agent will help in enhancement of the image and plays a prominent role in magnetic resonance imaging (Behzadi, Farooq, Newhouse, & Prince, 2018). In the recent years of advancement in nanotechnology it can be seen that these nanoparticles can be used as contrast agent in magnetic resonance imaging.

### 16.8.2 *Computed Tomography*

In this technique we use one X-ray source and a detector array which can detect and later convert it to images. During all these times computed tomography had been employed for imaging purposes in clinics. They can create an image with better spatial and temporal resolution. One limitation with CT is that its sensitivity toward contrast agents. Still this technique is a promising technique (Dong et al., 2019).

### 16.8.3 *Positron Emission Tomography*

This is a technique for nuclear biomedical imaging. PET is a noninvasive method. In this technique, radiotracers are used to create images. The tracers used in positron emission tomography can give us an idea on the biological pathways (Santos & Ferreira, 2019).

### 16.8.4 *Single Photon Emission Computerized Tomography*

This technique for nuclear imaging and employs gamma radiations to report the biochemical alterations. All these years, this technique has been the nuclear imaging technique (Chakravarty, Hong, & Cai, 2015).

### 16.8.5 *Optical Imaging*

Using this technique one can observe numerous types of structures which are involved in autophagy at both macro and micro levels. This technique involves fluorescence, chemiluminescence, and Raman imaging. This helps in getting a two dimensional or 3D image at microscopic and macroscopic dynamic levels. Fluorescence imaging provides intuitive outcomes, takes much lesser time and the image formed can be understood better when compared with the various other techniques. This is the sole reason why scientists prefer optical imaging for getting images (Wang, Li, Wei, & Duan, 2017).

### *16.8.6 Ultrasound*

This technique is a noninvasive imaging method which helps in observing the structure, morphology, direction, and limitations (Guo, Lu, Qin, & Fei, 2018). In the recent years, nanobubbles (NBs) have developed to be a promising agent. These are basically composed of gas cores and shells which are stabilized. Chitosan is a compound derived from chitin and it is reported to be one of the major materials on the earth. Nanobubbles have been created using chitosan and they have been proved to be potential in treating tumors since they are biocompatible and can carry the drug within. In one of the recent studies DOX-loaded chitosan nanobubbles were made. Later in the year of 2016 new kind of ultrasound imaging contrast agent was developed.

### *16.8.7 Photoacoustic Imaging*

In this technique it utilizes the photoacoustic effect. It creates images from the signals captured from the materials (Choi, Park, Jeon, & Kim, 2018). This method is also known as optoacoustic imaging (Steinberg et al., 2019).

### *16.8.8 Multimodality Imaging*

SPIONs, (Feraheme, FH) and [89Zr]Zr was utilized as a nanoplatform for PET and MRI. PET-MRI coordinates the phenomenal affectability of PET with the spatial goal and differentiation of delicate tissue by MRI. Feraheme can abbreviate the cross over unwinding time, T2, and is commonly utilized for dull differentiation improvement. Nonetheless, dull differentiation is frequently difficult to execute in clinical settings for applications, for example, identification and determination of metastases in the lymph hubs. FH radiolabelled with OET tracer can exploit exceptionally delicate brilliant signs from PET. It can distinguish the presence of FH in districts, where the MRI contrast is excessively low or loud. Test results demonstrated that FH is a truly appropriate SPION for without chelate marking of PET tracers, and can be utilized in half and half PET-MRI (Yuan et al., 2020). For consolidated magnetomotive ultrasound PET/CT and MRI for sentinel lymph hubs, 68Ga-marked SPIONs were proposed.

### 16.8.9 Image-Guided Therapy

A semiconducting plasmonic nanovesicle was proposed; this contained both semiconducting poly and PEG which were joined together to a gold nanoparticle (Au@ PPDI/PEG). The electromagnetic field increased the efficiency of absorbing the light. It can produce photothermal effect. These can also produce high photoacoustic signal. All together due to their various characteristics, these complexes have great importance as theranostic agents (Yang et al., 2017). Au nanorods in photoacoustic imaging and photothermal therapy have been examined. The beneficial properties of Au NPs, for example, behavior of biomaterials, tunable surface plasmonic reverberation, and controlled combination settle on them an incredible decision for theranostic applications. Photoacoustic imaging guided photothermal therapy is conceivable when the beat is utilized to crush the malignant growth cells. The nanotubes can be employed as tools in various therapies in PTT and PAI (Siregar, Oktamuliani, & Saijo, 2018).

The atomic number of gold nanoparticles is high. Due to this very high atomic number these nanoparticles, absorb low and medium energy X-rays very strongly (He & Chow, 2016). One limitation is when the electrons are released and these can cause damage (Wang et al., 2016). One of studies showed their capacity of active targeting and these can be used in CT imaging for enhancing the technique and can also be used as tools in medicine and therapies.

### 16.8.10 Combination Therapy

This therapy cures various malignancies and there by improve the test outcomes. Usually these have synergistic drug action and have been reported to be successful in cases when there is a problem of drug resistance. Liposomes are generally employed in these kind of therapies. Numerous liposomal formulation of DOX include DaunoXome and ONCO-TCS. These are very popularly employed in delivering drugs at various sites. Few other nanoparticles that are employed in these therapies are polymeric nanoparticles. These polymeric nanoparticles have high thermodynamic and kinetic characteristics which help in target based anticancer drug delivery to the specific cancer cells (Gurunathan, Kang, Qasim, & Kim, 2018). In recent studies it has been reported that studies conducted both in in vivo and in vitro, and outcomes showed the nanoprobe had great efficiency in targeting cancer tissues, the drug was retained for longer period of time, and appropriate therapy effect was there on the cancer cells (Zhang et al., 2019).These can get highly accumulated in the cancer cells because of their very small size, around 60 nm and modified outer surface. All of these characters make it a great agent to be applied in image-guided photodynamic therapy or photothermal therapy (Miao et al., 2016).

Conventionally cancer is treated with the help of chemotherapy but it has many limitations like it kills other neighboring healthy cells too. The recent theranostic nanoparticles are a combination of both detecting if there is a disorder and also treat the same disorder using the drug entrapped within the nanoparticle and there by increases the target site delivery thereby only the tumor cells gets treated with the drug and along with that helps in detecting and monitoring of the cancerous cells.

## 16.9 Summary

The very recent advancements in the field of nanotechnology have shown a prominent development, which have great benefits from the modifications of numerous nanoparticles. Another advantage with these materials is that they can be used in target delivery of various drugs. Otherwise these kind of target site based drug delivery would not have been possible and clinical results have shown that these nanomaterials provide a fundamental basis for some tumor treatments. Researchers believe that since there are continuous discoveries in the field of nanotechnology, there will be a great impact of nanotechnology in the fields of tumor therapies and medical imaging. There are some limitations like cytotoxicity and non-biodegradability. These have to be observed more to understand the side effects on human health. But still barriers like the nanomaterials' characteristics, metabolism of the drug, the effect on cells and their compatibility, still remain. Another limitation is that the mechanisms of action are not reported so far. Based on the above challenges, it is recommended to study the further optimizations and mechanisms between the cell and nanomaterials are also to be understood, this can lead to a better imaging and therapeutic effects.

## References


Abrahamse, H., & Hamblin, M. R. (2016). New photosensitizers for photodynamic therapy. *The Biochemical Journal, 473*, 347–364.

Abunahla, H., Mohammad, B., Alazzam, A., Jaoude, M. A., Al-Qutayri, M., Abdul Hadi, S., & Al-Sarawi, S. F. (2019). MOMSense: Metal-oxide-metal elementary glucose sensor. *Scientific Reports, 9*(1), 5524.

Ajima, K., Yudasaka, M., Murakami, T., Maigne, A., Shiba, K., & Iijima, S. (2005). Carbon nanohorns as anticancer drug carriers. *Molecular Pharmacology, 2*, 475–480.

Amulyavichus, A., Daugvila, A., Davidonis, R., & Sipavichus, C. (1998). Study of chemical composition of nanostructural materials prepared by laser cutting of metals. *Fizika Metallov i Metallovedenie, 85*, 111–117.

Arap, W. (1998). Cancer treatment by targeted drug delivery to tumor vasculature in a mouse model. *Science, 279*, 377–380.

Ashkarran, A. A. (2012). Synthesis and characterization of gold nanoparticles via submerged arc discharge based on a seed-mediated approach. *Journal of Theoretical and Applied Physics, 6*(14), 1–6.

Baetke, S. C., Lammers, T., & Kiessling, F. (2015). Applications of nanoparticles for diagnosis and therapy of cancer. *The British Journal of Radiology, 88*(1054), 20150207.

Banchereau, J., & Steinman, R. M. (1998). Dendritic cells and the control of immunology. *Nature, 392*, 245–252.

Bang, J. H., & Suslick, K. S. (2010). Applications of ultrasound to the synthesis of nanostructured materials. *Advanced Materials, 22*(10), 1039–1059.

Behlke, M. A. (2008). Chemical modification of siRNAs for in vivo use. *Oligonucleotides, 18*(4), 305–320.

Behzadi, A. H., Farooq, Z., Newhouse, J. H., & Prince, M. R. (2018). MRI and CT contrast media extravasation. *Medcine, 97*, e0055.

Bello, L. (2001). Alpha(v)beta3 and alpha(v)beta5 integrin expression in glioma periphery. *Neurosurgery, 49*, 380–389.

Benne, N., van Duijn, J., Kuiper, J., Jiskoot, W., & Slutter, B. (2016). Orchestrating immune responses: How size, shape and rigidity affect the immunogenicity of particulate vaccines. *Journal of Controlled Release, 234*, 124–134.

Beydoun, D., Amal, R., & Low, G. (1999). Role of nanoparticles in photocatalysis. *Journal of Nanoparticle Research, 1*, 439.

Bozzuto, G., & Molinari, A. (2015). Liposomes as nanomedical devices. *International Journal of Nanomedicine, 10*, 975.

Bretin, L., Pinon, A., Bouramtane, S., Ouk, C., Richard, L., Perrin, M., Chaunavel, A., & Carrion, C. (2019). Photodynamic therapy activity of new human colorectal cancer. *Cancers, 11*, 1474.

Brown, H. C., & Brown, C. A. (1962). New, highly active metal catalysts for the hydrolysis of borohydride. *Journal of the American Chemical Society, 84*, 1493–1494.

Burke, B. P., Cawthorne, C., & Archibald, S. J. (2017). Multimodal nanoparticle imaging agents: Design and applications. *Philosophical Transactions of the Royal Society A - Mathematical Physical and Engineering Sciences, 375*(2107), 20170261.

Buss, J. L., Torti, F. M., & Torti, S. V. (2003). The role of iron chelation in cancer therapy. *Current Medicinal Chemistry, 10*, 1021–1034.

Bystrom, L. M., Guzman, M. L., & Rivella, S. (2014). Iron and reactive oxygen species: Friends or foes of cancer cells? *Antioxidants and Redox Signaling, 20*(12), 1917–1924.

Chabra, V., Pillai, V., Mishra, B. K., Morrone, A., & Shah, D. O. (1995). Synthesis, characterization, and properties of microemulsion-mediated nanophase TiO2 particles. *Langmuir, 11*, 3307.

Chakravarty, R., Hong, H., & Cai, W. (2015). Image-guided drug delivery with single-photon emission computed tomography: A review of literature. *Current Drug Targets, 16*, 592–609.

Chandra, P., Singh, J., Singh, A., Srivastava, A., Goyal, R. N., & Shim, Y. B. (2013). Gold nanoparticles and nanocomposites in clinical diagnostics using electrochemical methods. *Journal of Nanoparticles, 2013*(12), 535901.

Chen, X., & Mao, S. S. (2007). Titanium dioxide nanomaterials: Synthesis, properties, modifications, and applications. *Chemical Reviews, 107*, 2891–2959.

Cheng, Y., Xu, Z., Ma, M., & Xu, T. (2008). Dendrimers as drug carriers: Applications in different routes of drug administration. *Journal of Pharmaceutical Sciences, 97*, 123–143.

Chiu, Y., & Rana, T. M. (2003). siRNA function in RNAi: A chemical modification analysis. *RNA, 9*, 1034–1048.

Choi, S. J., Lee, J. K., Jeong, J., & Choy, J. H. (2013). Toxicity evaluation of inorganic nanoparticles: Considerations and challenges. *Molecular & Cellular Toxicology, 9*, 205–210.

Choi, W., Park, E. Y., Jeon, S., & Kim, C. (2018). Clinical photoacoustic imaging platforms. *Biomedical Engineering Letters, 8*, 139–155.

Christian, S. (2003). Nucleolin expressed at the cell surface is a marker of endothelial cells in angiogenic blood vessels. *The Journal of Cell Biology, 163*, 871–878.

Culver, K. S., Shin, Y. J., Rotz, M. W., Meade, T. J., Hersam, M. C., & Odom, T. W. (2016). Shape-dependent relaxivity of nanoparticle-based T1 magnetic resonance imaging contrast agents. *The Journal of Physical Chemistry. C, Nanomaterials and Interfaces, 120*(38), 22103–22109.

Deepak, V., Umamaheshwaran, P. S., Guhan, K., Nanthini, R. A., Krithiga, B., Jaithoon, N. M., & Gurunathan, S. (2011). Synthesis of gold and silver nanoparticles using purified URAK. *Colloid Surface B, 86*, 353–358.

Desai, N. (2012). Challenges in development of nanoparticle-based therapeutics. *The AAPS Journal, 14*(2), 282–295.

Dimov, N., Kastner, E., Hussain, M., Perrie, Y., & Szita, N. (2017). Formation and purification of tailored liposomes for drug delivery using a module-based micro continuous-flow system. *Scientific Reports, 7*, 12045.

Dong, Y. C., Hajfathalian, M., Maidment, P. S. N., Hsu, J. C., Naha, P. C., Si-Mohamed, S., Breuilly, M., Kim, J., Chhour, P., & Douek, P. (2019). Effect of gold nanoparticle size on their properties as contrast agents for computed tomography. *Scientific Reports, 9*, 1–13.

Duan, H., Wang, D., & Li, Y. (2015). Green chemistry for nanoparticle synthesis. *Chemical Society Reviews, 44*, 5778–5792.

Duan, X., He, C., Kron, S. J., & Lin, W. (2016). Nanoparticle formulations of cisplatin for cancer therapy. *Wiley Interdisciplinary Reviews. Nanomedicine and Nanobiotechnology, 8*, 776–791.

Elsupikhe, R. F., Shameli, K., Ahmad, M. B., Ibrahim, N. A., & Zainudin, N. (2015). Green sonochemical synthesis of silver nanoparticles at varying concentrations of κ-carrageenan. *Nanoscale Research Letters, 10*, 302.

Emanet Ciofani, M., Şen, Ö., & Culha, M. (2020). Hexagonal boron nitride nanoparticles for prostate cancer treatment. *ACS Applied Nano Materials, 3*(3), 2364–2372.

Estelrich, J., & Antònia Busquets, M. (2018). Iron oxide nanoparticles in photothermal therapy. *Molecules, 23*, 1567.

Evans, E. R., Bugga, P., Asthana, V., & Drezek, R. (2018). Metallic nanoparticles for cancer immunotherapy. *Materials Today, 21*, 673–685.

Fang, J. (2011). The EPR effect: Unique features of tumor blood vessels for drug delivery, factors involved, and limitations and augmentation of the effect. *Advanced Drug Delivery Reviews, 63*, 136–151.

Farrow, B., Evers, B. M., Iwamura, T., Murillo, C., O'Connor, K. L., & Rychahou, P. (2003). Inhibition of pancreatic cancer cell growth and induction of apoptosis with novel therapies directed against protein kinase A. *Surgery, 134*(2), 197–205.

Felber, A. E., Dufresne, M. H., & Leroux, J. C. (2012). pH-sensitive vehicles, polymeric micelles, and nanospheres prepared with polycarboxylates. *Advanced Drug Delivery Reviews, 64*, 979–992.

Fire, A., Xu, S., Montgomery, M. K., Kostas, S. A., Driver, S. E., & Mello, C. C. (1998). Potent and specific genetic interference by double-stranded RNA in Caenorhabditis elegans. *Nature, 391*, 806.

Fisusi, F. A., & Akala, E. O. (2019). Drug combinations in breast cancer therapy. *Pharmaceutical Nanotechnology, 7*, 3–23.

Foy, S. P., & Labhasetwar, V. (2011). Oh the irony: Iron as a cancer cause or cure? *Biomaterials, 32*, 9155–9158.

Frappier, V., Duran, M., & Keating, A. E. (2018). PixelDB: Protein–peptide complexes annotated with structural conservation of the peptide binding mode. *Protein Science, 27*(1), 276–285.

Fujiwara, R., Luo, Y., & Sasaki, T. (2015). Cancer therapeutic effects of titanium dioxide nanoparticles are associated with oxidative stress and cytokine induction. *Pathobiology, 82*, 243–251.

Gallardo-Williams, M. T., Chapin, R. E., King, P. E., Moser, G. J., Goldsworthy, T. L., Morrison, J. P., & Maronpot, R. R. (2004). Boron supplementation inhibits the growth and local expression of IGF-1 in human prostate adenocarcinoma (LNCaP) tumors in nude mice. *Toxicologic Pathology, 32*(1), 73–78.

Ge, Y., Zhang, Y., Xia, J., et al. (2009). Effect of surface charge and agglomerate degree of magnetic iron oxide nanoparticles on KB cellular uptake in vitro. *Colloids and Surfaces. B, Biointerfaces, 73*(2), 294–301.

Ghafouri-Fard, S., & Ghafouri-Fard, S. (2012). siRNA and cancer immunotherapy. *Immunotherapy, 4*(9), 907–917.

Gopinath, P., Gogoi, S. K., Chattopadhyay, A., & Ghosh, S. S. (2008). Implications of silver nanoparticle induced cell apoptosis for in vitro gene therapy. *Nanotechnology, 19*, 075104.

Guo, R., Lu, G., Qin, B., & Fei, B. (2018). Ultrasound imaging technologies for breast cancer detection and management: A review. *Ultrasound in Medicine & Biology, 44*, 37–70.

Gurav, A. S., Kodas, T. T., Wang, L. M., Kauppinen, E. I., & Joutsensaari, J. (1994). Generation of nanometer-size fullerene particles via vapor condensation. *Chemical Physics Letters, 218*, 304–308.

Gurunathan, S., Kang, M. H., Qasim, M., & Kim, J. H. (2018). Nanoparticle-mediated combination therapy: Two-in-one approach for cancer. *International Journal of Molecular Sciences, 19*, 1–37.

Harada, A., Ono, M., & Yuba, E. (2013). Titanium dioxide nanoparticle-entrapped polyion complex micelles generate singlet oxygen in the cells by ultrasound irradiation for sonodynamic therapy. *Biomaterials Science, 1*, 65–73.

Harris, J. M., & Chess, R. B. (2003). Efect of pegylation on pharmaceuticals. *Nature Reviews. Drug Discovery, 2*(3), 214–221.

He, C., & Chow, J. C. (2016). Gold nanoparticle DNA damage in radiotherapy: A Monte Carlo study. *AIMS Bioengineering, 3*, 352.

Hemond, C. C., & Bakshi, R. (2018). Magnetic resonance imaging in multiple sclerosis. *Cold Spring Harbor Perspectives in Medicine, 8*, 1–21.

Hillen, F., & Griffioen, A. W. (2007). Tumour vascularization: Sprouting angiogenesis and beyond. *Cancer Metastasis Reviews, 26*(3–4), 489–502.

Hirschberg, H., & Madsen, S. J. (2019). Cell mediated photothermal therapy of brain tumors. *Journal of Neuroimmune Pharmacology, 12*, 99–106.

Holgate, S. T. (2010). Exposure, uptake, distribution and toxicity of nanomaterials in humans. *Journal of Biomedical Nanotechnology, 6*(1), 1–19.

Huang, H., Zhang, W. K., Li, M. C., Gan, Y. P., Ma, C. A., & Zhang, X. B. (2004). Electrochemical production of Sn-filled carbon nanotubes in molten salts. *Transactions of Nonferrous Metals Society of China, 14*, 441–445.

Huynh, E., & Zheng, G. (2015). Cancer nanomedicine: Addressing the dark side of the enhanced permeability and retention effect. *Nanomedicine (London, England), 10*, 1993–1995.

Jain, K., Gupta, U., & Jain, N. K. (2014). Dendronized nanoconjugates of lysine and folate for treatment of cancer. *European Journal of Pharmaceutics and Biopharmaceutics, 87*, 500–509.

Jørgensen, J. T., Norregaard, K., Tian, P., Bendix, P. M., Kjaer, A., & Oddershede, L. B. (2016). Single particle and PET-based platform for identifying optimal plasmonic nano-heaters for photothermal cancer therapy. *Scientific Reports, 2*(3), 201–210.

Kamaly, N. (2012). Targeted polymeric therapeutic nanoparticles: Design, development and clinical translation. *Chemical Society Reviews, 41*, 2971–3010.

Kerbel, R. S., & Kamen, B. A. (2004). The anti-angiogenic basis of metronomic chemotherapy. *Nature Reviews. Cancer, 4*(6), 423–436.

Kesharwani, P., Jain, K., & Jain, N. K. (2014). Dendrimer as nanocarrier for drug delivery. *Progress in Polymer Science, 39*, 268–307.

Kesharwani, P., Xie, L., Banerjee, S., Mao, G., Padhye, S., Sarkar, F. H., & Iyer, A. K. (2015). Hyaluronic acid-conjugated polyamidoamine dendrimers for targeted delivery of 3, 4-difluorobenzylidene curcumin to CD44 overexpressing pancreatic cancer cells. *Colloids and Surfaces B: Biointerfaces, 136*, 413–423.

Kim, J., Lee, N., & Hyeon, T. (2017). Recent development of nanoparticles for molecular imaging. *Philosophical Transactions of the Royal Society A - Mathematical Physical and Engineering Sciences, 375*(2107), 20170022.

Klabunde, K. J. (2001). *Nanoscale materials in chemistry*. New York, NY: Wiley-Interscience.

Klippstein, R., & Pozo, D. (2010). Nanotechnology-based manipulation of dendritic cells for enhanced immunotherapy strategies. *Nanomedicine: Nanotechnology, Biology and Medicine, 6*(4), 523–529.

Kobayashi, H. (2013). Improving conventional enhanced permeability and retention (EPR) effects; what is the appropriate target? *Theranostics, 4*, 81–89.

Kodiha, M., Wang, Y. M., Hutter, E., Maysinger, D., & Stochaj, U. (2015). Off to the organelles killing cancer cells with targeted gold nanoparticles. *Theranostics, 5*(4), 357–370.

Kong, F. Y., Zhang, J. W., Li, R. F., Wang, Z. X., Wang, W. J., & Wang, W. (2017). Unique roles of gold nanoparticles in drug delivery, targeting and imaging applications. *Molecules, 22*, 1445.

Kruis, F. E., Fissan, H., & Rellinghaus, B. (2000). Sintering and evaporation characteristics of gas-phase synthesis of size-selected PbS nanoparticles. *Materials Science and Engineering B, 69*, 329–334.
Kudr, J., Haddad, Y., Richtera, L., Heger, Z., Cernak, M., Adam, V., & Zitka, O. (2017). Magnetic nanoparticles: From design and synthesis to real world applications. *Nanomaterials, 7*, 243.
Kulthe, S. S., Choudhari, Y. M., Inamdar, N. N., & Mourya, V. (2012). Polymeric micelles: Authoritative aspects for drug delivery. *Designed Monomers and Polymers, 15*, 465–521.
Kumar, T. P., Ramesh, R., Lin, Y. Y., & Fey, G. T. K. (2004). Tin-filled carbon nanotubes as insertion anode materials for lithium-ion batteries. *Electrochemistry Communications, 6*, 520–525.
Kwiatkowski, G., Jähnig, F., Steinhauser, J., Wespi, P., Ernst, M., & Kozerke, S. (2017). Nanometer size silicon particles for hyperpolarized MRI. *Scientific Reports, 7*, 1–6.
Kyrychenko, A., Pasko, D. A., & Kalugin, O. N. (2017). Poly(vinyl alcohol) as a water protecting agent for silver nanoparticles: The role of polymer size and structure. *Physical Chemistry Chemical Physics, 19*(13), 8742–8756.
Lecuyer, T., Teston, E., Ramirez-Garcia, G., Maldiney, T., Viana, B., Seguin, J., Mignet, N., Scherman, D., & Richard, C. (2016). Chemically engineered persistent luminescence nanoprobes for bioimaging. *Theranostics, 6*, 2488.
Lee, M. S., Dees, E. C., & Wang, A. Z. (2017). Nanoparticle-delivered chemotherapy: Old drugs in new packages. *Oncology, 31*, 198–208.
Leonenko, Z., Finot, E., & Amrein, M. (2007). Adhesive interaction measured between AFM probe and lung epithelial type II cells. *Ultramicroscopy, 107*(10–11), 948–953.
Leserman, L. D. (1980). Targeting to cells of fluorescent liposomes covalently coupled with monoclonal antibody or protein A. *Nature, 288*, 602–604.
Li, X., Wang, X., Zhang, J., Hanagata, N., Wang, X., Weng, Q., Ito, A., Bando, Y., & Golberg, D. (2017). Hollow boron nitride nanospheres as boron reservoir for prostate cancer treatment. *Nature Communications, 8*, 13936.
Limbach, L. K., Li, Y., & Grass, R. N. (2005). Oxide nanoparticle uptake in human lung fibroblasts: Effects of particle size, agglomeration, and diffusion at low concentrations. *Environmental Science & Technology, 39*(23), 9370–9376.
Liu, J., Lécuyer, T., Seguin, J., Mignet, N., Scherman, D., Viana, B., & Richard, C. (2019). Imaging and therapeutic applications of persistent luminescence nanomaterials. *Advanced Drug Delivery Reviews, 138*, 193–210.
Liu, M. C., Lin, T. S., & Sartorelli, A. C. (1992). Synthesis and antitumor activity of amino derivatives of pyridine-2-carboxaldehyde thiosemicarbazone. *Journal of Medicinal Chemistry, 35*, 3672–3677.
Liyanage, P. Y., Hettiarachchi, S. D., Zhou, Y., Ouhtit, A., Seven, E. S., Oztan, C. Y., Celik, E., & Leblanc, R. M. (2019). Nanoparticle-mediated targeted drug delivery for breast cancer treatment. *Biochimica Et Biophysica Acta. Reviews on Cancer, 1871*, 419–433.
Longmire, M., Choyke, P. L., & Kobayashi, H. (2008). Clearance properties of nano-sized particles and molecules as imaging agents: Considerations and caveats. *Nanomedicine, 3*(5), 703–717.
Ma, J., Pulfer, S., Li, S., Chu, J., Reed, K., & Gallo, J. M. (2001). Pharmacodynamic-mediated reduction of temozolomide tumor concentrations by the angiogenesis inhibitor TNP470. *Cancer Research, 61*(14), 5491–5498.
Macák, M. (2008). *Growth of anodic self-organazed titanium dioxide nanotube layers [Ph.D. thesis]*. University Erlangen-Nürnberg-Germany.
Madaan, K., Kumar, S., Poonia, N., Lather, V., & Pandita, D. (2014). Dendrimers in drug delivery and targeting: Drug-dendrimer interactions and toxicity issues. *Journal of Pharmacy & Bioallied Sciences, 6*, 139.
Maeda, H. (2001). The enhanced permeability and retention (EPR) effect in tumor vasculature: The key role of tumor-selective macromolecular drug targeting. *Advances in Enzyme Regulation, 41*, 189–207.
Manchun, S., Dass, C. R., & Sriamornsak, P. (2012). Targeted therapy for cancer using pH-responsive nanocarrier systems. *Life Sciences, 90*, 381–387.

Mandal, A., Bisht, R., Rupenthal, I. D., & Mitra, A. K. (2017). Polymeric micelles for ocular drug delivery: From structural frameworks to recent preclinical studies. *Journal of Controlled Release, 248*, 96–116.
Mangal, S., Gao, W., Li, T., & Zhou, Q. T. (2017). Pulmonary delivery of nanoparticle chemotherapy for the treatment of lung cancers: Challenges and opportunities. *Acta Pharmacologica Sinica, 38*, 782–797.
Manikprabhu, D., Cheng, J., Chen, W., Sunkara, A. K., Mane, S. B., Kumar, R., Das, M. N., Hozzein, W., Duan, Y. Q., & Li, W. J. (2016). Sunlight mediated synthesis of silver nanoparticles by a novel actinobacterium (*Sinomonas mesophila* MPKL 26) and its antimicrobial activity against multi drug resistant *Staphylococcus aureus*. *Journal of Photochemistry and Photobiology A: Chemistry, B158*, 202–205.
Marcu, A., Pop, S., Dumitrache, F., Mocanu, M., Niculite, C., Gherghiceanu, M., Lungu, C., Fleaca, C., Ianchis, R., & Barbut, A. (2013). Magnetic iron oxide nanoparticles as drug delivery system in breast cancer. *Applied Surface Science, 281*, 60–65.
Mateti, S., Wong, C. S., Liu, Z., Yang, W., Li, Y., Li, L. H., & Chen, Y. (2017). Biocompatibility of boron nitride nanosheets. *Nanoscale Research Letters, 11*(1), 334–342.
Matsumura, S., Ajima, K., Yudasaka, M., Iijima, S., & Shiba, K. (2007). Dispersion of cisplatin-loaded carbon nanohorns with a conjugate comprised of an artificial peptide aptamer and polyethylene glycol. *Molecular Pharmacology, 4*, 723–729.
McNamara, K., & Tofail, S. A. (2015). Nanosystems: The use of nanoalloys, metallic, bimetallic, and magnetic nanoparticles in biomedical applications. *Physical Chemistry Chemical Physics, 17*, 27981–27995.
McNamara, K., & Tofail, S. A. (2017). Nanoparticles in biomedical applications. *Advances in Physics, 2*, 54–88.
Meng, F., Cheng, R., Deng, C., & Zhong, Z. (2012). Intracellular drug release nanosystems. *Materials Today, 15*, 436–442.
Meng, F., Han, N., & Yeo, Y. (2017). Organic nanoparticle systems for spatiotemporal control of multimodal chemotherapy. *Expert Opinion on Drug Delivery, 14*, 427–446.
Meng, F., Hennink, W. E., & Zhong, Z. (2009). Reduction-sensitive polymers and bioconjugates for biomedical applications. *Biomaterials, 30*, 2180.
Miao, W., Kim, H., Gujrati, V., Kim, J. Y., Jon, H., Lee, Y., Choi, M., Kim, J., Lee, S., & Lee, D. Y. (2016). Photo-decomposable organic nanoparticles for combined tumor optical imaging and multiple phototherapies. *Theranostics, 6*, 2367–2379.
Miyata, K., Christie, R. J., & Kataoka, K. (2011). Polymeric micelles for nano-scale drug delivery. *Reactive and Functional Polymers, 71*, 227–234.
Moghimi, S. M., Hunter, A. C., & Murray, J. C. (2001). Long-circulating and target-specific nanoparticles: Theory to practice. *Pharmacological Reviews, 53*(2), 283–318.
Mokoena, D. R., George, B. P., & Abrahamse, H. (2019). Enhancing breast cancer treatment using a combination of cannabidiol and gold nanoparticles for photodynamic therapy. *International Journal of Molecular Sciences, 20*, 4771.
Monthioux, M. (2002). Filling single-wall carbon nanotubes. *Carbon, 40*, 1809–1823.
Mou, Y., Xing, Y., & Ren, H. (2017). The effect of superparamagnetic iron oxide nanoparticle surface charge on antigen cross-presentation. *Nanoscale Research Letters, 12*(1), 52.
Muthukumar, T., Kumari, S., Sambandam, B., Aravinthan, A., Sastry, T. P., & Kim, J. (2016). Green synthesis of gold nanoparticles and their enhanced synergistic antitumor activity using HepG2 and MCF7 cells and its antibacterial effects. *Process Biochemistry, 51*, 384–391.
Nishida, N., Yano, H., Nishida, T., Kamura, T., & Kojiro, M. (2006). Angiogenesis in cancer. *Vascular Health and Risk Management, 2*(3), 213–219.
Noriega-Luna, B., Godínez, L. A., Rodríguez, F. J., Rodríguez, A., Larrea, G., Sosa-Ferreyra, C., Mercado-Curiel, R., Manríquez, J., & Bustos, E. (2014). Applications of dendrimers in drug delivery agents, diagnosis, therapy, and detection. *Journal of Nanomaterials, 2014*, 39.
Orme, M. E., & Chaplain, M. A. J. (1997). Two-dimensional models of tumour angiogenesis and anti-angiogenesis strategies. *IMA Journal of Mathematics Applied in Medicine and Biology, 14*(3), 189–205.

Park, W., Heo, Y. J., & Han, D. K. (2018). New opportunities for nanoparticles in cancer immunotherapy. *Biomaterials Research, 22*, 1–10.
Parveen, S., & Sahoo, S. K. (2008). Polymeric nanoparticles for cancer therapy. *Journal of Drug Targeting, 16*(2), 108–123.
Pasqualini, R. (2000). Aminopeptidase N is a receptor for tumor-homing peptides and a target for inhibiting angiogenesis. *Cancer Research, 60*, 722–727.
Pelaz, B., Charron, G., & Pfeiffer, C. (2013). Interfacing engineered nanoparticles with biological systems: Anticipating adverse nano-bio interactions. *Small, 9*(9–10), 1573–1584.
Pelt, J., Busatto, S., Ferrari, M., Thompson, E. A., Mody, K., & Wolfram, J. (2018). Chloroquine and nanoparticle drug delivery: A promising combination. *Pharmacology & Therapeutics, 191*, 43–49.
Peng, Y., Lu, B., Wang, N., Li, L., & Chen, S. (2017). Impacts of interfacial charge transfer on nanoparticle electrocatalytic activity towards oxygen reduction. *Physical Chemistry Chemical Physics, 19*(14), 9336–9348.
Pittella, F., Zhang, M., & Lee, Y. (2011). Enhanced endosomal escape of siRNA-incorporating hybrid nanoparticles from calcium phosphate and PEG-block charge-conversional polymer for efficient gene knockdown with negligible cytotoxicity. *Biomaterials, 32*(11), 3106–3114.
Pluym, T., Powell, Q., Gurav, A., Ward, T., Kodas, T., & Glicksman, H. (1993). Solid silver particle production by spray pyrolysis. *Journal of Aerosol Science, 24*, 383–392.
Pratsinis, S. E. (2011). History of manufacture of fine particles in high-temperature aerosol reactors. In D. S. Ensor (Ed.), *Aerosol science and technology: History and reviews* (1st ed.). Research Triangle Park: RTI International.
Press, D. (2018). Application of active targeting nanoparticle delivery system for chemotherapeutic drugs and traditional/herbal medicines in cancer therapy: A systematic review. *International Journal of Nanomedicine, 13*, 3921.
Prusty, K., & Swain, S. K. (2018). Nano silver decorated polyacrylamide/dextran nanohydrogels hybrid composites for drug delivery applications. *Materials Science and Engineering, 85*, 130–141.
Qi, G. B., Gao, Y. J., Wang, L., & Wang, H. (2018). Self-assembled peptide-based nanomaterials for biomedical imaging and therapy. *Advanced Materials, 30*(22), 1703444.
Qiao, Z. Y., Lin, Y. X., Lai, W. J., Hou, C. Y., Wang, Y., Qiao, S. L., Zhang, D., Fang, Q. J., & Wang, H. A. (2016). General strategy for facile synthesis and in situ screening of self-assembled polymer-peptide nanomaterials. *Advanced Materials, 28*(9), 1859–1867.
Quintanar-Guerrero, D. (1998). Preparation techniques and mechanisms of formation of biodegradable nanoparticles from preformed polymers. *Drug Development and Industrial Pharmacy, 24*(12), 1113–1128.
Rampersaud, S., Fang, J., & Wei, Z. (2016). The effect of cage shape on nanoparticle-based drug carriers: Anticancer drug release and efficacy via receptor blockade using dextran-coated iron oxide nanocages. *Nano Letters, 16*(12), 7357–7363.
Rao, V. A., Klein, S. R., Agama, K. K., Toyoda, E., Adachi, N., Pommier, Y., & Shacter, E. B. (2009). The iron chelator Dp44mT causes DNA damage and selective inhibition of topoisomerase IIA in breast cancer. *Cells American Association of Cancer Research, 69*, 948–957.
Riley, R. S., & Day, E. S. (2017). Gold nanoparticle-mediated photothermal therapy: Applications and opportunities for multimodal cancer treatment. *Wiley Interdisciplinary Reviews. Nanomedicine and Nanobiotechnology, 9*(4), e1449.
Riley, R. S., June, C. H., Langer, R., & Mitchell, M. J. (2019). Delivery technologies for cancer immunotherapy. *Nature Reviews Drug Discovery, 18*, 175–196.
Rosado-De-Castro, P. H., Morales, M. D. P., Pimentel-Coelho, P. M., Mendez-Otero, R., & Herranz, F. (2018). Development and application of nanoparticles in biomedical imaging. *Contrast Media & Molecular Imaging, 2018*, 1403826.
Saleh, T., & Shojaosadati, S. A. (2016). Multifunctional nanoparticles for cancer immunotherapy. *Human Vaccines and Immunotherapeutics, 12*, 1863–1875.
Santos, B. S., & Ferreira, M. J. (2019). Positron emission tomography in ischemic heart disease. *Revista Portuguesa de Cardiologia, 38*, 599–608.

Sapsford, K. E., Algar, W. R., Berti, L., Gemmill, K. B., Casey, B. J., Oh, E., Stewart, M. H., & Medintz, I. L. (2013). Functionalizing nanoparticles with biological molecules: Developing chemistries that facilitate nanotechnology. *Chemical Reviews, 113*, 1904–2074.

Sau, S., Alsaab, H. O., Bhise, K., Alzhrani, R., Nabil, G., & Iyer, A. K. (2018). Multifunctional nanoparticles for cancer immunotherapy: A groundbreaking approach for reprogramming malfunctioned tumor environment. *Journal of Controlled Release, 274*, 24–34.

Schlesinger, H. I., Brown, H. C., Finholt, A. E., Gilbreath, J. R., Hoekstra, H. R., & Hyde, E. K. (1953). Sodium borohydride, its hydrolysis and its use as a reducing agent and in the generation of hydrogen. *Journal of the American Chemical Society, 75*(1), 215–219.

Schwarz, D. S., Hutvágner, G., Haley, B., & Zamore, P. D. (2002). Evidence that siRNAs function as guides, not primers, in the Drosophila and human RNAi pathways. *Molecular Cell, 10*(3), 537–548.

Scorei, I., & Popa, R. (2012). Boron-containing compounds as preventive and chemotherapeutic agents for cancer. *Anti-Cancer Agents in Medicinal Chemistry, 10*(4), 346–351.

Sercombe, L., Veerati, T., Moheimani, F., Wu, S. Y., Sood, A. K., & Hua, S. (2015). Advances and challenges of liposome assisted drug delivery. *Frontiers in Pharmacology, 6*, 286.

Shah, M., Badwaik, V., Kherde, Y., Waghwani, H. K., Modi, T., Aguilar, Z. P., Rodgers, H., Hamilton, W., Marutharaj, T., Webb, C., Lawrenz, M. B., & Dakshinamurthy, R. (2014). Gold nanoparticles: Various methods of synthesis and antibacterial applications. *Frontiers in Bioscience, 1*(19), 1320–1344.

Shameli, K., Ahmad, M. B., Yunus, W. M. Z. W., Ibrahim, N. A., Gharayebi, Y., & Sedaghat, S. (2010). Synthesis of silver/montmorillonite nanocomposites using γ-irradiation. *International Journal of Nanomedicine, 5*, 1067–1077.

Siddique, S., & Chow, J. C. L. (2020). Application of nanomaterials in biomedical imaging and cancer therapy. *Nanomaterials, 10*, 1700.

Silva, A. L., Soema, P. C., Slutter, B., Ossendorp, F., & Jiskoot, W. (2016). PLGA particulate delivery systems for subunit vaccines: Linking particle properties to immunogenicity. *Human Vaccines & Immunotherapeutics, 12*(4), 1056–1069.

Silva, G. A. (2004). Introduction to nanotechnology and its applications to medicine. *Surgical Neurology, 61*(3), 216–220.

Simón, M., Norregaard, K., Jørgensen, J. T., Oddershede, L. B., & Kjaer, A. (2019). Fractionated photothermal therapy in a murine tumor model: Comparison with single dose. *International Journal of Nanomedicine, 14*, 5369–5379.

Singh, P., Pandit, S., Mokkapati, V. R. S. S., Garg, A., Ravikumar, V., & Mijakovic, I. (2018). Gold nanoparticles in diagnostics and therapeutics for human cancer. *International Journal of Molecular Sciences, 19*, 1979.

Siregar, S., Oktamuliani, S., & Saijo, Y. A. (2018). Theoretical model of laser heating carbon nanotubes. *Nanomaterials, 8*, 580.

Sitharaman, B., Kissell, K. R., Hartman, K. B., Tran, L. A., Baikalov, A., Rusakova, I., Sun, Y., Khant, H. A., Ludtke, S. J., Chiu, W., Laus, S., Toth, E., Helm, L., Merbach, A. E., & Wilson, L. J. (2005). Superparamagnetic gadonanotubes are high-performance MRI contrast agents. *Chemical Communications*, 3915–3917.

Steinberg, I., Huland, D. M., Vermesh, O., Frostig, H. E., Tummers, W. S., & Gambhir, S. S. (2019). Photoacoustic clinical imaging. *Photoacoustics, 8*(14), 77–98.

Tang, J., Li, L., Howard, C. B., Mahler, S. M., Huang, L., & Xu, Z. P. (2015). Preparation of optimized lipid-coated calcium phosphate nanoparticles for enhanced in vitro gene delivery to breast cancer cells. *Journal of Materials Chemistry. B, Materials for Biology and Medicine, 3*(33), 6805–6812.

Tao, A., Sinsermsuksakul, P., & Yang, P. (2006). Polyhedral silver nanocrystals with distinct scattering signatures. *Angewandte Chemie, International Edition, 45*, 4597–4601.

Terman, A., & Kurz, T. (2012). Lysosomal iron, iron chelation, and cell death. *Anitoxidants and Redox Signaling, 18*(8), 888–898.

Tian, Y., Guo, Z., Zhang, T., Lin, H., Li, Z., Chen, J., Deng, S., & Liu, F. (2019). Inorganic boron-based nanostructures: Synthesis, optoelectronic properties, and prospective applications. *Nanomaterials, 9*(4), 538.

Tien, D. C., Liao, C. Y., Huang, J. C., Tseng, K. H., Lung, J. K., Tsung, T. T., Kao, W. S., Tsai, T. H., Cheng, T. W., & Yu, B. S. (2008). Novel technique for preparing a nano-silver water suspension by the arc-discharge method. *Reviews on Advanced Materials Science, 18*, 750–756.
Tombácz, E., Turcu, R., Socoliuc, V., & Vékás, L. (2015). Magnetic iron oxide nanoparticles: Recent trends in design and synthesis of magneto responsive nanosystems. *Biochemical and Biophysical Research Communications, 468*, 442–453.
Torchilin, V. (2011). Tumor delivery of macromolecular drugs based on the EPR effect. *Advanced Drug Delivery Reviews, 63*, 131–135.
Tran, S., DeGiovanni, P. J., Piel, B., & Rai, P. (2017). Cancer nanomedicine: A review of recent success in drug delivery. *Clinical and Translational Medicine, 6*(1), 44.
Tripathy, S., & Das, M. (2013). Dendrimers and their applications as novel drug delivery carriers. *Journal of Applied Pharmaceutical Science, 3*, 142–149.
Vansant, E. F., Voort, P. V. D., & Vrancken, K. C. (1995). *Characterization and chemical modification of the silica surface*. New York, NY: Elsevier Science.
Veiseh, O., Gunn, J. W., & Zhang, M. (2010). Design and fabrication of magnetic nanoparticles for targeted drug delivery and imaging. *Advanced Drug Delivery Reviews, 62*(3), 284–304.
Ventola, C. L. (2017). Progress in nanomedicine: Approved and investigational nanodrugs. *Pharmacy and Therapeutics, 42*(12), 742–755.
Vines, J. B., Yoon, J. H., Ryu, N. E., Lim, D. J., & Park, H. (2019). Gold nanoparticles for photothermal cancer therapy. *Frontiers in Chemistry, 5*(7), 167.
Wakaskar, R. R. (2017). Polymeric micelles for drug delivery. *The International Journal of Drug Development and Research, 9*, 1–2.
Wang, B., Zhang, W., Zhou, X., Liu, M., Hou, X., Cheng, Z., & Chen, D. (2019). Development of dual-targeted nano-dandelion based on an oligomeric hyaluronic acid polymer targeting tumor-associated macrophages for combination therapy of non-small cell lung cancer. *Drug Delivery, 26*, 1265–1279.
Wang, X. (2014). The development of site-specific drug delivery nanocarriers based on receptor mediation. *Journal of Controlled Release, 193*, 139–153.
Wang, Y., Li, Y., Wei, F., & Duan, Y. (2017). Optical imaging paves the way for autophagy research. *Trends in Biotechnology, 35*, 1181–1193.
Wang, Y., Lin, Y. X., Qiao, Z. Y., An, H. W., Qiao, S. L., Wang, L., Rajapaksha, R. P. Y. J., & Wang, H. (2015). Self-assembled autophagy-inducing polymeric nanoparticles for breast cancer interference in-vivo. *Advanced Materials, 27*(16), 2627–2634.
Wang, Y., Strohm, E. M., Sun, Y., Wang, Z., Zheng, Y., Wang, Z., & Kolios, M. C. (2016). Biodegradable polymeric nanoparticles containing gold nanoparticles and Paclitaxel for cancer imaging and drug delivery using photoacoustic methods. *Biomedical Optics Express, 7*, 4125.
Watts, J. K., Deleavey, G. F., & Damha, M. J. (2008). Chemically modified siRNA: Tools and applications. *Drug Discovery Today, 13*(19–20), 842–855.
Wen, R., Umeano, A. C., Kou, Y., Xu, J., & Farooqi, A. A. (2019). Nanoparticle systems for cancer vaccine. *Nanomedicine, 8*, 627–648.
Wu, D., Gao, Y., Qi, Y., Chen, L., Ma, Y., & Li, Y. (2014). Peptide-based cancer therapy: Opportunity and challenge. *Cancer Letters, 351*(1), 13–22.
Wu, W., He, Q. G., & Jiang, C. Z. (2008). Magnetic iron oxide nanoparticles: Synthesis and surface functionalization strategies. *Nanoscale Research Letters, 3*(11), 397–415.
Xiao, Y., Shlyahovsky, B., Popov, I., Pavlov, V., & Willner, I. (2005). Shape and color of Au nanoparticles follow biocatalytic processes. *Langmuir, 21*(13), 5659–5662.
Xu, C., & Wang, J. (2015). Delivery systems for siRNA drug development in cancer therapy. *Asian Journal of Pharmaceutical Sciences, 10*(1), 1–12.
Xu, S., Zhang, R., Zhao, W., Zhu, Y., Wei, W., Liu, X., Luo, J. (2017). Self-assembled polymeric nanoparticles film stabilizing gold nanoparticles as a versatile platform for ultrasensitive detection of carcino-embryonic antigen. *Biosensors and Bioelectronics, 92*, 570–576.
Xu, W., Ling, P., & Zhang, T. (2013). Polymeric micelles, a promising drug delivery system to enhance bioavailability of poorly water-soluble drugs. *Journal of Drug Delivery, 2013*, 340315.

Yan, J., He, W., Yan, S., Niu, F., Liu, T., Ma, B., Shao, Y., Yan, Y., Yang, G., Lu, W., Du, Y., Lei, B., & Ma, P. X. (2017). Self-assembled peptide-lanthanide nanoclusters for safe tumor therapy: Overcoming and utilizing biological barriers to peptide drug delivery. *ACS Nano, 12*(2), 2017–2026.

Yang, Z., Song, J., Dai, Y., Chen, J., Wang, F., Lin, L., Liu, Y., Zhang, F., Yu, G., & Zhou, Z. (2017). Theranostics self-assembly of semiconducting-plasmonic gold nanoparticles with enhanced optical property for photoacoustic imaging and photothermal therapy. *Theranostics, 7*(8), 2177–2185.

Yoncheva, K., & Momekov, G. (2011). Antiangiogenic anticancer strategy based on nanoparticulate systems. *Expert Opinion on Drug Delivery, 8*(8), 1041–1056.

Yousaf, T., Dervenoulas, G., & Politis, M. (2018). Advances in MRI methodology. *International Review of Neurobiology, 141*, 31–76.

Yu, Z., Zhou, P., Pan, W., Li, N., & Tang, B. (2018). A biomimetic nanoreactor for synergistic chemiexcited photodynamic therapy and starvation therapy against tumor metastasis. *Nature Communications, 9*, 1–9.

Yuan, H., Wilks, M. Q., Maschmeyer, R., Normandin, M. D., Josephson, L., & Fakhri, G. E. (2020). Original research a radio-nano-platform for T1/T2 dual-mode PET-MR imaging. *International Journal of Nanomedicine, 15*, 1253.

Zhang, A., Pan, S., Zhang, Y., Chang, J., Cheng, J., Huang, Z., Li, T., Zhang, C., De La Fuentea, J. M., & Zhang, Q. (2019). Carbon-gold hybrid nanoprobes for real-time imaging, photothermal/photodynamic and nanozyme oxidative therapy. *Theranostics, 9*, 3443–3458.

Zhang, C., Ni, D., Liu, Y., Yao, H., Bu, W., & Shi, J. (2017). Magnesium silicide nanoparticles as a deoxygenation agent for cancer starvation therapy. *Nature Nanotechnology, 12*(4), 378–386.

Zhang, L., Gu, F., Chan, J., Wang, A., Langer, R., & Farokhzad, O. (2008). Nanoparticles in medicine: Therapeutic applications and developments. *Clinical Pharmacology and Therapeutics, 83*, 761–769.

Zhang, X., Lin, Y., & Gillies, R. J. (2010). Tumor pH and its measurement. *Journal of Nuclear Medicine, 51*, 1167–1170.

Zhu, J., & Shi, X. (2013). Dendrimer-based nanodevices for targeted drug delivery applications. *Journal of Materials Chemistry B, 1*, 4199–4211.

Zylberberg, C., & Matosevic, S. (2016). Pharmaceutical liposomal drug delivery: A review of new delivery systems and a look at the regulatory landscape. *Drug Delivery, 23*, 3319–3329.

# Chapter 17
# Progress in Nanomaterial Self-Assembly for Bio-scaffolds: Exclusive Biomedical Applications

**Gözde Koşarsoy Ağçeli, Kanika Dulta, Parveen Chauhan, P. K. Chauhan, and Kaushik Pal**

## 17.1 Introduction

Natural ageing, injury, congenital diseases and disorders are accountable for causing tissue injury and dysfunction in human body as reported by health science. Higher mortality rate, tissue damage, and organ dysfunctions are caused due to majority of diseases. Organ transplantation between one person to another or from animals was approved as the crucial therapeutic method that has been utilized since decades (Yang, Leong, Du, & Chua, 2001). In Tissue Engineering (TE) there are three basic stages, first phase involves having the donor's base cells to interact with. Second phase is linked with the development of cultured cells and finally third phase, dealing with the introduction of the transformed cells into scaffold for incubation of cells. Final step deals with the utilization of newly formed cells or organs by transplantation approach. Tissue Engineering is an integrated area dealing with multiple disciplines that combines the cellular activities and the methodology of increasing it over an artificial substrate classified as scaffolding with sufficient biochemical factors necessary for synthesizing engineered tissues and organs or regeneration of tissue damage (Langer & Vacanti, 1993; Liu, Xia, & Czernuszka, 2007). It comprises of cells seeded on a scaffold, followed by in vitro culturing for

G. Koşarsoy Ağçeli (✉)
Department of Biology, Hacettepe University, Ankara, Turkey
e-mail: gozdekosarsoy@hacettepe.edu.tr

K. Dulta · P. Chauhan · P. K. Chauhan
Faculty of Applied Sciences and Biotechnology, Shoolini University, Solan, Himachal Pradesh, India

K. Pal
Laboratório de Biopolímeros e Sensores, Instituto de Macromoléculas, Universidade Federal do Rio de Janeiro (LABIOS/IMA/UFRJ)–Cidade Universitária, Rio de Janeiro, Brazil

Wuhan University, Wuchang Dist., Hubei, China


K. Pal (ed.), *Bio-manufactured Nanomaterials*,
https://doi.org/10.1007/978-3-030-67223-2_17

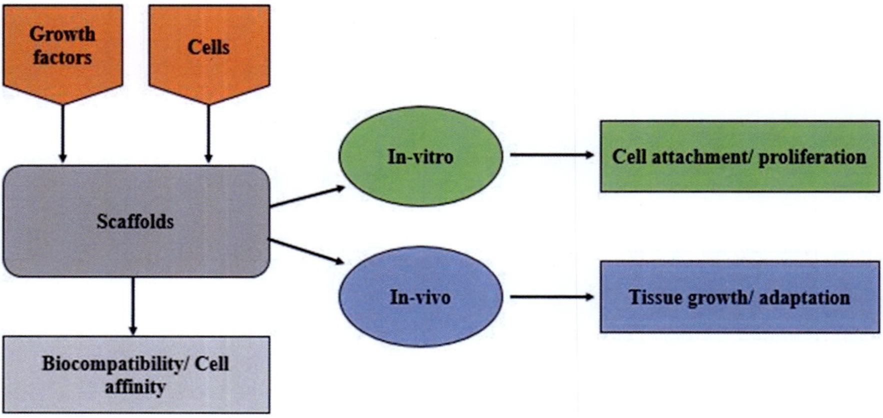


**Fig. 17.1** General concept of scaffold

formation of mature tissues. The injured parts are repaired, including fractured bones, skin or cartilage by implantation. The general mechanism of tissue regeneration occurs among scaffolds, whereby blood vessels penetrates in system and thus slowly weakening the scaffold when a tissue recently designed is in position as schematically illustrated in Fig. 17.1 (Liu et al., 2007; Liu & Czernuszka, 2006). Generally, TE scaffolds should meet three major intended objectives: (1) establishing well defined framework that forms the tissue regeneration; (2) providing transient role in a malfunction during regeneration of tissues; (3) promoting the development of tissues and possibly facilitating the incorporation of seeded cells, genes or proteins for promotes regeneration of tissue. Major advancements of TE for understanding the association of cell-scaffolds and also the technological development for synthesis and characterization of porous scaffolds facilitating emergence of bio-scaffolds of "third generation"; biodegradable and biologically active scaffolds engineered to serve cell and tissues with a transient three dimensional microenvironment and simultaneously directing the cellular processes associated with de novo tissue regeneration (Hutmacher, 2000). Scaffold offers cells with a structure and preliminary aid for attachment, proliferation, and differentiation for developing an extracellular matrix (ECM) (Agarwal & Ray, 2001). Scaffolds being biologically compatible in both degraded and embedded context must have adequate mechanical properties for providing the neo-tissues with required environmental stress. The framework should be engineered to sustain the stability of scaffold before the freshly grown tissues carried up by synthetic supports (Gomes, Ribeiro, Malafaya, Reis, & Cunha, 2001).

### *17.1.1 Importance of Scaffold*

Scaffold is a biodegradable extracellular 3D matrix which facilitates regeneration of tissues. An optimal scaffold should be biologically compatible and degradable, enabling complete spare of functioning tissue. Morphology, chemical makeup and biological signals are essential aspects for controlling cell activity and facilitating tissue regeneration. Biologically degradable polymeric materials are being studied as bone tissue-engineered scaffolds. However, there are certain limits to these technologies, including inadequate mechanical resistance. On other hand, nanoscale inorganic and organic materials combined into polymeric scaffolds can provide the most desirable properties required for structural support and discharge of biologically active components and cellular adhesion, integration differentiation, into the adjacent environment.

Scaffolding materials serves crucial part in regeneration of tissues as it offers a microenvironment appropriate for cellular attachment, propagation and variation. Generally, an optimal scaffolding material must be biologically compatible, with manageable degradation and sufficient physical and chemical characteristics to mimic the composition of the actual tissues as in extracellular matrix (ECM) (Walmsley et al., 2015). Various mixtures of materials with particular functions should be able to handle utilizing engineered membranes, encapsulating cells and controlled chemical release. In addition, it should support and monitor abnormal activities arising at cellular and tissues level (D'Amora et al., 2017; Hosseinpour et al., 2017).

The scaffold is required to perform many roles in scaffold-based tissue engineering. Suitable mechanical strength and stability should be given to exchange the structural functions with misplaced or wounded tissue. Most significantly, an effective scaffolding can promotes ingraining of newly formed tissues and gradual development and redesigning by delivering appropriate protection (Ferracane & Giannobile, 2014; Zhang, Sun, Song, Gu, & Sun, 2015). Kinetics of degradation and physical and chemical properties should also be taken into consideration.

### *17.1.2 Role of Nanomaterials*

The utilization and development of biologically compatible and applied structures depending on nanostructured systems has expanded scientific and industrial focus as the material surface displays a rapid expansion in this field that could greatly boost physical and chemical properties. Nano-sized particles (1–100 nm) are known to be an essential technique for pharmacological carriers, antimicrobial agents, and skin regenerating devices (Shi, Votruba, Farokhzad, & Langer, 2010; Vieira, Vial, Reis, & Oliveira, 2017).

Over the past decades, nanotechnology has been rigorously evolved; it is rapidly growing and making important impacts in materials science. The secret to its achievement is the exciting characteristics of nanostructured materials aimed for improved systems of efficiency on the basis of morphological dimension, shape, and composition. These products presently reflecting huge scope to consumer development and also have currently been commercialized as nanotechnology-enhanced goods. The existence of nanoparticles (NPs) is attributed to the incorporation of stabilizers, such as additives, and for the enhancement of physicochemical or antibacterial properties (Contado, 2015; Leydesdorff & Zhou, 2007).

Chemical processes such as lipids, polymers, nano gels, dendrimers, supramolecular compounds, nano emulsions etc., are part of the general grouping of NPs as additives. Inorganic NPs utilized in the regeneration of tissues, including carbon-based nanostructures (graphene and carbon nanotubes), metallic nanoparticles (like gold, titanium dioxide, silver, and copper), magnetic nanoparticles and quantum dots, in specific (Appel, Anastasio, Larson, & Brey, 2013). Recent advancements in utilization of nano-sized materials in pharmaceuticals include development of regulated drug delivery mechanisms, biomarkers for detection of disease, recognition of pathogen/protein, molecular separation/purification and methods for regenerative medicine. Subsequent experiments based on biomedicine to carry out research in multiple disciplines, incorporating areas such as, biology, physics, chemistry, and materials science and engineering; related to the formation of active structures in reaction to the regeneration of tissue in organisms (Laurencin & Nair, 2015).

### *17.1.3 Desirable Characteristics of Nanomaterial for Bio-scaffold Designing*

Recently a number of nanomaterials (NMs) are being utilized as tissue-engineered scaffolds of various tissues or organs such as bone, cartilage, vascular, skin, corneal, pancreatic, liver, spinal, etc. Tissue engineering circles primarily around three major foundations: material scaffolds, cells and growth factors/bioactive molecules circles primarily around three major foundations: cells, growth factors and material scaffolds/bioactive molecules. The features of nanomaterials accounted for tissue engineering involves various aspects, i.e., (1) nano-dimensions confer appropriate framework to cells or tissues for its attachment, differentiation and proliferation; (2) nanomaterials serves as similar systems for engineered extracellular matrix in the form of chemical composition and physical structure, as scaffolds for TE (Appel et al., 2013); (3) highly conductive carbon-based nanomaterials offers electric stimulus to bone engineering scaffolds, (4) micro or nano-encapsulation of essential growth factors allows the release at the target site slowly and sustainably, and (5) NMs offer improved biocompatibility, bioactivity, and improve scaffold interactions with cells or proteins (De et al., 2018; Edwards, Werkmeister, & Ramshaw, 2009; Guo, Lei, Li, & Ma, 2015; Shi, Votruba, Farokhzad, & Langer, 2010; Van & Sabine, 2017; Yi, Rehman, Zhao, Liu, & He, 2016).

### *17.1.4 Biomedical Applications of Nano-based Bio-scaffolds*

Designing biomaterials has grown rapidly over the last few decades. Preferences have recently been set for such products which may be used as major aspect of biomedical fields. Biomaterials for biomedical applications are intended at improving artificial materials which could be utilized to replace or restore diseased tissue functioning in the human body and improve the quality of life. After an initial experimental stage of preference of biomaterials based on its ease of access, the design efforts were concentrated mainly on either achieving structural or mechanical efficiency. Biomaterial serves another essential part in tissue engineering by serving as an artificial bone plates, joint substitutes, ligaments, intraocular lenses, vascular grafts, dental implants for heart valves, as well as other medical equipment including pacemakers, biosensors, etc.

## 17.2 Construction of 3D Tissue Engineering

Nanomaterials offer future prospects in tissue engineering. Nanoscale structures can regulate cellular processes such as conformity, proliferation and differentiation, and most notably, nanomaterials have special optical and magnetic properties and are thus ideal for observing in vivo cell output after transplantation. Nanomaterials have been successively designed as 3D tissue-engineered scaffolds for bone, skin, cardiac, liver, cartilage, vascular tissues, and pancreas; also, nanostructures were recognized to monitor and control the activity of primary stem cells, such as cell attachment, development, and differentiation. Magnetic nanoparticles have acquired much attention currently because of their remarkable magnetic properties and incredible morphology. Magnetic nanoparticles are usually smaller (20–300 nm) and their orientation of magnetization facilitates thermal variations at atmospheric or biological temperature (Fig. 17.2).

### *17.2.1 Construction of Liver Bio-scaffold*

In humans, liver is the major internal organ serving digestive functions. Acute liver failure develops unexpectedly and degree of progress of this disorder may often result in the malfunction of many other organs and even death of an individual. Hepatic transplantation can hardly cure end-stage liver failure. Patient infected with hepatic failure have only a restricted therapeutic choice thus growing need for liver donors relates to means of using tissue engineering for treatment and exploration of alternative therapies. Hepatocytes are the primary cellular cells which perform the main liver role (Birla, 2014; Stravitz & Kramer, 2009). Nanotechnology is in an early years developing and replacing nanomaterials for hepatocyte production and repair. Nanofibers/Nanoparticles have the morphological similarities to natural engineered extracellular matrix, and their mechanical properties are therefore advantageous for regeneration of the liver tissue. A research conducted by Chen et al. documented the active distribution of hepatocyte nuclear factor 3β plasmid DNA by positive charges of mesoporous silica NPs. The treated-induced pluripotent stem cells were differentiated into hepatocyte-like cells and thus suggested the

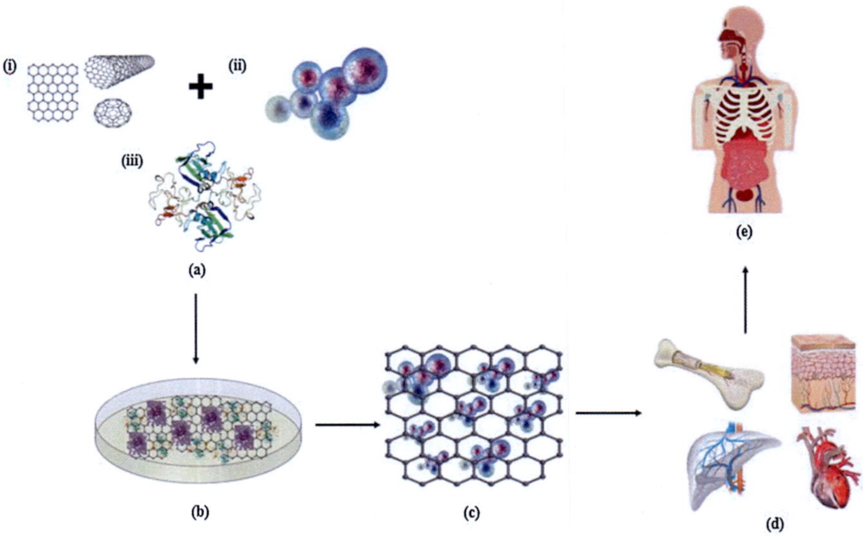


**Fig. 17.2** Construction of different bio-scaffolds, (**a**) includes, (i) Nanomaterial, (ii) Cells, (iii) Growth factors; (**b**) In vitro cultivation (cell seeding); (**c**) Scaffold with proliferated cells; (**d**) Cells differentiated to form tissues/organs; (**e**) Tissue/organ implantation into human body

significance of these NPs in the transfer of genes to facilitate the development of hepatocyte-like cells (Chen, Tsai, Hung, Chiou, & Mou, 2013). Likewise, polyethyleneimine-modified selenium NPs was reported for supplying important growth factors for transforming embryonic mouse stem cells into hepatocyte-like cells as well as for the restoration of liver damage, which plays a role in liver regeneration (Amin, Hashem, Alshehri, Awad, & Hassan, 2017). Stem cells transplant of the bone marrow with a conjunction of the hepatocyte growth factor added in chitosan nanoparticles has contributed to enhanced development of stem cells of the bone marrow into hepatocytes for the cure of liver cirrhosis (Pulavendran, Rose, & Mandal, 2011). Nanofiber have characteristics which proves these nanofiber a better choice for liver tissue engineering, such as high porosity, small pore size, and large surface area-to-volume ratio. Another literature reported utilization of polycaprolactone nanofiber scaffolds to induce differentiation of human cord blood via unregulated somatic stem cells along hepatogenic sections (Hashemi et al., 2008).

### *17.2.2 Construction of Skin Bio-scaffold*

Skin is the main human body organ that serves as a primary barrier for shielding internal organs from the pathogens that exist in the outside world. Skin also allows the entire body in homeostasis and to maintain moisture. Skin has a propensity to constant repairing and renewal based on many stem cells involved with it, in order to satisfy its job well. Skin injury related to burns, acute injuries, accidental or chronic foot ulcers (diabetes), and leads to skin deficiency. Then, tissue engineering came into play to develop ideal skin replacements, artificial engineered extracellular matrix, and an origin of cells by discovering the nature of weakened skin tissue and stem cells to recover the skin's incomplete structure and physiology. Skin tissue engineering involves cells transferred to the target by itself, cell delivery along with existing biomaterials, materials utilized to substitute the skin's dermal layer and substances that aid substitute the skin's epidermis and dermis (Groeber, Holeiter, Hampel, Hinderer, & Schenke-Layland, 2011; Sood, Granick, & Tomaselli, 2014). Substitutes of skin tissue must be readily compatible, must exhibit strong mechanical and physical properties and thus should be non-antigenic. The development of skin tissue engineering scaffolds involves sufficient understanding of wound healing activities, wound forms, patient medical conditions, recovery process and the appropriate properties of an optimal dressing. A number of biomaterials are used conventionally for skin tissue engineering. Hydrogels, nanocomposites, scaffolds, and dressings in various other ways were developed for skin recovery from various types of nanomaterials such as, polymeric carbon-based, solid lipid, and inorganic nanomaterials and regeneration owing to its durability, high surface area-to-volume ratio, excellent interactions with the biological target, ease of surface alteration, high chances of penetration into the skin tissue and the capacity to monitor physical and chemical properties (Babitha & Korrapati, 2017). Metal and metal oxide NPs (silver, zinc, gold, etc.) serve a significant part in inorganic nanomaterials used for wound healing. Silver (Ag) has been utilized for decades for wound healing agent but nano-Ag's medicinal effectiveness has been reported to enhance wound regeneration compared to conventional Ag or ionic Ag dressings. Because of the antimicrobial function of silver NPs (AgNPs), AgNPs-based dressings over a wide variety of pathogens include infection-free wound healing of acute and chronic wounds in burned wounds, incisions, and excisioned skin (Rai, Yadav, & Gade, 2009). The research indicates that the usage of nano-Ag (3–10 nm size) gel synthesized via whole cell and cell-free extract of different microbes has stimulated the healing of burnt wounds as it relates to silver ions and formulations supported. Manufacturers and distributors have confirmed a number of nano-Ag dressings available in market that such dressings demonstrate gradual and regulated release of Ag ions to wound site. These dressings are also considered healthy and non-cytotoxic for human use as the released silver nanoparticles does not gets absorbed deep into the skin while accumulating in the dermal portion of the skin, serving to protect the wounded area with development of keratinocytes (Tocco, Zavan, Bassetto, & Vindigni, 2012).

### 17.2.3 *Construction of Bone Bio-scaffolds*

Bone is a nanofibrous, hierarchically organized arrangement of calcified connective tissue consisting primarily of collagen strengthened with hydroxyapatite. Bone tissue gives muscular strength, strengthens body stability, and preserves bone marrow as well. Bones exhibit the ability of regeneration and repair. Significant bone abnormalities such as bone fracture, however, arising due to osteoporosis or arthritis, bone cancer, trauma, strong intensity of impact and tension, requires bone grafts. Bone tissue transplantation is very common and the treatment of non-healing bone defects involves replacement grafts (Zhang, Wu, Friis, & Xiao, 2010). Bone grafting technology for bone abnormalities suffers from limitations such as shortage of donor access as well as donor site morbidity. Utilizing tissue-engineered methods offers the development of 3D scaffolds with an analogous structure to organic bone tissues. The tremendous clinical demand of bone tissue engineering for restoring the normal functioning of defective bone tissues has increased dramatically in recent decades. Because of the bone's inherent nanostructure, several nanotechnology-based materials are engineered for imitating bone functions with the goal of assisting bone development and improved absorption into the host tissue. Nanomaterials carrier systems can improve accurate and selective distribution of bone regenerative factors to facilitate local repair at the site of injury (Giannoudis, Dinopoulos, & Tsiridis, 2005; Shi et al., 2015).

In TE, one of the chief research topics is to development of bio constructs that can be combined with the in vivo tissues. Current systems reflect shortcomings such as limited diffusion and irregular distribution of cell matrix. In order to overcome the disadvantages of existing materials, various types of scaffolds and bioreactors have been developed and tested over the last few years (D'Amora et al., 2017). Oral tissues regeneration is very challenging and nanomaterials may pose a great opportunity for future TE applications of numerous craniofacial and oral defects. The most widely studied nanomaterials for oral tissues can be categorized as nanofibers, nanoparticles, nanotubes, nanosheets, and nanospheres. Researchers have also recently examined multilayer nanoscaffolds for oral tissue regeneration (Kargozar & Mozafari, 2018; Neel, Chrzanowski, Salih, Kim, & Knowles, 2014).

### *17.2.4 Construction of Pancreas Bio-scaffolds*

Pancreas is an organ that performs both endocrine and exocrine roles. Endocrine components, essential for lifespan such as, α, β, δ, and ε cells secretes hormones in the body. β-cells are mainly responsible for insulin secretion in the liver. Deficiencies in β-cells contribute to type 1 diabetes mellitus. During last few decades different techniques using tissue engineering have been utilized to repair defective cells by further understanding the diverse β-cell microenvironment. The synthesis of a bio-engineered pancreas by pancreas tissue engineering has been established through a suitable mixture of cells, biologically active molecules and biomaterial scaffolds that could aid in islet transplantation and regeneration as an alternative means of diabetes therapy (Bowers et al., 2018). Islet transplant even includes some drawbacks, including lack of islet donors, islet isolation challenges, reduced viability during isolation, and intra-islet vasculature injury. Alternative transplant sites and supportive scaffolds have been utilized in the early days of transplantation to minimize immunological responses and enhance the viability of the islet, however there are still problems of viability. Nanotechnological methods are utilized for the manufacturing nanomaterials for the regeneration and reconstructing pancreatic islets. Nanotechnological initiatives have been made to research the efficacy of collagen I nanofibrils in conjunction with nicotinamide and exendin-4 for the repair of damaged pancreatic islets. The evolved system was capable to facilitate the differentiation of mesenchymal stem cells into insulin-producing cells as well as exhibit excellent activity in rats with type-2 diabetes mellitus for islet regeneration (Ma et al., 2016). Nicotinamide is a poly (ADPribose) synthetase inhibitor that converts pancreatic progenitor cells into insulin-secreting cells, while the exendin-4 can prevent lipotoxicity-induced apoptosis in the in vitro pancreatic cell line. Another research revelaed, heparin-binding peptide amphiphilic nanofibers were observed to transmit vascular endothelial growth factors and fibroblast growth factors, which dramatically increased the density of the blood vessels, as well as enhanced insulation to facilitate islet transplantation. Heparin mimetic peptide amphiphile nanofiber gels were observed to enhance the power of islet transplantation in the viability and activity of pancreatic islands. Peptide amphiphilic nanofibers form an aqueous 3D network structure that imitates natural engineered extracellular matrix to facilitate angiogenesis for islet transplantation (Zhao et al., 2010).

### *17.2.5 Construction of Cartilage Bio-scaffolds*

Cartilage is a flexible, solid, avascular, aneural, and alymphatic connective tissue consisting predominantly of chondrocytes enclosed in thick and strongly hydrated ECM components. Amongst the various forms of cartilage, the articular cartilage has a load in human joints and a restricted potential for self-regeneration due to lack of vasculature. Cartilage abnormalities occur as a result of damage, ageing and few conditions that lead to extreme pain and impairment. Avascular existence stops chondrocytes and progenitor cells from heading into the lesions to form matrix. Thereby, a great need was felt to develop TE-based materials, having functions similar to natural cartilage so as to make them use for cartilage repair and regeneration. The growth of cartilage TE solutions, such as biomaterial scaffolds or cell-impregnated scaffolds depends mainly on the suitable biomaterials, cell source, and stimulating factors. Nanofibers are known to be the most suitable biomaterial scaffolds because fibrous morphology is capable of stimulating natural collagen fibers, between 50 and 500 nm diameter. Electrospun nanofibers having high porosity, high surface to volume ratio, and variable pore size distribution resulted in promising applications in the area of TE (Shin et al., 2006). Polymeric PLGA nanofibrous scaffolds for joint cartilage regeneration have been identified as having superior cell proliferation and ECM growth compared to membrane type scaffolds. The nanofiber scaffold was adequately mechanically stable to resist implantation and to sustain regenerative cartilage. In another study, electrospun nanofibers scaffolds of protein tyrosine bound to modified gelatin has supported the growth and adhesion of chondrocytes. In vitro cell culture finding have shown that electrospun engineered protein scaffolds enable the attachment and development of chondrocytes. These nanofibers also showed cell proliferation showing their applicability in cartilage engineering (Agheb, Dinari, Rafienia, & Salehi, 2017).

### 17.2.6 Construction of Cardiac Bio-scaffolds

Heart failure due to myocardial infarction is the leading cause of deaths worldwide. Myocardium is an essential layer of heart muscle tissue that is made up of myocytes. Cardiac myocytes have a propensity to self-contract that is essential for maintaining pumping action of heart. Damage to cardiomyocytes can cause heart failure, so it is necessary to improve curative heart failure therapy. Heart transplantation is considered to be the best way to repair the injured heart. However, there is a significant drawback of the scarcity of organs for transplantation procedures owing to lack of graft donors and also immunosuppression (Castells-Sala et al., 2013; Chiu & Radisic, 2013). Cardiac TE is a promising technique with the goal of restoring and rebuild infarct cardiac tissues with 3D cell-based grafts, through the manufacture of in vitro implantation cardiac tissue engineering and the advancement of biomaterials to enable cell proliferation, growth, and vascularization. In addition, nanotechnology can be used to build TE cardiac myocytes NMs. Various nano-sized biomimetic assemblies have been used for the tissue regeneration by cell/drug delivery, such as electrospun scaffolds, peptide amphiphiles, self-assembling peptides and layer-by-layer (LbL) complexes (Amezcua, Shirolkar, Fraze, & Stout, 2016; Hirt, Hansen, & Eschenhagen, 2014). Another method using $Al/Al_2O_3$ nanowires has opened a new path for large-scale development of contractile myocardial grafts by maintaining orientation of cardiomyocytes for cardiovascular TE. Nanofibrous Polyurethane and ethyl cellulose-based scaffolds for cardiac TE have also been reported (Chen et al., 2015).

### 17.2.7 Construction of Vascular Bio-scaffolds

Cardiovascular disorders, in particular coronary artery disease or atherosclerosis may cause obstruction of blood vascular system. Atherosclerosis is another dangerous disease associated with the plaque development in the blood vessels that causes wall thickening and eventually restricts blood flow. Blood vessels (arteries and veins) passage blood to provide nutrients and oxygen to all the areas of the body. Blocking these boats is a cause of deaths worldwide (Li, Sengupta, & Chien, 2014). Vascular TE has arisen as an alternative area that includes multidisciplinary methods and technology to create biomaterials or natural or synthetic polymers scaffolds for modern vascular grafts to preserve post implant work. The advancement of nanotechnology will serve as a promising approach to increasing the performance of vascular TE by designing scaffolds that can imitate the nanoscale architecture of natural human tissues. To this end, ultra-thin fibers produced by the electrospinning process provide a desirable approach to the manufacture of scaffolds for tissue engineering applications. Nanofibrous scaffold has a high surface area-to-volume ratio that provides more substrate for cell attachment. All these characteristics are useful for tissue ingrowth (Stitzel, Pawlowski, Wnek, Simpson, & Bowlin, 2001). Electrospun PLA/PCL nanofibers with a diameter 400–800 nm have been synthesized to replicate the natural environment of ECM. Human smooth muscles and endothelial cells cultivated with these nanofiber scaffolds displayed cell proliferation and proper attachment and also preserved their form, which indicated the function of these scaffold as vessel substitutes. An additional attempt was made to create collagen blended PLA/PCL nanofiber scaffolds to imitate ECM in terms of morphology and chemistry. Human coronary artery endothelial cells (HCAEC) seeded onto these scaffolds demonstrated viability and cell adhesion indicated the promise of these NMs for TE channel (Wang, Ding, & Li, 2013).

## 17.3 Nanomaterials in Biomolecular Detection

The use of biomolecular detection and biosensors is increasingly gaining speed in the field of TE. Biosensors are biological instruments that identify change in molecules by the identification of physiochemical. Various types of biosensors have been used in TE, such as electrochemical, electrical, magnetic, and acoustic, for the detection of efficient protein molecules, glucose and pathogenic microbes. Most notably, though, there, is a growth in the usage of nanomaterials in biomolecular detection. Nanomaterials in biosensors have been used for the identification of DNA, nucleic acid and protein, among other uses, due to their high reactivity and useful chemical properties. The rising eminence of nanoparticle applications in biomolecular sensors and the importance of biosensors in TE highpoint the potential of nanomaterials in TE. Hopefully in the near future, researchers would be able to insert intelligent nanoscale biosensors within scaffolds, cell sheets or related locations to monitor the growth of engineered tissues after transplantation and grafting. It is also assumed that the integration of nanomaterials into 3D-engineered tissues could reduce inflammatory responses to transplanted engineered tissues by modulating immune cell adhesion and cell viability (Tansil & Gao, 2006; Thaxton, Rosi, & Mirkin, 2005).

### *17.3.1 Gene Delivery*

Gene delivery technologies targeting mature cells or stem cells has been a core concern within TE. Human mesenchymal stem cells (hMSCs) are multipotent cells possess immunosuppressive properties and are capable of distinguishing between various types of cells, including chondrocytes, osteoblasts, myocytes and adipocytes. It is important to create an adequate vector system with high gene transfection effectiveness, low cytotoxicity and high sensitivity to unhealthy cells for successful gene therapy applications (Wang, Castro, An, et al., 2012). Gene delivery is divided into two categories: viral and nonviral. The potential for gene transmission was shown by the use of nanoparticles and self-assembled nanomaterials as described below:

#### 17.3.1.1 DNA Transfection

The benefits of nonviral approaches are their flexibility and lack of immune response, while the drawback is little efficacy due to low transfection rate. Magnetofection is a modem gene transmission process for the transfection of gene using MNPs. In order to achieve magnetofection using plasmid DNA, cationic lipids or polymers with DNA complexes interact with magnetic beads and then by magnetic energy, are drawn to target cells so that they can aggregate on the surface. Substantial study has been performed on various methods of magnetofection in combination with polyethyleneimine or dioleoyltrimethylammonium and Lipofectamine. Furthermore, in endothelial cells and embryonic stem cells, which are typically immune to conventional transfection approaches, magnetofection has demonstrated a high level of expression in target cells. In fibroblasts and keratinocytes using reporter genes, transfection efficiencies using magnetofection techniques were found to be 36- and 10-fold higher compared with cationic liposomes.

Some experiments have shown that iron oxide magnetic particles can bind to the gene and increase the efficiency of transfection. These particles have been dispersed inside a polymer matrix or internalized in a polymer or metallic environment, which binds DNA via charge contact. Another research looked at enhancing the absorption of mesoporous silica nanoparticles by target cells (Cai, Mataraza, Qin, et al., 2005). The composition combined with *N*-{1-(2,3-dioleoyloxy)propyl}-*N*,*N*,*N* trimethylammonium chloride (DOTAP chloride), helped to prevent aggregation and facilitated absorption. CNTs, which have revealed numerous applications and can be synthesized through a variety of different methods, have also been shown to be unexpectedly useful as nonviral gene delivery agents. A method called as nanotube spearing was used to prepare nickel-embedded, magnetic nanotubes where DNA is bound.

#### 17.3.1.2 Viral Transduction

Viral transduction approaches have also been used in TE applications; however, one of the issues associated with this approach is the difficulties in the preparation of high titer viral vectors. In one scenario, retroviral vectors were captured using magnetite cationic liposomes (MCLs) by adding the MCLs to the retroviral vector pseudo type solution of the vesicular stomatitis virus glycoprotein. MCLs were then obtained using magnetic force and injected into mouse neuroblastoma cells, raising the viral titer by 55-fold. Strong interactions due to magnetic force have been shown to improve the cellular absorption of MNPs. As magnetofection transduction efficiency with MCLs was contrasted with traditional methods using polybrene, the efficacy for MCLs was 6.7-fold greater.

No coxsackievirus and adenovirus receptor (CAR) were stated by using adenoviral vectors conjugated with polyethyleneimine. These vectors were coated with MNPs, which allowed a number of cell line to be transduced. Magnetofection has been used to transfect a variety of cell types, including epithelial cells from lungs and endothelial cells from blood vessels (Akiyama, Ito, Kawabe, & Kamihira, 2010). These particles were also been used to efficiently distribute antisense oligonucleotides and siRNA to suppress gene expression. Further production of magnetically field vectors can provide methods for the successful treatment of various types of diseases (Krötz, de Wit, Sohn, et al., 2003).

#### 17.3.1.3 Magnetic Cell Patterning

Efficient TE needs the production of tissue structures that are identical to in vivo environments. One technique to do this is to regulate cell adhesion within a detailed design sequence. Present confines of traditional approaches include the need for substantial additional infrastructure or need for the use of chemically adapted surfaces. The use of MNPs for the purpose of cell patterning can help to overcome these barriers by offering an efficient and practical alternative. A cell-patterning technique was developed using MCLs where a magnet with a magnetic field concentrator was placed under a cell culture surface. Various cell patterns have been successfully developed using this method by controlling the traces of the magnetic field concentrators. When human umbilical vein endothelial cells were used, it was observed that, with the same number of cells, the cells were associated and capillary-like structures formed when arranged in a line. In the development of vascular grafts, it was observed that the tubular geometry did not permit the successful transfer of the cells to the scaffold. Another analysis using porcine decellularized common carotid artery found that when a cylindrical magnet was placed into the lumen of the artery and then submerged in a suspension of NIH3T3 fibroblasts (3T3s) that had been magnetically labeled using initial MCLs with a positive surface charge, almost all the 3T3s were bound to the artery relative to the very low seeding efficiency that was detected in the absence of a magnet. The reseeding of porcine decellularized common carotid artery with both human dermal fibroblasts and smooth human muscle cells effectively built the vascular graft using the Mag-seeding technique (Ino, Ito, & Honda, 2007; Shimizu, Ito, Arinobe, et al., 2007).

#### 17.3.1.4 Stimulation of Cells for Mechanotransduction

It is well known that several bioactive particles and growth factors control the role of cell in the human body. Mechanical factors, play a key role in defining cell functions by disturbing mechanotransduction pathways. Several methods have been used, including the use of bioreactors of shear stress and rigidity patterned substrates for mechanically regulation of cell functions. However, MNPs have proved to both of these approaches, as they have the ability to be globally, spatially and temporally controlled by a magnetic field. At the microscopic stage, the mechanism takes place as follows. Initial, the MNPs are coated with a particular attacking antibody. When the magnetic field has been applied, the cells are grouped in the direction of the magnetic field. Dependent on the antibody used, the cell structure of the receptor is affected. Mannix et al. were able to obtain a rise in intracellular calcium ion levels, while Gopinath et al. were able to trigger an apoptotic cascade using these methods. These two findings, along with several other studies have demonstrated how the applications of MNPs can provide complete descriptions of cell structure and differentiation in regenerative medicine in a remote manner (Gopinath et al., 2010; Mannix, Kumar, Cassiola, et al., 2008).

#### 17.3.1.5 Antibacterial Applications

Centers for Disease Control (CDC), USA, has recently estimated more deaths from antibiotic-resistant bacteria than from all cancers. This has definitely emphasized the increasing need to find alternative way of destroying bacteria without use of antibiotics, and nanoparticles can provide a response. Some metal oxides, mostly silver NPs, have shown great disinfectant and wound healing properties. Xing et al. measured the antibacterial efficacy of poly(3-hydroxybutyrate-*co*-3-hydroxyvalerate) (PHBV) nanofibrous silver-containing scaffolds (Xing et al., 2010). It was observed that silver-containing PHBV nanofibrous scaffolds had better antibacterial properties and had exhibited excellent in vitro cell viability. This indicated that PHBV nanofibrous scaffolds incorporating silver nanoparticles have potential for usage in joint arthroplasty and, thus, should be further examined. Another research deals with the classification and disinfectant activity of a nanosilver-based biocomposite scaffold for bone TE (Saravanan et al., 2011). Biocomposite nanosilver-containing scaffolds may prevent bacterial contamination during reconstructive bone surgery and that the existence of silver nanoparticles in scaffolds were used as a coating to guard against contamination, sepsis and malfunctioning of implants.

## 17.4 Conclusion

TE is a research division covering interaction, physical science, material science, engineering, and biomedicine. The key purpose of using NMs such as nanofibers, NPs, nanotubes, and nanocomposites for TE applications is to accomplish the structural and functional reconstruction of weakened or dysfunctional tissues/organs which functions abnormally. Strategies have been built to provide growth factors in a regulated manner to satisfy the needs of tissue regeneration. Nanotechnology has an enormous probable to create biomaterials for the production of major cells building block of ECM of cells as NMs provide support for cell binding, variation, and production. The major obstacle for NM-based TE is the potential of NMs to monitor nanoscale vasculogenesis. There is also a need to implement more techniques to render cost-effective NMs containing scaffolds to satisfy the rising demands of future regenerative medicine.

## References

Agarwal, C. M., & Ray, R. B. (2001). Biodegradable polymeric scaffolds for musculoskeletal tissue engineering. *Journal of Biomedical Materials Research, 55*, 141–150.

Agheb, M., Dinari, M., Rafienia, M., & Salehi, H. (2017). Novel electrospun nanofibers of modified gelatin-tyrosine in cartilage tissue engineering. *Materials Science and Engineering: C, 71*, 240–251.

Akiyama, H., Ito, A., Kawabe, Y., & Kamihira, M. (2010). Cell-patterning using poly (ethylene glycol)-modified magnetite nanoparticles. *Journal of Biomedical Materials Research. Part A, 92*(3), 1123–1130.

Amezcua, R., Shirolkar, A., Fraze, C., & Stout, D. A. (2016). Nanomaterials for cardiac myocyte tissue engineering. *Nanomaterials, 6*, 133–147.

Amin, K. A., Hashem, K. S., Alshehri, F. S., Awad, S. T., & Hassan, M. S. (2017). Antioxidant and hepatoprotective efficiency of selenium nanoparticles against acetaminophen-induced hepatic damage. *Biological Trace Element Research, 175*, 136–145.

Appel, A. A., Anastasio, M. A., Larson, J. C., & Brey, E. M. (2013). Imaging challenges in biomaterials and tissue engineering. *Biomaterials, 34*, 6615–6630.

Babitha, S., & Korrapati, P. S. (2017). Biodegradable zein–polydopamine polymeric scaffold impregnated with TiO2 nanoparticles for skin tissue engineering. *Biomedical Materials, 12*, 055008.

Birla, R. (2014). *Liver tissue engineering. Introduction to tissue engineering: applications and challenges* (pp. 295–321). Houstan: Wiley.

Bowers, D. T., Claire, E. O., Chhabra, P., Linda, L., Parker, H. M., Ritu, S. L., Michael, L. T., Dan, L., Kenneth, L. B., & Edward, A. B. (2018). An engineered microencapsulation membrane releasing FTY720 to precondition pancreatic islet transplantation. *Journal of Biomedical Materials Research. Part B, Applied Biomaterials, 10*(6), 555–568.

Cai, D., Mataraza, J. M., Qin, Z. H., et al. (2005). Highly efficient molecular delivery into mammalian cells using carbon nanotube spearing. *Nature Methods, 2*(6), 449–454.

Castells-Sala, C., Alemany-Ribes, M., Fernandez-Muinos, T., Recha- Sancho, L., Lopez-Chicon, P., Aloy-Reverte, C., & Semino, C. E. (2013). Current applications of tissue engineering in biomedicine. *Journal of Biochips and Tissue Chips, 4*, 1–14.

Chen, P. H., Liao, H. C., Hsu, S. H., Chen, R. S., Wu, M. C., Yang, Y. F., & Su, W. F. (2015). A novel polyurethane/cellulose fibrous scaffold for cardiac tissue engineering. *RSC Advances, 5*, 6932–6939.

Chen, W., Tsai, P. H., Hung, Y., Chiou, S. H., & Mou, C. Y. (2013). Nonviral cell labelling and differentiation agent for induced pluripotent stem cells based on mesoporous silica nanoparticles. *ACS Nano, 7*, 8423–8440.

Chiu, L. L., & Radisic, M. (2013). Cardiac tissue engineering. *Current Opinion in Chemical Engineering, 2*, 41–52.

Contado, C. (2015). Nanomaterials in consumer products: A challenging analytical problem. *Frontiers in Chemistry, 3*, 48.

D'Amora, U., Russo, T., Gloria, A., Rivieccio, V., D'Ant, V., Negri, G., Ambrosio, L., & De Santis, R. (2017). 3D additive-manufactured nanocomposite magnetic sca_olds: E_ect of the application mode of a time-dependent magnetic field on hMSCs behavior. *Bioactive Materials, 2*, 138–145.

De, W., Tinke, M., Lidy, E., Fratila, A., Amir, A. Z., & Nicholas, A. P. (2018). Bone tissue engineering via growth factor delivery: From scaffolds to complex matrices. *Regenerative Biomaterials, 5*, 197–211.

Edwards, S. L., Werkmeister, J. A., & Ramshaw, J. A. (2009). Carbon nanotubes in scaffolds for tissue engineering. *Expert Review of Medical Devices, 6*, 499–505.

Ferracane, J. L., & Giannobile, W. V. (2014). Novel biomaterials and technologies for the dental, oral, and craniofacial structures. *Journal of Dental Research, 93*, 1185–1186.

Giannoudis, P. V., Dinopoulos, H., & Tsiridis, E. (2005). Bone substitutes: An update. *Injury, 36*, 20–27.

Gomes, M. E., Ribeiro, A. S., Malafaya, P. B., Reis, R. L., & Cunha, A. M. (2001). A new approach based on injection moulding to produce biodegradable starch-based polymeric scaffolds morphology. *Biomaterials, 22*, 883–889.

Gopinath, P., Gogoi, S. K., Sanpui, P., Paul, A., Chattopadhyay, A., & Ghosh, S. S. (2010). Signaling gene cascade in silver nanoparticle induced apoptosis. *Colloids and Surfaces. B, Biointerfaces, 77*(2), 240–245.

Groeber, F., Holeiter, M., Hampel, M., Hinderer, S., & Schenke-Layland, K. (2011). Skin tissue engineering-in vivo and in vitro applications. *Advanced Drug Delivery Reviews, 63*, 352–366.

Guo, B., Lei, B., Li, P., & Ma, P. X. (2015). Functionalized scaffolds to enhance tissue regeneration. *Regenerative Biomaterials, 2*, 47–57.

Hashemi, S. M., Soleimani, M., Zargarian, S. S., Haddadi-Asl, S. V., Ahmadbeigi, N., Soudi, S., & Mohammadi, Y. (2008). In vitro differentiation of human cord blood-derived unrestricted somatic stem cells into hepatocyte-like cells on poly (ε-caprolactone) nanofiber scaffolds. *Cells, Tissues, Organs, 190*, 135–149.

Hirt, M. N., Hansen, A., & Eschenhagen, T. (2014). Cardiac tissue engineering. *Circulation Research, 114*, 354–367.

Hosseinpour, S., Ahsaie, M. G., Rad, M. R., Baghani, M., Motamedian, S. R., & Khojasteh, A. (2017). Application of selected sca_olds for bone tissue engineering: A systematic review. *Oral and Maxillofacial Surgery, 21*, 109–129.

Hutmacher, D. W. (2000). Scaffolds in tissue engineering bone and cartilage. *Biomaterials, 21*, 2529–2543.

Ino, K., Ito, A., & Honda, H. (2007). Cell patterning using magnetite nanoparticles and magnetic force. *Biotechnology and Bioengineering, 97*(5), 1309–1317.

Kargozar, S., & Mozafari, M. (2018). Nanotechnology and nanomedicine: Start small, think big. *Materials Today Proceedings, 5*, 15492–15500.

Krötz, F., de Wit, C., Sohn, H. Y., et al. (2003). Magnetofection—A highly efficient tool for antisense oligonucleotide delivery in vitro and in vivo. *Molecular Therapy, 7*(5), 700–710.

Langer, R., & Vacanti, J. P. (1993). Tissue engineering. *Science, 260*, 920–927.

Laurencin, C. T., & Nair, L. (2015). Nanotechnology and regenerative engineering: The Scaffold. In: C. N. R. Rao (Eds.), (p. 459). Boca Raton: CRC Press.

Leydesdorff, L., & Zhou, P. (2007). Nanotechnology as a field of science: Its delineation in terms of journals and patents. *Scientometrics, 70*, 693–713.

Li, S., Sengupta, D., & Chien, S. (2014). Vascular tissue engineering: From in vitro to in situ. *Wiley Interdisciplinary Reviews. Systems Biology and Medicine, 6*, 61–76.

Liu, C., Xia, Z., & Czernuszka, J. T. (2007). Design and development of three-dimensional scaffolds for tissue engineering. *Chemical Engineering Research and Design, 86*(7), 1051–1064.

Liu, C. Z., & Czernuszka, J. T. (2006). Development of biodegradable scaffolds for tissue engineering: a perspective on emerging technology. *Materials Science and Technology*. 23:4, 379–391, https://doi.org/10.1179/174328407X177027

Ma, C. T., Wu, Y. J., Huang, H. H., Kang, P. L., Hsiao, K. Y., Lo, D. Y., & Kuo, S. M. (2016). In vitro and in vivo evaluation of the effect of nano-sized collagen molecules and nicotinamide on mesenchymal stem cell differentiation. *Journal of Materials Chemistry B, 4*, 3892–3902.

Mannix, R. J., Kumar, S., Cassiola, F., et al. (2008). Nanomagnetic actuation of receptor-mediated signal transduction. *Nature Nanotechnology, 3*(1), 36–40.

Neel, E. A. A., Chrzanowski, W., Salih, V. M., Kim, H. W., & Knowles, J. C. (2014). Review tissue engineering in dentistry. *Journal of Dentistry, 42*, 915–928.

Pulavendran, S., Rose, C., & Mandal, A. B. (2011). Hepatocyte growth factor incorporated chitosan nanoparticles augment the differentiation of stem cell into hepatocytes for the recovery of liver cirrhosis in mice. *Journal of Nanbiotechnology, 9*, 15–25.

Rai, M., Yadav, A., & Gade, A. (2009). Silver nanoparticles as a new generation of antimicrobials. *Biotechnology Advances, 27*, 76–83.

Saravanan, S., Nethala, S., Pattnaik, S., Tripathi, A., Moorthi, A., & Selvamurugan, N. (2011). Preparation, characterization and antimicrobial activity of a bio-composite scaffold containing chitosan/nano-hydroxyapatite/nano-silver for bone tissue engineering. *International Journal of Biological Macromolecules, 49*(2), 188–193.

Shi, J., Votruba, A. R., Farokhzad, O. C., & Langer, R. (2010). Nanotechnology in drug delivery and tissue engineering: From discovery to applications. *Nano Letters, 10*, 3223–3230.

Shi, S., Jiang, W., Zhao, T., Aifantis, K. E., Wang, H., Lin, L., Fan, Y., Feng, Q., Cui, F. Z., & Li, X. (2015). The application of nanomaterials in controlled drug delivery for bone regeneration. *Journal of Biomedical Materials Research. Part A, 103*, 3978–3992.

Shimizu, K., Ito, A., Arinobe, M., et al. (2007). Effective cell-seeding technique using magnetite nanoparticles and magnetic force onto decellularized blood vessels for vascular tissue engineering. *Journal of Bioscience and Bioengineering, 103*(5), 472–478.

Shin, H. J., Lee, C. H., Cho, I. H., Kim, Y. J., Lee, Y. J., Kim, I. A., & Shin, J. W. (2006). Electrospun PLGA nanofiber scaffolds for articular cartilage reconstruction: Mechanical stability, degradation and cellular responses under mechanical stimulation in vitro. *Journal of Biomaterials Science. Polymer Edition, 17*, 103–119.

Sood, A., Granick, M. S., & Tomaselli, N. L. (2014). Wound dressings and comparative effectiveness data. *Advances in Wound Care, 3*, 511–529.

Stitzel, J. D., Pawlowski, K. J., Wnek, G. E., Simpson, D. G., & Bowlin, G. L. (2001). Arterial smooth muscle cell proliferation on a novel biomimicking, biodegradable vascular graft scaffold. *Journal of Biomaterials Applications, 16*, 22–33.

Stravitz, R. T., & Kramer, D. J. (2009). Management of acute liver failure. *Nature Reviews. Gastroenterology & Hepatology, 6*, 542–553.

Tansil, N. C., & Gao, Z. (2006). Nanoparticles in biomolecular detection. *Nano Today, 1*(1), 28–37.

Thaxton, C. S., Rosi, N. L., & Mirkin, C. A. (2005). Optically and chemically encoded nanoparticle materials for DNA and protein detection. *MRS Bulletin, 30*(5), 376–380.

Tocco, I., Zavan, B., Bassetto, F., & Vindigni, V. (2012). Nanotechnology-based therapies for skin wound regeneration. *Journal of Nanomaterials, 4*, 1–11.

Van, R., & Sabine, P. H. (2017). Enhancing regenerative approaches with nanoparticles. *Journal of the Royal Society Interface, 14*, 20170093–201700102.

Vieira, S., Vial, S., Reis, R. L., & Oliveira, J. M. (2017). Nanoparticles for bone tissue engineering. *Biotechnology Progress, 33*, 590–611.

Walmsley, G. G., Mc Ardle, A., Tevlin, R., Momeni, A., Atashroo, D., Hu, M S., Feroze, A. H., Wong, V. W., Lorenz, P. H., Longaker, M. T., et al. (2015). Nanotechnology in bone tissue engineering. *Nanomedicine, 11*, 1253–1263.

Wang, S., Castro, R., An, X., et al. (2012). Electrospun laponite-doped poly (lactic-co-glycolic acid) nanofibers for osteogenic differentiation of human mesenchymal stem cells. *Journal of Materials Chemistry, 22*(44), 23357–23367.

Wang, X., Ding, B., & Li, B. (2013). Biomimetic electrospun nanofibrous structures for tissue engineering. *Materials Today, 16*, 229–241.

Xing, Z. C., Chae, W. P., Baek, J. Y., Choi, M. J., Jung, Y., & Kang, I. K. (2010). In vitro assessment of antibacterial activity and cytocompatibility of silver-containing PHBV nanofibrous scaffolds for tissue engineering. *Biomacromolecules, 11*(5), 1248–1253.

Yang, S., Leong, K. F., Du, Z., & Chua, C. K. (2001). The design of scaffolds for use in tissue engineering. Part I. Traditional factors. *Tissue Engineering, 7*, 679–689.

Yi, H., Rehman, F., Zhao, C., Liu, B., & He, N. (2016). Recent advances in nano scaffolds for bone repair. *Bone Research, 4*, 16050–16060.
Zhang, Y., Sun, H., Song, X., Gu, X., & Sun, C. (2015). Biomaterials for periodontal tissue regeneration. *Reviews in Advanced Sciences and Engineering, 40*, 209–214.
Zhang, Y., Wu, C., Friis, T., & Xiao, Y. (2010). The osteogenic properties of CaP/- silk composite scaffolds. *Biomaterials, 31*, 2848–2856.
Zhao, M., Song, C., Zhang, W., Hou, Y., Huang, R., Song, Y., & Song, C. (2010). The three-dimensional nanofiber scaffold culture condition improves viability and function of islets. *Journal of Biomedical Materials Research Part B-Applied Biomaterials, 94*, 667–672.

# Chapter 18
# An Impact of Antibacterial Efficacy of Metal Oxide Nanoparticles: A Promise for Future

**Suma Sarojini and Saranya Jayaram**

## 18.1 Introduction

The field of biomedicine has seen immense growth in technology on one hand and plenty of challenges on the other. While highly efficient therapeutic practices have been able to provide protection against various microbial diseases, the rapid adaptation of microbes to drugs has caused limitations in the efficacy of these treatments. Microbial drug resistance caused by adaptation to selective pressure has increased mortality and morbidity due to resistant pathogens and opportunistic infections. Attributing to the misuse and overuse of antibiotics along with inadequacy of other novel therapeutic drugs, there has been an incremental rise in drug resistant microbes. Hence, research has garnered attention to develop novel strategies to overcome multidrug resistant bacteria (Vega-Jiménez, Vázquez-Olmos, Acosta-Gío, & Álvarez-Pérez, 2019). Owing to their small size, varying shapes, large surface area, high electronegativity and other important physicochemical properties, NP have found utility as antimicrobial agents with high efficiency and stability. While these methods are cost-effective, displaying promising results *in vitro*, there remains a lack of research in understanding their nanotoxicology in biomedicine. Hence, understanding their biochemical and regulatory modalities would further expand their potential as antimicrobial drugs. Presently, a lot of research is being conducted worldwide on metal oxide NP as antimicrobial agents; with popular examples being metal oxide NP of copper, silver and titanium. One of the accepted beliefs is that nanoparticles complement and support traditional antibiotics by functioning as 'good carriers' for drug delivery thereby enhancing their efficiencies (Zhang,

S. Sarojini (✉) · S. Jayaram
Department of Life Sciences, CHRIST (Deemed to be University), Bangalore, India
e-mail: suma@christuniversity.in


K. Pal (ed.), *Bio-manufactured Nanomaterials*,
https://doi.org/10.1007/978-3-030-67223-2_18

Pornpattananangkul, Hu, & Huang, 2010), disrupting biofilm formation and exhibiting various antimicrobial mechanisms, thus giving less probability for host microbes to develop antimicrobial resistance (Wang, Hu, & Shao, 2017). The present chapter describes the concept of metal oxide NP used as antimicrobial agents with reference to their types, mechanisms of action, advantages, limitations and future scope.

### *18.1.1 General Mechanisms of Action of Metal Oxide Nanoparticles*

Metal oxide NPs have been extensively used as antimicrobial agents due to their high efficiency, enhanced stability, cost-effectiveness and longer shelf-life. Additionally, since they do not target any specific protein in microbes, there exist fewer chances of microbes gaining resistance against these therapeutic agents. Transition metal oxide NPs has been extensively studied for their antimicrobial properties *in vitro* and their mechanisms of action have been proposed to be in lieu of their advantageous physicochemical properties (Vega-Jiménez et al., 2019). While the success of conventional antimicrobial therapies get compromised due to multi drug resistance, metal oxide NP exhibit high microbial toxicity and help circumvent problems of drug resistance.

In general, the antimicrobial mechanisms of NP can be categorised as: oxidative stress induction or generation of ROS, metal ion release or non-oxidative mechanisms, penetration and degradation of bacterial cell membrane and disruption of bacterial proteins and DNA. While they may adhere to any one of these categories, NP-based cytotoxicity can occur also due to these mechanisms simultaneously. This feature of NP contributing towards antimicrobial properties through multiple mechanisms makes them superior to traditional antibiotics that work based on simple pathways thus leading to frequent cases of antibiotic drug resistance. Interestingly, apart from bactericidal properties, the unique physicochemical properties of NP enable them to circumvent biofilm formation. The general mode of action of NP is as follows: NP only relies on interaction and cellular contact to perform their toxic activities. Through electrostatic interactions (Li, Chen, Zhao, & Urmila, 2015), Van der Waals forces (Armentano et al., 2014), receptor-ligand bonds (Gao, Thamphiwatana, Angsantikul, & Zhang, 2014) or hydrophobic interactions (Luan, Huynh, & Zhou, 2016), these NP bind to and accumulate on bacterial cells. Subsequent internalisation of these particles leads to their intracellular interaction with bacterial DNA, proteins and other organelles. Cumulatively, NP-based bacterial cytotoxicity is then mediated by oxidative stress with ROS formation, heterogeneous alterations in cellular membranes and factors, disruptions in cell wall structures and permeabilities, deactivation of signalling pathways and changes in genetic material (Wang et al., 2017).

#### 18.1.1.1 Promoting ROS Formation

Reactive Oxygen Species (ROS) induced oxidative stress is an important mode of action of metal oxide NP due to oxygen vacancies in their crystal structures. In general, the four types of ROS produced are: superoxide radical ($O^{2-}$), hydroxyl radical ($OH^-$), hydrogen peroxide ($H_2O_2$) and singlet oxygen ($O_2$). Upon entering bacterial cells, these metal oxide NP promote ROS formation and cause increased intracellular oxidative stress, damage to DNA and oxidation of PUFA (Polyunsaturated Fatty Acids). Ultimately they either cause less acute oxidative stress due to generation of $O_2$ and $H_2O_2$ or highly acute oxidative stress due to generation of $O_2$ and $OH^-$. Various studies have also reported that metal oxide NP induced intracellular oxidative stress also contributes towards damage to permeability of bacterial cell membranes (Wang et al., 2017). Hence, metal oxide NP induced ROS formation leads to bactericidal and bacteriostatic effects.

#### 18.1.1.2 Electrostatic Interactions Leading to Cell Wall Damages

The metal cations of metal oxide NP interact with electronegative groups on bacterial cell walls and accumulate on these cells. Prolonged accumulation leads to altered structure and permeability of bacterial cell membranes. Gram negative bacteria consisting of more electronegative cell walls are thus more susceptible to microbicidal actions of these metal oxides NP. The small size and large surface area of NP coupled with microporosity of microbial cell membranes enhances accumulation and efficiency of NP. Upon getting accumulated on bacterial cell walls, metal oxide NP inhibits DNA replication and also creates pit-like structures which change cell membrane permeability by propagating the release of membrane proteins, internal factors and lipopolysaccharides (Vega-Jiménez et al., 2019).

#### 18.1.1.3 Loss of Homeostasis by Metal Ions

Microbes contain a host of essential metal ions that generally function as coenzymes, cofactors and catalysts but cause deleterious cellular effects when present in excess. The gradual release of metal ions from metal oxide NP and adsorption onto bacterial cells leads to their internalisation and reactions with host proteins and nucleic acid, especially through functional groups like mercapto (-SH), amino (-NH) and carboxyl (-COOH) (Wang et al., 2017). Metal oxide NP internalised by microbial cells causes cell damage through high intracellular metal ion concentrations, altered helical structure of DNA by cross-linking and intercalation, increased membrane permeability by neutralisation of charges on LPS and high oxidative stress by generation of large amounts of $OH^-$ radicals.

#### 18.1.1.4 Dysfunction of Proteins and Enzymes

Apart from altering homeostasis of microbial cells leading to altered DNA and proteins, metal oxide NP cause carboxylation of protein side chains of amino acids and render them dysfunctional by causing loss of catalytic activity and degradation (Vega-Jiménez et al., 2019).

#### 18.1.1.5 Inhibition of Transduction Signals

The electrochemical properties of metal oxide NP alter the structure of microbial nucleic acid, suppress cell division by hindering plasmid and chromosomal replication and dephosphorylate phosphotyrosine residues thus impairing one of the key molecules in bacterial cell signal transductions. Metal oxide NP follows different routes of entry into microbial cells through diffusion, adsorption or biosorption (through carboxyl and phosphate functional groups present in negatively charged cell membranes) (Wang et al., 2017). The antimicrobial mechanisms of metal oxide NP have been found to be influenced by their physicochemical properties like: size, shape, dimensions of <30 nm, high surface area, higher surface reactivity and surface/zeta potential of metal oxide NP which influence their surface area: mass ratio thereby influencing rate of accumulation on bacterial cells. Various studies have reported that needle shaped metal oxide NP had higher antibacterial activity and that negatively charged surfaces had low affinity for adhesion to NP. Few modification techniques of NP like chemical doping with antibiotics help increase their efficiency, promote ROS formation, enhance their surface area: volume ratio and prevent their agglomeration in solvents (Gold, Slay, Knackstedt, & Gaharwaret, 2018). Research has also been performed on functionalising metal oxide NP with other compounds to create biocomposites with enhanced antimicrobial activities (like reduced graphene oxide–zinc oxide (rGO-ZnO) nanocomposites with enhanced antibacterial activity and PEGylated rGO-ZnO nanocomposites with increased biocompatibility) (Rajaura et al., 2017).

## 18.2 Zinc Metal Oxide Nanoparticles and Their Scope

ZnO NP are used extensively in various industrial products but in the past two decades their use in biological applications has become popular due to their excellent biocompatibility, high economic value and low toxicity. Zinc oxide based NP with small particle sizes makes Zinc to be easily absorbed by cells. ZnO is also graded as a 'GRAS' by the USFDA, which is corroborated by research, exhibiting ZnO NP to be less toxic, more biocompatible and relatively less expensive than other metal oxide NP (Rasmussen, Martinez, Louka, & Wingett, 2010). Properties of increased specific surface area, reduced particle size, enhanced surface reactivity and photo-oxidising/ photo-catalysing properties have been shown to confer extremely high antimicrobial activities to ZnO NP.

Numerous studies have reported ZnO NP induced ROS generation in host bacterial cells to be predominated by free radicals like superoxide anion ($O^{2-}$), hydrogen peroxide ($H_2O_2$) and hydroxide ($OH^-$). ROS formation induced by internalisation of ZnO NP leads to destruction of cellular components like lipids, DNA and proteins in host bacterial cells. While earlier studies had reported this to be exclusive to the presence of UV and visible light, recent studies have exhibited ROS induced nanotoxicity even in dark conditions. Hence extensive research is being performed to further investigate the mechanism of ZnO NP induced ROS formation in bacterial cells as a function of light exposure. The photoactivity of ZnO NP is resulted by transitioning of valence electrons upon accepting incident light and producing highly reactive reactants on the surface of and inside the crystal material (Wang et al., 2017). Upon excitation, ZnO electrons interact with water and produce $OH^*$ free radicals and $H^+$ ions. The oxygen molecules suspended in the mixture of bacteria and ZnO NP yield $O^{2-}$ that reacts with $H^+$ ions to produce $HO^{2-}$ free radicals. Subsequently this interferes with electrons generating $HO^{2*}$ free radicals and combines with $H^+$ to form $H_2O_2$ molecules (Yu et al., 2014). While free radicals cannot enter bacterial cells and only remain suspended in the growth medium, $H_2O_2$ molecules get adsorbed on the cellular surfaces and pass through the bacterial cells to exhibit their toxicities (Matai et al., 2014).

Various studies have reported that $Zn^{2+}$ ions also leak into growth media leading to host cell adsorption, intracellular percolation and inhibition of bacterial active transport mechanisms, amino acid metabolisms and enzyme system functions. This mechanism postulates that ZnO NP nanotoxicity is induced by $Zn^{2+}$ solubility in the media and its physicochemical properties like: porosity, concentration, size, morphology, functionalisation with other compounds, pH, UV illumination, exposure time and existence of other elements. Adsorption and subsequent particle internalisation of $Zn^{2+}$ ions have also been attributed to damage bacterial cell membrane permeability leading to progressive release of LPS, membrane proteins and internal factors. The nanosized ZnO particles have emerged as functional, strategic, promising and versatile inorganic materials with broad ranging applications in lieu of their efficient antimicrobial properties, least toxicity to human cells, high cost-effectiveness and longer shelf-life. Recent research has depicted toxicity induced by ZnO NP to vary among study type and test conditions. Hence this calls for further elucidation of exact mechanisms of action of ZnO NP because their extensive use in biomedicine has shown promising results as antimicrobial and anticancer agents. By studying the underlying effects of structural, optical and electrical properties of ZnO, its efficacy as antimicrobial agents can be further enhanced for potential use as drugs against multidrug-resistant bacteria. Interestingly, in the field of food technology, the use of metal oxide NP conjugated with heterogeneous composites (like zeolites) to be used as antimicrobial agents is gaining increased attention. Specifically, the use of ZnO NP conjugated with zeolite membranes along with UV light exposure has exhibited enhanced antimicrobial properties, a phenomenon also known as photoactivity of ZnO NP. For this reason, UV exposure has been used to enhance ZnO NP stability, antimicrobial efficacy and substrate absorption in composite surfaces leading to $Zn^{2+}$ ion release, accumulation and internalisation into host cells (Azizi-Lalabadi, Ehsani, Divband, & Alizadeh-Sani, 2019).

## 18.3 Titanium Metal Oxide Nanoparticles and Their Scope

Most studies have reported antimicrobial Titanium oxide NP to be potent agents producing ROS in microbial cells, with an emphasis on Green synthesised $TiO_2$ NP. Apart from bactericidal properties, titania metal oxide NP have been shown to possess other properties like photocatalysis, ability to induce degeneration of bacterial cell wall and dysfunction of signalling pathways while functioning as antimicrobial agents (Nikolova & Chavali, 2020; Wang et al., 2017). The high antibacterial activity of $TiO_2$ NP is brought about by photooxidation decomposition and mineralisation peroxidation of saturated and unsaturated fatty acids of the bacterial plasma membrane. $TiO_2$ NP have been shown to react with the -SH (thiol) groups of bacterial cell wall proteins thereby inactivating transport proteins and nutrients causing reduced cell wall permeability. The bactericidal properties of UV light activated $TiO_2$ NP has been exhibited to be dependent on the amount of dissolved $O_2$ and NP-cell contact which enhance $TiO_2$ NP translocation into host cells. Photocatalysis of $TiO_2$ NP has been shown to affect the cell signalling pathways rendering them dysfunctional by also decreasing coenzyme activities in the respiratory chain, lowering the ability for heme biosynthesis and transporting Iron and Phosphorous. These photocatalytic properties were enhanced by conjugation with metal composites and plant extracts (e.g.: *G. zeylanica* extracts). Recent studies have shown that $TiO_2$ NP cause bacterial DNA compression, degeneration and fragmentation through their affinity to bind to G-C rich regions in bacterial genetic material (Iram et al., 2015).

Unique surface chemistry and morphologies enable the wide use of $TiO_2$ NP for various biomedical applications. However, their chemical synthesis has disadvantages of requiring high temperatures, pressures and toxic chemicals. Hence recent research in the Green Synthesis of $TiO_2$ NP (from plants, microorganisms and other biological derivatives) has gained attention owing to being environmentally safe and biologically less toxic (to non-host organisms). Bio-mediated titanium oxides NP have enhanced physicochemical properties and find various biomedical applications. Additionally, Green synthesised $TiO_2$ NP has been proved to have enhanced photocatalytic activity, solubility and stability when compared to their chemically synthesised counterparts, along with greater ability to degrade toxic compounds by oxidation. However the exact mechanism for this antimicrobial phenomenon still remains to be deciphered.

## 18.4 Copper Metal Oxide Nanoparticles and Their Scope

Being far cheaper than Silver oxide NP, CuO NP hold greater potential in Biomedicine, but require higher concentrations to achieve their nanotoxicities. Their antimicrobial effects are shown to be mediated by release of $Cu^{2+}$ ions and are more effective towards bacterial cell walls rich in amine and carboxyl groups such as those of Gram positive *Bacillus subtilis*. This is attributed to the dense peptidoglycan layer of Gram positive cell walls that bind to $Cu^{2+}$ ions released from the CuO nanoparticles. The subsequent cytotoxicity is mediated by ROS induction in bacterial cells. Various studies have reported that smaller sized CuO NP (20 nm) induced strong bactericidal activities against Gram positive and Gram negative bacteria (Gilbertson et al., 2016).

With growing concerns of antibiotic resistant bacteria, tremendous research is being carried out on antimicrobial applications of metal oxide based NP. CuO NP exerts their antimicrobial effects through ROS formation by dissolution into bacterial cells and interaction with cellular ligands. Amino acids display high affinity towards CuO NP and hence facilitate their adsorption on bacterial surface, complex formation and intracellular dissolution/release of $Cu^{2+}$ ions. Specifically, studies revealed that the participating amino acid residues have profound effects on the antimicrobial activity of CuO NP through their ability to promote copper ion dissolution after adsorption to bacterial cells. While Val and Arg residues having hydrophobic and positive side chains, respectively, form soluble Cu(II) complexes allowing diffusion of copper ions, poorly soluble Cu(II) complexes are formed upon the interaction between these NP and residues like Asp, Leu, Phe and Tyr. This significant feature is currently being explored to functionalize copper metal oxide NP with amino acid residues that enhance their biosorption and antimicrobial properties (Badetti et al., 2019).

## 18.5 Silver Metal Oxide Nanoparticles and Their Scope

Contrary to other metals, silver has the advantage of being effective at lower concentrations and hence is popularly utilised in different biomedical applications. Silver oxide NP act as antimicrobial agents by enabling ROS mediated cell lysis through interaction of silver ions with thiol groups (-SH) of bacterial proteins leading to increased permeability and disintegration of bacterial cell membranes (Walder, Pittet, & Tramer, 2002). Various studies have also reported these redox reactions to cause increased bacterial lysis (Nikolova & Chavali, 2020). Silver oxide-based NP exert antimicrobial effects through adsorption to bacterial surfaces causing microbial membrane damage; ROS induced oxidative stress and creation of pits. Morphology, size, oxidation, dissolution states, surface charges, colloidal state, molar absorptivity, surface coatings, aggregation properties and dispersion capability are features that influence the antimicrobial properties of silver oxide NP (Burduşel et al., 2018).

In recent years, silver oxide-based NP have been extensively researched for their prospective biomedical applications. Given their broad spectrum of useful features, silver based NP are being utilised as antimicrobials, for drug delivery and diagnostics, in healthcare devices and in other industries. Interestingly, Ag NP were the first metal oxide based nanoparticles that attracted worldwide attention as unconventional antimicrobial agents despite insufficient information regarding their toxicities. Their major advantages include facile and safe synthesis techniques using physical, chemical and biological processes with reduced economic implications and enhanced repeatability and reproducibility. Additionally, their physicochemical properties help silver oxide NP elicit antimicrobial properties which can be enhanced by functionalising them with other compounds or capping agents. In this regard, extensive research is being carried out to design cost-effective, efficient, eco-friendly, safe and simple methods to translate the applications of silver oxide NP into areas of biomedical products, therapeutics and unconventional antimicrobial agents (especially against drug resistant bacteria) (Burduşel et al., 2018). Silver NP, being the most favoured and widely used nanoparticles, make for efficient antimicrobial agents especially since studies have shown their potency against Gram positive and negative bacteria. However, this would mandate enhanced understanding of their bioregulatory mechanisms and toxicological regulations.

## 18.6 Magnesium Metal Oxide Nanoparticles and Their Scope

The mean zeta potential of MgO NP exhibit positive charge in the pH ranges of 4 to 8 which favour the electrostatic interactions between these NP and bacterial cells. Similar to most other antimicrobial mechanisms of action of metal oxide NP, MgO NP react with bacterial cells and induce ROS formation ($OH^-$ free radicals) leading to cytotoxicity. Additionally, their interaction with bacterial cells has been reported to lead to destruction of peptide linkages of bacterial cell membrane proteins through the action of superoxide ions (Huang et al., 2005). The other antimicrobial mechanism of action is proposed to be through destruction and disintegration of bacterial cell walls leading to leakage of intracellular contents and cell lysis. While MgO NP possess the advantage of not being toxic to human cells at lower concentrations (Huang et al., 2005), their bactericidal activity has been shown to increase proportionately with their concentrations (Nikolova & Chavali, 2020).

Various studies have reported magnesium oxide NP to have different antimicrobial properties in different conditions (like altered shapes and sizes) but only few studies have reported their exact antimicrobial potency. In biomedical coating devices, MgO NP has been used as additives to provide significant antimicrobial applications without side effects. Currently, MgO NP are being extensively used as coating additives for next generation biodegradable medical devices in clinical practice to combat infections caused by drug resistant bacteria (Nguyen, Grelling, Wetteland, Rosario, & Liu, 2018). Further research is essential to develop the standard concentrations and dosages of MgNP to be used as antimicrobial agents by also establishing standards required to mitigate toxicities caused due to excessive magnesium ion concentrations.

## 18.7 Calcium Oxide Nanoparticles and Their Scope

There lies a close similarity between the mechanisms of action of Calcium and Magnesium oxide NP, in that they both favour close proximity with bacterial cells for enhanced ROS induced bactericidal properties. CaO NP upon internalisation into bacterial cells generates superoxide ions which react with carbonyl groups of peptides in bacterial cell walls leading to destruction of these cells (Makhluf et al., 2005).

Among the inorganic metal oxide nanoparticles like MgO, $TiO_2$, CuO, ZnO and CaO NP (Ramola, Joshi, Ramola, Chhabra, & Singh, 2019), applications of CaO NP in biomedical research has been due to its biosafety in terms of being nontoxic to humans and animals. Research is also underway to promote the biogenic synthesis of CaO NP using environmentally safe procedures (Anantharaman, Ramalakshmi, & George, 2016). Green synthesis of CaO NP has been shown to enhance their photocatalytic and antimicrobial properties (even at low nanoparticle concentrations), along with being highly economic, cost-effective and highly efficient. When compared to conventional chemical syntheses, these Green processes minimise the use of harmful chemicals and varied instrumentations. CaO NP has multiple features like unique structural and optical properties, significant and stable antimicrobial properties and low environmental toxicities. Due to their specific structural and optical properties CuO NPs have been employed as drug delivery agents in photothermal therapy, photodynamic therapy and synaptic delivery (Abraham and Saranthy, 2018).

## 18.8 Aluminium Oxide Nanoparticles and Their Scope

Contrary to most of the previously cited examples of metal oxide NP, the distinguishing factor of $Al_2O_3$ NP is that they exhibit bactericidal properties only at high concentrations. Their mechanism of action was proposed to primarily be based upon distortion of bacterial cell walls by virtue of agglomeration, creation of pits and intracellular percolation (Ansari et al., 2014). In the quest for finding potent antimicrobial drugs against drug resistant pathogens, metal oxide nanoparticles have been efficaciously utilised due to their antimicrobial efficiencies and advantages of being less toxic, having high specificity and lowering the cost of production. Being one of the most important ceramic materials $Al_2O_3$ possess various commercial, industrial and biological applications. Specific to the biomedical field, $Al_2O_3$ NP have exhibited high bacterial cytotoxicity that has been attributed to their small size which mediates bacterial cell electrostatic interactions leading to distortion of bacterial membrane morphologies, altered permeability, disrupted transport of bacterial proteins and intracellular $Al_2O_3$ NP dissolution causing oxidative damage. L-alpha-Phosphatidyl-Ethnolamine (PE) of LPS in bacterial cell wall proteins has been shown to interact with $Al_2O_3$ NP and mediate adhesion and internalisation of these nanoparticles (Ansari et al., 2014; Ansari, Khan, Khan, Pal, & Cameotra, 2013). There remains the need for further studies on ecotoxicity evaluation and biological safety regulations of $Al_2O_3$ NP to further propel their use in biomedical applications.

## 18.9 Iron Oxide Nanoparticles and Their Scope

Magnetite is of great importance in biomedical research due to its high magnetism, biocompatibility, low cost and reduced toxicities. Iron oxide NP have great antimicrobial effects mediated by ROS formation in bacterial cells through the interplay between oxygen of reduced iron species ($Fe^{3+}$ or $Fe^{2+}$) and the distributed electronic or ionic transport chains in bacterial cells (Ismail, Sulaiman, Abdulrahman, & Marzoog, 2015). Various studies have also reported a synergistic relationship between phytochemical alkaloids and generation of ROS by iron oxide NP. Corroborating these claims, research has shown enhancement of antibacterial effects of iron oxide NP brought about by phytochemicals like those in *Cynomet*ra *ramiflora* leaf extracts and the negative charges on these NP favouring strong electrostatic interactions with bacterial cells thus gaining entry through adsorption and internalisation due to their small size and large surface area (Groiss, Selvaraj, Varadavenkatesan, & Vinayagam, 2017).

Magnetic nanoparticles, due to their biocompatibility, chemical stability and magnetic behaviour, are widely used more than other types of NP for biomedical applications. In this category, FeO NP come in different types like magnetite ($Fe_2O_3$), maghemite ($\gamma$-$Fe_2O_3$), hematite ($\alpha$-$Fe_2O_3$) and goethite (FeO(OH)) (Bucak, Yavuztürk, & Sezer, 2012). Iron oxide NP possess advantageous properties like small size and high surface area:volume ratio that result in very high free energy content due to which they interact with cellular surfaces to become stable entities. In this course of interactions photocatalytic iron oxide NP overcome the energy gap electron barrier and produce free ROS which induce host cell damage that induces antimicrobial effects (Arakha et al., 2015). Currently, research is being done to functionalize iron oxide based nanoparticles with natural or synthetic polymers and perform size controlled synthesis to create more stable and hydrophilic structures providing variable functional groups facilitating bacterial cell interactions. Supermagnetic iron oxide NP loaded with antioxidant enzymes (like gallic acid, catalase) are also used for targeted enzyme delivery, preparation of biomedical devices and carriers for drug delivery. These nanocarriers coupled with antioxidants have shown increased efficacies by revealing an inverse relationship between their particle size and antimicrobial efficiency. Biomedical applications always favour nanosized particles with high magnetic values and narrow particle size distribution (Shah et al., 2017). However, on the downside, there has been conflicting evidence on the toxicity of Iron oxide NP. Furthermore, electropositivity and their surface charges have also been identified as factors affecting genotoxicity and cytotoxicity (Arias et al., 2018). These highly promising features of magnetic iron oxide NP in

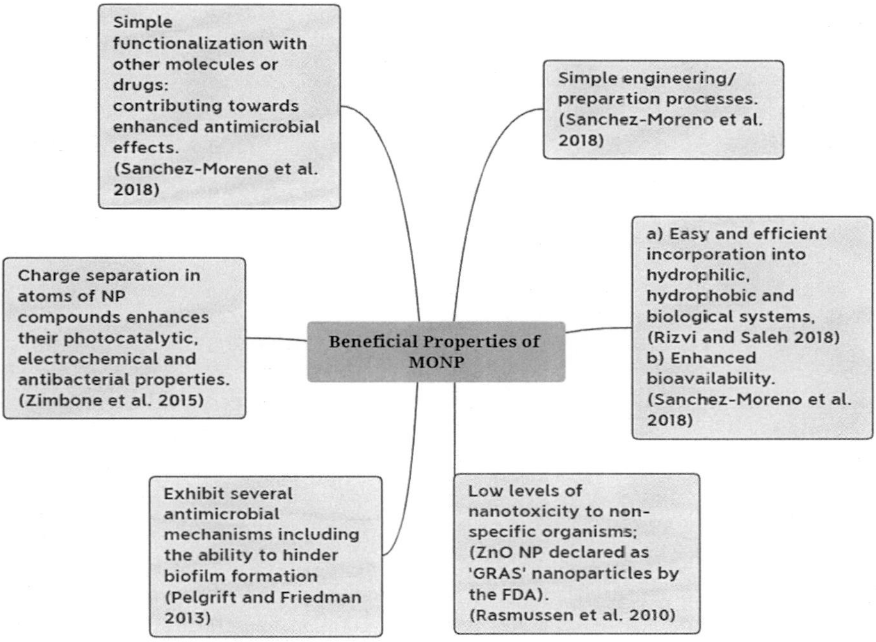


**Fig. 18.1** Beneficial properties of metal oxide nanoparticles

nanotoxicology which are being utilised to develop antimicrobial agents need further biochemical elucidation so as to further their research and utility in biomedical applications. Few beneficial properties of MONP are depicted in Fig. 18.1.

## 18.10 Conclusions, Outlook and Future Aspects

The biomedical applications of metal oxide-based nanoparticles have brought great promise especially in the management of recalcitrant microbes. They tailor to various promising and wide ranging biomedical applications due to their preferential size of about >200 nm that enhances nanoparticle surface area and colloidal stability (Biswas, Islam, Choudhury, Mostafa, & Kadir, 2014). Furthermore, nanoparticle based antimicrobial agents have shown promise against cases of drug resistant microbes that have evolved due to rampant use of antibiotics, inappropriate administration of antibiotics without relevant medical prescriptions and excessive switching of antibiotics without adequate medical supervision (Wang et al., 2017). Based on their dimensions MONP are classified as 0-dimension (nanoparticles, nanoclusters and so on), 1-dimension (nanorods, nanotubes and so on), 2-dimension/ planar (nanosheets, nanocoatings and so on) and 3-dimension (dendrimers, nanopillars and so on) (Rafiei-Sarmazdeh, Zahedi-Dizaji, & Kang, 2019). Their unique physicochemical properties have enabled controlled synthesis of MONP tailoring to specific configurations, distributions and biomedical requirements. While being employed as antimicrobial agents, MONP are also being excessively utilised in tissue therapy or regeneration, wound healing, diagnostics, medical imaging (due to their fluorescence and magnetic properties enabling precise localisation and reliable quantification) and as biosensors (Nikolova & Chavali, 2020). Microbial biofilm formation that functions as an embedding zone aggregating microbes causing chronic infections, relaying drug resistance and generating superantigens, has been shown to be hindered by metal oxide NP (Pelgrift & Friedman, 2013). The crucial aspect of deciphering their nanotoxicity remains highly dependent on distinguishing the NP specific effects from other concurrent ionic effects in organisms (Nikolova & Chavali, 2020). Research has been limited in elucidating the in vivo biological toxicities of MONP, comparative analysis with different types of metal oxide NP and in conducting evidence based randomised trials on different organisms. Due to these reasons, predominant research has been performed in vitro and only on selected microbes, whose results cannot be extrapolated into the complex in vivo biological milieu (Vega-Jiménez et al., 2019). This mandates for standardising specific nanoparticle synthesis parameters, testing and regulations. While the use of antimicrobial MONP has elicited encouraging responses, there still remains more scope in deciphering their mechanisms of action and standardising their toxicities, syntheses and efficiencies. Antimicrobial applications of MONP would benefit in great proportions with further randomised and regulated research (emphasising on animal trials) in biological systems.

## References


Abraham, S., & Saranthy, V. P. (2018). Biomedical applications of calcium oxide nanoparticles—A spectroscopic study. *International Journal of Pharmaceutical Sciences Review and Research, 49*, 121.

Anantharaman, A., Ramalakshmi, S., & George, M. (2016). Green synthesis of calcium oxide nanoparticles and its applications. *International Journal of Engineering Research and Applications, 6*, 27–31.

Ansari, M. A., Khan, H. M., Khan, A. A., Ahmad, M. K., Mahdi, A. A., Pal, R., et al. (2014). Interaction of silver nanoparticles with *Escherichia coli* and their cell envelope biomolecules. *Journal of Basic Microbiology, 54*, 905–915.

Ansari, M. A., Khan, H. M., Khan, A. A., Pal, R., & Cameotra, S. S. (2013). Antibacterial potential of $Al_2O_3$ nanoparticles against multidrug resistance strains of Staphylococcus aureus isolated from skin exudates. *Journal of Nanoparticle Research, 15*, 1970.

Arakha, M., Pal, S., Samantarrai, D., Panigrahi, T. K., Mallick, B. C., Pramanik, K., et al. (2015). Antimicrobial activity of iron oxide nanoparticle upon modulation of nanoparticle-bacteria interface. *Scientific Reports, 5*, 14813.

Arias, L. S., Pessan, J. P., Vieira, A. P. M., Lima, T. M. T. D., Delbem, A. C. B., & Monteiro, D. R. (2018). Iron oxide nanoparticles for biomedical applications: A perspective on synthesis, drugs, antimicrobial activity, and toxicity. *Antibiotics, 7*, 46.

Armentano, I., Arciola, C. R., Fortunati, E., Ferrari, D., Mattioli, S., Amoroso, C. F., et al. (2014). The interaction of bacteria with engineered nanostructured polymeric materials: A review. *The Scientific World Journal, 2014*, 1–18.

Azizi-Lalabadi, M., Ehsani, A., Divband, B., & Alizadeh-Sani, M. (2019). Antimicrobial activity of Titanium dioxide and Zinc oxide nanoparticles supported in 4A zeolite and evaluation the morphological characteristic. *Scientific Reports, 9*, 1–10.

Badetti, E., Calgaro, L., Falchi, L., Bonetto, A., Bettiol, C., Leonetti, B., et al. (2019). Interaction between copper oxide nanoparticles and amino acids: Influence on the antibacterial activity. *Nanomaterials, 9*, 792.

Biswas, A. K., Islam, M. R., Choudhury, Z. S., Mostafa, A., & Kadir, M. F. (2014). Nanotechnology based approaches in cancer therapeutics. *Advances in Natural Sciences: Nanoscience and Nanotechnology, 5*, 043001.

Bucak, S., Yavuztürk, B., & Sezer, A. D. (2012). Magnetic nanoparticles: Synthesis, surface modifications and application in drug delivery. *Recent Advances in Novel Drug Carrier Systems, 2*, 165–200.

Burduşel, A. C., Gherasim, O., Grumezescu, A. M., Mogoantă, L., Ficai, A., & Andronescu, E. (2018). Biomedical applications of silver nanoparticles: An up-to-date overview. *Nanomaterials, 8*, 681.

Gao, W., Thamphiwatana, S., Angsantikul, P., & Zhang, L. (2014). Nanoparticle approaches against bacterial infections. *Wiley Interdisciplinary Reviews: Nanomedicine and Nanobiotechnology, 6*, 532–547.

Gilbertson, L. M., Albalghiti, E. M., Fishman, Z. S., Perreault, F., Corredor, C., Posner, J. D., et al. (2016). Shape-dependent surface reactivity and antimicrobial activity of nano-cupric oxide. *Environmental Science & Technology, 50*, 3975–3984.

Gold, K., Slay, B., Knackstedt, M., & Gaharwar, A. K. (2018). Antimicrobial activity of metal and metal-oxide based nanoparticles. *Advanced Therapeutics, 1*, 1700033.

Groiss, S., Selvaraj, R., Varadavenkatesan, T., & Vinayagam, R. (2017). Structural characterization, antibacterial and catalytic effect of iron oxide nanoparticles synthesised using the leaf extract of Cynometra ramiflora. *Journal of Molecular Structure, 1128*, 572–578.

Huang, L., Li, D. Q., Lin, Y. J., Wei, M., Evans, D. G., & Duan, X. (2005). Controllable preparation of Nano-MgO and investigation of its bactericidal properties. *Journal of Inorganic Biochemistry, 99*, 986–993.

Iram, N. E., Khan, M. S., Jolly, R., Arshad, M., Alam, M., Alam, P., et al. (2015). Interaction mode of polycarbazole–titanium dioxide nanocomposite with DNA: Molecular docking simulation and *in vitro* antimicrobial study. *Journal of Photochemistry and Photobiology B: Biology, 153*, 20–32.

Ismail, R. A., Sulaiman, G. M., Abdulrahman, S. A., & Marzoog, T. R. (2015). Antibacterial activity of magnetic iron oxide nanoparticles synthesized by laser ablation in liquid. *Materials Science and Engineering: C, 53*, 286–297.

Li, H., Chen, Q., Zhao, J., & Urmila, K. (2015). Enhancing the antimicrobial activity of natural extraction using the synthetic ultrasmall metal nanoparticles. *Scientific Reports, 5*, 11033.

Luan, B., Huynh, T., & Zhou, R. (2016). Complete wetting of graphene by biological lipids. *Nanoscale, 8*, 5750–5754.

Makhluf, S., Dror, R., Nitzan, Y., Abramovich, Y., Jelinek, R., & Gedanken, A. (2005). Microwave-assisted synthesis of nanocrystalline MgO and its use as a bactericide. *Advanced Functional Materials, 15*, 1708–1715.

Matai, I., Sachdev, A., Dubey, P., Kumar, S. U., Bhushan, B., & Gopinath, P. (2014). Antibacterial activity and mechanism of Ag–ZnO nanocomposite on *S. aureus* and GFP-expressing antibiotic resistant E. coli. *Colloids and Surfaces B: Biointerfaces, 115*, 359–367.

Nguyen, N. Y. T., Grelling, N., Wetteland, C. L., Rosario, R., & Liu, H. (2018). Antimicrobial activities and mechanisms of magnesium oxide nanoparticles (nMgO) against pathogenic bacteria, yeasts, and biofilms. *Scientific Reports, 8*, 1–23.

Nikolova, M. P., & Chavali, M. S. (2020). Metal oxide nanoparticles as biomedical materials. *Biomimetics, 5*, 27.

Pelgrift, R. Y., & Friedman, A. J. (2013). Nanotechnology as a therapeutic tool to combat microbial resistance. *Advanced Drug Delivery Reviews, 65*, 1803–1815.

Rafiei-Sarmazdeh, Z., Zahedi-Dizaji, S. M., & Kang, A. K. (2019). Two-dimensional nanomaterials. In S. Ameen, M. S. Akhtar, & H. S. Shin (Eds.), *IntechOpen two-dimensional nanomaterials, nanostructures*. London: IntechOpen.

Rajaura, R. S., Sharma, V., Ronin, R. S., Gupta, D. K., Srivastava, S., Agrawal, K., et al. (2017). Synthesis, characterization and enhanced antimicrobial activity of reduced graphene oxide–zinc oxide nanocomposite. *Materials Research Express, 4*, 025401.

Ramola, B., Joshi, N. C., Ramola, M., Chhabra, J., & Singh, A. (2019). Green synthesis, characterisations and antimicrobial activities of CaO nanoparticles. *Oriental Journal of Chemistry, 35*, 1154–1157.

Rasmussen, J. W., Martinez, E., Louka, P., & Wingett, D. G. (2010). Zinc oxide nanoparticles for selective destruction of tumor cells and potential for drug delivery applications. *Expert Opinion on Drug Delivery, 7*, 1063–1077.

Rizvi, S. A., & Saleh, A. M. (2018). Applications of nanoparticle systems in drug delivery technology. *Saudi Pharmaceutical Journal, 26*, 64–70.

Sanchez-Moreno, P., Ortega-Vinuesa, J. L., Peula-Garcia, J. M., Marchal, J. A., & Boulaiz, H. (2018). Smart drug-delivery systems for cancer nanotherapy. *Current Drug Targets, 19*, 339–359.

Shah, S. T., Yehya, A. W., Saad, O., Simarani, K., Chowdhury, Z., Alhadi, A. A., et al. (2017). Surface functionalization of iron oxide nanoparticles with gallic acid as potential antioxidant and antimicrobial agents. *Nanomaterials, 7*, 306.

Vega-Jiménez, A. L., Vázquez-Olmos, A. R., Acosta-Gío, E., & Álvarez-Pérez, M. A. (2019). In vitro antimicrobial activity evaluation of metal oxide nanoparticles. In K. S. Koh & V. L. Wong (Eds.), *Nanoemulsions—Properties, fabrications and application*. London: IntechOpen.

Walder, B., Pittet, D., & Tramer, M. R. (2002). Prevention of bloodstream infections with central venous catheters treated with anti-infective agents depends on catheter type and insertion time: Evidence from a meta-analysis. *Infection Control & Hospital Epidemiology, 23*, 748–756.

Wang, L., Hu, C., & Shao, L. (2017). The antimicrobial activity of nanoparticles: Present situation and prospects for the future. *International Journal of Nanomedicine, 12*, 1227.

Yu, J., Zhang, W., Li, Y., Wang, G., Yang, L., Jin, J., et al. (2014). Synthesis, characterization, antimicrobial activity and mechanism of a novel hydroxyapatite whisker/nano zinc oxide biomaterial. *Biomedical Materials, 10*, 015001.

Zhang, L., Pornpattananangkul, D., Hu, C. M., & Huang, C. M. (2010). Development of nanoparticles for antimicrobial drug delivery. *Current Medicinal Chemistry, 17*, 585–594.

Zimbone, M., Buccheri, M. A., Cacciato, G., Sanz, R., Rappazzo, G., Boninelli, S., et al. (2015). Photocatalytical and antibacterial activity of $TiO_2$ nanoparticles obtained by laser ablation in water. *Applied Catalysis B: Environmental, 165*, 487–494.

# Chapter 19
# Structural Analysis and Thermal Properties of Graphene and Biocomposite Potential Application in Various Sensors

**Kehkashan Alam, Faiz Warsi, Arshi Khan, and Ishaat M. Khan**

## 19.1 Graphene and Biocomposites: An Overview of Smart Electronic Materials

The existence of carbon in our everyday experiences has a remarkable capability to bond together powerfully, forming a covalent bond in various types, i.e., in different states of hybridization (sp, $sp^2$, and $sp^3$). Due to the presence of a carbon atom in multiple states, it results in the formation of molecules ranging between small to a long chain. Centuries ago, it was seen that the first element present in organic compounds and biomolecules is the carbon atom. Due to the catenation property of carbon, many organic compounds are known. Carbon is also present in a natural carbon material known as the allotropes of carbon, such as graphite and diamond. These allotropic forms contain carbon, but they possess very different properties. Graphite is physically soft and appears as an opaque black material with good electrical conductivity, while diamond is known as the most rigid material with a transparent electrical insulator. The differences observed in the allotropes of carbon are due to the attached carbon atom in both cases. The crystal systems in which the graphite ($sp^2$ carbon atom) and diamond (tetrahedral $sp^3$ carbon atoms) crystallize are hexagonal and cubic.

Moreover, graphite is comprised of stacked graphene monolayer which stacked together by van der Waals interactions. Graphene is packed in a 2D sheet of $sp^2$-hybridized carbon organized in a hexagonal lattice. Its basic structure is like the network of honeycomb for the formation of other allotropes such as one-dimensional (1D) carbon nanotubes (CNTs), stacked three-dimensional (3D) graphite and zero-dimensional (0D) fullerenes, and C-dots (Lu, Zu, Byun, Kim, & Chou, 2012; Jianfeng Wang, Cheng, & Tang, 2012).

K. Alam (✉) · F. Warsi · A. Khan · I. M. Khan
Department of Chemistry, Aligarh Muslim University, Aligarh, Uttar Pradesh, India
e-mail: gc9837@myamu.ac.in; drishaatamu@gmail.com


K. Pal (ed.), *Bio-manufactured Nanomaterials*,
https://doi.org/10.1007/978-3-030-67223-2_19

Over the previous decade, the graphene's 2D crystal has sprouted as an appealing building block with its significant properties, creating and designing high-quality, multifunctional materials (Castro Neto, Guinea, Peres, Novoselov, & Geim, 2009; Compton & Nguyen, 2010). The impressive successive efforts have made rich graphene macroscopic materials using liquid assembly methodology up to this point, including 2D films/papers and 3D structures (aerogels) with flexible utilization in broad applications (Compton & Nguyen, 2010; Han, Xu, & Gao, 2013; Kou & Gao, 2011; Sun, Xu, & Gao, 2013). By combining 0D, 1D, and 2D carbon nanostructures, we can form 1D nanowires, 2D thin films, 3D microspheres, carbon aerogels, etc.

Nowadays, the graphene study seeks many researchers' attention in different research areas, which involves energy, electronics, photonics, and composites (Servant, Bianco, Prato, & Kostarelos, 2014). Additionally, graphene produces great desires in the biomedical field because of its significant mechanical and electrochemical properties (Liu, Lee, & Im, 2013), joined by biocompatibility, transparency, high structural stability, and electrical conductivity. Its surface area was found to be near 2630 $m^2/g$. Collectively, all these properties are highly favorable for thermal and switchable sensors (Mao, Hatton, & Rutledge, 2013). Also, graphene has many applications in scaffold production, drug release, and sensing. Besides these, the stretchable matrix has been intensively explored for their new applications like electronic medical care and smart clothing (Ahn & Je, 2012; Kim & Rogers, 2008). Its polymeric or organic light-emitting diode (OLED) shows given flexible composite electrodes that consist of silver wires (Liang, Li, Niu, Yu, & Pei, 2013) or carbon nanotubes (CNTs) (Sekitani et al., 2009), and conductive hydrogel (Larson et al., 2016; Wang, Yan, Chee, & Lee, 2015) have seen the utility in stretchable electroluminescent (EL) are illustrated.

This chapter portrays the recently grown idea to acknowledge graphene macroscopic ordered materials, going through graphene chemistry and graphene liquid crystals (LCs) with macroscopic categorizing in the liquid state. The entire framework comprises of three legitimate areas: (1) physical properties of graphene; (2) its thermal properties; and (3) applications of graphene as sensors ranging between gas and vapor sensor to the biomedical devices.

## 19.2 Physical Studies of Graphene

### *19.2.1 Structural Properties*

Of all the elements that exist in nature, carbon has the most unique and remarkable property. It is also the very first element to be known by humans. The application of carbon in various areas is due to its diverse structure, and properties like bonding between carbon atoms can vary to nanostructures, microstructures, and crystalline alignment. Carbon exists in two allotropic forms, i.e., diamond and graphite both exist in three-dimensional structures. Graphite undergoes $sp^2$ hybridisation, whereas diamond form $sp^3$ hybridisation. The carbon arrangement in a diamond's 3D crystal lattice is a face-centered cubic (fcc) (Nazare & Neves, 2018). A strong covalent bond between carbon atoms ($sp^3$ hybridization) governs immense physical properties. Graphite exists in a thin, flat scale (planar) structure (Graphite and Precursors - Google Books, n.d.). Each carbon is arranged in a hexagonal lattice in its crystal, and each carbon atom is separated by a distance of 0.142 nm. The distance between the two carbon layers is 0.335 nm (Fig. 19.1a). Graphite exists in two forms, α and β, and their arrangement in the crystal lattice is hexagonal and rhombohedral. Both states have similar properties. The thin, flat scale has a stronger attraction between lateral planes than between the aircraft. As the graphitic carbon has $sp^2$ hybridization and hexagonal arrangement, all the carbon nanomaterials can be regarded as in the same group. We mean that they possess striking similarities with the same group and have considerable differences in different shapes and sizes. Graphene structural unit of all allotrope (Hirsch, 2010) like carbon nanotube (CNT), carbon fiber (CF), and amorphous carbon (AC), and fullerene. Fullerenes have unique chemistry and technological applications in nanotechnology and electronics. They were part of intense research in the 1990s. Fullerenes are large balls consisting of entirely carbon and have the shape of an ellipsoid or hollow sphere (Kroto, Heath, O'Brien, Curl, &

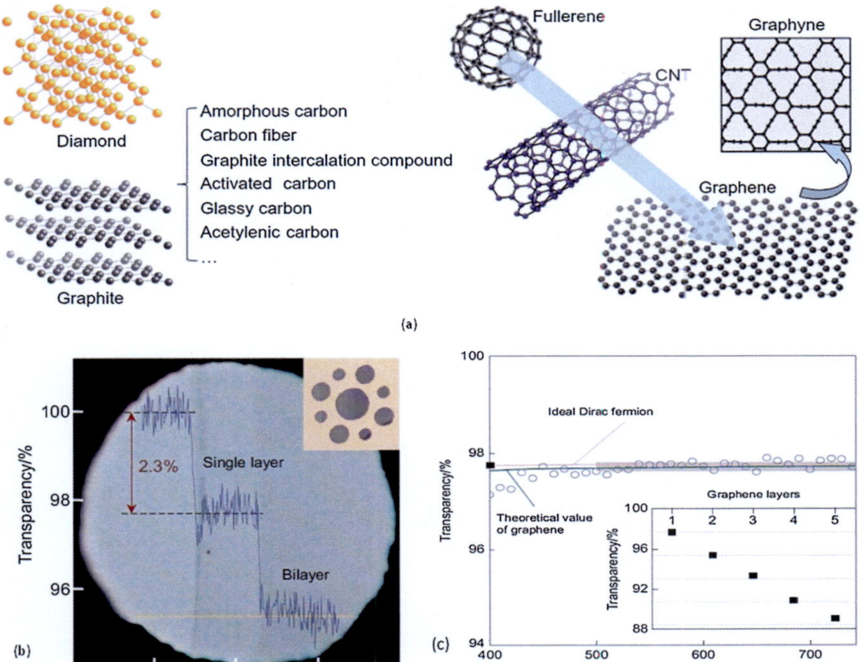


**Fig. 19.1** (**a**) Allotropes of carbon and its related materials, (**b**) difference in transparency between a substrate and single-layer graphene, and (**c**) Theoretical transparency of graphene reproduced with permission from reference (Zhen & Zhu, 2017)

Smalley, 1985). When the graphene sheet is rolled in cylindrical form, CNT is formed. They have similarities as to fullerene. They have high strength efficient electrical properties and a better conductor of heat (Baughman, Zakhidov, & De Heer, 2002). Multi-walled carbon nanotubes (MWNTs) and single-walled carbon nanotubes (SWNTs) diameters are 0.8–2 nm and 5–20 nm, respectively. Walls of CNT can be metals or semiconductors, depending on the lattice movement concerning the principal axis, called Chirality. Carbon fiber (CF) is in 5–10μm in diameter and consists mainly of a carbon atom joined with each other in crystal and aligned parallel to the fiber axis. CFs are long, and its molecular chain is twisted (Hiremath, Mays, & Bhat, 2017). These fibers are firm for its small size. CFs molded with composite materials have immense applications in car bodies, bicycles, automobiles, and fishing rods.

Graphene is planar in nature, super light, and possess a density of 0.77 mg/m$^2$. Hexagonal carbon is the unit structure of graphene, having a unit area of 0.052nm$^2$. The ring contains only two carbon atoms, as three rings share each carbon atom. Graphene is the super thin and ultra-light as it consists of one layer of carbon atoms. This property makes graphene high transparent 97.7%. It absorbs only 2.3% of light (Nair et al., 2008b) shown in Fig. 19.1b. Single-layer graphene, bilayer graphene, and substrate have a transparency difference of 2.3%. The number of graphene layers can be effectively obtained by transparency difference of graphene; this is also confirmed by using the Dirac fermion theory represented in Fig. 19.1c. Various graphene layers have different colors and contrast, which is due to interference and refraction of light. Due to different colors, these layers of graphene can be easily identified (Ni et al., 2007). Graphene can replace other membranes such as a fluorine-doped tin oxide (FTO) and indium tin oxide (ITO), as it is a very high transparent conductive membrane. This can solve different problems concerning lesser indium availability and environmental pollution. Another characteristic property of graphene is the mobility of its electron, by which it is highly conducting, having a conductivity of $10^6$ S/m and resistance of 31 Ω/sq. (Kim et al., 2009). Another factor that enhances graphene conductivity is its π-electron, which is free to move. Its displacement adds up in conductance.

## 19.2.2 Electrochemical Optical Switching

The graphene's current build-up is due to the rich electronic property with free path and high motilities at room temperature (Geim & Novoselov, 2007; Novoselov, 2004). Its application in nanotechnology for the starting material field effect transistor (FED), but it is concerning problems at neutral (Dirac) point graphene have minimum conductivity, least electronic carriers by which $I_{on}/I_{off}$ ratios become considerably low and hamper most electronic applications. Using graphene nanoribbons (GNRs), width less than 5 nm, having the capacity to open band gap larger than 500Mev (Fiori & Iannaccone, 2007; Liang, Neophytou, Lundstrom, & Nikonov, 2007), be used to overcome this problem. Even if we use state-of-the-art techniques to control the optical scale (Maile et al., 2000), the smallest GNRs we get are width 10 nm (Avouris, Chen, & Perebeinos, 2009; Geim & Novoselov, 2007; Özyilmaz et al., 2007). This demanding challenge to clarify GNR-transistors that can be given to integrated circuits without wrapped electronic quality. Szafranek and coworkers have shown that attachment or detachment of active radicals to graphene alters graphene property. This process depends upon the electrochemical mechanism where $I_{on}/I_{off}$ ratios are greater than$10^6$ (Echtermeyer et al., 2007). The extremists are hydrogen ($H^+$) and hydroxyl ($OH^-$), which are catalytically generated from water molecules. The addition of radical makes graphene non-conductive, but the process can flip by a small current that strengthens its conductive behavior.

### *19.2.3 Nonlinear Optical Response*

Nonlinear optical methods have an essential ability to be used in different photonic operations. Hence the study of such material is highly recommended. These photonic materials provide better performance in industries for artificial growth and fabrication of material, transparency, high threshold, and fast response.

Carbon-based material, graphene is used to have rank nonlinear optical response over the varying spectral range. Bloch equation is used to observe visual effects in graphene, including field interaction, current density, and photon generation. Nonlinear effects are generated when exposed to an electric field. Electrons in graphene are coupled to an applied electric field. The analysis of nonlinearity is due to induced electronic current and harmonic generation. Graphene consists of two bands, the valence and conduction. Doping of graphene is not done in the valence band, which is filled, whereas the conduction band is empty, graphene subjected to the electric field flow of electrons takes place from valence to conductance through electron valley. The movement of charge carriers is followed by a time-dependent Dirac equation (TDDE) (Ishikawa, 2010). Xin jin (Jin, 2015) has shown from (TDDE) studies that easy visualization of coupling of light with field interaction under applied magnetic field was observed in graphene. Photo signals generated when graphene is subjected to 90-degree excitation pulse and 180-degree refocusing pulse.

## 19.3 Thermal Properties

### *19.3.1 Interconnects*

Graphene was found to be of significant advantages in nanoscale devices and interconnects due to its higher thermal conductivity. The temperature rise and heat sinking are significantly less due to this property during the operation of devices shown in Fig. 19.2. In 2020, a group of researchers have reported the effect of size and temperature for the monolayer graphene (MLG) based interconnects (Debroy,

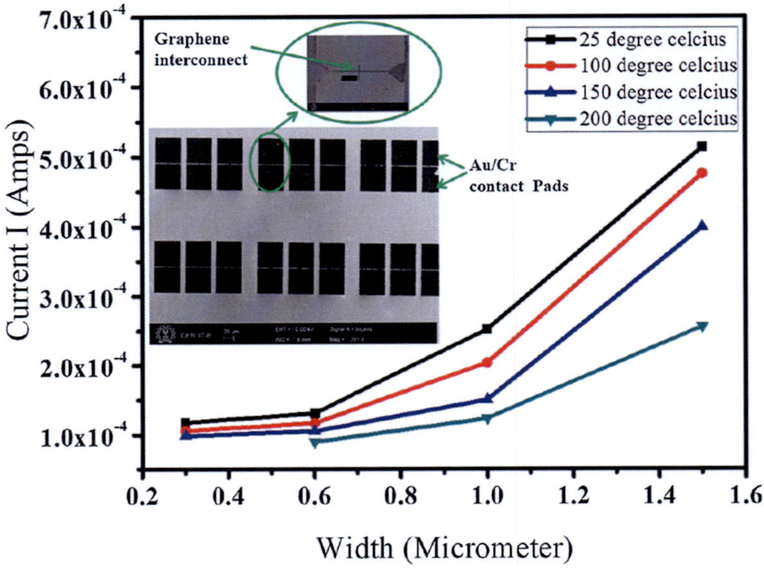


**Fig. 19.2** Current vs width plot for monolayer graphene interconnect at different temperatures respectively. The inset shows the SEM image of the patterned graphene interconnects between the metal pads (reprinted with permission from reference Debroy et al., 2020)

Sivasubramani, Vaidya, Acharyya, & Acharyya, 2020), which have investigated for the copper interconnects and result obtained that graphene interconnects were found to have a 100 times more current density $1.18 \times 10^8$ A/cm$^2$ than the copper (Cu). So, copper wires show less potential than graphene. The multilayer graphene nanoribbons (MLGNR) were investigated to replace Cu-based interconnects for future generation on-chip interconnects, which have been explained by many researchers. Further, experts in 2012 described that the MLGNR interconnect the better performance as compared to single-layer GNR interconnects (SLGNR). They reported that, when the edges are smooth, the delay of MLGNRs is less than the Cu interconnects at 9.5-nm node (Kumar, Rakheja, & Naeemi, 2012). Another group of scientists explained in 2020 that the Cu-MLG showed the energy delay product (EDP), power delay product and reduced power dissipation interconnects there an improvement in the performance up to 30. 67% in Cu-MLG as compared with Cu interconnects (Badugu & S., 2020).

### *19.3.2 Conductivity*

The thermal conductivity of graphene was found up to 5000 W/mk (Naik & Krishnaswamy, 2017). Its high thermal conductivity has been used to tackle the severe issues of heat dispersal in microelectronics, which has become a great

**Table 19.1** Thermal conductivity of graphene

| Sample | K (W/mk) | Method | Comments | References |
|---|---|---|---|---|
| Graphene | 2000–5000 | Theory: VFF,BTE, $\gamma$ (q) | Strong width dependence | Nika, Pokatilov, et al. (2009) |
| Graphene | 1000–5000 | Theory: RTA, $\gamma_{TA}$, $\gamma_{LA}$ | Strong size dependence | Nika, Ghosh, et al. (2009) |
| Graphene | 8000–10,000 | Theory: MD Tersoff | Square graphene sheet | Evans et al. (2010) |
| Graphene | 1400–2400 | Theory: BTE | Length dependence | Lindsay et al. (2010) |
| Graphene | ~4000 | Theory: Ballistic | Strong width dependence | Muñoz et al. (2010) |
| CNT | ~6600 | Theory: MD | $K_{CT} < K_{graphene}$ | Berber et al. (2000) |
| CNT | ~3000 | Theory: MD | Strong defect dependence | Che et al. (2000) |

interest to many researchers (Table 19.1). An excellent report on the significant thermal conductivity to modify graphene oxide paper was explained by Yu et al. utilizing alkaline earth metals (Yu, Xie, Li, Zhao, & Zhang, 2013). Carbon-based nanoscale materials like carbon nanotubes (CNTs) and graphene has shown excellent thermal conductivity. Because of this property, they are regarded as the future generation of thermal interface materials. The thermal conductivity was limited when CNTs are used as fillers in polymeric matrices (Deng, Zheng, Wang, & Nan, 2007; Nan, Shi, & Lin, 2003). Also, graphene has been used in polymeric composites, and it was noted that there is significantly less interface resistance between the polymeric matrix and graphene than CNTs. It was seen that with increasing the layer of graphene, thermal conductivity also decreases (Ghosh et al., 2010; Zhong, Zhang, Ai, & Zheng, 2011). Moreover, it was noticed that graphene's most thermal properties had been derived from graphite. In a review, Renteria et al. (2014) have explained the uses of graphene in thermal management and energy storage. They review the few-layer graphene (FLG) and GNRs, which gives an example of graphene in thermal phase change materials (PCMs).

### *19.3.3 Graphene Derivatives of Potential Utilities*

It was explained by the scientists in recent researches, the graphene derivative such as graphene oxide (GO) and reduced graphene oxide (rGO) was investigated broadly for the synthesis of graphene-based fibers (GBFs) (Xu, Zhang, & Qu, 2020) (Cheng, Hu, Zhao, & Qu, 2014; Kim et al., 2009; Meng et al., 2015; Xu & Gao, 2015; Ye, Zhang, Zhao, & Qu, 2018). There is much application of CNTs and carbon fibers such as in batteries (Wang et al., 2018) actuator, supercapacitors and solar cells (Fang et al., 2014; Yang et al., 2013). The previous reviews, such as Chou and coworkers (Meng et al., 2015), recently gave a complete survey for the formation, potential uses of GBFs. In another study, the microscopic mechanism was discussed to form liquid

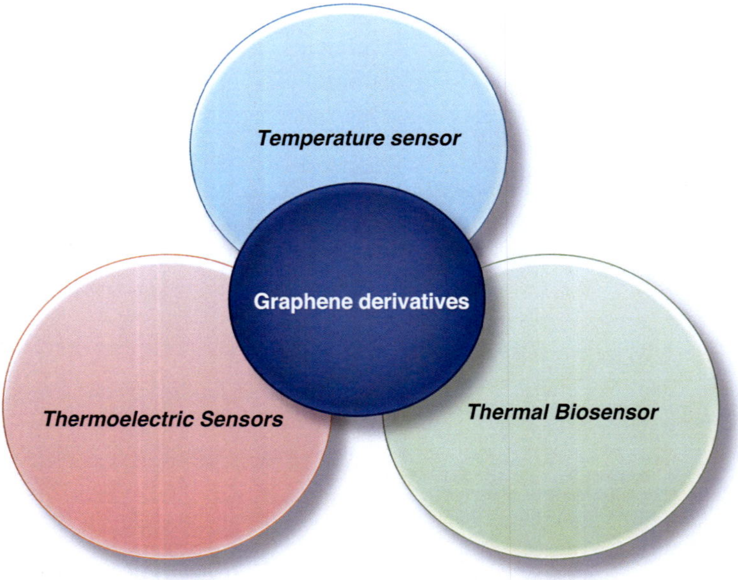


**Fig. 19.3** Presenting different potential utilities for derivatives of graphene

crystals and summed up its impact in the making of continuous fibers, explained by Gao and his colleagues (Liu, Xu, Gao, Cheng, & Gao, 2017). Also, Huang et al. (2018) examined precisely for the uses of flexible and wearable super capacitors and Cheng and coworkers (Wan, Peng, Jiang, & Cheng, 2016), explained the uses of bioinspires nanocomposites. Some of the thermal uses are as follows (Fig. 19.3):

#### 19.3.3.1 Temperature Sensor

The transparent, stretchable graphene-based multifunctional E-skin sensor, which shows 90% transmittance in the range of 400–1000 nm reported by (Ho et al., 2016). A technique was investigated to make up the graphene/poly (3,4-ethylene dioxythiophene): poly(styrene sulfonate) (PEDOT: PSS) based skin-conformable inkjet-printed temperature sensors (Vuorinen et al. 2016). Also, it was explained that graphene-based sensors on silicon nitride for highly sensitive and fast application (SiNx) membranes could be utilized (Davaji et al., 2017).

#### 19.3.3.2 Thermoelectric Sensors

The direct conversion between the electrical voltage and temperature difference is the thermoelectric effect, which is reversible through a thermocouple, and the sensors based on this effect are passive. The use of graphene is not limited to temperature sensors only, but also many essential parameters. Cai et al. reported that the impressive sensing ability was shown by graphene thermoelectric terahertz photodetector ranging between 10 | V | $W^{-1}$ to 700 | V | $W^{-1}$ at 25 °C (Cai et al., 2014). The graphene-based sensor procedure is identical to that of state-of-the-art at 25 °C terahertz detectors (Sizov & Rogalski, 2010). Lunderberg et al. suggest converting the Plasmon's natural decay product of voltage through the thermoelectric effect by an all-graphene-based mid-infrared Plasmon detector (Lundeberg et al., 2017).

#### 19.3.3.3 Thermal Biosensor

The thermal energy is released or absorbed by biochemical reactions measured by thermal biosensor and has applications in nanomedicine and clinical biomedical fields (Chen, Roy, Yang, & Prasad, 2016; Li et al., 2016; Ma, Huang, Song, Chen, & Zhang, 2016). Further, it was reported that the multifunctional theranostic system coupling diagnostic and therapeutic functions with graphene quantum dots (GQD's) and derivative of porphyrin (P) (Cao et al., 2017). Also, explained that a multifunctional platform with GQD-capped magnetic mesoporous silica nanoparticle (MMSNs) for successful and safe treatment. The nanoparticle of MMSN/GQD shows excellent potential for efficient and accurate cancer therapy.

## 19.4 Diversities of Sensor Applications

### *19.4.1 Gas Sensor and Vapors Sensor*

Gas sensors have a significant role in environmental pollution control. They are used in diverse fields of application, such as environmental monitoring, process control, and medical diagnosis. They detect a low concentration of combustible, flammable, and toxic gases in an ambient atmosphere. Various gas sensors have been used, but 2D layered nanomaterials achieved great success, especially the graphene-based sensors. Graphene-based sensors have special features, such as atom-thick 2D conjugated structure, high conductivity, large specific surface area, high temperature, electric tolerance properties that make them a suitable candidate in gas-sensing applications. Graphene is known to have a unique chemical structure with a heterogeneous electronic system ($sp^2$ & $sp^3$). This makes graphene offers exciting, mechanical, optical, and electrical properties, which enable it to be emerging as a new generation of sensors (Yang, Gan, Li, & Zhai, 2016).

Semiconductivity is an essential property for nanomaterials to be used in gas-sensing applications. A material must possess a band gap so that it can be used as sensors. The sensing ability of graphene-based sensors depends on the physical and chemical adsorption of a gas. Adsorption will be achieved greater in graphene material with the defective site. Different types of gas molecules are adsorbed on graphene sensing layers that give similar conductivity changes. Various graphene-based gas—sensors have been reported in the literature, including gas sensors based on defective and functionalized graphene materials, graphene/polymer composites, and two-layered graphene materials. A gas sensor possesses several properties for high response and sensibilities to target gases such as ultrahigh surface to volume ratio, intense surface activities, and work effectively at low temperatures.

Further, it was noticed that problems appeared in the selective detection of gas molecules. These are contamination by pollutants. The conductance of graphene-based materials increases or decreases due to the adsorption of an oxidizing or reducing gas. The proper fabrication of methods can solve this problem.

Chemiresistor is the most generally utilized configurations of gas sensors (Fig. 19.4b). For this situation, the vaporous analytes were identified by estimating the resistance changes of detecting layers prompted by adsorbing the gas atoms. The upsides of this sort of sensors are their direct manufacture and direct measurements. This sensor can be applied to distinguish $NO_2$, $NH_3$, dinitrotoluene (DNT), and the sensors' performances firmly rely upon the temperature. Field-effect transistors (FETs) have additionally been applied for detecting gases. Field-effect transistors (FETs) shown in Fig. 19.4a have likewise been applied for detecting gases.

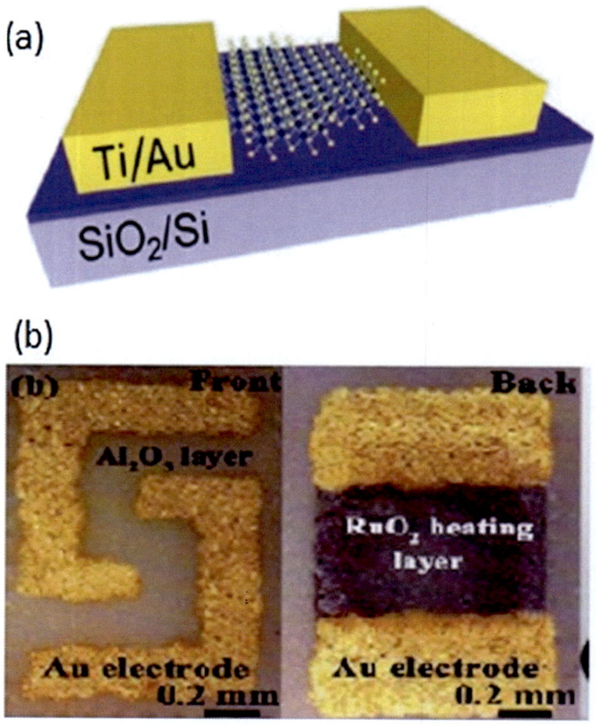


**Fig. 19.4** Gas-sensing devices (**a**) Field-effect transistor (reproduced from reference Zhang, Feng, Fei, Liu, & Zhang, 2014); (**b**) chemiresistor (reproduced from reference Liu et al., 2014)

Bulk graphene materials have likewise been applied for creating gas-sensing devices. For instance, graphene foam with consistent 3D networks was set up by CVD procedure. Also, nickel foam was utilized as the format. These foam have huge porosities, and gas particles can promptly diffuse to inside graphene dividers' surface to add to the detecting signals. To find out the principal limit of detection (LOD) of graphene-based gas sensors, they further upgraded the detecting device by utilizing the hall geometry to give the most grounded reaction to the adjustment in control transporter thickness close to the Dirac point (Yuan & Shi, 2013). Some of the sensors have been prepared by Yoon et al. an elite graphene sheet.

Based on FET carbon dioxide ($CO_2$) sensors (Yoon et al., 2011), Li et al. built up a graphene-based FET sensor with a palladium-adjusted (rGO) adsorbed on a Si/$SiO_2$ wafer for NO identification (Li et al. 2011). The conductance of a device expanded with the adsorption of NO, showing that NO go about as accepter, which concurs with Chen et al.'s. study (Chen, Gao, Zhao, Cai, & Fu, 2012). The response time was taken 240 s by Dan et al. manufactured a graphene sensor FET vapor sensor similar to Yoon et al.'s. concentrate aside from that they utilized as a spotless cycle, which was accomplished by putting the device in a reducing climate ($H_2$/Air) at 400 °C for 1 h to eliminate 1 nm thick defilement layer, which was a build-up of opposing material on the delivered during the EBL process (Dan, Lu, Kybert, Luo, & Johnson, 2009). Some of the graphene materials for sensing gas are shown in Table 19.2.

### 19.4.2 Modulation of Biosensor Activities

The discovery of naturally dynamic particles is of fundamental significance for biomedical, environmental, and security purposes. Such findings can be done by biosensor or by bioanalytical conventions. Schematic diagram of biosensor is shown in Fig. 19.5. A chemical sensor is a device that quantitatively or semiquantitatively changes over chemical information about a chemical species' presence to a valuable signal. Sensors comprise of two components: a receptor and a transducer. A receptor can be any organic or inorganic material with (ideally) a particular connection with one analyte or gathering of analytes. On account of biosensors, the acknowledgment component is a biomolecule. The second key component of the detecting stage is the transducer, which changes over chemical information into a quantifiable signal (Hulanicki, Glab, & Ingman, 1991). The fascinating and energizing properties of the single-layer graphene sheet, for example, high mechanical quality, increased flexibility, and thermal conductivity exhibition of the room temperature quantum hall impact too high room temperature electron velocity, tunable optical properties, and a tunable bandgap have energized mainstream researchers particularly in the region of material, physics, and chemistry. Different, yet correspondingly graphene shows staggering properties with multilayer design (Bunch et al., 2007).

**Table 19.2** Graphene materials for gas-sensing applications (*G* graphene, *DNT* 2, 4 dinitrotoluene, *DMMP* dimethyl methylphosphonate (reprinted with permission from reference Yang et al., 2016)

| No | Graphene materials | Sensor type | Detected gas | Sensitivity | LOD |
|---|---|---|---|---|---|
| 1 | Mechanically exfoliated G | Hall geometry | $NO_2$ | $\Delta R = 2.5\ \Omega$ (1 electron) | 1 molecule |
| 2 | CVD-grown G | FET | $NO_2$ | $\Delta R/R = 12.49\%$ (5 ppm) | 6.87 ppb |
| 3 | CVD-grown G | FET | $NO_2$ | $\Delta I/I_0 = 1.67\ ppm^{-1}$ | 300 ppt |
| 4 | Epitaxially grown G | Chemiresistor | $NO_2$ | $\Delta G/G = 9\%$ (18 ppm, Si-face) | |
| | | | | $\Delta G/G = 4\%$ (18 ppm, C-face) | |
| 5 | Hydrazine-reduced GO | Chemiresistor | DNT | $\Delta R/R = -0.028\%$ (52 ppb) | |
| 6 | Vc-reduced GO | Chemiresistor | $NO_2$ | $\Delta R/R = 6\%$ (10 ppm) | 400 ppt |
| 7 | PPD-reduced GO | Chemiresistor | DMMP | $\Delta R/R = 8\%$ (20 ppm) | |
| 8 | $NaBH_4$-reduced GO | Chemiresistor | $NH_3$ | $\Delta R/R = 5.5\%$ (200 ppm) | |
| 9 | Sulfonated rGO | Chemiresistor | $NO_2$ | $\Delta G/G = 0.443\ ppm^{-1}$ | 3.6 ppm |
| 10 | Ag-S-RGO | Chemiresistor | $NO_2$ | $\Delta R/R = -45\%$ (50 ppm) | |
| 11 | Pt-G | Chemiresistor | $H_2$ | $\Delta R/R = 16\%$ (4%) | |
| 12 | Pd-RGO | Chemiresistor | NO | $\Delta G/G = 4\%$ (2 ppb) | |
| 13 | AgNWs coated rGO | Chemiresistor | $NH_3$ | $\Delta R/R = 3\%$ (15 ppm) | |
| 14 | $SnO_2$-RGO | FET | $NO_2$ | $I_g/I_a = 1.11$ (1 ppm) | |
| 15 | $Cu_2O$-rGO | Chemiresistor | $NO_2$ | $\Delta I/I_0 = 67.8\%$ (2 ppm) | 64 ppb |
| 16 | CuO-ZnO/G | Chemiresistor | Acetone | $R_a/R_g = 9.4$ (10 ppm) | |
| 17 | ZnO QDs/G | Chemiresistor | Formaldehyde | $\Delta G/G = 1.1$ (100 ppm) | |
| 18 | Pt-$SnO_2$/G | Chemiresistor | $H_2$ | $\Delta R/R = 2$ (1%) | |
| 19 | CNT/rGO | FET | $NO_2$ | $\Delta R/R = 20\%$ (20 ppm) | |
| 20 | PANI/G | Chemiresistor | $NH_3$ | $\Delta R/R = 32.5\%$ (10 ppm) | |
| 21 | PSS-PANI/G | Chemiresistor | $H_2S$ | $\Delta R/R = 60\%$ (20 ppm) | 1 ppm |
| 22 | PEDOT/G | FET | $NO_2$ | $\Delta R/R = 20\%$ (2 ppm) | 500 ppb |
| 23 | P NFs/G | Chemiresistor | $NO_2$ | $\Delta R/R = 1.03\ ppm^{-1}$ | 150 ppb |

#### 19.4.2.1 Protein Biosensors

The human body's response to treatment can be studied by measuring the number of disease biomolecules in the body. These are the protein found in body fluids, and their abnormal level in the human body is used for determining the bodily response. Various life-threatening diseases can be detected in the early stages by making use of this technique. On detecting a single or an assay of protein biomolecules in the body fluids, e.g., blood or urine on a biosensor device. The disease can be timely controlled has opened new research avenues (Nair et al., 2008a).

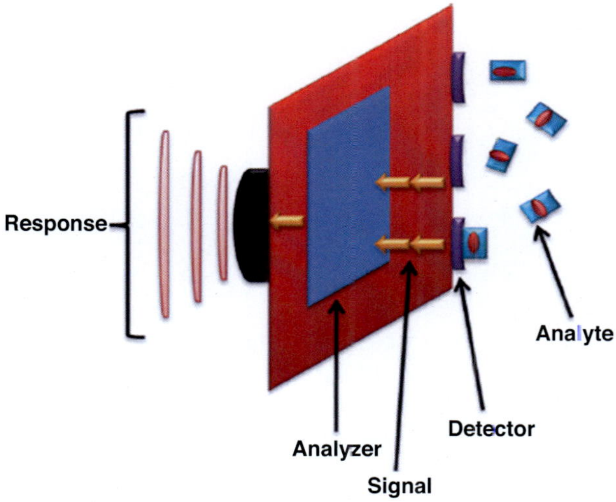


**Fig. 19.5** Schematic representation of a biosensor (reprinted with permission from reference Kuila et al., 2011)

### 19.4.2.2 Electrochemical Biosensor

An electrochemical DNA sensor based on graphene was reported by Zhon. The current signal of the four free bases of DNA (G, A, T, and C) CR-GO/GC electrode when separated efficiently leads to the conclusion that CR-GO/GC can detect the bases all at the same time. Still, graphite and glassy carbon cannot be used for it. The electrochemical DNA sensor provides high sensitivity, high selectivity at a low cost for identifying selected DNA sequences or mutated genes. The mutated gene leads to various diseases among human beings. The sensors come up with a simple, accurate, and affordable way for diagnosis (Shi, Fang, Zhang, Zhang, & Li, 2009). Moreover, graphene oxide-based fluorescent sensors were initially introduced to disclose DNA, target DNA, and protein was made possible by it. The detection is made due to dye's specific binding nature labeled ssDNA and target DNA (Geim & Novoselov, 2007).

### 19.4.2.3 Glucose Biosensor

The first Graphene–Shan reported based glucose biosensor. This Biosensor had graphene polyethyleneimine functionalizes ionic liquid nanocomposite. These nanocomposites acted as an electrode and could exhibit a complete linear glucose response ranging from 2 Mm to 14 mM at $R = 0.994$. It showed good reproducibility and high stability. The credit to make us familiar with the glucose biosensor based on chemically reduced graphene oxide [CR-GP] goes to Zhou. Graphene-Based biosensors showed enhanced electronic signals, which helped sense glucose. These signals include wide linear range, high sensitivity, and low detection limit 2.00 nm. Glucose biosensor has a wide linear range from detection than other carbon material–based detection like carbon tubes and carbon nanofibers.

#### 19.4.2.4 DNA Biosensor

Electrochemical Immunobiosensor

A novel electrochemical immunosensor for the touchy discovery of malignancy biomarkers alpha-fetoprotein in (AFP) is portrayed that utilizes a graphene sheet sensor scaffold, and carbon functionalized nanospheres. Based on the double intensification strategy of a graphene sheet and multienzyme labeling, the developed immunosensor indicated a seven overlay expanded in the identification signal contrasted with the immunosensor without graphene alteration and carbon nanospheres labeling. The following approach has been created for signal enhancement nanoparticle-based electrochemical biosensor, metal and semiconductor nanoparticles are legitimately used as an electroactive mark to intensify the electrochemical reaction of DNA and protein. The nanoparticle is utilized as transporters to stack many electroactive species, such as ferrocene, to enhance biomolecules' recognition. Recently, Lui and associates revealed (HRP) functionalized silica-based nanoparticle as liable for identifying alpha-fetoprotein (Zeng et al., 2018). The appealing and energizing properties of the single-layer graphene sheet, for example, high mechanical strength, high elasticity, and thermal conductivity demonstration of the room temperature quantum hall effect very high room electron mobility, tunable optical properties, and a tunable bandgap have excited the scientific community especially in the area of material, physics, and chemistry, Different, but similarly, graphene shows stunning properties with multilayer architecture.

### *19.4.3 Biomedical Devices*

Graphene Quantum Dots (GQD's) are carbon-based anisotropic nanomaterials with (0-D) comprising a texture like structure homologous to graphene. The morphological highlights of GQD's emulate the two CDs, just as graphene (Mansuriya & Altintas, 2020). In the course of the most recent decade, various nanomaterials have been consistently contemplated and utilized as sign enhancing species, for example, nanoparticles (NPs), graphene, nanowires, carbon nanotubes (CNTs), attractive dabs, and quantum dots (QDs). Among these nanomaterials, QDs, for example, graphene quantum dots (GQD's) and carbon dots (CDs), are turning out to be very notable for their various properties, signal gainer attributes, excellent biocompatibility, tunable size, electrocatalytic execution just as their ability for the simultaneous and different identification of biomolecules as portrayed in Fig. 19.6.

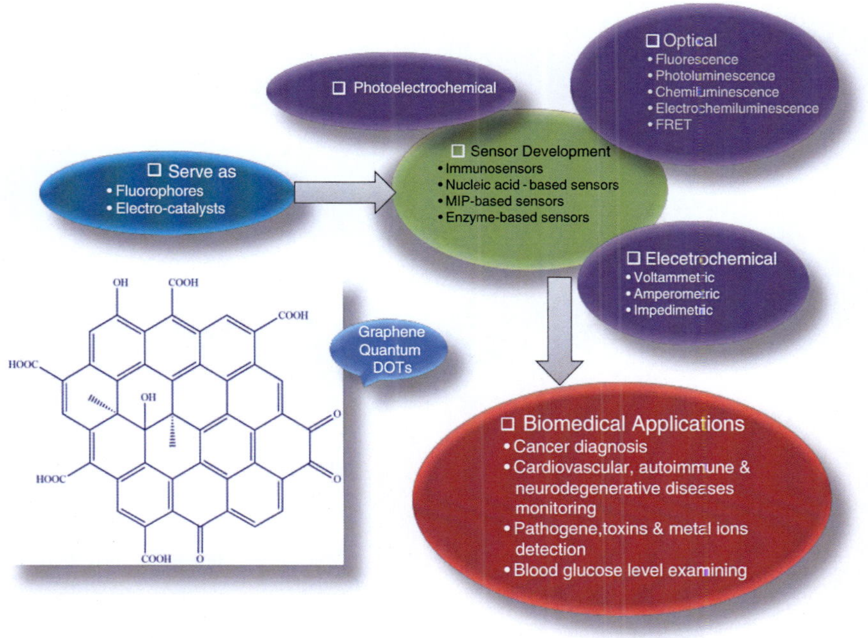


**Fig. 19.6** Biomedical applications in sensor development and the chemical structure of graphene quantum dots (GQDs) reprinted with permission from reference (Mansuriya & Altintas, 2020)

Additionally, their strength, dormancy, non-poisonous, long chain compound steadiness, and water dissolvability, just as their photograph dependability against photograph blanching and flickering, are a portion of the ordinary qualities that are considered for their applications in biomedicine. GQD's can likewise be united with proteins, antibodies, and little nucleic acids because of their dimensional likeness to such atoms. They can viably improve the outside of biosensors for retaining an observable number of receptors (Ma, Li, & Zhang, 2018). GQD's can go about as enzymes or electro-stimulants for catalyzing hydrogen peroxide ($H_2O_2$) to decide target analytes through a mark-free methodology (Savas & Altintas, 2019). GQD's fill in as alluring fluorophores, i.e., as fluorescent names, quenchers, and energy just as charge contributors (Zheng & Wu, 2017).

### 19.4.3.1 Fluorescence-Based GQDs

A biosensor was announced in 2016 by Laurenti et al. that the miRNA sequences was determined by combining single-stranded DNA (ssDNA) and GQDs with the up-conversion nanoparticles (UCNPs) and silica ($SiO_2$) (Laurenti et al., 2016). Further, the perceptive fluorescence reaction was noted for this biosensor towards miRNA in human serum within the calibration series of $1 \times 10^{-18}$ to $10^{-12}$ M and

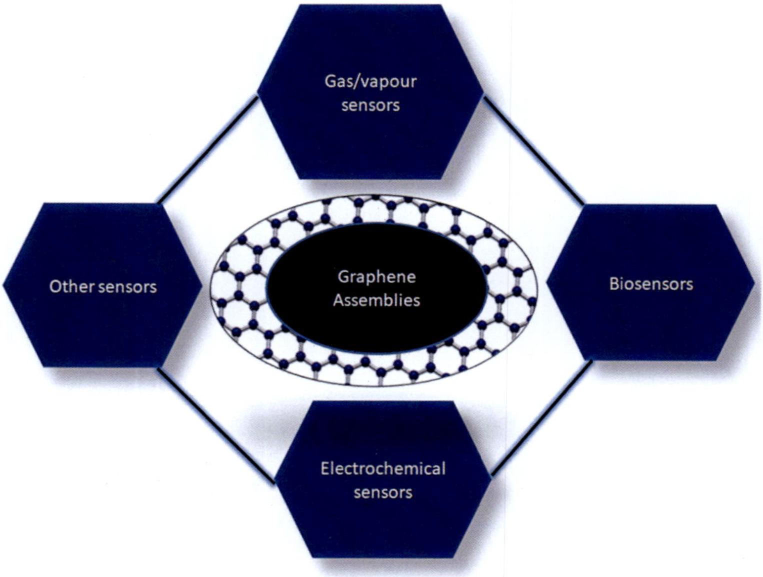


**Fig. 19.7** Schematic representation for graphene assemblies for various applications

might accomplish $4.3 \times 10^{-19}$ M as the detection limit (LOD). Some other applications based on GQDs-sensors are electroluminescence (EL), chemiluminescence (CL), photoluminescence (PL). Schematic representation has been given for graphene assemblies for some applications shown in Fig. 19.7.

## 19.5 Conclusions and Future Prospects

In summary, here, we have discussed the basics of graphene. Graphene is a single–thick layer of $sp^2$ hybridized carbon atoms arranged in a two-dimensional (2D) honeycomb crystal lattice consisting of carbon allotropes containing 0D fullerenes, 1D carbon annotates, and 3D graphite. We also presented the structural and thermal properties of graphene, which includes thermal conductivity, interconnects, and recent uses. It was found that graphene explores good thermal conductivity which is nearly about 5000 W/mk. Due to this property, it has been used to bear with various issues related to heat dispersal in microelectronics. Graphene was discovered to be of huge preferences in nanoscale gadgets and interconnects because of its high thermal conductivity. Besides these, there are various thermal uses of graphene derivative like graphene oxide (GO) and reduced graphene oxide (rGO) such as in temperature and thermoelectric sensor and thermal biosensor. Further, there are several applications of graphene-based materials such as based quantum dots (GQDTs) and carbon dots (CDs) display application in biomedical devices. Also, it was reported that graphene have shown a significant role for environmental pollution controlling in gas and vapor sensors. Besides these graphene has application in biosensors which is divided into different types such as protein, electrochemical, glucose and DNA biosensors.

It is normal that the amazing electronic property of graphene will get another era of nanoelectronics. The most direct application for graphene is presumably its utilization in composite materials. The composites direct power and can withstand a lot higher temperatures than the polymers. The graphene-based polymer composites could be ideal to make lightweight gas tanks and plastic compartments. These composites can be conceivably used to make lighter, more eco-friendly airplane and vehicle parts, more grounded wind turbines, clinical embeds, and sports types of gear. High conductivity and high optical straightforwardness make graphene reasonable for manufacturing straightforward directing covering in LCDs and sun oriented cells.

Further, scaling down and creation of conservative biosensors for indicative objects is an emanant need in sensor innovation since it requires advancement of solid, reproducible, and financially savvy sensors with high precision, affectability, and particularity. Bringing down the expense of a portion of these sensors is important to expand convenience in far off zones for crisis employments. Besides, scaling down of the sensors can permit fast discovery of infection and bacterial microorganisms, just as use in self-checking biological implants to identify genuine ailments. The previously mentioned applications in the everyday routine sciences will serve to secure experiences and improve individuals' health. Notwithstanding, extensive work should even now been done to ensure, and confirm the biocompatibility and non-poisonousness of graphene-based nanomaterials with the end goal that their long-lasting use doesn't represent any health hazard.

**Acknowledgments** The authors wish to thank the Chairman, Department of Chemistry, Aligarh Muslim University (AMU), Aligarh, India, for providing research facilities. The authors are thankful for the support of DST-FIST & UGC DRS-SAP (II) programs for providing necessary research facilities.

**Conflict of Interest** All the authors declared that there is no conflict of interest for this book chapter publishrng.

## References

Ahn, J. H., & Je, J. H. (2012). Stretchable electronics: Materials, architectures and integrations. *Journal of Physics D: Applied Physics, 45*(10), 103001.

Avouris, P., Chen, Z., & Perebeinos, V. (2009). Carbon-based electronics. *Nanoscience and Technology, 376*, 174–184.

Badugu, D. M., & S., S (2020). Crosstalk reduction in copper on-chip interconnects with graphene barrier for ternary logic applications. *International Journal of Circuit Theory and Applications, 48*(12), 2097–2110.

Baughman, R. H., Zakhidov, A. A., & De Heer, W. A. (2002). Carbon nanotubes—The route toward applications. *Science, 297*(5582), 787–792.

Berber, S., Kwon, Y.-K., & Tománek, D. (2000). Unusually high thermal conductivity of carbon nanotubes. *Physical Review Letters, 84*(20), 4613–4616.

Bunch, J. S., van der Zande, A. M., Verbridge, S. S., Frank, I. W., Tanenbaum, D. M., Parpia, J. M., Craighead, H. G., & McEuen, P. L. (2007). Electromechanical resonators from graphene sheets. *Science, 315*(5811), 490–493.

Cai, X., Sushkov, A. B., Suess, R. J., Jadidi, M. M., Jenkins, G. S., Nyakiti, L. O., Myers-Ward, R. L., Li, S., Yan, J., Gaskill, D. K., Murphy, T. E., Drew, H. D., & Fuhrer, M. S. (2014). Sensitive room-temperature terahertz detection via the photothermoelectric effect in graphene. *Nature Nanotechnology, 9*(10), 814–819.

Cao, Y., Dong, H., Yang, Z., Zhong, X., Chen, Y., Dai, W., & Zhang, X. (2017). Aptamer-conjugated graphene quantum dots/porphyrin derivative theranostic agent for intracellular cancer-related MicroRNA detection and fluorescence-guided photothermal/photodynamic synergetic therapy. *ACS Applied Materials & Interfaces, 9*(1), 159–166.

Castro Neto, A. H., Guinea, F., Peres, N. M. R., Novoselov, K. S., & Geim, A. K. (2009). The electronic properties of graphene. *Reviews of Modern Physics, 81*(1), 109–162.

Che, J., Çagin, T., & Goddard, W. A. (2000). Thermal conductivity of carbon nanotubes. *Nanotechnology, 11*(2), 65–69.

Chen, G., Roy, I., Yang, C., & Prasad, P. N. (2016). Nanochemistry and nanomedicine for nanoparticle-based diagnostics and therapy. *Chemical Reviews, 116*(5), 2826–2885.

Chen, Y., Gao, B., Zhao, J.-X., Cai, Q.-H., & Fu, H.-G. (2012). Si-doped graphene: An ideal sensor for NO- or NO2-detection and metal-free catalyst for N2O-reduction. *Journal of Molecular Modeling, 18*(5), 2043–2054.

Cheng, H., Hu, C., Zhao, Y., & Qu, L. (2014). Graphene fiber: A new material platform for unique applications. *NPG Asia Materials, 6*(7), 113.

Compton, O. C., & Nguyen, S. T. (2010). Graphene oxide, highly reduced graphene oxide, and graphene: Versatile building blocks for carbon-based materials. *Small, 6*(6), 711–723.

Dan, Y., Lu, Y., Kybert, N. J., Luo, Z., & Johnson, A. T. C. (2009). Intrinsic response of graphene vapor sensors. *Nano Letters, 9*(4), 1472–1475.

Davaji, B., Cho, H. D., Malakoutian, M., Lee, J. K., Panin, G., Kang, T. W., & Lee, C. H. (2017). A patterned single layer graphene resistance temperature sensor. *Scientific Reports, 7*, 8811.

Debroy, S., Sivasubramani, S., Vaidya, G., Acharyya, S. G., & Acharyya, A. (2020). Temperature and size effect on the electrical properties of monolayer graphene based interconnects for next generation MQCA based nanoelectronics. *Scientific Reports,*10, 6240.

Deng, F., Zheng, Q.-S., Wang, L.-F., & Nan, C.-W. (2007). Effects of anisotropy, aspect ratio, and nonstraightness of carbon nanotubes on thermal conductivity of carbon nanotube composites. *Applied Physics Letters, 90*(2), 021914.

Echtermeyer, T. J., Lemme, M. C., Baus, M., Szafranek, B. N., Geim, A. K., & Kurz, H. (2007). A graphene-based electrochemical switch. *ArXiv*, 1–12.

Evans, W. J., Hu, L., & Keblinski, P. (2010). Thermal conductivity of graphene ribbons from equilibrium molecular dynamics: Effect of ribbon width, edge roughness, and hydrogen termination. *Applied Physics Letters, 96*(20).

Fang, X., Yang, Z., Qiu, L., Sun, H., Pan, S., Deng, J., Luo, Y., & Peng, H. (2014). Core-sheath carbon nanostructured fibers for efficient wire-shaped dye-sensitized solar cells. *Advanced Materials*. https://doi.org/10.1002/adma.201305241

Fiori, G., & Iannaccone, G. (2007). Simulation of graphene nanoribbon field-effect transistors. *IEEE Electron Device Letters, 28*(8), 760–762.

Geim, A. K., & Novoselov, K. S. (2007). The rise of graphene. *Nature Materials, 6*(3), 183–191.

Ghosh, S., Bao, W., Nika, D. L., Subrina, S., Pokatilov, E. P., Lau, C. N., & Balandin, A. A. (2010). Dimensional crossover of thermal transport in few-layer graphene. *Nature Materials, 9*(7), 555–558.

Graphite and Precursors - Google Books. (n.d.). Retrieved October 30, 2020, from https://books.google.co.in/books?hl=en&lr=&id=7p2pgNOWPbEC&oi=fnd&pg=PP9&dq=%5B18%5D.+P.+Delhaes,+Graphite+and+Precursors,+CRC+Press,+2000&ots=1P5adIsYQv&sig=uEUxuvnQE6HMJUPJpL_UkBgGRc&redir_esc=y#v=onepage&q=%5B18%5D. P. Delhaes%2C Graphite and Precurso.

Han, Y., Xu, Z., & Gao, C. (2013). Ultrathin graphene nanofiltration membrane for water purification. *Advanced Functional Materials*. https://doi.org/10.1002/adfm.201202601

Hiremath, N., Mays, J., & Bhat, G. (2017). Recent developments in carbon fibers and carbon nanotube-based fibers: A review. *Polymer Reviews, 57*(2), 339–368.

Hirsch, A. (2010). The era of carbon allotropes. *Nature Materials, 9*(11), 868–871.

Ho, D. H., Sun, Q., Kim, S. Y., Han, J. T., Kim, D. H., & Cho, J. H. (2016). Stretchable and multimodal all graphene electronic skin. *Advanced Materials, 28*(13), 2601–2608.

Huang, L., Santiago, D., Loyselle, P., & Dai, L. (2018). Graphene-based nanomaterials for flexible and wearable supercapacitors. *Small, 14*(43), 1800879.

Hulanicki, A., Glab, S., & Ingman, F. (1991). Chemical sensors definitions and classification. *Pure and Applied Chemistry, 9*(63) 1247–1250.

Ishikawa, K. L. (2010). Nonlinear optical response of graphene in time domain. *Physical Review B - Condensed Matter and Materials Physics*.

Jin, X. (2015). *Nonlinear optical response in graphene*. University of Iowa.

Kim, D. H., & Rogers, J. A. (2008). Stretchable electronics: Materials strategies and devices. *Advanced Materials, 20*(24), 4887–4892.

Kim, K. S., Zhao, Y., Jang, H., Lee, S. Y., Kim, J. M., Kim, K. S., Ahn, J.-H., Kim, P., Choi, J.-Y., & Hong, B. H. (2009). Large-scale pattern growth of graphene films for stretchable transparent electrodes. *Nature, 457*(7230), 706–710.

Kou, L., & Gao, C. (2011). Making silicananoparticle-covered graphene oxide nanohybrids as general building blocks for large-area superhydrophilic coatings. *Nanoscale, 3*(2), 519–528.

Kroto, H. W., Heath, J. R., O'Brien, S. C., Curl, R. F., & Smalley, R. E. (1985). C60: Buckminsterfullerene. *Nature, 318*(6042), 162–163.

Kuila, T., Bose, S., Khanra, P., Mishra, A. K., Kim, N. H., & Lee, J. H. (2011). Recent advances in graphene-based biosensors. *Biosensors and Bioelectronics, 26*(12), 4637–4648.

Kumar, V., Rakheja, S., & Naeemi, A. (2012). Performance and energy-per-bit modeling of multilayer graphene nanoribbon conductors. *IEEE Transactions on Electron Devices, 59*(10), 2753–2761.

Larson, C., Peele, B., Li, S., Robinson, S., Totaro, M., Beccai, L., Mazzolai, B., & Shepherd, R. (2016). Highly stretchable electroluminescent skin for optical signaling and tactile sensing. *Science, 351*(6277), 1071–1074.

Laurenti, M., Paez-Perez, M., Algarra, M., Alonso-Cristobal, P., Lopez-Cabarcos, E., Mendez-Gonzalez, D., & Rubio-Retama, J. (2016). Enhancement of the upconversion emission by visible-to-near-infrared fluorescent graphene quantum dots for miRNA detection. *ACS Applied Materials & Interfaces, 8*(20), 12644–12651.

Li, L., Chen, C., Liu, H., Fu, C., Tan, L., Wang, S., Fu, S., Liu, X., Meng, X., & Liu, H. (2016). Multifunctional carbon–silica nanocapsules with gold core for synergistic photothermal and chemo-cancer therapy under the guidance of bimodal imaging. *Advanced Functional Materials, 26*(24), 4252–4261.

Li, W., Geng, X., Guo, Y., Rong, J., Gong, Y., Wu, L., Zhang, X., Li, P., Xu, J., Cheng, G., Sun, M., & Liu, L. (2011). Reduced graphene oxide electrically contacted graphene sensor for highly sensitive nitric oxide detection. *ACS Nano, 5*(9), 6955–6961.

Liang, G., Neophytou, N., Lundstrom, M. S., & Nikonov, D. E. (2007). Ballistic graphene nanoribbon metal-oxide-semiconductor field-effect transistors: A full real-space quantum transport simulation. *Journal of Applied Physics, 102*(5), 054307.

Liang, J., Li, L., Niu, X., Yu, Z., & Pei, Q. (2013). Elastomeric polymer light-emitting devices and displays. *Nature Photonics, 7*(10), 817–824.

Lindsay, L., Broido, D. A., & Mingo, N. (2010). Diameter dependence of carbon nanotube thermal conductivity and extension to the graphene limit. *Physical Review B - Condensed Matter and Materials Physics, 82*, 161402(R).

Liu, B., Chen, L., Liu, G., Abbas, A. N., Fathi, M., & Zhou, C. (2014). High-performance chemical sensing using Schottky-contacted chemical vapor deposition grown monolayer MoS2 transistors. *ACS Nano 8*(5), 5304–5314.

Liu, F., Lee, C. W., & Im, J. S. (2013). Graphene-based carbon materials for electrochemical energy storage. *Journal of Nanomaterials, 2013*. 8(5), 5304–5314.

Liu, Y., Xu, Z., Gao, W., Cheng, Z., & Gao, C. (2017). Graphene and other 2D colloids: Liquid crystals and macroscopic fibers. *Advanced Materials, 29*(14), 1606794.

Lu, W., Zu, M., Byun, J.-H., Kim, B.-S., & Chou, T.-W. (2012). State of the art of carbon nanotube fibers: Opportunities and challenges. *Advanced Materials, 24*(14), 1805–1833.

Lundeberg, M. B., Gao, Y., Woessner, A., Tan, C., Alonso-González, P., Watanabe, K., Taniguchi, T., Hone, J., Hillenbrand, R., & Koppens, F. H. L. (2017). Thermoelectric detection and imaging of propagating graphene plasmons. *Nature Materials, 16*(2), 204–207.

Ma, F., Li, C. C., & Zhang, C. Y. (2018). Development of quantum dot-based biosensors: Principles and applications. *Journal of Materials Chemistry B, 6*, 6173–6190.

Ma, Y., Huang, J., Song, S., Chen, H., & Zhang, Z. (2016). Cancer-targeted nanotheranostics: Recent advances and perspectives. *Small, 12*(36), 4936–4954.

Maile, B. E., Henschel, W., Kurz, H., Rienks, B., Polman, R., & Kaars, P. (2000). Sub-10 nm linewidth and overlay performance achieved with a fine-tuned EBPG-5000 TFE electron beam lithography system. *Japanese Journal of Applied Physics, Part 1: Regular Papers and Short Notes and Review Papers, 39*(12 B), 6836–6842.

Mansuriya, B. D., & Altintas, Z. (2020). Applications of graphene quantum dots in biomedical sensors. *Sensors, 20*(4), 1072.

Mao, X., Hatton, T., & Rutledge, G. (2013). A review of electrospun carbon fibers as electrode materials for energy storage. *Current Organic Chemistry*.

Meng, F., Lu, W., Li, Q., Byun, J. H., Oh, Y., & Chou, T. W. (2015). Graphene-based fibers: A review. *Advanced Materials, 27*(35), 5113–5131.

Nazare, M. H., & Neves, A. J. (Eds.). (2018). Properties, growth and applications of diamond, institution of engineering and technology, 2000. *Industrial and Engineering Chemistry Research, 57*(24), 8220–8229.

Muñoz, E., Lu, J., & Yakobson, B. I. (2010). Ballistic thermal conductance of graphene ribbons. *Nano Letters, 10*(5), 1652–1656.

Naik, G., & Krishnaswamy, S. (2017). Photoreduction and thermal properties of graphene-based flexible films. *Graphene, 6*(2), 27–40.

Nair, R. R., Blake, P., Grigorenko, A. N., Novoselov, K. S., Booth, T. J., Stauber, T., Peres, N. M. R., & Geim, A. K. (2008a). Fine structure constant defines visual transparency of graphene. *Science, 320*(5881), 1308.

Nair, R. R., Blake, P., Grigorenko, A. N., Novoselov, K. S., Booth, T. J., Stauber, T., Peres, N. M. R., & Geim, A. K. (2008b). Fine structure constant defines visual transparency of graphene. *Science, 320*(5881), 1308–1308.

Nan, C. W., Shi, Z., & Lin, Y. (2003). A simple model for thermal conductivity of carbon nanotube-based composites. *Chemical Physics Letters, 375*, 666–669.

Ni, Z. H., Wang, H. M., Kasim, J., Fan, H. M., Yu, T., Wu, Y. H., Feng, Y. P., & Shen, Z. X. (2007). Graphene thickness determination using reflection and contrast spectroscopy. *Nano Letters, 7*(9), 2758–2763.

Nika, D. L., Ghosh, S., Pokatilov, E. P., & Balandin, A. A. (2009). Lattice thermal conductivity of graphene flakes: Comparison with bulk graphite. *Applied Physics Letters, 94*(20), 203103.

Nika, D. L., Pokatilov, E. P., Askerov, A. S., & Balandin, A. A. (2009). Phonon thermal conduction in graphene: Role of Umklapp and edge roughness scattering. *Physical Review B, 79*(15), 155413.

Novoselov, K. S. (2004). Electric field effect in atomically thin carbon films. *Science, 306*(5696), 666–669.

Özyilmaz, B., Jarillo-Herrero, P., Efetov, D., Abanin, D. A., Levitov, L. S., & Kim, P. (2007). Electronic transport and quantum hall effect in bipolar graphene p-n-p junctions. *Physical Review Letters, 99*(16), 166804.

Renteria, J., Nika, D., & Balandin, A. (2014). Graphene thermal properties: Applications in thermal management and energy storage. *Applied Sciences, 4*(4), 525–547.

Savas, S., & Altintas, Z. (2019). Graphene quantum dots as nanozymes for electrochemical sensing of yersinia enterocolitica in milk and human serum. *Materials, 12*(13), 2189.

Sekitani, T., Nakajima, H., Maeda, H., Fukushima, T., Aida, T., Hata, K., & Someya, T. (2009). Stretchable active-matrix organic light-emitting diode display using printable elastic conductors. *Nature Materials, 8*(6), 494–499.

Servant, A., Bianco, A., Prato, M., & Kostarelos, K. (2014). Graphene for multi-functional synthetic biology: The last "zeitgeist" in nanomedicine. *Bioorganic and Medicinal Chemistry Letters, 24*(7), 1638–1649.

Shi, Y., Fang, W., Zhang, K., Zhang, W., & Li, L. J. (2009). Photoelectrical response in single-layer graphene transistors. *Small, 5*(17), 2005–2011.

Sizov, F., & Rogalski, A. (2010). THz detectors. *Progress in Quantum Electronics, 34*(5), 278–347.

Sun, H., Xu, Z., & Gao, C. (2013). Multifunctional, ultra-flyweight, synergistically assembled carbon aerogels. *Advanced Materials, 25*(18), 2554–2560.

Vuorinen, T., Niittynen, J., Kankkunen, T., Kraft, T. M., & Mäntysalo, M. (2016). Inkjet-printed graphene/PEDOT:PSS temperature sensors on a skin-conformable polyurethane substrate. Scientific Reports, 6(October), 1–8.

Wan, S., Peng, J., Jiang, L., & Cheng, Q. (2016). Bioinspired graphene-based nanocomposites and their application in flexible energy devices. *Advanced Materials, 28*(36), 7862–7898.

Wang, J., Cheng, Q., & Tang, Z. (2012). Layered nanocomposites inspired by the structure and mechanical properties of nacre. *Chemical Society Reviews, 41*(3), 1111–1129.

Wang, J., Yan, C., Chee, K. J., & Lee, P. S. (2015). Highly stretchable and self-deformable alternating current electroluminescent devices. *Advanced Materials, 27*(18), 2876–2882.

Wang, L., Pan, J., Zhang, Y., Cheng, X., Liu, L., & Peng, H. (2018). A Li–air battery with ultralong cycle life in ambient air. *Advanced Materials, 30*(3), 1704378.

Xu, T., Zhang, Z., & Qu, L. (2020). Graphene-based fibers: Recent advances in preparation and application. *Advanced Materials, 32*(5), 1901979.

Xu, Z., & Gao, C. (2015). Graphene fiber: A new trend in carbon fibers. *Materials Today, 18*(9), 480–492.

Yang, W., Gan, L., Li, H., & Zhai, T. (2016). Two-dimensional layered nanomaterials for gas-sensing applications. *Inorganic Chemistry Frontiers, 3*(4), 433–451.

Yang, Z., Sun, H., Chen, T., Qiu, L., Luo, Y., & Peng, H. (2013). Photovoltaic wire derived from a graphene composite fiber achieving an 8.45% energy conversion efficiency. *Angewandte Chemie, 125*(29), 7693–7696.

Ye, M., Zhang, Z., Zhao, Y., & Qu, L. (2018). Graphene platforms for smart energy generation and storage. *Joule, 2*(2), 245–268.

Yoon, H. J., Jun, D. H., Yang, J. H., Zhou, Z., Yang, S. S., & Cheng, M. M. C. (2011). Carbon dioxide gas sensor using a graphene sheet. *Sensors and Actuators, B: Chemical, 157*, 301–313.

Yu, W., Xie, H., Li, F., Zhao, J., & Zhang, Z. (2013). Significant thermal conductivity enhancement in graphene oxide papers modified with alkaline earth metal ions. *Applied Physics Letters,103*(14), 141913.

Yuan, W., & Shi, G. (2013). Graphene-based gas sensors. *Journal of Materials Chemistry A, 1*(35), 10078–10091.

Zeng, Z., Xiao, F. X., Phan, H., Chen, S., Yu, Z., Wang, R., Nguyen, T. Q., & Yang Tan, T. T. (2018). Unraveling the cooperative synergy of zero-dimensional graphene quantum dots and metal nanocrystals enabled by layer-by-layer assembly. *Journal of Materials Chemistry A, 6*, 1700–1713.

Zhang, H., Feng, J., Fei, T., Liu, S., & Zhang, T. (2014). SnO2 nanoparticles-reduced graphene oxide nanocomposites for NO2 sensing at low operating temperature. *Sensors and Actuators, B: Chemical, 190*, 472–478.

Zhen, Z., & Zhu, H. (2017). Structure and properties of graphene. In H. Zhu, Z. Xu, D. Xie, & Y. Fang (Eds.), *Graphene: Fabrication, characterizations, properties and applications* (pp. 1–12). Amsterdam: Elsevier.

Zheng, P., & Wu, N. (2017). Fluorescence and sensing applications of graphene oxide and graphene quantum dots: A review. *Chemistry - An Asian Journal, 12*(18), 2343–2353.

Zhong, W.-R., Zhang, M.-P., Ai, B.-Q., & Zheng, D.-Q. (2011). Chirality and thickness-dependent thermal conductivity of few-layer graphene: A molecular dynamics study. *Applied Physics Letters, 98*(11), 113107.

# Chapter 20
# Conclusion, Outlook, Future Aspects, and Utilization of Functional Bio-engineered Materials

**Alka Gupta**

## 20.1 Overviews of Nano/Bio-engineered Materials

Nano-biotechnology is a multidisciplinary technique in which bio-engineered nanomaterials are synthesized and implicated in nanodevices in health, agriculture and food science etc. applications. A specially designed nano system (functional biomolecules like small molecules, genes and proteins etc.) has such a dimensional control which can tune the properties for various useful applications. Specifically, in the medical field the nano-biotechnology provides efficacies for theranostics (A type of diagnostic testing system employed for selecting targeted therapy) of numerous diseases like cancer by fine and precision modification of the morphology, size, and surface property. The ability of systematically modification of the properties of bio-engineered nanomaterials via controlling its structure and surface properties makes them extremely striking candidates for use in biological contexts, not only for important scientific studies but also for commercially feasible modern technologies. The Fig. 20.1 presents the schematic diagram showing the methods used for constructing bio-engineered nanomaterials (Salata, 2004).

A. Gupta (✉)
Department of Materials Science and Metallurgical Engineering (MSME), University Institute of Engineering and Technology; Chatrapati Shahu Ji Maharaj Kanpur University Kalyanpur, Kanpur, UP, India
e-mail: alkagupta@csjmu.ac.in


K. Pal (ed.), *Bio-manufactured Nanomaterials*,
https://doi.org/10.1007/978-3-030-67223-2_20

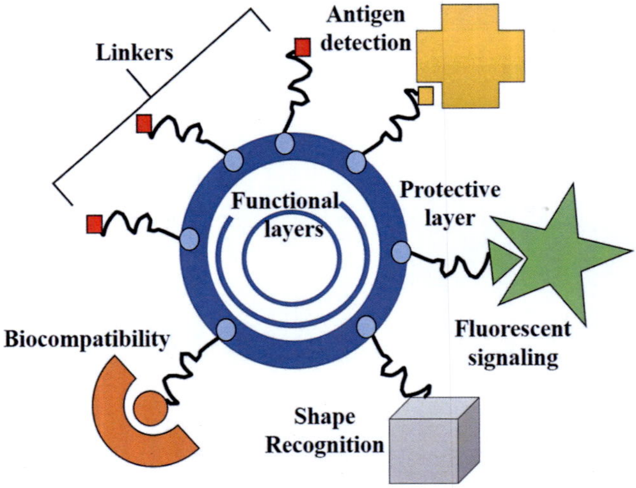


**Fig. 20.1** Configurations depicting the nano-biomaterial's applications in medical or biological fields (Salata, 2004)

## 20.2 Theranostics and Biomedical Engineering Applications

Amalgamation of functionalized bio-engineered nanomaterials in several biomedical applications has posed enumerable research attention in last few decades. The major applications of nano-biotechnology in biomedical engineering, theranostics (therapeutics and diagnostics), tissue engineering and agricultural fields, etc. are vast areas of research (Sahoo & Labhasetwar, 2003). The field of nanotechnology met a challenge to channel or bridge the barriers of the biological and physical sciences by designing and implication of nanomaterials and nanotools for solving biological problems. The evolution of new field nano-biotechnology, which more specifically deals with nanorobotics for targeted drug delivery and nanomedicines for curing deadly diseases like cancer. This chapter summarizes recent developments in nano-biotechnology specifically focusing on health, medical science and tissue engineering. Innovative research approaches in these fields and their main challenges toward practical applicability are also emphasized in a chronological way.

## 20.3 Hazardous Environmental Impacts of Energy Applications

The future implications of nano-biotechnology include a choice of numerous new materials and devices possibly advantageous in the field of electronics, medicine, biomaterials and energy materials. These vast applicability of nano-biotechnology comes with a package of many issues including problems of toxicity and hazardous environmental impact of nanomaterials (Buzea et al., 2007) and their consequences on global economics. These apprehensions need attention by governments to put statutory warranted regulations of nano-biotechnology commercialization and R&D sectors (Milunovich & Roy, 2001). In spite of the many arguments, this nano-biotechnology presents enormous prospects for the future. Nano-biotechnology might lead to novel innovations in the various distinct fields of biomedical applications extending from gene therapy to molecular imaging, disease diagnosis to drug delivery and, effective biomarkers to biosensors etc. Some of these specific fields and their applications are already discussed in detail in the following chapters such as bio imaging in Chap. 2, Quantum dots based bio sensors in Chap. 3, Antimicrobial capabilities of bio engineered nanoparticles and antibiofilm and nanomaterials based bio scaffolds in Chaps. 4–8, Major issues of toxicity and environmental impact in Chaps. 9, 12 and tissue engineering, industrial and agricultural application are covered in Chaps. 10, 13, 15 also commercialization and therapeutic drug delivery aspects are touched upon in Chap. 16 respectively. Applications of nano-biotechnology in health and medical fields are briefly depicted in Fig. 20.2 in the form of conclusive and precise flow chart. As listed in Fig. 20.2, the applications of nanomaterials to biology or medicine are reported by many research groups such as in fluorescent biological labels (Bruchez et al., 1998; Chan & Nie, 1998; Wang et al., 2002), in drug and gene delivery (Mah et al., 2002; Pantarotto et al., 2003), in bio detection of pathogens (Edelstein et al., 2000), in detection of proteins (Nam et al., 2003), in probing of DNA structure (Mahtab et al., 1995), in Tissue engineering (de la Isla et al., 2003; Ma et al., 2003), in tumor destruction via heating (Shinkai et al., 1999), in separation and purification of biological molecules and cells (Molday & Mackenzie, 1982), in MRI contrast enhancement (Weissleder et al., 1990), in Phagokinetic studies (Parak et al., 2002).

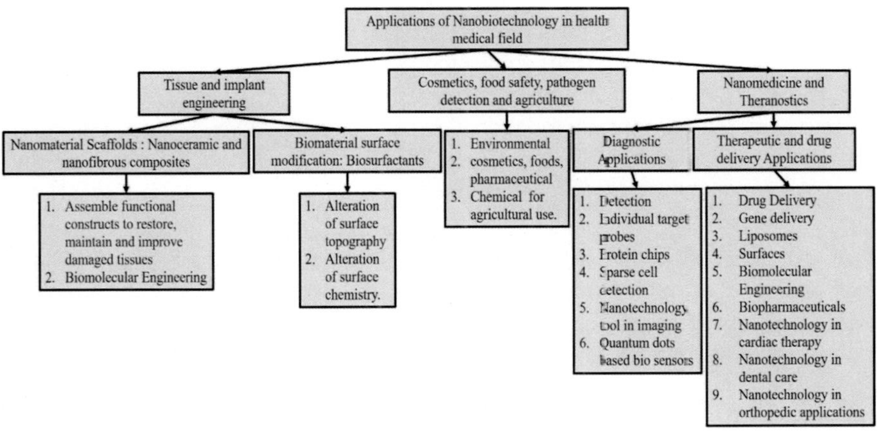


**Fig. 20.2** Applications of nano-biotechnology in health and medical fields

## 20.4 Electrical, Optical and Analytical Tools of Applications

Besides the vast application of bio-engineered nanomaterial in medical and health science, nano-biotechnology find their use is various other fields like electronics, food industries, environmental sciences, and defense services too. In spite of numerous scopes, vast varieties of applicability and future prospects of these bio manufactured nanomaterials have lack of reproducibility and regulations. For an example, there are many diseases driven from the food related items, signifying the importance of monitoring food borne pathogens throughout. Conventional detection methods are not sufficient for quality control of semi perishable foods due to laborious methods and high time consumption (ranging from 24 h for *E. coli* and 7 days for Listeria monocytogenes, etc.), also pathogens numbers were underestimated by these methods indicating lack of quantification and qualification in accuracy. Advancements in manipulation of new bio engineered functionalized nanomaterials, allow to bind more accurately different molecules (such as toxins, Bacteria and proteins and nucleus acids) for biosensing due to their large surface to volume ratio (permeable large surface area exposure) making them to be immobilized by increasing the number of reaction sites of targeted specimen consequently, these properties coupled with electronics and optical analytical tools leads to biosensors with enhanced sensitivity and improved accurate response time determination. Another versatile use of Functional bio-engineered nanomaterials in cancer theranostics application restrains by many challenges that make it difficult in practical clinical translations. Among all one of the toughest challenges in this field is to induce multiple functionalities into a single nanoparticle to make a complex system. The, next problem is encountered during the production process on an industrial scale and economical clinical translations in the future. Scientists and research engineers of concerned field must refine the way of choosing the appropriate imaging techniques and contrast modalities for exact clinical implications. An interdisciplinary approach must be followed to facilitate the functional bio-engineered nanomaterials for targeted cancer theranostics in near future.

No single person can provide the answers to challenges that nanotechnology brings, nor can any single group or intellectual discipline. The five main challenges encountered so far, by the researchers and scientists are first, to develop tools for evaluation of bio-engineered functionalized nanomaterials in the air and water exposure to fairly understand that exposure for animals, plants and humans keeping in mind the potential environment contamination by these nanomaterials which might lead to adverse consequences. The second challenge is to develop an appropriate technique to detect and quantify the toxicity level of bio engineered nanomaterials in the near future, as this challenge needs to propose specific models by the researchers, for forecasting inevitable adverse effects of nanomaterials and nanodevices on the environment and human health. To make an advancement of reverse systems for the assessment of accurate impact of bio engineered nanomaterials and devices (including their total life span) on health and the environment, would be the third challenge. The fourth challenge would be to invent the precise technique which could appropriately evaluate the risk of these nanomaterials on human health and to the environment. Commercialization (innovation efficacy, scalability, proper funding, unusual resources, etc.) and reproducibility (development of new products and the upgrading of existing products with ability of productivity) would be the fifth important challenges of nano-biotechnology. The most fundamental interrogations raised about nano-biotechnology is to develop new regulations by authorities all over the world to evaluate possible hazards and risks by the extensive applications of this unconventionally advanced technology for controlling its optimum usage (Li et al., 2007). This nano-dimensional bio engineered functional particles and nano-molecular devices cause various cardiovascular, respiratory and gastrointestinal system irregularities (Nijhara & Balakrishnan, 2006). For an example, various lung pathologies like, interstitial inflammation, peribronchial inflammation, epitheloid granuloma and necrosis of lung are caused by injecting (Intra-tracheal instillation) carbon nanotubes (nano sized allotrope of carbon) in mice (Lam et al., 2004). It has been revealed by research analysis that human body can be exposed to nanomaterials during production most likely via the lungs and rapidly translocate to other vital organs of human body through the blood circulation even nanoparticles can enter the central nervous system too (Oberdörster et al., 2005; Williams, 2005; Elder et al., 2006). This results in more careful usage of nanoparticle-mediated drug delivery and diagnostic as a tool for diagnosis and therapeutic medication of diseases. Radomski et al. has detected the pro-aggregators side effects (like acceleration of vascular thrombosis) of intra-tracheal instillation of nanotubes on the platelets counts (in the in vitro experiments) on rat specimen. It is also reported that fullerenes are safer in comparison to nanotubes for making nanoparticle-based drug delivery system (Radomski et al., 2005). Currently, nanosized devices are designed by using carbon nanotubes, fullerenes, silica and silica-based derivatives.

## 20.5 Nanoparticle Toxicity and Nanomedicine

The toxicity of nanoparticles can also affect the gastrointestinal system, its inflammatory response leads to organ damage as well. In case of overdoes of these nanomedicine over and above human tolerance limit, would lead to many toxic health consequences like jaundice, hemolysis and even organs failure (Chen et al., 2006; Medina et al., 2007). Actually, these effects have been experimented in vitro in animal so their actual effects on human are far difficult to extrapolate. As soon as, nanomaterials of different size, shape, functionalities, hydrophilic or hydrophobic kinds of reactive surfaces, charge, etc. react with a biological system then various types of bio physicochemical interactions (positive or negative) may possibly take place. As these bio functionalized nanoparticles are widely used in different biomedical applications, like drug delivery, therapeutics, and imaging etc., their cumulative exposure to the human body and their internal organs too leads many potential hazards on human health due to their easy penetration capacity through skin and through inhalation leading to very adverse effects on exposed organs. Because of their small particle size, sometimes phagocytic defense system of the human body may not recognize them and these tiny particles may reach to the blood circulation or brain nerves to directly affect the normal cardiac or cerebral functions leads to toxication and abnormalities.

In near future, probably these products would be trailed and tested in vivo (including humans) by academia and governmental owned laboratories leads to economical commercialization. In current scenario most of the medical devices and therapeutics diagnostic tools are still far away from the market and only are at trial stage. Therefore, target drug delivery systems using bio engineered nanomaterials and nanodevice implantations necessitates a more complex infrastructural techniques and more accessible regulatory management system (Hamad-Schifferli et al., 2002). Continuous progressions in the field of nanomedicine have presented new future prospects for variety of applications such as diagnostic and regenerative medicines. By the use of nanomedicine and devices precise diagnosis and detection of particular diseased cells become faster, more specifically targeted drug delivery at the exact point of a sick cell is possible and assuring only the diseased cells to be cured without affecting the surrounded healthy cell.

## 20.6 Multidisciplinary Challenges of Nano-biosciences and Technology

Although, the multidisciplinary field of nano-biotechnology is in early stages of developments. The consequences of these developments will be so enormous that it affects almost all the fields of science and engineering. In future today's uncurable diseases may be fully cured by using nano-biotechnology. However, the potentials of nano-biotechnology in health and medicine field are high enough with enlisted enormous benefits, but the safety issues of nanomedicine are still a big challenge. It is possible that nano-biotechnology would play a crucial role in diagnosis and treatment of human diseases in the future and become an inevitable part of our day today's life. In the upcoming decade, these nanodevices (nanorobots for targeted drug delivery, nanomedicines, etc.) based on nano-biotechnology, make thousands of treatments very rapidly, accurately and very inexpensively. Future directions in diagnostics lead to miniaturization from bio-chip technology to the nanoscale technology. For an example the most common clinical diagnostic the "blood test" (blood protein analysis) which reflects the health state or disease related to most organs, can become as small as detecting blood molecular fingerprints by using nanotechnology leads to more sensitive evaluate of disease. Similarly, another field of this technology, Molecular electronics make it possible to design the microscopic sensors that are able to detect the patterns of chemicals in a body fluid and deconvolute the information from the bloodstream to estimate the characteristics of small chemical sources present in tiny microscopic volume of the tissues (Hogg & Kuekes, 2006). The tissue engineering via quantum dots based high-resolution bio sensors indicate that these devices can have capability of sensing in both time and space. The new research innovation trend, build the nano-diagnostic devices using the bottom up approach, starting from the smallest building blocks as atoms and molecules. Nano-biotechnology will facilitate the new invention of non-PCR diagnostic methods, replacing the fluorescent labeling or marking conventional techniques. Further modification in diagnostic tools by using, nanotechnology enables the analysis of a single cell for genetic findings. Moreover, the use of nano-diagnostics tools must reduce the time of waiting for the pathological test results in the near future so that the patients could then get treatment immediately without any delay making the future healthcare systems more economic, precise and accurate. Another significant application of nano-biotechnology is cancer diagnostics. The Molecular diagnosis of cancer by genetic profiling is an important tool for detection of this deadly disease at a much early stage which would assure the possibility to cure it completely. Nanorobots may be frequently used in the future for targeted therapy and drug delivery for the treatment of cancer. A nanodevice for theranostics (the combination of diagnosis and therapeutics) could be implanted in persons who have hereditary history of cancer, although they do not show any obvious symptoms of cancer, consequently cancer surveillance would be possible by external monitoring system lead to suitable therapeutic intervention if necessary, at an early stage. These early detections of disease would definitely increase the chances of cure and survival (Jain, 2005).

## 20.7 Biochemical Reactions of Drug and Gene Delivery

The present research emphasis on bio-derived nanomaterials in the filed tissue engineering and medical science and technology for drug and gene delivery using different types of nanoparticles extracted from the green sources such as bio-inspired nanomaterials are nontoxic in nature and can be used for extensive variety of applications. In this context Elia et al. synthesized gold nanoparticles by various types of plant extracts which are very significantly useful for biochemical reactions (De Adhikari et al., 2015). In the same line, biopolymers and bio nanocomposites are another type of important nanomaterials having great advantage of biocompatibility and enhanced reactive sites. Adhikari and his coworkers reported the applicability of graphene-based cellulose nanofibers used in supercapacitors, which was extracted from the waste papers. Other novel composite materials are synthesized by using, dextran and cellulose, which were extracted from the natural resources. It has proven that plant-based bio engineered nanomaterials have an incredible possibility in health, medical, food and agricultural applications. Though, the main challenges for the use of these materials are the difficulty in their mass production, reproducibility, process of purification and their consumptions with additional matrices. Researchers, scientists, engineers, chemists and also biotechnologist around the world are trying hard to explore possible explanations and potential solutions for the above pointed challenges to increase the applicability of nano-biomaterials from the laboratories to the market. For better future prospective of these materials there is a requirement of a strong collaboration between research and innovative marketing strategies to make this technology economic and affordable for everybody. It could be predicted that this need of affordable bio nanotechnology will be fulfilled and be available with much innovations within 5–10 years in the market for the public usage.

Starting from the nanotechnological tools to nanodevice design, the next important field of nano-biotechnology is DNA and protein-based nanotechnology, in which proteins and nucleic acids are used as a building block for constructing nanostructures, for example DNA origami and nanorobots. This new endeavor of nanotechnology certainly has potential to challenge miniaturization of diagnostic tools for drug delivery to extremely small level and give opportunity to develop the replacements for antibodies. DNA nanotechnology can be supported by computational methods which permit the researchers to draw the coherent design of specific patterns and complex structures on the screen for further analysis.

## 20.8 Future Aspects of Clinical Applications and Bio-implants

Bio engineered nanomaterials results in a new future direction of the modern biomedical applications starting from biomedicine to targeted drug delivery, bio imaging to biosensors for health and life sciences. The constant progression in synthesis techniques of the nanomaterials offers a full control of their physicochemical properties. Furthermore, modern therapeutics also discover bio-engineered functionalized nanomaterials (such as bio-implants) to mimic natural biological entities (like body organs) with increased biocompatibility into the human body. Currently, an engineered nanomaterial has extremely advanced the imaging tools and bio-sensing techniques results in ultra-precise and accurate therapeutics. Although, the toxicological effects of these nanomaterials like bio-implants and nanomedicines on the human body are still not fully understood and justified till date. Further in-depth research and analysis are essentially needed to create an explicit knowledge of toxic effect of these materials on human health. The future of these bio-engineered nanomaterials in medical and clinical applications is promising, but improved understanding of their toxic side effects and techniques to fix them are the future areas of research interest. The final summary is that the nano-biotechnology potentially promises to encompass the boundaries of present molecular diagnostics by ensuring to the point diagnosis, combining diagnostics with therapeutics, and advancements of more personalized medicine for health care. The most vital applications of bio engineered functionalized nanomaterials and devices are in the field of biomarker, cancer diagnosis tools, and early diagnosis of transmittable microorganisms.

**Acknowledgments** The author acknowledges the Springer journal for publishing this book chapter.

**Conflict of Interest** There are no conflicts of interest to declare.

## References

Bruchez, M., Moronne, M., et al. (1998). Semiconductor nanocrystals as fluorescent biological labels. *Science, 281*(5385), 2013.

Buzea, C., Pacheco, I. I., et al. (2007). Nanomaterials and nanoparticles: Sources and toxicity. *Biointerphases, 2*(4), MR17–MR71.

Chan, W. C. W., & Nie, S. (1998). Quantum dot bioconjugates for ultrasensitive nonisotopic detection. *Science, 281*(5385), 2016.

Chen, Z., Meng, H., et al. (2006). *Toxicology Letters, 163*(2), 109–120.

De Adhikari, A., Oraon, R., et al. (2015). Effect of waste cellulose fibres on the charge storage capacity of polypyrrole and graphene/polypyrrole electrodes for supercapacitor application. *RSC Advances, 5*(35), 27347–27355.

de la Isla, A., Brostow, W., et al. (2003). Nanohybrid scratch resistant coatings for teeth and bone viscoelasticity manifested in tribology. *Materials Research Innovations, 7*(2), 110–114. Acute toxicological effects of copper nanoparticles in vivo.

Edelstein, R. L., Tamanaha, C. R., et al. (2000). The BARC biosensor applied to the detection of biological warfare agents. *Biosensors and Bioelectronics, 14*(10), 805–813.

Elder, A., Gelein, R., et al. (2006). Translocation of inhaled ultrafine manganese oxide particles to the central nervous system. *Environmental Health Perspectives, 114*(8), 1172–1178.

Hamad-Schifferli, K., Schwartz, J. J., et al. (2002). Remote electronic control of DNA hybridization through inductive coupling to an attached metal nanocrystal antenna. *Nature, 415*(6868), 152–155.

Hogg, T., & Kuekes, P. J. (2006). Mobile microscopic sensors for high resolution in vivo diagnostics. *Nanomedicine: Nanotechnology, Biology and Medicine, 2*(4), 239–247.

Jain, K. K. (2005). Role of nanobiotechnology in developing personalized medicine for cancer. *Technology in Cancer Research & Treatment, 4*(6), 645–650.

Lam, C.-W., James, J. T., et al. (2004). Pulmonary toxicity of single-wall carbon nanotubes in mice 7 and 90 days after intratracheal instillation. *Toxicological Sciences, 77*(1), 126–134.

Li, Z., Hulderman, T., et al. (2007). Cardiovascular effects of pulmonary exposure to single-wall carbon nanotubes. *Environmental Health Perspectives, 115*(3), 377–382.

Ma, J., Wong, H., et al. (2003). Biomimetic processing of nanocrystallite bioactive apatite coating on titanium. *Nanotechnology, 14*(6), 619–623.

Mah, C., Fraites, T. J., Jr., Zolotukhin, I., Song, S., Flotte, T. R., Dobson, J., et al. (2002). Improved method of recombinant AAV2 delivery for systemic targeted gene therapy. *Molecular Therapy, 6*(1), 106–112.

Mahtab, R., Rogers, J. P., et al. (1995). Protein-sized quantum dot luminescence can distinguish between "straight", "bent", and "kinked" oligonucleotides. *Journal of the American Chemical Society, 117*(35), 9099–9100.

Medina, C., Santos-Martinez, M. J., et al. (2007). Nanoparticles: Pharmacological and toxicological significance. *British Journal of Pharmacology, 150*(5), 552–558.

Milunovich, S., & Roy, J. (2001). The next smaller thing—An introduction to nanotechnology. Technical report, Merril Lynch, Sept. 4, 2001.

Molday, R. S., & Mackenzie, D. (1982). Immunospecific ferromagnetic iron-dextran reagents for the labeling and magnetic separation of cells. *Journal of Immunological Methods, 52*(3), 353–367.

Nam, J.-M., Thaxton, C. S., et al. (2003). Nanoparticle-based bio-bar codes for the ultrasensitive detection of proteins. *Science, 301*(5641), 1884.

Nijhara, R., & Balakrishnan, K. (2006). Bringing nanomedicines to market: Regulatory challenges, opportunities, and uncertainties. *Nanomedicine: Nanotechnology, Biology and Medicine, 2*(2), 127–136.

Oberdörster, G., Oberdörster, E., et al. (2005). Nanotoxicology: An emerging discipline evolving from studies of ultrafine particles. *Environmental Health Perspectives, 113*(7), 823–839.

Pantarotto, D., Partidos, C. D., et al. (2003). Immunization with peptide-functionalized carbon nanotubes enhances virus-specific neutralizing antibody responses. *Chemistry & Biology, 10*(10), 961–966.

Parak, W. J., Boudreau, R., et al. (2002). Cell motility and metastatic potential studies based on quantum dot imaging of phagokinetic tracks. *Advanced Materials, 14*(12), 882–885.

Radomski, A., Jurasz, P., et al. (2005). Nanoparticle-induced platelet aggregation and vascular thrombosis. *British Journal of Pharmacology, 146*(6), 882–893.

Sahoo, S. K., & Labhasetwar, V. (2003). Nanotech approaches to drug delivery and imaging. *Drug Discovery Today, 8*(24), 1112–1120.

Salata, O. V. (2004). Applications of nanoparticles in biology and medicine. *Journal of Nanobiotechnology, 2*(1), 3.

Shinkai, M., Yanase, M., et al. (1999). Intracellular hyperthermia for cancer using magnetite cationic liposomes. *Journal of Magnetism and Magnetic Materials, 194*(1), 176–184.

Wang, S., Mamedova, N., et al. (2002). Antigen/antibody immunocomplex from CdTe nanoparticle bioconjugates. *Nano Letters, 2*(8), 817–822.

Weissleder, R., Elizondo, G., Wittenberg, J., Rabito, C. A., Bengele, H. H., & Josephson, L. (1990). Ultrasmall superparamagnetic iron oxide: Characterization of a new class of contrast agents for MR imaging. *Radiology, 175*(2), 489–493.

Williams, D. (2005). The risks of nanotechnology. *Medical Device Technology, 16*(9), 6, 9–10.

# Index




K. Pal (ed.), *Bio-manufactured Nanomaterials*,
https://doi.org/10.1007/978-3-030-67223-2

**O**



**P**

**Q**

The manufacturer's authorised representative in the EU is Springer Nature Customer Service Centre GmbH, Europaplatz 3, 69115 Heidelberg, Germany. If you have any concerns regarding our products, please contact ProductSafety@springernature.com

Printed and bound by CPI Group (UK) Ltd, Croydon, CR0 4YY
30/09/2026
02224059-0001